MW01492866

THERMAL DESIGN AND OPTIMIZATION

Adrian Bejan

Department of Mechanical Engineering and Material Science
Duke University

George Tsatsaronis

Institut für Energietechnik
Technische Universität Berlin

Michael Moran

Department of Mechanical Engineering
The Ohio State University

A WILEY-INTERSCIENCE PUBLICATION

JOHN WILEY & SONS, INC.

New York / Chichester / Brisbane / Toronto / Singapore

This text is printed on acid-free paper.

Copyright © 1996 by John Wiley & Sons, Inc.

All rights reserved. Published simultaneously in Canada.

No part of this publication may be reproduced, stored in a retrieval system or transmitted in any form or by any means, electronic, mechanical, photocopying, recording, scanning or otherwise, except as permitted under Sections 107 or 108 of the 1976 United States Copyright Act, without either the prior written permission of the Publisher, or authorization through payment of the appropriate per-copy fee to the Copyright Clearance Center, 222 Rosewood Drive, Danvers, MA 01923, (978) 750-8400, fax (978) 750-4470. Requests to the Publisher for permission should be addressed to the Permissions Department, John Wiley & Sons, Inc., 111 River Street, Hoboken, NJ 07030, (201) 748-6011, fax (201) 748-6008.

To order books or for customer service please, call 1(800)-CALL-WILEY (225-5945).

This publication is designed to provide accurate and
authoritative information in regard to the subject
matter covered. It is sold with the understanding that
the publisher is not engaged in rendering
professional services. If legal advice or other
expert assistance is required, the services of a competent
professional person should be sought.

Library of Congress Cataloging in Publication Data:
Bejan, Adrian, 1948–
    Thermal design and optimization / Adrian Bejan, George
Tsatsaronis, Michael Moran.
        p.   cm.
    Includes index.
    ISBN 0-471-58467-3
    1. Heat engineering.   I. Tsatsaronis, G. (George)   II. Moran,
Michael J.   III. Title.
TJ260.B433   1996
621.402—dc20   95-12071

Printed in the United States of America

10 9 8 7 6

PREFACE

This book provides a comprehensive and rigorous introduction to thermal system design and optimization from a contemporary perspective. The presentation is intended for engineering students at the senior or first-year graduate level and for practicing engineers and technical managers working in the energy field. The book is appropriate for use in a capstone design course, in a technical elective course, and for self-study. Sufficient end-of-chapter problems are provided for these uses. In class testing, the material has been found to work well with the intended audience.

We assume readers have had introductory courses in engineering thermodynamics and heat transfer and are familiar with the basics of fluid mechanics. Some background in engineering economics is also desirable but not required. For readers with limited backgrounds in engineering thermodynamics, heat transfer, and engineering economics, reviews are provided in Chapters 2, 4, and 7, respectively. Our presentation does not provide a detailed discussion of component design or extensive operating and cost data. Information on these topics is available in various standard references, handbooks, and manufacturers' catalogs. Readers should refer to such sources as needed; we have provided extensive reference lists to facilitate this. The book has been written to allow flexibility in the use of units. It can be studied using International System (SI) units only or a mix of SI and English units.

In the area of thermal systems, engineering curricula are largely component and design analysis oriented. Students initially learn to apply mass and energy balances and, increasingly, entropy and exergy balances. Then, on the basis of known engineering descriptions and specifications, students learn to calculate the size, performance, and cost of heat exchangers, turbines, pumps, and other components. These activities are important, but the scope of engineering design is much wider. Design is primarily system oriented and the objective is to effect a design solution: to devise a means for accomplishing a stated purpose subject to real-world constraints. Design requires synthesis: selecting and putting together components to form a smoothly working whole. Design also often requires that principles from different disciplines be applied within a single problem, for example, principles from engineering thermodynamics and heat transfer. Moreover, design usually requires explicit consideration of engineering economics, for cost is almost invariably a key issue. Finally, design requires optimization techniques that are not typically encountered elsewhere in the engineering curricula. Synthesis, engineering economics, and optimization are among several design topics discussed in this book.

The current presentation departs from those of previous books on the subject of thermal system design in three important respects: (1) A concerted effort has been made to include material drawn from the best of contemporary thinking about design and design methodology. Thus, Chapter 1 provides discussions of concurrent design, quality function deployment, and other contemporary design ideas. (2) This book includes current developments in engineering thermodynamics, heat transfer, and engineering economics relevant to design. Many of these developments are based on the second law of thermodynamics. In particular, we feature the use of exergy analysis and entropy generation minimization. We employ the term thermoeconomics to denote exergy-aided cost minimization. (3) A case study is considered throughout the book for continuity of the presentation. The case study involves the design of a cogeneration system.

The presentation of design topics initiated in Chapter 1 continues in Chapter 2 with the development of a thermodynamic model for the case study cogeneration system and a discussion of piping system design. Chapter 3 provides a discussion of design guidelines evolving from reasoning using the second law of thermodynamics and, in particular, the exergy concept.

Chapters 4, 5, and 6 all contain design-related material, including heat exchanger design. These presentations are intended to illuminate the design process by gradually introducing first-level design notions such as degrees of freedom, design constraints, and thermodynamic optimization. In these chapters the role of second-law reasoning in design is further emphasized. Examples familiar from previous courses in thermodynamics and heat transfer are used to illustrate principles. Chapters 5 and 6 also illustrate the effectiveness of elementary modeling in design. Such modeling is often an important element of the concept development stage. Elementary models can highlight key design variables, the relations among them, and fundamental trade-offs. In some instances, such models can lead directly to design solutions, as for example in the case of electronic package cooling considered in Chapter 5.

With Chapter 7 detailed engineering economic evaluations enter the presentation explicitly and are featured in the remainder of the book. Chapter 8 presents a powerful and systematic design approach that combines the exergy concept of engineering thermodynamics with principles of engineering economics. Exergy costing methods are introduced and applied in this chapter. These methods identify the real cost sources at the component level: the capital investment costs, the operating and maintenance costs, and the costs associated with the destruction and loss of exergy. The optimization of thermal systems is based on a careful consideration of these cost sources. Chapter 9 provides discussions of the pinch method for the design of heat exchanger networks and the iterative optimization of complex systems. Results of the thermoeconomic optimization of the cogeneration system case study are also presented.

This book has been developed to be flexible in use: to satisfy various instructor preferences, curricular objectives, course durations, and self-study needs. Some instructors will elect to present material in depth from all nine chapters. Shorter presentations are also possible. For example, courses might be built around Chapters 1–6 plus topics from Chapter 7, or formed from Chapters 1–3 plus Chapters 7–9. Some instructors may want to provide a highly-focused presentation by simply tracking the cogeneration system case study or another case study drawn from the references provided or the individual instructor's professional experience. Other course arrangements are also possible, and instructors are encouraged to contact the authors directly concerning alternatives.

The use of the second law of thermodynamics in thermal system design and optimization is still a novelty in U.S. industry, but not in Europe and elsewhere. This approach is featured here because an increasing number of engineers and engineering managers worldwide agree that it has considerable merit and are advocating its use. We offer it in an evolutionary spirit as a worthy alternative. Our aim is to contribute to the education of the next generation of thermal system designers and to the background of currently active designers who feel the need for more effective design methods. We welcome constructive comments and criticism from readers. Such feedback is essential for the further development of the design approaches presented in this book.

Several individuals have contributed to this book. A. Özer Arnas and Gordon M. Reistad reviewed the manuscript and provided several helpful suggestions. We appreciate their input. Additionally, we owe special thanks to faculty colleagues, staff and students at our respective institutions for their support and assistance: at Duke University, Kathy Vickers, Linda Hayes, Jose V. C. Vargas, Oana Craciunescu, and Gustavo Ledezma; at Tennessee Technological University, Kenneth Purdy, David Price, Helen Haggard, Agnes Tsatsaronis, and the 126 student members of the thermal design class in the academic year 1993–1994; at Technische Universität Berlin, Yanzi Chen, Frank Cziesla, Andreas Krause, and Christine Gharz; and at The Ohio State University, Kenneth Waldron, Margaret Drake, and Carol Bird.

ADRIAN BEJAN
GEORGE TSATSARONIS
MICHAEL MORAN

June, 1995

CONTENTS

1

INTRODUCTION TO THERMAL SYSTEM DESIGN

Engineers not only must be well versed in the scientific fundamentals of their disciplines, but also must be able to design and analyze components typically encountered in their fields of practice: mechanisms, turbines, reactors, electrical circuits, and so on. Moreover, engineers should be able to synthesize something that did not exist before and evaluate it using criteria such as economics, safety, reliability, and environmental impact. The skills of synthesis and evaluation are at the heart of *engineering design*. Design also has an important creative component not unlike that found in the field of art.

This book concerns the design of *thermal systems*. Thermal systems typically experience significant *work* and/or *thermal* interactions with their surroundings. The term thermal interactions is used generically here to include heat transfer and/or the flow of hot or cold streams of matter, including reactive mixtures. Thermal systems are widespread, appearing in diverse industries such as electric power generation and chemical processing and in nearly every kind of manufacturing plant. Thermal systems are also found in the home as food freezers, cooking appliances, furnaces, and heat pumps. Thermal systems are composed of compressors, pumps, turbines, heat exchangers, chemical reactors, and a variety of related devices. These components are interconnected to form networks by conduits carrying the *working substances,* usually gases or liquids.

Thermal system design has two branches: *system design* and *component design*. The first refers to overall thermal systems and the second to the individual components (heat exchangers, pumps, reactors, etc.) that make up the overall systems. System design may refer to new-plant design or to design associated with plant expansion, retrofitting, maintenance, and refurbishment.

The principles introduced in this book are relevant to each of these application areas.

The design of thermal systems requires principles drawn from *engineering thermodynamics, fluid mechanics, heat and mass transfer,* and *engineering economics.* These principles are developed in subsequent chapters. The objective of this chapter is to introduce some of the fundamental concepts and definitions of thermal system design.

1.1 PRELIMINARIES

In developed countries worldwide, the effectiveness of using oil, natural gas, and coal has improved markedly over the last two decades. Still, usage varies widely even within nations, and there is room for improvement. Compared to some of its principal international trading partners, for example, U.S. industry as a whole has a higher energy resource consumption on a per unit of output basis *and* generates considerably more waste. These realities pose a challenge to continued U.S. competitiveness in the global marketplace. One way that this challenge can be addressed is through better industrial plant design and operation.

For industries where energy is a major contributor to operating costs, an opportunity exists for improving competitiveness through more effective use of energy resources. This is a well-known and largely accepted principle today. A related but less publicized principle concerns the waste and effluent streams of plants. The waste from a plant is often not an unavoidable result of plant operation but a measure of its inefficiency: The less efficient a plant is, the more unusable by-products it produces, and conversely. Effluents not produced owing to a more efficient plant require no clostly cleanup and do not impose a burden on the environment. Such *clean technology* considerations provide another avenue for improving long-term competitiveness.

Economic, fuel-thrifty, environmentally benign plants have to be carefully engineered. Careful engineering calls for the rigorous application of principles and practices that have withstood the test of time. It also requires the best current thinking on the subject of design. Accordingly, recognizing the linkage between efficient plants and the goals of better energy resource use with less environmental impact, we stress the use in design of the *second law of thermodynamics,* including the *exergy methods* discussed in Chapter 3.

We also stress the use of principles of good design practice that have been formulated in recent years [see, e.g., 1–11]. Witnessing the relationship between superior engineering design and enhanced competitiveness in world markets, many industries have adopted innovative design and management principles that were largely unknown just a few years ago. The *product realization process* concept is one. This is the procedure whereby new and improved products are conceived, designed, produced, brought to market, and supported. The product realization process is a firm's strategy for product excellence through *continuous improvement*. As such, the product realization process is not static, but always evolving in response to new information and conditions in the marketplace. Another concept employed nowadays involves DFX strategies. In this acronym, DF denotes *design for* and X identifies a characteristic to be optimized, such as DFA (design for assembly), DFM (design for manufacturing), and DFE (design for the environment). Many other relatively new concepts are now being used by industry, as for example *just-in-time* manufacturing, where parts and assemblies are delivered or manufactured as needed, thereby reducing inventory costs.

Despite their relatively recent development, innovative practices for better design engineering have taken on such a wide variety of forms with sometimes overlapping functions that it is not always a simple matter to select from among them. Moreover, not all of these contemporary design practices are applicable to or useful in the design of every system or product. Accordingly, in this chapter we refer only to those practices that are particularly relevant to thermal systems design. More extended discussions are found in the references cited.

1.2 WORKABLE, OPTIMAL, AND NEARLY OPTIMAL DESIGNS

A critical early step in the design of a system is to pin down what the system is required to do and to express these requirements quantitatively, that is, to formulate the *design specifications* (Section 1.3). A *workable* design is simply one that meets all the specifications. Among the workable designs is the *optimal* design: the one that is in some sense the "best." Several alternative notions of best can apply depending on the application: optimal cost, size, weight, reliability, and so on. Although the term optimal can be defined as the most favorable or most conducive to a given end, especially under fixed conditions, in practice the term can take on many guises depending on the ends and fixed conditions that apply. To avoid ambiguity in subsequent discussions, therefore, we will make explicit in each different context how the term optimal is being used. Still, the design that minimizes cost is usually the one of interest.

Owing to system complexity and uncertainties in data and information about the system, a true optimum is generally impossible to determine. Accordingly, we often accept a design that is close to optimal—a *nearly optimal* design. To illustrate, consider a common component of thermal systems, a counterflow heat exchanger. A key *design variable* for heat exchangers is $(\Delta T)_{min}$, the minimum temperature difference between the two streams. Let us consider the variation of the total annual cost associated with the heat exchanger as $(\Delta T)_{min}$ varies, holding other variables fixed [8].

From engineering thermodynamics we know that the temperature difference between the two streams of a heat exchanger is a measure of nonideality (irreversibility) and that heat exchangers approach ideality as the temperature difference approaches zero (Section 2.1.3). This source of nonideality exacts a penalty on the fuel supplied to the overall plant that has the heat exchanger as a component: A part of this fuel is used to *feed* the source of nonideality. A larger temperature difference corresponds to a thermodynamically less ideal heat exchanger, and so the *fuel cost* increases with $(\Delta T)_{min}$, as shown in Figure 1.1.

To reduce operating costs, then, we would seek to reduce the stream-to-stream temperature difference. From the study of heat transfer, however, we know that as the temperature difference between the streams narrows more heat transfer area is required for the same rate of energy exchange. More heat transfer area corresponds to a larger, more costly heat exchanger. Accordingly, the *capital cost* increases with the heat transfer area and varies inversely with $(\Delta T)_{min}$, as shown in Figure 1.1. All costs shown in the figure are *annual levelized costs* (Section 7.4).

The *total cost* associated with the heat exchanger is the sum of the capital cost and the fuel cost. As shown in Figure 1.1, the variation of the total cost with $(\Delta T)_{min}$ is concave upward. The point labeled *a* locates the design with minimum total annual cost: the optimal design for a system where the heat exchanger is the sole component. However, the variation of total cost with temperature difference is relatively flat in the portion of the curve whose end points are *a'* and *a"*. Heat exchangers with $(\Delta T)_{min}$ values giving total cost on this portion of the curve may be regarded as nearly optimal.

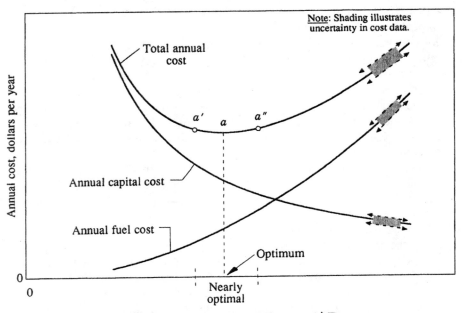

Figure 1.1 Cost curves for a single heat exchanger.

The design engineer would be free to specify, without a significant difference in total cost, any $(\Delta T)_{min}$ in the interval designated nearly optimal in Figure 1.1. For example, by specifying $(\Delta T)_{min}$ smaller than the economic optimum, near point a', the designer would save fuel at the expense of capital *but* with no significant change in total cost. At the higher end of the interval near point a'', the designer would save capital at the expense of fuel. Specifying $(\Delta T)_{min}$ outside of this interval would be a *design error* for a stand-alone heat exchanger, however: With too low a $(\Delta T)_{min}$, the extra capital cost would be much greater than what would be saved on fuel. With too high a $(\Delta T)_{min}$, the extra fuel cost would be much greater than what would be saved on capital.

Thermal design problems are typically far more complex than suggested by this example. For one thing, costs cannot be predicted as precisely as implied by the curves of Figure 1.1. Owing to uncertainties, cost data should be shown as bands and not lines. This is suggested by the shaded intervals on the figure. The annual levelized fuel cost, for example, requires a prediction of future fuel prices and these vary, sometimes widely, with the dictates of the world market for such commodities. Moreover, the cost of a heat exchanger is not a fixed value but often the consequence of a bidding procedure involving many factors. Equipment costs also do not vary continuously, as shown in Figure 1.1, but in a stepwise manner with discrete sizes of the equipment.

Considerable additional complexity enters because thermal systems typically involve several components that interact with one another in complicated ways. Because of component interdependency, a unit of fuel saved by reducing the nonideality of a heat exchanger, for example, may lead to the waste of one or more units of fuel elsewhere in the system, resulting in no net change, or even an increase, in fuel consumption for the overall system. Moreover, unlike the heat exchanger example that involved just one design variable, several design variables usually must be considered and optimized *simultaneously*. Still further, the objective is normally to optimize a system consisting of several components. Optimization of components individually, as was done in the example, usually does not guarantee an optimum for the overall system.

Thermal system design problems also may admit several alternative, fundamentally different solutions. Workable designs might exist, for example, that require one, two, or perhaps three or more heat exchangers. The most significant advances in design generally come from identifying the best design configuration from among several alternatives and not from merely polishing one of the alternatives. Thus, as a final cautionary note, it should be recognized that applying a mathematical optimization procedure to a particular design configuration can only consider trade-offs for that configuration (e.g., the trade-off between capital cost and fuel cost). No matter how rigorously formulated and solved an optimization problem may be, such procedures are generally unable to identify the existence of alternative design configurations or discriminate between fundamentally different solutions of the same problem. Additional discussion of optimization is provided in Chapter 9.

1.3 LIFE-CYCLE DESIGN

Engineering design requires highly structured *critical thinking* and *active communication* among the members of the *design team,* that is, the group of individuals whose responsibility is to design a product or system. In this section the design process is considered broadly and the role of the design team discussed generally. Further discussion is provided in Sections 1.4 and 1.5.

1.3.1 Overview of the Design Process

Figure 1.2 shows a flow chart of the design process for thermal systems. Five distinct stages are indicated:

1. Understanding the problem
2. Concept development
3. Detailed design
4. Project engineering
5. Service

Alternative labels can be employed for this sequence and a different number of stages can be used. Those given in Figure 1.2 are by no means the only possible choices. Other design process flow charts are provided in the literature; see [2], for example.

Since the flow chart proceeds from a general statement of use or opportunity, the *primitive problem,* through placing the system into service and its ultimate retirement from service, the sequence may be called *life-cycle design.* Although the flow of activities is shown as sequential from the top of the chart to the bottom, left and right arrows are used at certain locations, and at others reversed arrows, labeled *iterate,* are shown. The left–right arrows suggest both project work proceeding in parallel by individuals or groups having distinctive expertise *and* the interactions that normally occur between these individuals or groups. An effective design process necessarily blends series and parallel activities synergistically. The desirability of iterating as more knowledge is obtained and digested should be evident. Although iteration does involve costs, there is normally an economic advantage in identifying and correcting problems as early as possible. Iteration continues at any level until the amount and quality of information allows for an informed decision.

The design process involves a wide range of skills and experience. Normally these are well beyond the capabilities of a single individual and call for a group effort. This is the role played by the design team. A strategy known as *concurrent design* addresses both the makeup of the design team and the interactions between team members. A central tenet of concurrent design is that all departments of a company be involved in the design process

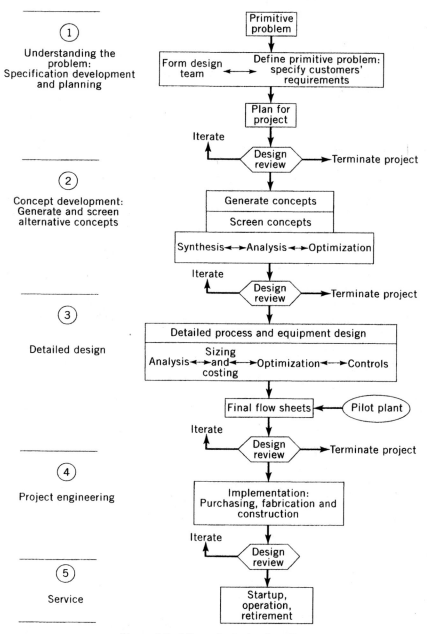

Figure 1.2 Life-cycle design flow chart.

from the very beginning so that decisions can be made earlier and with better knowledge, avoiding delay and errors and shortening the design process.

Concurrent design seeks to combine the efforts of process engineering, controls engineering, manufacturing and construction, service and maintenance, cost accounting, environmental engineering, production engineering, marketing and sales, and so on, into one integrated procedure. To broaden the experience base and widen the scope of thinking, design teams should include nonengineers whenever appropriate, and all participants, technical and nontechnical, should have a voice in fitting their differing, partial contributions into the whole. Active communication among design team members is essential to ensure that problems are identified and dealt with as early as possible. The real concurrency of the approach resides in interactions between groups that historically may not have communicated well with one another, process design and marketing, for instance. Concurrent design aims to overcome the weaknesses inherent in traditional design approaches where evolving designs are passed from department to department with minimal communication, little appreciation of the significance of design decisions made earlier, and a poor understanding of the impact of subsequent changes on such decisions.

Let us now consider briefly the design stages shown in Figure 1.2. The typical design project begins with an idea that something might be worth doing. Thus, in the first stage is the recognition of a need or an economic opportunity. In the second stage the development of concepts for implementing the idea takes place. This is a crucially important stage because decisions made here can determine up to 80% of the total capital cost of a project. In the best design practice, all of the X's in the applicable DFX strategies (Section 1.1) should be identified at these early stages and considered throughout the life-cycle design process. Critical factors that influence the X's should be identified and related to their impact on the X's throughout the life cycle. Such meticulous attention contributes significantly in the realization of the X's in the final design.

In the third stage, the component parts and their interconnections are detailed. As suggested by Figure 1.2, a number of simultaneous, parallel activities occur at this stage. These might deal with further analysis, sizing and costing of equipment, optimization, and controls engineering. The objective is to combine several pieces of equipment into one smoothly running system. When accurate design data are lacking or there is uncertainty about features of the system, laboratory testing, prototyping, or pilot plant tests may be necessary to finalize the design. At stage 4 the effort moves into *project engineering,* which is where the detailed design is turned into a list of actual equipment to be purchased or fabricated. A set of blueprints or the equivalent are developed for the construction phase, and piping and instrumentation diagrams are prepared. This stage also includes the detailed mechanical design of the equipment items and their supports. The end result of the design process is stage 5, the commissioning and safe operation of the system.

As indicated on the flow chart, periodic *design reviews* occur. The design review or design audit is an important aspect of good industrial design practice. Design reviews usually take the form of presentations to managers and other design team members. Reviews may require both written and oral communications. Design reviews are key elements of the concurrent design approach because they provide formal opportunities to share information and solicit input that may improve the design. It is also an opportune time to make necessary schedule adjustments and reallocate resources between budget items. One possible outcome of a design review is the decision to terminate the project. An important question to be asked throughout the design process and especially at design reviews is "Why?" By frequently asking this question, design team members challenge one another to think more deeply about underlying assumptions and the viability of proposed approaches to problems arising during the design process.

To conclude this survey of the design process, note that a fundamental aspect of the process makes it mandatory that knowledge about the project be generated as early as possible: At the beginning of the design process knowledge about the problem and its potential solutions is a minimum, but designers have the greatest freedom of choice because few decisions have been made. As time passes, however, knowledge increases but design freedom decreases because earlier decisions tend to constrain what is allowable later. Indeed, as discussed next, most of the early design effort should focus on establishing an understanding of and the need for the problem.

1.3.2 Understanding the Problem: "What?" not "How?"

Engineers typically deal with ill-defined situations in which the problems are not crisply formulated as in textbooks, the data required are not immediately at hand, decisions about one variable can affect a number of other variables, and economic considerations are of overriding importance. A design assignment also rarely takes the form of a specific system to design. The engineer often receives only a general statement of need or opportunity. This statement constitutes the primitive problem. The following provides an example:

Sample Problem. To provide for a plant expansion, additional power and steam are required. Determine how the power and steam are best supplied.

We will return to this sample problem in subsequent discussions to illustrate various concepts.

The first stage of the design process is to understand the problem. At this stage it must be recognized that the object is to determine what qualities the system should possess and not how a system can be engineered to satisfy the requirements. The question of how enters at the concept development stage. Specifically, at this initial stage the primitive problem must be defined and the requirements to which the system must adhere determined.

One approach to achieving these objectives is a design strategy known as *quality function deployment.* Quality function deployment has several features including the following:

- Identify the customers.
- Determine the customers' requirements:
    a. Classify the requirements as *musts* and *wants.*
    b. Express the requirements in measurable terms.

A premise of quality function deployment is that quality must be *designed in* during the life-cycle design process and not merely tested for at the end. The importance of using the quality function deployment approach at the outset of the design process cannot be overemphasized. The time is well spent, for the methodology aims at making the objectives understood and forces such in-depth thinking that potential design solutions may evolve serendipitously.

It might seem that the manager who relayed the primitive problem is the customer. However, the integrated approach of concurrent design mandates the identification of other customers. If the system is intended for the marketplace, the ultimate consumer would evidently be counted as a customer. But manufacturing personnel, marketing and sales personnel, service personnel, and other individuals from management might all be considered as customers. This also applies to systems destined for in-house use: Customers normally can be identified from among various departments of the company. That the concept of customer should include various internal segments of the company organization stems mainly from the realization that poor communications among the various constituencies involved in a design project not only lengthens the design process making it costlier, but often yields a lower quality outcome as well.

Once the *internal* and *external* customers have been identified, the next step is to determine what requirements the system should satisfy. The key question now for design team members to be asking is "Why?" The team must examine critically every proposed requirement, seeking justification for each. Functional performance, cost, safety, and environmental impact are just a few of the issues that may be among the customers' requirements. *Function analysis,* which aims at expressing the overall function for the design in terms of the conversion of *inputs* to *outputs,* might be applied usefully at this juncture [2].

Successful design teams pay great attention to formulating objectives that are appropriate and attainable. The team must be thorough to reduce the possibility of a new requirement being discovered late in the design process when it may be very costly to accommodate. Still, the list of requirements should be regarded as dynamic and not static. If there is good reason for modifying the requirements during the design process, it is appropriate to do so through negotiations between the design team and customers. An effective and frequently elegant approach to clarifying and ordering the objectives is provided by an *objective tree* [2].

Not all of the requirements are equally important or even absolutely essential. The list of requirements normally can be divided into hard and soft constraints: musts and wants. As suggested by the name itself, the musts are essential and have to be met by the final design. The wants vary in importance and may be placed in rank order. This can be done simply by having each team member rank the wants on a 10-point scale, say, forming a composite ranking, and then finalizing the rank-ordered list by discussion and negotiation. Although the wants are not mandatory, some of the top-ranked ones may be highly desirable and useful in screening alternatives later.

Next, the design team must translate the customers' requirements into quantitative specifications. Requirements should be expressed, where appropriate, in *measurable terms,* for the lack of a measure usually means that the requirement is not well understood. Moreover, it is only with numerical values that the design team can evaluate the design as it evolves. To illustrate, let us return to the sample problem stated earlier. The problem calls for additional power and steam. As these are needed to accommodate a plant expansion, the customers would include the departments requiring the additional power and steam, plant operations, management, and possibly other segments of the company organization. Beginning with the statement of the primitive problem, the design team should question the need and seek justification for such an undertaking. Each suggested requirement should be scrutinized.

Let us assume that eventually the following musts are identified: (1) the power developed must be at least 30 MW, (2) the steam must be saturated or slightly superheated at 20 bars and have a mass flow rate of at least 14 kg/s, and (3) all federal, state, and local environmental regulations must be observed. These environmental-related musts also would be expressed numerically. A plausible want in this case is that any fuel required by the problem solution be among the fuels currently used by the company. Normally many other requirements—both musts and wants—would apply; the ones mentioned here are only representative. We will return to this sample problem later.

1.3.3 Project Management

For success in design, it is necessary to be systematic and thorough. Meticulous attention to detail should be given from the beginning to the end. This requires careful project management. Accordingly, once the customers' requirements have been expressed quantitatively, a plan for the overall project is developed. The plan aims at keeping the project under control and allows for the monitoring of progress relative to goals. The plan typically includes a *work statement, time schedule,* and *budget.* The work statement defines the tasks to be completed, the goal of each task, the time required for each task, and the sequencing of the tasks. These must be expressed with specificity. The personnel requirements for each task have to be detailed, and it is necessary to identify the individual who will provide oversight for each task and be responsible for meeting the goals.

A *Gantt chart,* a form of bar chart, is commonly used in time scheduling. On such a chart each task is plotted against a time schedule in weeks or months, the personnel committed to each task are indicated, and other information such as the timing of design reviews is shown. *The schedule represented on a Gantt chart should be updated as conditions change.* Some tasks may require more time than originally envisioned, and this may dictate changes in the time periods assigned to other tasks. When updating the chart, it is recommended that a different bar shading than used initially be employed to indicate the revised time period for a task. This allows schedule changes to be seen at a glance. A simple Gantt chart is shown in Figure 1.3. As a Gantt chart does not show the interdependencies among the tasks, some special techniques have been developed for planning, scheduling, and control of projects. The most commonly used are the *critical path method* (CPM) and *program evaluation and review technique* (PERT) [11, 12].

The design project budget has a number of components. The largest component is usually personnel salaries. Since personnel may divide their time among two or more projects, this budget component should be keyed to the project work statement and schedule. Another component of the budget accounts for equipment and materials for construction and testing, travel, computer services, and other nonpersonnel expenses. Finally, the *indirect costs* cover personnel benefits, charges for offices, meeting rooms and other facilities, and so on.

1.3.4   Project Documentation

Thorough documentation throughout the project is compulsory. The documents become part of the project *design file.* The design file provides valuable reference material for future projects involving the system designed, documentation to·support patent applications, and documentation to demonstrate that regulatory requirements were fulfilled. In the event of a lawsuit owing to an injury or some loss related to the system, the design file also provides information to demonstrate that professional design methods were employed. Most companies require careful record keeping by the design team. Forms that this takes include the personal *design notebook, technical reports,* and *process flow sheets.*

Design Notebook. The designer's notebook provides a complete record of the work done on the project by that individual in the sequence that the work was completed. The notebook is a personal diary of the individual's contri-

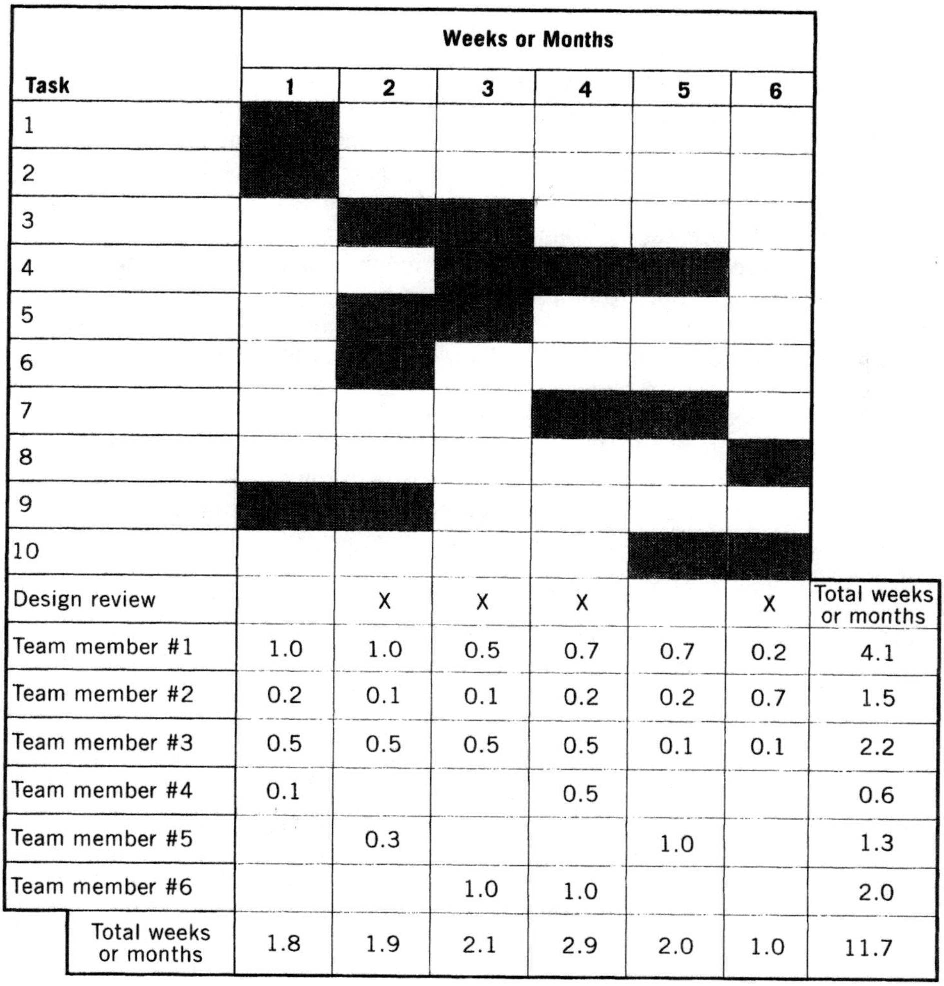

Task	Weeks or Months						
	1	**2**	**3**	**4**	**5**	**6**	
1	■						
2	■						
3		■					
4				■			
5		■					
6		■					
7				■			
8						■	
9	■						
10					■		
Design review		X	X	X		X	Total weeks or months
Team member #1	1.0	1.0	0.5	0.7	0.7	0.2	4.1
Team member #2	0.2	0.1	0.1	0.2	0.2	0.7	1.5
Team member #3	0.5	0.5	0.5	0.5	0.1	0.1	2.2
Team member #4	0.1			0.5			0.6
Team member #5		0.3			1.0		1.3
Team member #6			1.0	1.0			2.0
Total weeks or months	1.8	1.9	2.1	2.9	2.0	1.0	11.7

Figure 1.3 Gantt chart.

butions and should contain all calculations, notes, and sketches that concern the design. Completeness is essential. Although notebook requirements vary with the company, a few guidelines can be suggested: Normally only firmly bound notebooks are used; loose-leaf notebooks are unacceptable. Pages should be numbered consecutively and the notebook should be filled in without leaving blank pages. As entries are made in the notebook, the pages should be signed and dated. Companies also may require that each page be signed by a witness. Photocopied information, data from the literature, computer programs and printouts, and other such information should be glued or stapled into the notebook at appropriate locations. For information that is too lengthy for inserting in the notebook, a note should be entered indicating the nature of the information and where it can be located. There should be no erasures. Previously recorded work found to be incorrect should be crossed out but not obliterated in the event the work would have to be accessed later. A note should be inserted indicating both the source of error and the page where corrected work can be found. The first few pages of the notebook are normally reserved for a table of contents or index, which will allow all work done on a particular topic to be easily located.

Technical Reports. A final technical report is invariably required. Interim reports supporting design reviews and reports to agencies monitoring worker safety, compliance to environmental standards, and so on, are also usually required. Reports should be clear and concise, yet complete. They should be well organized and contain only relevant graphical and tabular data. Schematics and sketches should effectively communicate important features of the design. Technical reports should be written from an outline, both an outline of the overall report and, in finer detail, each subsection of the report. A number of drafts are normally required to achieve a smoothly flowing presentation. Ample time should be scheduled for this purpose.

Technical reports can be structured in various ways, as for example the following:

Summary. This part of the report is an expanded abstract stating the objectives and applied procedures and giving the principal results, conclusions, and recommendations for future action. The summary should be written last after the rest of the report is in final form. The summary should be concise and written in plain language. Since the summary is frequently the only part of the report circulated to and studied by management and other interested parties, it should be self-contained and highly polished.

Main Body. The main body provides details. This part of the report includes an introduction, a survey of the state of the art and the pertinent literature, and a discussion of the procedures used in solving the design problem. The principal results should be presented and discussed. A final section should state and discuss important conclusions and recommendations for future action.

End Matter. Collected at the end of the report are the literature reference list, graphical and tabular data, process flow sheets, layout drawings, cost analysis details, supporting analyses, computer codes and printouts, and other essential supporting information. Complete reference citations should be provided to facilitate subsequent retrieval of the literature.

Proficiency in technical writing requires practice. But the effort is worthwhile because good communications skills are a prerequisite for success in engineering and in nearly every other field. Further discussion of technical report writing relating specifically to thermal system design is provided in the literature [10, 11].

Process Flow Sheets (Flow Diagrams). As indicated in Figure 1.2, the third stage of the design process culminates in the final flow sheets (also known as process flow diagrams). Process flow sheets, analogous to the circuit diagrams of electrical engineering, are conventional and convenient ways to represent process concepts. In a rudimentary form the flow sheet may be no more than a block diagram showing inputs and outputs. More detailed flow sheets include the general types of major equipment required: pressure vessels, compressors, heat exchangers, and so on. Complete flow sheets provide details on the devices to be used, operating temperatures and pressures at key locations, compositions of flow streams, utilities, feed and product stream details, and other important data. As every language, including the language of mathematics, has rules and conventions to foster communication, so does flow sheet preparation. There are standard symbols for various types of equipment and conventions for numbering equipment and designating utility and process streams. Special symbols or conventions are commonly employed to label the temperature, pressure, and chemical composition at various locations. Further details concerning flow sheet preparation are provided elsewhere [10].

Other Documentation. The life-cycle design of a thermal system normally involves documentation in addition to what already has been discussed. This documentation includes, for instance, construction drawings, installation instructions, operating instructions (including how to operate the system in its normal service range and various operating modes: startup, standby, shutdown, and emergency), information about how to cope with equipment failure, diagnostic and maintenance procedures, quality control and quality assurance procedures, and retirement instructions (procedures for decommissioning and disposing of the system). All in all, the documentation required is considerable and underscores the value to industry of engineers with good communication skills.

1.4 THERMAL SYSTEM DESIGN ASPECTS

An integrated, well-structured design process engaging a design team with a broad range of experience and featuring good communications is an approach that can be recommended generally and is not limited to the thermal systems serving as the present focus. Accordingly, much of the discussion of Section

1.3 applies to engineering design of all kinds. In the present section we take a closer look at some key design aspects within the context of thermal systems.

1.4.1 Environmental Aspects

Compliance with governmental environmental regulations has customarily featured an *end-of-the-pipe* approach that addresses mainly the pollutants emitted from stacks, ash from incinerators, thermal pollution, and so on. Increasing attention is being given today to what goes *into the pipe,* however. This is embodied in the concept of *design for the environment* (DFE), in which the environmentally preferred aspects of a system are treated as design objectives rather than as constraints.

In DFE, designers are called on to anticipate negative environmental impacts throughout the life cycle and engineer them out. In particular, efforts are directed to reducing the creation of waste and to managing materials better, using methods such as changing the process technology and/or plant operation, replacing input materials known to be sources of toxic waste with more benign materials, and doing more in-plant recycling. Concurrent design with its multifaceted approach is well suited for considering environmental objectives at every decision level. Moreover, the quality function deployment design strategy naturally allows for environmental quality to be one of many quality factors taken into account.

Compliance with environmental regulations should be considered throughout the design process and not deferred to the end when options might be foreclosed owing to earlier decisions. Addressing such regulations early may result in fundamentally better process choices that reduce the size of the required cleanup. Costs to control pollution are generally much higher if left for resolution after the facility has begun operation. Still, under the best of circumstances some end-of-the-pipe cleanup might be required to meet federal, state or local environmental regulations. The cost of this may be considerable.

Design engineers should keep current on what is legally required by the federal EPA (Environmental Protection Agency) and OSHA (Occupational Safety and Health Administration) and corresponding state and local regulatory groups. An important reporting requirement is the formulation of an *environmental impact statement* providing a full disclosure of project features likely to have an adverse environmental effect. The report includes the identification of the specific environmental standards that require compliance by the project, a summary of all anticipated significant effluents and emissions, and the specification of possible alternative means to meet standards.

Specification of appropriate pollution control measures necessitates consideration of the type of pollutant being controlled and the features of the available control equipment. The size of the equipment needed is generally related to the quantity of pollutant being handled, and so equipment costs can be reduced by decreasing the volume of effluents. Depending on the nature of the processes taking place in the system, several types of pollution control may be needed: air, water, thermal, solid waste, and noise pollution.

Air pollution control equipment falls into two general types: *particulate removal* by mechanical means, such as cyclones, filters, scrubbers, and precipitators, and *gas component removal* by chemical and physical means, including absorption, adsorption, condensation, and incineration. For *liquid waste* effluents, physical, chemical, and biological waste treatment measures can be used. To avoid costly waste treatment facilities or reduce their cost, it is advisable to consider the recovery of valuable liquid-borne products prior to waste treatment. *Thermal pollution* resulting from the direct discharge of warm water into lakes, rivers, and streams is commonly ameliorated by cooling towers, cooling ponds, and spray ponds.

Solid wastes can be handled by incineration, pyrolysis, and removal to a sanitary landfill adhering to state-of-the-art waste management practices. As for liquid wastes, it is advisable to recover valuable substances from the solid waste before treating it. Coupling waste incineration with steam or hot water generation may provide an economic benefit. For effective and practical *noise* control it is necessary to understand the individual equipment and process noise sources, their acoustic properties and characteristics, and how they interact to cause the overall noise problem.

1.4.2 Safety and Reliability

Safety should be designed in from the beginning of the life-cycle design process. As for environmental considerations, the concurrent design approach is well suited for considering safety at every decision level, and quality function deployment naturally allows safety to be one of the quality factors taken into account.

The service life of a system will not be trouble free and the occasional failure of some piece of equipment is likely. The design team is responsible for anticipating such events and designing the system so that a local failure cannot mushroom into an overall system failure or even disaster. A tolerance to failure is an important feature of every system. One approach for instilling such a tolerance involves testing the response of each component via computer simulation at extreme conditions that are not part of the normal operating plan.

Personnel safety is an area where there can be no compromise. Safety studies should be undertaken throughout the design process. Deferring safety issues to the end is unwise because decisions might have been made earlier that foreclose effective alternatives. Hazards have to be anticipated and dealt with; exposure to toxic materials should be prevented or minimized; machinery must be guarded with protective devices and placarded against unsafe uses; and first-aid and medical services must be planned and available when needed. The design team should use safety checklists for identifying hazards that are provided in the literature [1, 11]. The design team must be aware of the requirements of the federal Occupational Safety and Health Act and applicable state and local requirements.

Published codes and standards must also be considered. For some types of systems both the required design calculations and performance levels to be achieved are specified in design codes and standards promulgated by government, professional societies, and manufacturing associations. The ASME (American Society of Mechanical Engineers) *Performance Test Codes* are well-known examples. The NRC (Nuclear Regulatory Commission) regulations apply to nuclear power generation technologies. The U.S. armed forces standard MIL-STD 882B (System Safety Program Requirements) aims at ensuring safety in military equipment. For shell-and-tube heat exchangers there are the TEMA (Tubular Exchanger Manufacturers Association) and APA (American Petroleum Association) standards. The ANSI (American National Standards Institute) standards may also be applicable. At the outset of the design process it is advisable that relevant codes be identified and accessed for subsequent use.

Reliability is a crucially important feature of systems and products of all kinds. As for other qualities discussed, reliability must be designed in from the beginning and considered at each decision level. Reliability is closely related to *maintainability* and *availability*. Reliability is the probability that a system will successfully perform specified functions for specified environmental conditions over a prescribed operating life. Since no system will exhibit absolute reliability, it is important that the system can be repaired and maintained easily and economically and within a specified time period. Maintainability is the probability that these features will be exhibited by the system. Availability is a measure of how often a system will be, or was, available (operational) when needed. Reliability, availability, and maintainability (RAM) contribute importantly to determining the overall system cost. Further discussion of these important design qualities is found elsewhere [13]. Of particular interest to thermal systems design is UNIRAM, a personal computer software package that has been developed to perform RAM analyses of thermal systems [14].

Ideally, a system is designed so that its performance is insensitive to factors out of the control of designers: external factors such as ambient temperature and air quality, internal factors such as wear, and imperfections related to manufacturing and assembly. The relative influence of different internal and external factors on reliability can sometimes be investigated via computer simulations, but in many cases an experimental approach is required. As traditional experimental approaches are often time consuming and costly, more effective testing strategies have been developed for determining statistically how systems may be designed to allow them to be robust (reliable) in the face of disturbances, variations, and uncertainties. Such methods, including the popular *Taguchi methods* [7], aim to achieve virtually flawless performance economically.

Each piece of equipment of a thermal system should be specified to carry out its intended function. Still, to perform reliably, protect from uncertainty in the design data, and allow for increases in capacity, reasonable *safety factors* are usually applied. It is important to engineer on the safe side; but an indiscriminate application of safety factors is not good practice because it can result in so much overdesign that the system becomes uneconomical.

1.4.3 Background Information and Data Sources

Design engineers should make special efforts to keep current on advances in their fields and allied fields. This includes reading the technical literature, attending industrial expositions and professional society meetings, and developing a network of professional contacts with individuals having kindred interests. Considerable background information and data normally exist for just about every type of design situation. The difficulty is to locate such material and have it available when needed. Both private and public sources of information and data may be available to support a design project.

Private sources of information consist of the usually closely guarded proprietary information accumulated by individual companies. Little of this is ever made available to outsiders. The project file containing correspondence leading up to the formulation of the primitive design problem is an example of a private source of information. Companies also may have a collection of files, reports, and data compilations related to the project under consideration. Special questionnaires soliciting input about the project may be available from respondents both inside and outside the company. *Internal* design standards may fix limits on the choices that can be made throughout the design process. Personal contact with experts within the company is especially recommended, for a wealth of useful information often can be obtained simply by asking.

Public sources of information include the open technical literature. This potentially rich source should be explored, but owing to the rapid growth of the literature it is increasingly difficult to search effectively for specific types of information. This has given rise to commercial *online databases* as a way to facilitate searches. A database is an organized collection of information on particular topics. An online database is one accessible by computer, normally for a fee. A survey of online databases relevant to thermal system design is provided in Reference 15. Included in the survey are thermal property databases and databases on environmental protection, safety and health, and patents.

Handbooks are another valuable source of information [16–18]. Textbooks can also provide background information and data. Review articles and articles describing current technology are published in technical periodicals such as *Energy—The International Journal, Chemical Engineering Progress, ASHRAE Transactions, Chemical Engineering, Hydrocarbon Processing,* and *Power.* Considerable useful information often can be obtained by contacting the authors of pertinent recent publications.

The *Thomas Register* [19] provides an exhaustive listing of manufactured items. An excellent way to obtain information on components and materials is to locate sources from the *Thomas Register* and telephone company representatives. Design engineers often spend considerable time on the telephone with vendor sales representatives or engineering departments to obtain current performance and cost information.

The patent literature is a potentially high payoff public source of information. Though frequently underutilized by industry, the return on time spent studying patents is probably as great as for any other engineering activity. The patent literature not only can provide ideas that can assist in achieving a design solution but also can help in avoiding approaches that will not work. Patents are classified alphabetically by class and subclass in the *Index to Classification.* A weekly publication of the U.S. patent office, the *Official Gazette,* lists in numerical order an abstract of each patent issued in an earlier week. The *Index* and *Gazette* are available in many libraries. To find the numbers of the specific patents that have been filed in a particular class, a computer index can be used: CASSIS (Classification and Search Support Information System).

Codes and standards have been mentioned in Section 1.4.2. The wisdom of adhering to codes and standards cannot be overemphasized. Using them can shorten design time, reduce uncertainty in performance, and improve product quality and reliability. Federal, state, and local codes and standards provide information in the form of allowable limits on the performance of various systems. Useful background information is also available in standards published by professional associations and manufacturers' groups. Courts of law normally consider it a sign of good engineering practice if designs adhere to applicable standards, even if there is no legal compulsion to do so.

1.4.4 Performance and Cost Data

Equipment performance and cost data are required at various stages of the design process. These data might be obtained from vendors located via the *Thomas Register* or other sources mentioned in Section 1.4.3. Detailed cost estimates are conducted by specialists, usually in a cost-estimating department. Working from the details of a completely designed system, this group is normally able to estimate costs within ±5% for a plant without major new components. Design engineers frequently make approximate cost estimates at various stages. Quick, back-of-the-envelope calculations may have an inaccuracy of estimate of ±50% or more. For more detailed calculations, estimates may be in the ±10 to ±30% range.

To facilitate equipment cost estimating, relatively easy-to-use tabular and graphical compilations are provided in References 9–11 and the accompanying reference lists. The design engineer must also estimate the final product cost. The total product cost is the sum of costs related to investment, resources and utilities, and labor. The subject of cost calculation in design is detailed in Chapter 7.

1.5 CONCEPT CREATION AND ASSESSMENT

In this section we consider the most important stage of the design process, that of creating and evaluating alternative design concepts. Although success at this stage is crucial, only general guidelines can be suggested. Conceptual designers rely heavily on their practical experience and innate creativity, and these qualities are not readily transferred to others. Concept creation and evaluation are actually ongoing design activities not limited to any particular stage of the design process. Accordingly, the present discussion provides guidance for all such activities whenever they may occur.

1.5.1 Concept Generation: "How?" not "What?"

The focus of the first stage of the design process (Figure 1.2) is on building the design team's knowledge base. Now, in the second stage, that knowledge is used to develop alternative solutions to the primitive problem. Emphasis shifts from the "What to do" phase to the "How to do it" phase. We now want to consider specifically how the primitive problem can be solved. In some cases the objective is to design something similar to an existing design, and so there are previous solutions to access. In the most favorable outcome, only minor modifications of a previous design may be needed. However, the search for alternatives should not only consider adapting solutions of closely similar problems or combining special features of previous solutions in a fresh way to meet the current need, but also the possibility of achieving novel solutions.

There is generally a relation between the number of alternative solutions generated and their quality: To uncover a few good concepts, several have to be generated initially. Since the methods of engineering analysis and optimization can only polish specific solutions, it is essential that alternatives be generated having sufficient quality to merit polishing. Major design errors more often occur because a flawed alternative has been selected than because of subsequent shortcomings in analysis and optimization. And by the time a better quality alternative is recognized, it may be too late to take advantage of any alternative.

The second stage of the design process opens, then, by identifying as many alternative concepts as possible. Imagination should be given free rein and uninhibited thinking should be encouraged. New technology should be carefully considered; in addition, concepts that may have been unusable in the past should not be passed over uncritically because they may now be viable owing to changes in technology or costs. The sources of background information cited in Section 1.4.3 should be explored for promising alternatives.

The commonly used *brainstorming* approach may be effective on an individual or group basis. Another approach is based on *analogical thinking*. The objective of this approach is to express the problem in terms of more familiar analogies that are then used to open up avenues of inquiry [2]. There is some evidence that individuals working independently generate more ideas, and frequently more creative ideas, than do groups. Still, design team brainstorming, analogical thinking, and other group-interactive strategies can be effective in refining a list of alternatives, developing new ideas, and reducing the chance that a good approach is overlooked.

All ideas generated during a brainstorming session should be recorded, even if at first glance some may look impractical or unworkable. The goal should be to generate as many ideas as possible but not evaluate them. Evaluation occurs later. Thus, no team member should be allowed to comment critically until after the brainstorming session ends. Ideas may come quickly at first, but the initial surge is often followed by a fallow period when few ideas are forthcoming. The design team should keep working, however, until there is a consensus that the group has run out of ideas and exhausted the possibilities.

The concept creation phase just considered can lead to a number of plausible alternative solutions to the primitive problem differing from one another in basic concept and detail. For example, alternative concepts for the sample problem introduced in Section 1.3.2 might include the following:

- Generate all of the steam required in a boiler and purchase the required power from the local utility.
- Cogenerate steam and power. Generate all of the steam required and:
  a. Generate the full electricity requirement.
  b. Generate a portion of the electricity requirement. Purchase the remaining electricity needed from the utility.
  c. Generate more than the electricity requirement. Sell the excess electricity to the utility.

Each cogeneration alternative may be configured using a steam turbine system, a gas turbine system, or a combined steam and gas turbine system. Moreover, the type of fuel (coal, natural gas, oil) introduces still more flexibility. Thus, there are many alternative concepts to explore.

As the design of a thermal system is a significant undertaking involving considerable time and expense, no design effort should be wasted on alternatives lacking merit. Accordingly, the alternative concepts must be screened to eliminate all but the most worthwhile. Concept screening is considered next.

1.5.2 Concept Screening

In this section, we discuss concept screening and related issues, beginning with the observation that there are perils associated with screening. One is that an inferior concept will be retained only to be discarded later after considerable additional effort has been expended on it. Another more serious peril is that the best alternative is unwittingly screened out. Still, it is necessary to carefully make choices among alternatives.

Decision Matrix. The decision matrix is a formal procedure for evaluating alternative concepts. This method can be used for screening at any stage of the design process. To illustrate its use, let us consider the preliminary screening of alternative design concepts. In such an application, each team member would rate the alternatives numerically against criteria related to the musts and wants discussed in Section 1.3.2. Ideally, the quantitative measures established for each of these would be the basis for comparison. However, when only sketchy alternative concepts lacking in specifics are being compared, this would not be feasible. The basis for comparison might then be the customers' requirements expressed verbally and not quantitatively.

A sample decision matrix as might be prepared by an individual team member is shown in Figure 1.4. Note that the first alternative has been selected as a reference case and each of the other alternatives is evaluated on how well it competes against the reference case in satisfying the criteria. In the figure, +1 is used to signify that the alternative is better in satisfying a given criterion than the reference case, 0 is used if the alternative and reference case are about the same in satisfying the criterion, and −1 is used if the alternative is not as good as the reference. A positive total score for an alternative indicates that it is better overall in satisfying the criteria than the reference case. The reference case might be selected independently by each individual, or all design team members might use the same alternative concept as the reference case.

Since the criteria employed are not equally important, a variation of the matrix method uses different weighting factors for the criteria. Another variation employs a wider scoring range (+5 to −5, say) to allow for finer judgments. During discussion of the decision matrices prepared by several evaluators, some of the alternative concepts might be eliminated, new concepts might be suggested, the criteria used for evaluation might be clarified or revised, and so on. The procedure would then be repeated and continued iteratively until all but the best of the alternatives remain. This would conclude the preliminary screening of alternatives. The *decision tree* is another method that can be used to evaluate alternatives [2, 13]. It should be stressed that such methods do not make the decisions but only allow for orderly interaction and decision making by the participants.

Screening Issues. An important screening issue is *appropriate technology*. Mature, state-of-the-art technologies are generally favored. Unproven technology may introduce many uncertainties about safe and reliable operation, design data, costs, and so on. An alternative is likely to be discarded if it is inferior to another concept under consideration. An alternative might also be replaced when on scrutiny it suggests a modification leading to an improved version. Concepts judged as fundamentally unsafe or requiring highly toxic

			ALTERNATIVES			
			1[a]	2	3	4
C R I T E R I A	M U S T S	1	0	+1	−1	0
		2	0	0	0	−1
		3	0	−1	−1	+1
		4	0	+1	+1	0
	W A N T S	1	0	0	−1	0
		2	0	+1	0	+1
		3	0	0	−1	0
		4	0	−1	+1	−1
TOTAL			0	+1	−2	0

[a] Alternative 1 is used as the reference

Figure 1.4 Sample decision matrix.

or hazardous materials would be eliminated. Cost is another important way to decide among alternatives. These are just a few of the general considerations that may apply for screening alternatives. In addition, some project-specific considerations may apply.

Consider again the alternative cogeneration concepts for the sample problem listed in Section 1.5.1. Let us suppose that after preliminary screening the following alternatives have been retained for further screening and evaluation:

- Produce all the steam required in a natural gas-fired boiler. Purchase the electricity from the local utility.
- Employ a coal-fired steam turbine cogeneration system.

- Employ a natural gas-fired gas turbine cogeneration system.
- Employ a natural gas-fired combined steam and gas turbine cogeneration system.

Each cogeneration system might be sized to provide the required steam. If less than the required power would be produced, supplementary electricity would be purchased; excess power produced would be sold to the utility.

1.5.3 Concept Development

In the concept development stage (stage 2 of Figure 1.2), each alternative passing the screening procedure discussed in Section 1.5.2 would be subject to further screening until the preferred design serving as the focus of the detailed design stage (stage 3 of Figure 1.2), emerges. This is known as the *base-case design*. Then, using many of the same methods as in stage 2, concept development would continue in the detailed design stage until the final flow sheets for the base-case design are obtained. The final flow sheets fully specify the equipment items and the interconnections among them required to meet all specifications.

The concept development stage is idealized in Figure 1.5 in terms of three interrelated steps: synthesis, analysis, and optimization. *Synthesis* is concerned with putting together separate elements into a whole. In this step the particular equipment items making up the overall thermal system and their interconnections are specified. The schematics of Figure 1.6 show possible outcomes of the synthesis step for the three cogeneration alternatives listed at the close of Section 1.5.2. Synthesis is considered in more detail later in the present section.

The objective of the analysis and optimization steps is to identify the preferred configuration from among the configurations synthesized. *Analysis* generally entails thermal analysis (solving mass, energy, and exergy balances, as required), costing and sizing equipment on at least a preliminary basis, and considering other key issues quantitatively. *Optimization* can take two general forms: structural optimization and parameter optimization.[1] In *structural optimization* the equipment inventory and/or interconnections among the equipment items are altered to achieve a superior design. Structural optimization is indicated in Figure 1.5 by the return arrow linking optimization to synthesis. In *parameter optimization* the pressures, temperatures, and chemical compositions at various state points and/or other key system variables are determined, at least approximately, with the aim of satisfying some desired objective—minimum total cost of the system product(s), for example. At this juncture, the term optimization means only improvement toward the desired

---

[1]The term *parameter* is used here generically to denote independent variables.

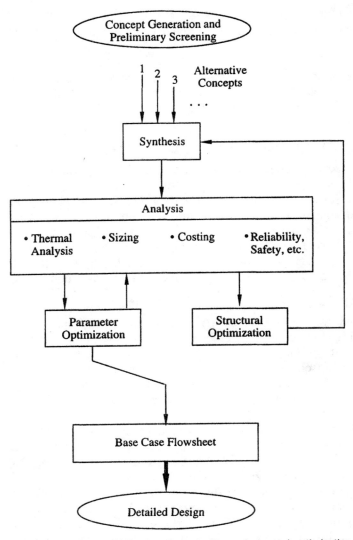

Figure 1.5 Concept development: synthesis, analysis, and optimization.

objective, however. As shown in Figure 1.2, the detailed design stage (stage 3) normally includes a more thorough optimization effort aimed at finalizing the design.

Although the goal of the concept development stage is clear, means for achieving it are not, for there is no generally accepted step-by-step procedure for arriving at a preferred design beginning from one or more alternative concepts. There are, however, several considerations that apply broadly. When

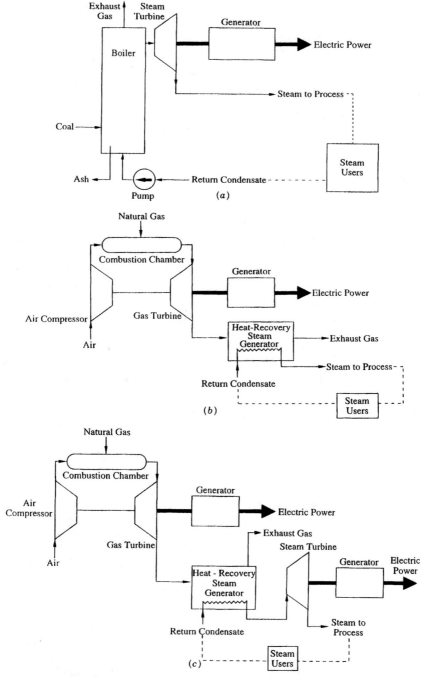

Figure 1.6   Cogeneration system alternatives: (a) coal-fired steam turbine cogeneration system; (b) Gas turbine cogeneration system; (c) combined steam and gas turbine cogeneration system.

synthesizing a design, it is wise to start simply and add detail as decisions are made and avenues of inquiry open up. Ideally a *hierarchy of decision levels* is traversed, with the amount of detail and the accuracy of the calculations increasing with each successive decision level. With this approach a flow sheet evolves in a step-by-step manner, beginning with the most important details and continuing in ever finer detail. Each decision level should involve an economic evaluation so that later decisions rest on and are guided by the economic evaluations at earlier levels.

Proces synthesis has an inherently *combinatorial* nature. Take, for example, the design of a thermal system in which several liquid or gas streams enter the system and several such streams exit. Within the system the streams may interact in various ways, reacting chemically, exchanging energy by heat transfer, undergoing pressure changes with an associated development of or requirement for power, and so on. The number of possible flow sheets for such a system might be considerable: For complex systems the number of flow sheets might be of the order of 10^6, for example. Each flow sheet would be described in terms of numerous equations, perhaps hundreds of equations, and there may be a large number of optimization variables associated with each alternative. For systems of practical interest this combinatorial aspect soon becomes challenging, and an approach based on the complete enumeration and evaluation of all alternatives is not viable, even with the use of computers.

Engineers have traditionally approached such daunting design problems using experience, inventiveness, and extensive iteration as alternatives are evaluated against various criteria. To avoid the need to consider all possible alternatives, plausible but fallible screening rules are typically used to eliminate cases lacking merit and determine feasible alternatives relatively quickly. Such design guidelines are drawn from the experience of designers who have solved similar problems and recorded common features of their solutions. The use of design guidelines does not assure the discovery of a satisfactory design, let alone an optimal design, but only provides a reasonable exploratory approach to design.

A great many design guidelines have been reported in the literature. Table 1.1 provides a sampling. Additional design guidelines are provided in Sections 3.6 and 9.3. The table entries are organized under two headings: general guidelines and process/system-specific guidelines. The process/system-specific guidelines are based on reasoning from the second law of thermodynamics. In selecting processes and equipment, it is not enough that the second law merely be satisfied. Such selections should not introduce glaring design errors: gratuitous sources of irreversibility or needless capital expenditures. To reduce the chance of a fundamentally flawed design, the design team must be fully aware of second-law constraints and opportunities. Accordingly, at each decision level second-law reasoning, as developed in this book, should complement economic evaluation in selecting processes and establishing conditions.

Table 1.1 Design guidelines[a]

General

Keep it simple.

Consider state-of-the-art technology, at least initially.

Consider standard equipment whenever possible.

When assessing the possibility of improving performance, consider the overall system and not just individual components or processes.

When assessing the possibility of improving a particular process, first check whether the process is necessary. Do not overlook the impact of a modification of that process on other processes.

Process/System Specific[b]

Avoid processes requiring excessively large or excessively small thermodynamic driving forces (differences in temperature, pressure, composition).

Maximize the use of cogeneration of power and process steam (or hot water).

Minimize the use of throttling. Consider the use of expanders if the power available is greater than 100 kW.

Minimize the mixing of streams with different temperatures, pressures, or chemical compositions.

Minimize the use of combustion. When using combustion, try to preheat the reactants and minimize the use of excess air.

Use efficient pumps, compressors, turbines, and motors.

Avoid unnecessary heat transfer:

1. Avoid heat transfer at high temperatures directly to the ambient or cooling water.
2. Do not heat refrigerated streams with a stream at a temperature above ambient.
3. For heat exchanger networks, consider use of the pinch method[c] [20] and exergy-related methods [8].

---

[a] See Sections 3.6 and 9.3 for additional guidelines and discussion.

[b] These guidelines are based on second-law reasoning.

[c] See Section 9.3 for further discussion.

1.5.4 Sample Problem Base-Case Design

During the concept development stage, the sample problem cogeneration alternatives listed in Section 1.5.2 have led to the simple schematics of Figure 1.6. Additionally, there is the option of producing the steam in a natural gas-fired boiler and purchasing the electricity from the local utility. Let us suppose that appropriate component descriptions and preliminary costing evaluations also have been obtained. The objective now is to identify the alternative that serves as the focus of the detailed design stage. This crucial concept development step requires, however, an economic evaluation of each alternative using methods such as discussed in Chapter 7 or elsewhere [e.g., 21].

By comparing the results from such economic evaluations, the gas turbine cogeneration system emerges as the best option. The concept development stage then continues for this alternative until the base-case design provided in Figure 1.7 and Table 1.2 is obtained. Notice that in accordance with one

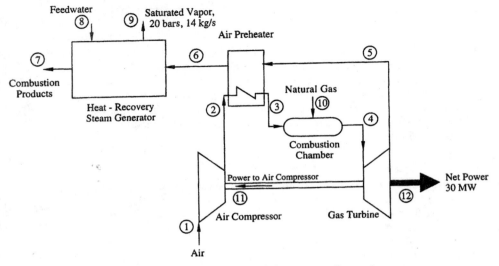

Figure 1.7   Base-case design of the cogeneration system.

Table 1.2   Mass flow rate, temperature, and pressure data for the cogeneration system of Figure 1.7[a]

State	Substance	Mass Flow Rate (kg/s)	Temperature (K)	Pressure (bars)
1	Air[b]	91.2757	298.150	1.013
2	Air	91.2757	603.738	10.130
3	Air	91.2757	850.000	9.623
4	Combustion products[c,d]	92.9176	1520.000	9.142
5	Combustion products	92.9176	1006.162	1.099
6	Combustion products	92.9176	779.784	1.066
7	Combustion products	92.9176	426.897	1.013
8	Water	14.0000	298.150	20.000
9	Water	14.0000	485.570	20.000
10	Methane[e]	1.6419	298.150	12.000[f]

[a] Data are shown with more significant figures than justifiable in practice to allow for ease of checking computer-generated values.
[b] Molecular weight M = 28.649. Molar analysis (%): 77.48 N_2, 20.59 O_2, 0.03 CO_2, 1.90 H_2O(g).
[c] Molecular weight M = 28.254. Molar analysis (%): 75.07 N_2, 13.72 O_2, 3.14 CO_2, 8.07 H_2O(g).
[d] Heat transfer from combustion chamber is estimated as 2% of lower heating value of fuel (Section 2.5).
[e] Molecular weight M = 16.043.
[f] Nominal value. See Sections 2.5 and 9.5 for discussion of p_{10}.

of the design guidelines of Table 1.1 an air preheater has been incorporated in the design. The configuration of Figure 1.7 then serves as the focus of the detailed design stage, namely stage 3 of Figure 1.2. This configuration is analyzed from the exergy viewpoint in Section 3.5.1 and from the perspective of thermoeconomics in Section 8.3.1.

1.6 COMPUTER-AIDED THERMAL SYSTEM DESIGN

As thermal system design involves considerable analysis and computation, including the mathematical modeling of individual components and the entire system, use of computers can facilitate and shorten the design process. Although not realized uniformly in each instance, benefits of computer-aided thermal system design may include increased engineering productivity, reduced design costs, and results exhibiting greater accuracy and internal consistency. This section provides a brief overview of computer-aided thermal system design.

1.6.1 Preliminaries

Information on computer-aided thermal system design can be found in the current engineering literature. For example, computer hardware and software developments are surveyed in the Chemputer and Software sections of the periodicals *Chemical Engineering* and *Chemical Engineering Progress,* respectively. The CPI Software and Software Exchange sections of *Chemical Processing* and *Mechanical Engineering,* respectively, also report on software of value to computer-aided thermal design.

The extent to which computer-aided thermal design can be applied is limited by the availability of property data in suitable forms. Accordingly, such data are vitally important. One respected source of property data is a program developed by the Design Institute for Physical Properties under the auspices of the American Institute of Chemical Engineers. Other noteworthy programs include those developed by the Institution of Chemical Engineers (London) and Germany's DECHEMA.

Computer-aided design also relies heavily on suitable process equipment design programs. Libraries of programs are available for designing or rating one of the most common thermal system components: heat exchangers. Widely used heat exchanger programs stem from the large-scale research efforts of Heat Transfer Research, Inc. (United States) and the Heat Transfer and Fluid Flow Service (United Kingdom). Software for numerous other types of equipment, including piping networks, are available in the literature. Many companies also have developed proprietary software.

1.6.2 Process Synthesis Software

In light of the explosively combinational nature of process synthesis noted previously, efforts have been directed in recent years to making process synthesis more systematic, efficient, and rapid. With the advent of very high speed computers and rapidly improving software, engineers are increasingly resorting to computer-oriented means for this purpose, and a considerable body of literature has been developed [22]. Though still in its infancy, one direction that this has taken is the development of *expert systems* for synthesizing flow sheets based on the experience of specialists and using principles from the field of artificial intelligence.

The *process invention procedure* is a hierarchical expert system for the synthesis of process flow sheets for a class of petrochemical processes [23]. It combines qualitative knowledge in the form of heuristics with quantitative knowledge in the form of design and cost models. A *knowledge-based* approach to flow sheet synthesis of thermal systems with heat–power–chemical transformations is presented in Reference 24. Another expert system for the design of thermal systems is discussed in Reference 25. This procedure aims at synthesizing a flow sheet starting from a list of components stored in a database. The design guidelines are drawn from the second law of thermodynamics and correspond closely to those listed in Reference 8.

In their present states of development such expert systems provide plausible means for synthesizing flow sheets. They aim at inventing feasible designs but not necessarily a final design. They allow for rapid screening of alternatives and obtaining first estimates of design conditions. The output of these procedures can be used as part of the input to one of the conventional simulators considered next.

1.6.3 Analysis and Optimization: Flowsheeting Software

Greater success has been achieved thus far in applying computer aids to analysis and parameter optimization than to process synthesis. This type of application is commonly called *flowsheeting* or *process simulation*. Flowsheeting software allows the engineer to model the behavior of a system, or system components, under specified conditions and do the thermal analysis, sizing, costing, and optimization required for concept development. *Spreadsheet software* is a less sophisticated but still effective approach for a wide range of applications. Such software has become popular because of its availability for microcomputers at reasonable cost, ease of use, and flexibility [11, 26].

Flowsheeting has developed along two lines: the sequential-modular approach and the equation-solving approach. In the sequential-modular approach, library models associated with the various components of a specified flow sheet are called in sequence by an executive program using the output stream data for each component as the input for the next component.

In the equation-solving approach all of the equations representing the individual flow sheet components and the links between them are assembled as a set of equations for simultaneous solution. Most of the more widely used flowsheeting programs: ASPEN PLUS, PROCESS, and CHEMCAD are of the sequential-modular type. SPEEDUP is one of the more widely adopted programs of the equation-solver type. A survey of the capabilities of 15 commercially available process simulators is reported in Reference 27. Of these, 10 are sequential-modular and 5 are equation solvers. A brief description of the features of each simulator is given, including options for sizing and costing equipment, computer operating systems under which the simulator is available, and optimization capabilities (if any).

Optimization deserves a special comment, for one of the main reasons for developing a flow sheet simulation is system improvement. And for complex thermal systems described in terms of a large number of equations, including nonlinear equations and nonexplicit variable relationships, the term optimization implies improvement rather than calculation of a global mathematical optimum. Many of the leading sequential-modular and equation-solving programs have optimization capabilities, and this technology is rapidly improving. Vendors should be contacted for up-to-date information concerning the features of flowsheeting software, including optimization capabilities.

Conventional optimization procedures may suffice for relatively simple thermal systems; though even for such systems cost and performance data are seldom in the form required for optimization. Moreover, with increasing system complexity these conventional methods can become unwieldy, time consuming, and costly. The method of *thermoeconomics* may then provide a better approach, particularly when chemical reactions are involved, regardless of the amount of information on cost and performance. Thermoeconomics aims to facilitate feasibility and optimization studies during the design phase of new systems and process improvement studies of existing systems; it also aims to facilitate decision making related to plant operation and maintenance. Knowledge developed via thermoeconomics assists materially in improving system efficiency and reducing the product costs by pinpointing required changes in structure and parameter values. Chapters 8 and 9 present the fundamentals of thermoeconomics.

1.7 CLOSURE

In the remainder of the book we build on the presentation of this chapter to provide a comprehensive and rigorous introduction to thermal system design and optimization from a contemporary perspective. The presentation of design topics initiated in this chapter continues in Chapter 2 ("Thermodynamics, Modeling, and Design Analysis") with the development of a thermodynamic model for the case study cogeneration system of Figure 1.7 and a dicsussion of piping system design. Chapter 3 ("Exergy Analysis") reinforces the discussion of design guidelines presented in Table 1.1 by introducing additional design guidelines evolving from reasoning using the second law of thermodynamics and, in particular, the exergy concept.

Chapter 4 ("Heat Transfer, Modeling, and Design Analysis"), Chapter 5 ("Applications with Heat and Fluid Flow"), and Chapter 6 ("Applications with Thermodynamics and Heat and Fluid Flow") all contain design-related material, including heat exchanger design. These presentations are intended to illuminate the design process by gradually introducing first-level design notions such as degrees of freedom, design constraints, and optimization. Chapters 5 and 6 also illustrate the effectiveness of elementary modeling in design. Such modeling is often an important element of the concept development stage of design. Elementary models can also highlight key design variables and relations among them. In some instances, they can lead directly to design solutions.

With Chapter 7 ("Economic Analysis") detailed engineering economic evaluations enter the presentation explicitly and are featured in the remainder of the book. In Chapter 8 ("Thermoeconomic Analysis and Evaluation") an effective and systematic design approach is presented that combines the exergy concept of engineering thermodynamics with principles of engineering economics. Exergy costing methods are introduced and applied in this chapter. These methods identify the real cost sources at the component level. The optimization of the design of thermal systems is based on a careful consideration of these cost sources. In Chapter 9 ("Thermoeconomic Optimization") there is a discussion of the *pinch method* for the design of heat exchanger networks and of the iterative optimization of complex systems. Results of the thermoeconomic optimization of the cogeneration system case study are also presented.

REFERENCES

1. J. M. Douglas, *Conceptual Design of Chemical Processes*, McGraw-Hill, New York, 1988.

2. N. Cross, *Engineering Design Methods*, Wiley, New York, 1989.

3. J. L. Nevins and D. E. Witney, *Concurrent Design of Products and Processes*, McGraw-Hill, New York, 1989.

4. S. F. Love, *Achieving Problem Free Project Management*, Wiley, New York, 1989.

5. S. Pugh, *Total Design: Integrated Methods for Successful Product Engineering*, Addison-Wesley, Wokingham, UK, 1991.

6. *Improving Engineering Design, Designing for Competitive Advantage* (U.S. National Research Council Committee on Engineering Design Theory and Methodologies), National Academy Press, Washington, D.C., 1991.

7. G. Taguchi, *Taguchi on Robust Technology Development*, ASME Press, New York, 1993.

8. D. A. Sama, The use of the second law of thermodynamics in the design of heat exchangers, heat exchanger networks and processes, J. Szargut, Z. Kolenda, G. Tsatsaronis, and A. Ziebik, eds., in *Proc. Int. Conf. Energy Sys. Ecol.,* Cracow, Poland, July 5–9, 1993, pp. 53–76.

9. D. E. Garrett, *Chemical Engineering Economics,* Van Nostrand Reinhold, New York, 1989.

10. G. D. Ulrich, *A Guide to Chemical Engineering Process Design and Economics,* Wiley, New York, 1984.

11. M. S. Peters and K. D. Timmerhaus, *Plant Design and Economics for Chemical Engineers,* 4th ed., McGraw-Hill, New York, 1991.

12. F. J. Gould and G. D. Eppen, *Introductory Management Science,* Prentice-Hall, Englewood Cliffs, NJ, 1984.

13. A. Ertas and J. C. Jones, *The Engineering Design Process,* Wiley, New York, 1993.

14. *User's Guide for the UNIRAM Availability Assessment Methodology: Version 3.0,* EPRI ET-7138s, Electric Power Research Institute, Palo Alto, CA, 1991.

15. A Wealth of Information Online, *Chem. Eng.,* Vol. 96, No. 6, 1989, pp. 112–127.

16. E. A. Avallone and T. Burmeister, *Marks' Standard Handbook for Mechanical Engineers,* 9th ed., McGraw-Hill, New York, 1987.

17. R. H. Perry and D. Green, *Chemical Engineers' Handbook,* 6th ed., McGraw-Hill, New York, 1984.

18. *ASHRAE Handbook 1993 Fundamentals,* American Society of Heating, Refrigerating, and Air Conditioning Engineers, Atlanta, 1993.

19. *Thomas Register of American Manufacturers and Thomas Register Catalog File,* Thomas, New York.

20. B. Linhoff, et. al., *A User Guide on Process Integration for the Efficient Use of Energy,* Institution of Chemical Engineers, Rugby, UK, 1982.

21. H. G. Stoll, *Least-Cost Electric Utility Planning,* Wiley-Interscience, New York, 1989.

22. P. Winter, Computer-aided process engineering: The evolution continues, *Chem. Eng. Prog.,* February, 1992, pp. 76–83.

23. R. L. Kirkwood, M. H. Locke, and J. M. Douglas, A prototype expert system for synthesizing chemical process flow sheets, *Comput. Chem. Eng.,* Vol. 12, 1988, pp. 329–343.

24. A. S. Kott, J. H. May, and C. C. Hwang, An autonomous artificial designer of thermal energy systems: Part 1 and Part 2, *J. Eng. Gas Turbines and Power,* Vol. 111, 1989, pp. 728–739.

25. R. Melli, B. Paoletti, and E. Sciubba, Design and functional optimization of thermo-mechanical plants via an interactive expert system, G. Tsatsaronis et al., eds., in *Computer-Aided Energy System Analysis,* AES-Vol. 21, ASME, New York, 1990.

26. E. M. Julian, Process modeling on spreadsheet, *Chem. Eng. Prog.,* October, 1989, pp. 33–40.

27. L. T. Biegler, Chemical process simulation, *Chem. Eng. Prog.,* October, 1989, pp. 50–61.

PROBLEMS

1.1 Consider the problem of determining the pipe diameter to use when pumping water at a specified volumetric flow rate from one fixed point to another. Sketch curves giving qualitatively (a) the cost for pumping the water, (b) the cost for the installed piping system, and (c) the total cost, each on an annual basis, versus pipe diameter. Identify the pipe diameter giving the least total cost per year. Discuss.

1.2 An engineering college at a large university is considering a major revision of its curricula, including the possibility of consolidating and/ or eliminating existing departments. Who are the customers that should be consulted? How might a design team be set up for this purpose?

1.3 The program evaluation and review technique (PERT) and critical path method (CPM) are often used for project planning and scheduling. Using the format listed in Section 1.3.4, prepare a brief report describing and contrasting PERT and CPM.

1.4 Together with two co-workers you have agreed to refinish the exterior of a two-story family dwelling. For this project, develop separate lists of (a) what needs to be done and (b) how to do it. Also prepare a Gantt chart and budget.

1.5 Contact a company regularly doing engineering design work. Request sample pages from the company's standard design notebook (photocopies will do) and its policy on the use of design notebooks by engineers.

1.6 List five words you often misspell and three grammatical errors you occasionally make in report writing.

1.7 Obtain federal, state, or local occupational safety and health regulations concerning (a) asbestos use and asbestos removal and (b) mercury use and cleanup of mercury spills. What are the health risks associated with these substances?

1.8 Contact your state environmental protection agency for regulations relating to liquid effluents discharged from industrial plants into lakes, rivers, and streams.

1.9 Using your library's computerized information retrieval system, obtain a printout listing references appearing in the last three years concerning a thermal systems aspect of your choice or one assigned to you.

1.10 Referring to U.S. patent office publications, obtain a copy of a patent granted within the last five years dealing with cogeneration of power and process steam. Outline the claims presented in the patent. Explain how the inventor presented proof of his or her claims.

1.11   Using the *Thomas Register,* locate vendors of gas turbines suitable for electric power generation in the 20–50-MW range. Obtain two vendor quotes for the installed cost of a gas turbine–electric generator system for this application.

1.12   A utility advertises that it is less expensive to heat water for domestic use with natural gas than with electricity. Is this claim correct? Discuss.

1.13   For the purchase of a new automobile for your personal use, develop a list of musts and wants. Use a decision matrix to evaluate three alternative automobiles. Repeat using a decision tree approach.

1.14   Referring to the engineering literature, list three additional design guidelines that do not overlap any of the entries appearing in Table 1.1.

1.15   Energy system simulation procedures include the Newton–Raphson and Hardy–Cross methods. Using the format listed in Section 1.3.4, prepare a brief report describing each method, its strengths and weaknesses, and its typical realm of application.

1.16   The method of Lagrange multipliers is often used to solve constrained optimization problems. Using the format listed in Section 1.3.4, prepare a brief report describing the Lagrange method together with an elementary illustrative example.

1.17   Using the format listed in Section 1.3.4, write a report on engineering ethics and the legal responsibilities of engineers.

2

THERMODYNAMICS, MODELING, AND DESIGN ANALYSIS

In this chapter we present the fundamental principles of engineering thermodynamics required for study of this book and illustrate their use for modeling and design analysis in applications involving engineering thermodynamics and fluid flow. The term *model* refers here to a description, invariably involving idealization to some degree, of a thermal system in terms of a set of mathematical relations. Design analysis simply refers to the reasoning and evaluations that are a normal adjunct of engineering design. The current presentation is introductory. Later chapters of the book provide further illustrations of modeling and design analysis, including applications in the heat transfer realm.

Fundamentals are surveyed in Sections 2.1–2.4. A premise underlying this presentation is that the reader has had an introduction to engineering thermodynamics and fluid flow. Most concepts are discussed only briefly in the belief that this is adequate to spark recall. If further elaboration is required, readers should consult References 1 and 2. In Section 2.5 we illustrate the fundamentals surveyed by presenting a thermodynamic model for the case study cogeneration system of Figure 1.7. In Section 2.6 we consider the modeling and design of piping systems.

2.1 BASIC CONCEPTS AND DEFINITIONS

Classical thermodynamics is concerned primarily with the macrostructure of matter. It addresses the gross characteristics of large aggregations of molecules and not the behavior of individual molecules. The microstructure of matter is studied in kinetic theory and statistical mechanics (including quan-

39

tum thermodynamics). In this book, the classical approach to thermodynamics is used.

2.1.1 Preliminaries

System. In a thermodynamic analysis, the system is whatever we want to study. Normally the system is a specified region that can be separated from everything else by a well-defined surface. The defining surface is known as the *control surface* or *system boundary*. The control surface may be movable or fixed. Everything external to the system is the surroundings. A system of fixed mass is referred to as a *control mass* or as a *closed system*. When there is flow of mass through the control surface, the system is called a *control volume* or *open system*.

State, Property. The condition of a system at any instant of time is called its *state*. The state at a given instant of time is described by the properties of the system. A *property* is any quantity whose numerical value depends on the state but not the history of the system. The value of a property is determined in principle by some type of physical operation or test.

Extensive properties depend on the size or extent of the system. Volume, mass, and energy are examples of extensive properties. An extensive property is additive in the sense that its value for the whole system equals the sum of the values for its parts. *Intensive* properties are independent of the size or extent of the system. Pressure and temperature are examples of intensive properties.

A *mole* is a quantity of substance having a mass numerically equal to its molecular weight. Designating the molecular weight by M and the number of moles by n, the mass m of the substance is $m = nM$. One kilogram mole, designated kmol, of oxygen is 32.0 kg and one pound mole (lbmol) is 32.0 lb. When an extensive property is reported on a unit mass or a unit mole basis, it is called a *specific* property. An overbar is used to distinguish an extensive property written on a per mole basis from its value expressed per unit mass. For example, the volume per mole is $\bar{v}$ whereas the volume per unit mass is v, and the two specific volumes are related by $\bar{v} = Mv$.

Process, Cycle. Two states are identical if, and only if, the properties of the two states are identical. When any property of a system changes in value, there is a change in state, and the system is said to undergo a *process*. When a system in a given initial state goes through a sequence of processes and finally returns to its initial state, it is said to have undergone a *cycle*.

Phase and Pure Substance. The term *phase* refers to a quantity of matter that is homogeneous throughout in both chemical composition and physical structure. Homogeneity in physical structure means that the matter is all *solid* or all *liquid* or all *vapor* (or equivalently all *gas*). A system can contain one or more phases. For example, a system of liquid water and water vapor (steam) contains two phases. A *pure substance* is one that is uniform and invariable in chemical composition. A pure substance can exist in more than one phase, but its chemical composition must be the same in each phase. For example, if liquid water and water vapor form a system with two phases, the system can be regarded as a pure substance because each phase has the same composition.

Equilibrium. *Equilibrium* means a condition of balance. In thermodynamics the concept includes not only a balance of forces but also a balance of other influences. Each kind of influence refers to a particular aspect of thermodynamic, or complete, equilibrium. *Thermal* equilibrium refers to an equality of temperature, *mechanical* equilibrium to an equality of pressure, and *phase* equilibrium to an equality of chemical potentials. *Chemical* equilibrium is also established in terms of chemical potentials. For complete equilibrium the several types of equilibrium must exist individually.

To determine if a system is in thermodynamic equilibrium, one may think of testing it as follows: Isolate the system from its surroundings and watch for changes in its observable properties. If there are no changes, it may be concluded that the system was in equilibrium at the moment it was isolated. The system can be said to be at an *equilibrium state.* When a system is *isolated,* it cannot interact with its surroundings; however, its state can change as a consequence of spontaneous events occurring internally as its intensive properties, such as temperature and pressure, tend toward uniform values. When all such changes cease, the system is in equilibrium. At equilibrium, temperature is uniform throughout the system. Also, pressure can be regarded as uniform throughout as long as the effect of gravity is not significant; otherwise a pressure variation with height can exist, as in a vertical column of liquid.

Temperature. The familiar mercury-in-glass thermometer relates the variation in length of a mercury column with the variation in temperature, thus determining a scale of temperature. The dependence of temperature measurements on a *thermometric substance* such as mercury is not satisfactory, however, for mercury remains a liquid only over a relatively narrow temperature interval. Moreover, the calibration of one thermometer between standard points will not necessarily yield intervals of temperature equal to those of a second thermometer using a different substance. The establishment of a scale of temperature independent of the working substance is clearly desirable. Such a scale is called a *thermodynamic* temperature scale. As discussed in Section 2.1.3, a thermodynamic scale, the Kelvin scale, can be elicited from the second law of thermodynamics. The definition of an absolute temperature following from the second law is valid over all temperature ranges and provides an essential connection between the several empirical measures of temperature.

Among several empirical temperature scales that have been devised, the empirical *gas scale* is based on the experimental observations that (1) at a given temperature level all gases exhibit the same value of the product $p\bar{v}$ (p is pressure and $\bar{v}$ the specific volume on a molar basis) if the pressure is low enough and (2) the value of the product $p\bar{v}$ increases with temperature level. With these points in mind the gas temperature scale is defined by

$$T = \frac{1}{\bar{R}} \lim_{p \to 0} (p\bar{v})$$

where T is temperature and $\bar{R}$ is a constant called the *universal gas constant*. The absolute temperature at the *triple point of water* is fixed by international agreement to be 273.16 K on the Kelvin temperature scale. Then $\bar{R}$ is evaluated experimentally as $\bar{R} = 8.314$ kJ/kmol·K.

2.1.2 The First Law of Thermodynamics, Energy

Energy is a fundamental concept of thermodynamics and one of the most significant aspects of engineering analysis. Energy can be *stored* within systems in various macroscopic forms: kinetic energy, gravitational potential energy, and internal energy. Energy can also be *transformed* from one form to another and *transferred* between systems. For closed systems, energy can be transferred by *work* and *heat transfer.* The total amount of energy is *conserved* in all transformations and transfers. We now organize these ideas into forms suitable for engineering analysis.

Work. In thermodynamics, the term *work* denotes a means for transferring energy. Work is an effect of one system on another, which is identified and measured as follows: Work is done by a system on its surroundings if the sole effect on everything external to the system *could have been* the raising of a weight. Notice that the raising of a weight is in effect a force acting through a distance, and so the work concept of mechanics is included. However, the test of whether a work interaction has taken place is not that the elevation of a weight is actually changed, or that a force actually acted through a distance, but that the sole effect *could be* the change in elevation of a mass. The magnitude of the work is measured by the number of standard weights that could have been raised.

Work done by a system is considered positive in value; work done on a system is considered negative. Using the symbol W to denote work, we have

$$W > 0: \quad \text{work done } by \text{ the system}$$

$$W < 0: \quad \text{work done } on \text{ the system}$$

The time rate of doing work, or power, is symbolized by $\dot{W}$.

Energy. A closed system undergoing a process that involves only work interactions with its surroundings experiences an *adiabatic* process. On the basis of experimental evidence, it can be postulated that when a closed system is altered adiabatically, the amount of work W_{ad} is fixed by the end states of the system and is independent of the details of the process. This postulate, which is one way the *first law of thermodynamics* can be stated, can be made regardless of the type of work interaction involved, the type of process, or the nature of the system.

As the work in an adiabatic process depends on the initial and final states only, it can be concluded that an extensive property can be defined for a system such that its change in value between two states is equal to the work in an adiabatic process that has these as the end states. This property is called *energy.* In engineering thermodynamics the change in the energy of a system is considered to be made up of three macroscopic contributions. One is the change in *kinetic energy* (KE) associated with the motion of the system as a whole relative to an external coordinate frame. Another is the change in gravitational *potential energy* (PE) associated with the position of the system as a whole in Earth's gravitational field. All other energy changes are lumped together in the *internal energy* (U) of the system. Like kinetic energy and gravitational potential energy, internal energy is an extensive property.

Collecting results, the change in energy between two states in terms of the work in an adiabatic process between these states is

$$(KE_2 - KE_1) + (PE_2 - PE_1) + (U_2 - U_1) = -W_{ad} \qquad (2.1)$$

where 1 and 2 denote the initial and final states, respectively, and the minus sign before the work term is in accordance with the previously stated sign convention for work. Since any arbitrary value can be assigned to the energy of a system at a given state 1, no particular significance can be attached to the value of the energy at state 1 or at *any* other state. Only *changes* in the energy of a system have significance.

The specific internal energy is symbolized by u or $\bar{u}$, respectively, depending on whether it is expressed on a unit mass or per mole basis. The specific energy (energy per unit mass) is the sum of the specific internal energy u, the specific kinetic energy $V^2/2$, and the specific gravitational potential energy gz. That is,

$$\text{Specific energy} = u + \tfrac{1}{2}V^2 + gz \qquad (2.2)$$

where V is the velocity and z is the elevation, each relative to a specified datum, and g is the acceleration of gravity.

A property related to internal energy u, pressure p, and specific volume v is *enthalpy,* defined by

$$h = u + pv \tag{2.3a}$$

or on an extensive basis

$$H = U + pV \tag{2.3b}$$

Energy Balance. Closed systems can also interact with their surroundings in a way that cannot be categorized as work, as for example a gas (or liquid) contained in a closed vessel undergoing a process while in contact with a flame. This type of interaction is called a *heat interaction,* and the process can be referred to as a *nonadiabatic* process.

A fundamental aspect of the energy concept is that energy is conserved. Thus, since a closed system experiences precisely the same energy change during a nonadiabatic process as during an adiabatic process between the same end states, it can be concluded that the *net* energy transfer to the system in each of these processes must be the same. It follows that heat interactions also involve energy transfer. Further, the amount of energy Q transferred *to* a closed system in such interactions must equal the sum of the energy change of the system and the amount of energy transferred *from* the system by work. That is,

$$Q = [(KE_2 - KE_1) + (PE_2 - PE_1) + (U_2 - U_1)] + W$$

This expression can be rewritten as

$$(U_2 - U_1) + (KE_2 - KE_1) + (PE_2 - PE_1) = Q - W \tag{2.4}$$

Equation 2.4, called the *closed system energy balance,* summarizes the conservation of energy principle for closed systems of all kinds.

Heat. The quantity denoted by Q in Equation 2.4 accounts for the amount of energy transferred to a closed system during a process by means other than work. On the basis of experiment it is known that such an energy transfer is induced only as a result of a temperature difference between the system and its surroundings and occurs only in the direction of decreasing temperature. This means of energy transfer is called an *energy transfer by heat.* The following sign convention applies:

$$Q > 0: \quad \text{heat transfer } to \text{ system}$$

$$Q < 0: \quad \text{heat transfer } from \text{ system}$$

Methods based on experiment are available for evaluating energy transfer by heat. These methods recognize two basic transfer mechanisms: *conduction* and *thermal radiation.* In addition, theoretical and empirical relationships are available for evaluating energy transfer involving combined modes such as *convection.* Further discussion of heat transfer fundamentals is provided in Chapter 4.

The quantities symbolized by W and Q account for transfers of energy. The terms work and heat denote different *means* whereby energy is transferred and not *what* is transferred. Work and heat are not properties, and it is improper to speak of work or heat "contained" in a system. However, to achieve economy of expression in subsequent discussions, W and Q are often referred to simply as work and heat transfer, respectively. This less formal approach is commonly used in engineering practice.

Power Cycles. Consider a closed system undergoing a thermodynamic cycle. Since energy is a property, the net change in energy over one cycle is zero. It follows from Equation 2.4 that for the cycle

$$Q_{cycle} = W_{cycle}$$

That is, over the cycle the net amount of energy received through heat interactions is equal to the net energy transferred out in work interactions. A *power cycle,* or *heat engine,* is one for which a net amount of energy is transferred out in work interactions: $W_{cycle} > 0$. This equals the net amount of energy received through heat interactions.

From experience it is found that power cycles are characterized both by an addition of energy by heat transfer and an inevitable rejection of energy by heat transfer:

$$Q_{cycle} = Q_A - Q_R$$

where Q_A denotes the total energy added by heat transfer and Q_R the total energy rejected by heat transfer. Combining the last two equations

$$W_{cycle} = Q_A - Q_R$$

The *thermal efficiency* of a heat engine is defined as the ratio of the net work developed to the total energy added by heat transfer:

$$\eta = \frac{W_{cycle}}{Q_A} = 1 - \frac{Q_R}{Q_A} \tag{2.5}$$

Experience with power cycles shows that the thermal efficiency is invariably less than 100%. That is, some portion of the energy Q_A supplied is rejected: $Q_R \neq 0$. Moreover, this condition is one that must be met by all power cycles, however idealized they may be. In other words, no power cycle, real or ideal, can have a thermal efficiency of 100%. Given this fact, it is of interest to determine the maximum theoretical efficiency. These considerations involve the *second law of thermodynamics,* considered next.

2.1.3 The Second Law of Thermodynamics

Many statements of the second law of thermodynamics have been proposed. Each of these can be called a statement of the second law or a corollary of the second law: If one is not valid, all are invalid. In every instance where a consequence of the second law has been tested directly or indirectly by experiment it has been verified. Accordingly, the basis of the second law, like every other physical law, is experimental evidence.

Kelvin–Planck Statement. Among the many alternative statements of the second law, the *Kelvin–Planck statement* is a convenient point of departure for further study. The Kelvin–Planck statement refers to the concept of a *thermal reservoir.* A thermal reservoir is a system that always remains at a constant temperature even though energy is added or removed by heat transfer. A reservoir is an idealization, of course, but such a system can be approximated in a number of ways—by Earth's atmosphere, large bodies of water (lakes, oceans), and so on. Extensive properties of thermal reservoirs, such as internal energy, can change in interactions with other systems even though the reservoir temperature remains constant, however.

The Kelvin–Planck statement of the second law may now be given as follows: It is impossible for any system to operate in a thermodynamic cycle and deliver a net amount of energy by work to its surroundings while receiving energy by heat transfer from a single thermal reservoir. In other words, a perpetual-motion machine of the second kind is impossible. Expressed analytically, the Kelvin–Planck statement is

$$W_{\text{cycle}} \leq 0 \quad \text{(single reservoir)} \tag{2.6}$$

where the words *single reservoir* emphasize that the system communicates thermally only with a single reservoir as it executes the cycle. The less than sign of Equation 2.6 applies when *internal irreversibilities* are present as the system of interest undergoes a cycle and the equal to sign applies only when no irreversibilities are present. The concept of irreversibilities is considered next.

Irreversibilities. A process is said to be *reversible* if it is possible for its effects to be eradicated in the sense that there is some way by which both the system and its surroundings can be exactly restored to their respective initial states. A process is *irreversible* if there is no way to undo it. That is, there is no means by which the system and its surroundings can be exactly restored to their respective initial states. A system that has undergone an irreversible process is not necessarily precluded from being restored to its initial state. However, were the system restored to its initial state, it would not also be possible to return the surroundings to the state they were in initially.

There are many effects whose presence during a process renders it irreversible. These include but are not limited to the following:

- Heat transfer through a finite temperature difference
- Unrestrained expansion of a gas or liquid to a lower pressure
- Spontaneous chemical reaction
- Mixing of matter at different compositions or states
- Friction—sliding friction as well as friction in the flow of fluids
- Electric current flow through a resistance
- Magnetization or polarization with hysteresis
- Inelastic deformation

The term *irreversibility* is used to identify effects such as these.

Irreversibilities can be divided into two classes, *internal* and *external*. Internal irreversibilities are those that occur within the system, while external irreversibilities are those that occur within the surroundings, normally the immediate surroundings. As this division depends on the location of the boundary, there is some arbitrariness in the classification (note that by locating the boundary to take in the immediate surroundings, all irreversibilities are internal). Nonetheless, valuable insights can result when this distinction between irreversibilities is made. When internal irreversibilities are absent during a process, the process is said to be *internally reversible*.

Engineers should be able to recognize irreversibilities, evaluate their influence, and develop cost-effective means for reducing them. However, the need to achieve profitable rates of production, high heat transfer rates, rapid accelerations, and so on invariably dictates the presence of significant irreversibilities. Furthermore, irreversibilities are tolerated to some degree in every type of system because the changes in design and operation required to reduce them would be too costly.

Although improved thermodynamic performance can accompany the reduction of irreversibilities, steps taken in this direction are normally constrained by a number of practical factors often related to costs. As an illustration, consider two bodies at different temperatures and able to communicate thermally. With a *finite* temperature difference between them, a spontaneous heat transfer would take place and, as noted previously, this would be a source of irreversibility. It might be expected that the importance of this irreversibility diminishes as the temperature difference narrows, and this is the case.

As the difference in temperature between the bodies approaches zero, the heat transfer would approach *ideality*. From the study of heat transfer we know that the transfer of a finite amount of energy by heat between bodies whose temperatures differ only slightly requires a considerable amount of time, a large heat transfer surface area, or both. To approach ideality, therefore, a heat transfer would require an infinite amount of time and/or an infinite surface area. Each of these options clearly have cost implications.

Carnot Corollaries. Since no power cycle can have a thermal efficiency of 100%, it is of interest to determine the maximum theoretical efficiency. The maximum theoretical efficiency for systems undergoing power cycles while communicating thermally with two thermal reservoirs at different temperature levels can be evaluated with reference to the following two corollaries of the second law, called the *Carnot corollaries*.

Corollary 1. The thermal efficiency of an irreversible power cycle is always less than the thermal efficiency of a reversible power cycle when each operates between the same two thermal reservoirs.

Corollary 2. All reversible power cycles operating between the same two thermal reservoirs have the same thermal efficiency.

A cycle is considered reversible when there are no irreversibilities within the system as it undergoes the cycle and heat transfers between the system and reservoirs occur ideally (i.e., with a vanishingly small temperature difference). The Carnot corollaries can be demonstrated using the Kelvin–Planck statement of the second law [1].

Kelvin Temperature Scale. Carnot Corollary 2 suggests that the thermal efficiency of a reversible power cycle operating between two thermal reservoirs depends only on the temperatures of the reservoirs and not on the nature of the substance making up the system executing the cycle or the series of processes. Using Equation 2.5, it can be concluded that the ratio of the heat transfers is related only to the temperatures, and is independent of the substance and processes. That is

$$\left(\frac{Q_C}{Q_H}\right)_{\text{rev cycle}} = \psi(T_C, T_H)$$

where Q_H is the energy transferred to the system by heat transfer from a hot reservoir at temperature T_H on a temperature scale to be defined and Q_C is the energy rejected from the system to a cold reservoir at temperature T_C. The words *rev cycle* emphasize that this expression applies only to systems undergoing reversible cycles while operating between the two reservoirs.

The *Kelvin temperature scale* is based on $\psi(T_C, T_H) = T_C/T_H$, [1, 2]. Then

$$\left(\frac{Q_C}{Q_H}\right)_{\text{rev cycle}} = \frac{T_C}{T_H} \qquad (2.7)$$

This equation defines only a ratio of temperatures. The specification of the Kelvin scale is completed by assigning a numerical value to one standard reference state. The state selected is the same used to define the gas scale: At the triple point of water the temperature is specified to be 273.16 K. If a reversible cycle is operated between a reservoir at the reference state temperature and another reservoir at an unknown temperature T, then the latter temperature is related to the value at the reference state by

$$T = 273.16 \frac{Q}{Q'}$$

where Q is the energy received by heat transfer from the reservoir at temperature T and Q' is the energy rejected to the reservoir at the reference state. Thus, a temperature scale is defined that is valid over all ranges of temperature and that is independent of the thermometric substance. Over their common range of definition the Kelvin and gas scales are equivalent.

The *Rankine scale,* the unit of which is the degree Rankine (°R), is proportional to the Kelvin scale: $T(\text{°R}) = 1.8\, T(\text{K})$. Thus, the Rankine scale is also a thermodynamic temperature scale with an absolute zero that coincides with the absolute zero of the Kelvin scale.

Carnot Efficiency. For the special case of a reversible power cycle operating between thermal reservoirs at temperatures T_H and T_C, combination of Equations 2.5 and 2.7 results in

$$\eta_{\text{max}} = 1 - \frac{T_C}{T_H} \qquad (2.8)$$

called the *Carnot efficiency.* By invoking the two Carnot corollaries, it should be evident that this is the efficiency for all reversible power cycles operating between thermal reservoirs at T_H and T_C. Moreover, it is the *maximum theoretical efficiency* that any power cycle, real or ideal, could have while operating between the same reservoirs. As temperatures on the Rankine scale differ from Kelvin temperatures only by the factor 1.8, Equation 2.8 may be applied with either scale of temperature.

Clausius Inequality. Primary consideration has been given thus far to the case of systems undergoing cycles while communicating thermally with two reservoirs, a hot reservoir and a cold reservoir. In the present discussion a corollary of the second law known as the Clausius inequality is introduced that is applicable to any cycle without regard for the body, or bodies, from which the cycle receives energy by heat transfer or to which the cycle rejects energy by heat transfer. The Clausius inequality provides the basis for introducing two ideas instrumental for quantitative evaluations of systems from a second-law perspective: the *entropy* and *entropy generation* concepts.

The Clausius inequality states that

$$\oint \left(\frac{\delta Q}{T} \right)_b \leq 0 \tag{2.9a}$$

where δQ represents the heat transfer at a part of the system boundary during a portion of the cycle, and T is the absolute temperature at that part of the boundary. The symbol δ is used to distinguish the differentials of *nonproperties,* such as heat and work, from the differentials of properties, written with the symbol d. The subscript b serves as a reminder that the integrand is evaluated at the boundary of the system executing the cycle. The symbol $\oint$ indicates that the integral is to be performed over all parts of the boundary and over the entire cycle. The Clausius inequality can be demonstrated using the Kelvin–Planck statement of the second law [1]. The significance of the inequality of Equation 2.9a is the same as in Equation 2.6: The equality applies when there are no internal irreversibilities as the system executes the cycle, and the inequality applies when internal irreversibilities are present.

For subsequent applications, it is convenient to rewrite the Clausius inequality as

$$\oint \left(\frac{\delta Q}{T} \right)_b = -S_{gen} \tag{2.9b}$$

where S_{gen} can be viewed as representing the "strength" of the inequality. The value of S_{gen} is positive when internal irreversibilities are present, zero when no internal irreversibilities are present, and can never be negative. Accordingly, S_{gen} is a measure of the effect of the irreversibilities present within the system executing the cycle. In the next section, S_{gen} is identified as the entropy generated by internal irreversibilities during the cycle.

2.1.4   Entropy and Entropy Generation

Up to this point, the discussion of the second law has been concerned primarily with what it says about systems undergoing thermodynamic cycles. Means are now introduced for analyzing systems from the second-law perspective as they undergo processes that are not necessarily cycles. The property entropy and the entropy generation concept play prominent parts in these considerations.

Entropy. Consider two cycles executed by a closed system. One cycle consists of an internally reversible process A from state 1 to state 2, followed by an internally reversible process C from state 2 to state 1. The other cycle consists of an internally reversible process B from state 1 to state 2, followed by the same process C from state 2 to state 1 as in the first cycle. For these cycles, Equation 2.9b takes the form

$$\left(\int_1^2 \frac{\delta Q}{T}\right)_A + \left(\int_2^1 \frac{\delta Q}{T}\right)_C = -S_{gen} = 0$$

$$\left(\int_1^2 \frac{\delta Q}{T}\right)_B + \left(\int_2^1 \frac{\delta Q}{T}\right)_C = -S_{gen} = 0$$

where S_{gen} has been set to zero since the cycles are composed of internally reversible processes. Subtracting these equations leaves

$$\left(\int_1^2 \frac{\delta Q}{T}\right)_A = \left(\int_1^2 \frac{\delta Q}{T}\right)_B$$

This shows that the integral of $\delta Q/T$ is the same for both processes. Since A and B are arbitrary, it follows that the integral of $\delta Q/T$ has the same value for any internally reversible process between the two states. In other words, the value of the integral depends on the end states only. It can be concluded, therefore, that the integral defines the change in some property of the system. Selecting the symbol S to denote this property, its change is given by

$$S_2 - S_1 = \left(\int_1^2 \frac{\delta Q}{T}\right)_{int\ rev} \tag{2.10}$$

where the subscript int rev is added as a reminder that the integration is carried out for any internally reversible process linking the two states. This property is called *entropy*. Entropy is an extensive property.

Since entropy is a property, the change in entropy of a system in going from one state to another is the same for all processes, both internally reversible and irreversible, between these two states. Thus, Equation 2.10 allows the determination of the change in entropy; once it has been evaluated, this is the magnitude of the entropy change for all processes of the system between these two states. The evaluation of entropy changes is discussed further in Sections 2.3 and 2.4.

As a closed system undergoes an internally reversible process, its entropy can increase, decrease, or remain constant. This can be brought out using the definition of entropy change on a differential basis

$$dS = \left(\frac{\delta Q}{T}\right)_{int \; rev} \tag{2.11}$$

Equation 2.11 indicates that when a closed system undergoing an internally reversible process *receives* energy by heat transfer, it experiences an *increase* in entropy. Conversely, when energy is *removed* from the system by heat transfer, the entropy of the system *decreases*. We can interpret this to mean that an entropy transfer is associated with (or accompanies) heat transfer. The direction of the entropy transfer is the same as that of the heat transfer. In an adiabatic internally reversible process of a closed system the entropy would remain constant. A constant-entropy process is called an *isentropic* process.

On rearrangement, Equation 2.11 becomes

$$(\delta Q)_{int \; rev} = T \, dS$$

Integrating from an initial state 1 to a final state 2

$$Q_{int \; rev} = \int_1^2 T \, dS \tag{2.12}$$

From Equation 2.12 it can be concluded that an energy transfer by heat to a closed system during an internally reversible process can be represented as an area on a plot of temperature versus entropy. On such a plot, the temperature must be in Kelvin or degrees Rankine, and the area is the entire area under the curve representing the process. The area interpretation of heat transfer is not valid for irreversible processes.

Entropy Balance. Consider next a cycle consisting of process *I* from state 1 to state 2, during which internal irreversibilities may be present, followed by an internally reversible process from state 2 to state 1. For this cycle, Equation 2.9b takes the form

$$\int_1^2 \left(\frac{\delta Q}{T}\right)_b + \int_2^1 \left(\frac{\delta Q}{T}\right)_{int \; rev} = -S_{gen} \tag{2.13}$$

where the first integral is for process *I* and the second is for the internally reversible process. The subscript *b* in the first integral emphasizes that the integrand is evaluated at the system boundary. The subscript is not required in the second integral because temperature is uniform throughout the system at each intermediate state of an internally reversible process. Since no irreversibilities are associated with the second process, the term S_{gen}, which accounts for the effect of irreversibilities during the cycle, refers only to process *I*, as this is the only site of irreversibilities.

Applying the definition of entropy change, the second integral of Equation 2.13 can be expressed as

$$S_1 - S_2 = \int_2^1 \left(\frac{\delta Q}{T}\right)_{int\ rev}$$

With this, Equation 2.13 becomes

$$\int_1^2 \left(\frac{\delta Q}{T}\right)_b + (S_1 - S_2) = -S_{gen}$$

On rearrangement, the *closed system entropy* balance results:

$$\underbrace{S_2 - S_1}_{\substack{\text{entropy} \\ \text{change}}} = \underbrace{\int_1^2 \left(\frac{\delta Q}{T}\right)_b}_{\substack{\text{entropy} \\ \text{transfer}}} + \underbrace{S_{gen}}_{\substack{\text{entropy} \\ \text{generation}}} \qquad (2.14)$$

When the end states are fixed, the entropy change on the left side of Equation 2.14 can be evaluated independently of the details of the process. However, the two terms on the right side depend explicitly on the nature of the process and cannot be determined solely from knowledge of the end states. The first term on the right side is associated with heat transfer to or from the system during the process. This term can be interpreted as the entropy transfer associated with (or accompanying) heat transfer. The direction of entropy transfer is the same as the direction of the heat transfer, and the same sign convention applies as for heat transfer: A positive value means that entropy is transferred into the system, and a negative value means that entropy is transferred out.

The entropy change of a system is not accounted for solely by the entropy transfer but is due in part to the second term on the right side of Equation 2.14 denoted by S_{gen}. The term S_{gen} is positive when internal irreversibilities are present during the process and vanishes when internal irreversibilities are absent. This can be described by saying that entropy is generated (or produced) within the system by action of irreversibilities. The second law of thermodynamics can be interpreted as specifying that entropy is generated by irreversibilities and conserved only in the limit as irreversibilities are reduced to zero. Since S_{gen} measures the effect of irreversibilities present within a system during a process, its value depends on the nature of the process and not solely on the end states. The variable S_{gen} is not a property.

When applying the entropy balance, the objective is often to evaluate the entropy generation term. However, the value of the entropy generation for a given process of a system usually does not have much significance by itself. The significance is normally determined through comparison: The entropy generation within a given component might be compared to the entropy generation values of the other components included in an overall system formed by these components. By comparing entropy generation values, the components where appreciable irreversibilities occur can be identified and rank ordered. This allows attention to be focused on the components that contribute most heavily to inefficient operation of the overall system.

To evaluate the entropy transfer term of the entropy balance requires information regarding both the heat transfer and the temperature on the boundary where the heat transfer occurs. The entropy transfer term is not always subject to direct evaluation, however, because the required information is either unknown or not defined, such as when the system passes through states sufficiently far from equilibrium. In practical applications, it is often convenient, therefore, to enlarge the system to include enough of the immediate surroundings that the temperature on the boundary of the *enlarged system* corresponds to the ambient temperature, T_{amb}. The entropy transfer term is then simply Q/T_{amb}. However, as the irreversibilities present would not be just those for the system of interest but those for the enlarged system, the entropy generation term would account for the effects of internal irreversibilities within the original system and external irreversibilities present within that portion of the surroundings included within the enlarged system.

The entropy balance can be expressed in alternative forms that may be convenient for particular analyses. One of these is the *rate form*, given by

$$\frac{dS}{dt} = \sum_j \frac{\dot{Q}_j}{T_j} + \dot{S}_{gen} \qquad (2.15)$$

where dS/dt is the time rate of change of entropy of the system. The term $\dot{Q}_j/T_j$ represents the time rate of entropy transfer through the portion of the boundary whose instantaneous temperature is T_j. The term $\dot{S}_{gen}$ accounts for the time rate of entropy generation due to irreversibilities within the system.

For a system *isolated* from its surroundings, the entropy balance for a process from state 1 to state 2 reduces to

$$(S_2 - S_1)_{isol} = S_{gen}$$

where S_{gen} is the total amount of entropy generated within the isolated system.

Since entropy is generated in all actual processes, the only processes that can occur are those for which the entropy of the isolated system increases. This is known as the *increase of entropy principle*. The increase of entropy principle is sometimes adopted as a statement of the second law. As systems left to themselves tend to undergo processes until a condition of equilibrium is attained, the increase of entropy principle suggests that the entropy of an isolated system increases as the state of equilibrium is approached, with the equilibrium state being attained when the entropy reaches a maximum.

2.2 CONTROL VOLUME CONCEPTS

The engineering applications considered in this book are analyzed on a *control volume* basis. Accordingly, the control volume formulations of the mass, energy, and entropy balances presented in this section play important roles. These are given in the present section in the form of overall balances. Equations of change for mass, energy, and entropy in the form of differential equations are also available in the literature [3].

2.2.1 Mass, Energy, and Entropy Balances

Conservation of Mass. When applied to a control volume, the principle of mass conservation is expressed as follows: The time rate of accumulation of mass within the control volume equals the difference between the total rates of mass flow in and out across the boundary.

An important case for subsequent developments is one for which inward and outward flows occur, each through one or more ports. For this case the conservation of mass principle takes the form

$$\frac{dm_{cv}}{dt} = \sum_i \dot{m}_i - \sum_e \dot{m}_e \tag{2.16}$$

The left side of this equation represents the time rate of change of mass contained within the control volume, $\dot{m}_i$ denotes the mass flow rate at an inlet port, and $\dot{m}_e$ is the mass flow rate at an outlet port.

The *volumetric flow rate* through a portion of the control surface with area dA is the product of the velocity normal to the area times the area: $V_n \, dA$ where the subscript n signifies that the velocity component normal to the area is intended. The *mass flow rate* through dA is found by multiplying the volume flow rate by the local fluid density: $\rho(V_n dA)$. The mass rate of flow through a port of area A is then found by integration over the area

$$\dot{m} = \int_A \rho V_n \, dA$$

One-dimensional flow is frequently assumed. One-dimensional flow means that the flow is normal to the boundary at locations where mass enters or exits the control volume, and the velocity and all other intensive properties do not vary with position across inlet and outlet ports. For one-dimensional flow, the last equation becomes

$$\dot{m} = \rho V A = \frac{VA}{v} \tag{2.17}$$

where v denotes the specific volume.

Control Volume Energy Balance. When applied to a control volume, the principle of energy conservation is expressed as follows: The time rate of accumulation of energy within the control volume is equal to the excess of the incoming rate of energy over the outgoing rate of energy. Energy can enter and exit a control volume by work and heat transfer. Energy also enters and exits with flowing streams of matter. Accordingly, for a control volume with one-dimensional flow at a single inlet and a single outlet

$$\frac{d(U + KE + PE)_{cv}}{dt} = \dot{Q} - \dot{W} + \dot{m}\left(u_i + \tfrac{1}{2}V_i^2 + gz_i\right)$$

$$- \dot{m}\left(u_e + \tfrac{1}{2}V_e^2 + gz_e\right) \tag{2.18}$$

where the terms in brackets account for the specific energy of the incoming and outgoing streams. The terms $\dot{Q}$ and $\dot{W}$ account, respectively, for the net rates of energy transfer by heat and work.

Because work is always done on or by a control volume where matter flows across the boundary, it is convenient to separate the term $\dot{W}$ of Equation 2.18 into two contributions: One contribution is the work associated with the force of the fluid pressure as mass is introduced at the inlet and removed at the exit. The other contribution, denoted as $\dot{W}_{cv}$, includes all other work effects, such as those associated with rotating shafts, displacement of the boundary, and electrical effects. The work rate concept of mechanics allows the first of these contributions to be evaluated in terms of the product of the pressure force and velocity at the point of application of the force. Thus, at outlet e

$$\begin{bmatrix} \text{Time rate of energy transfer} \\ \text{by work } \textit{from} \text{ the control} \\ \text{volume at outlet } e \text{ associated} \\ \text{with the fluid pressure} \end{bmatrix} = (p_e A_e) \mathbf{V}_e$$

A similar expression can be written for the rate of energy transfer by work into the control volume at inlet i. With these considerations, the work term $\dot{W}$ of Equation 2.18 can be expressed as

$$\dot{W} = \dot{W}_{cv} + (p_e A_e)\mathbf{V}_e - (p_i A_i)\mathbf{V}_i$$

or with Equation 2.17 as

$$\dot{W} = \dot{W}_{cv} + \dot{m}_e(p_e v_e) - \dot{m}_i(p_i v_i) \tag{2.19}$$

where $\dot{m}_i$ and $\dot{m}_e$ are the mass flow rates and v_i and v_e are the specific volumes evaluated at the inlet and outlet, respectively. The terms $\dot{m}_i(p_i v_i)$ and $\dot{m}_e(p_e v_e)$ account for the work associated with the pressure at the inlet and outlet, respectively, and are commonly referred to as *flow work*. The previously stated sign convention for work applies to Equation 2.19.

Substituting Equation 2.19 into Equation 2.18, and collecting all terms referring to the inlet and the outlet into separate expressions, the following form of the control volume energy rate balance results:

$$\frac{d(U + KE + PE)_{cv}}{dt} = \dot{Q} - \dot{W}_{cv} + \dot{m}_i \left(u_i + p_i v_i + \tfrac{1}{2}\mathbf{V}_i^2 + gz_i \right)$$

$$- \dot{m}_e \left(u_e + p_e v_e + \tfrac{1}{2}\mathbf{V}_e^2 + gz_e \right)$$

On introducing the specific enthalpy h (Equation 2.3a), the energy rate balance becomes simply

$$\frac{d(U + KE + PE)_{cv}}{dt} = \dot{Q}_{cv} - \dot{W}_{cv} + \dot{m}_i \left(h_i + \tfrac{1}{2}\mathbf{V}_i^2 + gz_i \right)$$

$$- \dot{m}_e \left(h_e + \tfrac{1}{2}\mathbf{V}_e^2 + gz_e \right) \tag{2.20}$$

where the subscript cv has been added to $\dot{Q}$ to emphasize that this is the heat transfer rate over the boundary (control surface) of the control volume.

To allow for applications where there may be several locations on the boundary through which mass enters or exits, the following expression is appropriate:

$$\frac{d(U + KE + PE)_{cv}}{dt} = \dot{Q}_{cv} - \dot{W}_{cv} + \sum_i \dot{m}_i \left(h_i + \tfrac{1}{2}V_i^2 + gz_i \right)$$

$$- \sum_e \dot{m}_e \left(h_e + \tfrac{1}{2}V_e^2 + gz_e \right) \qquad (2.21)$$

Equation 2.21 is an *accounting* balance for the energy of the control volume. It states that the time rate of accumulation of energy within the control volume equals the difference between the total rates of energy transfer in and out across the boundary. The mechanisms of energy transfer are heat and work, as for closed systems, and the energy accompanying the entering and exiting mass.

Control Volume Entropy Balance. Entropy, like mass and energy, is an extensive property, so it too can be transferred into or out of a control volume by streams of matter. Since this is the principal difference between the closed system and control volume forms, the control volume entropy rate balance can be obtained by modifying Equation 2.15 to account for these entropy transfers. The result is

$$\underbrace{\frac{dS_{cv}}{dt}}_{\substack{\text{rate of} \\ \text{entropy} \\ \text{change}}} = \underbrace{\sum_j \frac{\dot{Q}_j}{T_j} + \sum_i \dot{m}_i s_i - \sum_e \dot{m}_e s_e}_{\substack{\text{rates of} \\ \text{entropy} \\ \text{transfer}}} + \underbrace{\dot{S}_{gen}}_{\substack{\text{rate of} \\ \text{entropy} \\ \text{generation}}} \qquad (2.22)$$

where dS_{cv}/dt represents the time rate of change of entropy within the control volume. The terms $\dot{m}_i s_i$ and $\dot{m}_e s_e$ account, respectively, for rates of entropy transfer into and out of the control volume associated with mass flow. One-dimensional flow is assumed at locations where mass enters and exits. The term $\dot{Q}_j$ represents the time rate of heat transfer at the location on the boundary where the instantaneous temperature is T_j. The ratio $\dot{Q}_j/T_j$ accounts for the associated rate of entropy transfer. The term $\dot{S}_{gen}$ denotes the time rate of entropy generation due to irreversibilities *within* the control volume. When a control volume comprises a number of components, $\dot{S}_{gen}$ is the sum of the rates of entropy generation of the components.

2.2.2 Control Volumes at Steady State

Engineering systems are often idealized as being at *steady state,* meaning that all properties are unchanging in time. For a control volume at steady state, the identity of the matter within the control volume changes continuously, but the total amount of mass remains constant. At steady state, Equation 2.16 reduces to

$$\sum_i \dot{m}_i = \sum_e \dot{m}_e \tag{2.23a}$$

The energy rate balance at steady state is

$$0 = \dot{Q}_{cv} - \dot{W}_{cv} + \sum_i \dot{m}_i \left(h_i + \tfrac{1}{2}V_i^2 + gz_i \right) - \sum_e \dot{m}_e \left(h_e + \tfrac{1}{2}V_e^2 + gz_e \right) \tag{2.23b}$$

obtained by reducing Equation 2.21.

At steady state, the entropy rate balance is

$$0 = \sum_j \frac{\dot{Q}_j}{T_j} + \sum_i \dot{m}_i s_i - \sum_e \dot{m}_e s_e + \dot{S}_{gen} \tag{2.23c}$$

obtained by reducing Equation 2.22.

Mass and energy are conserved quantities, but entropy is not generally conserved. Equation 2.23a indicates that the total rate of mass flow into the control volume equals the total rate of mass flow out of the control volume. Similarly, Equation 2.23b states that the total rate of energy transfer into the control volume equals the total rate of energy transfer out of the control volume. However, Equation 2.23c shows that the rate at which entropy is transferred out must *exceed* the rate at which entropy enters, the difference being the rate of entropy generation within the control volume owing to irreversibilities.

2.2.3   Ancillary Concepts

Many important applications involve single-inlet, single-outlet control volumes at steady state. Several concepts related to this class of applications are now presented.

Isentropic Efficiencies. Consider first the case of an adiabatic control volume: a control volume for which there is no heat transfer. As there is a single inlet and a single outlet, Equation 2.23a reduces to $\dot{m}_i = \dot{m}_e$. Denoting the common mass flow rate by $\dot{m}$, Equation 2.23c gives

$$s_e - s_i = \frac{\dot{S}_{\text{gen}}}{\dot{m}} \quad \text{(no heat transfer)} \tag{2.24}$$

Accordingly, when irreversibilities are present within the control volume, the specific entropy increases as mass flows from inlet to outlet. In the ideal case in which no internal irreversibilities are present, mass passes through the control volume with no change in its entropy—that is, isentropically.

Isentropic efficiencies for turbines, compressors, and pumps involve a comparison between the actual performance of a device and the performance that would be achieved under idealized circumstances for the same inlet state and the same outlet pressure, with heat transfer between the device and its surroundings not occurring to any significant extent. The *isentropic turbine efficiency* η_{st} compares the actual turbine power $\dot{W}_{cv}$ to the power that would be developed in an isentropic expansion from the specified inlet state to the specified outlet pressure, $(\dot{W}_{cv})_s$:

$$\eta_{st} = \frac{\dot{W}_{cv}}{(\dot{W}_{cv})_s} \tag{2.25}$$

The *isentropic compressor efficiency* η_{sc} compares the actual power input to the power that would be required in an isentropic compression from the specified inlet state to the specified outlet pressure:

$$\eta_{sc} = \frac{(\dot{W}_{cv})_s}{\dot{W}_{cv}} \tag{2.26}$$

An *isentropic pump efficiency* is defined similarly.

Special Cases. Next, consider the control volume of Figure 2.1 where heat transfer occurs only at the temperature T_b. Denoting the common mass flow rate by $\dot{m}$, Equations 2.23b and 2.23c read, respectively

$$0 = \dot{Q}_{cv} - \dot{W}_{cv} + \dot{m}\left((h_i - h_e) + \tfrac{1}{2}(V_i^2 - V_e^2) + g(z_i - z_e)\right)$$

$$0 = \frac{\dot{Q}_{cv}}{T_b} + \dot{m}(s_i - s_e) + \dot{S}_{gen}$$

Eliminating the heat transfer term from these expressions, the work developed per unit of mass flowing through the control volume is

$$\frac{\dot{W}_{cv}}{\dot{m}} = [(h_i - h_e) - T_b(s_i - s_e) + \tfrac{1}{2}(V_i^2 - V_e^2) + g(z_i - z_e)] - T_b\frac{\dot{S}_{gen}}{\dot{m}}$$

The term in brackets is fixed by the states at the control volume inlet and outlet and the temperature T_b at which heat transfer occurs. Accordingly, an expression for the maximum value of the work developed per unit of mass flowing through the control volume, corresponding to the absence of internal irreversibilities, is obtained when the entropy generation term is set to zero:

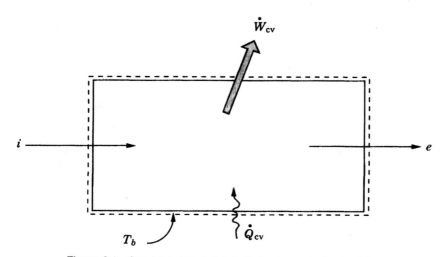

Figure 2.1 One-inlet, one-outlet control volume at steady state.

$$\left(\frac{\dot{W}_{cv}}{\dot{m}}\right)_{int\ rev} = (h_i - h_e) - T_b(s_i - s_e) + \tfrac{1}{2}(V_i^2 - V_e^2) + g(z_i - z_e) \quad (2.27)$$

(all heat transfer at temperature T_b)

This expression is applied in Section 3.3.2.

Using similar reasoning, expressions can be developed for the energy transfers by heat and work for a single-inlet, single-outlet control volume at steady state in the absence of internal irreversibilities, but without the restriction that heat transfer occurs only at the temperature T_b. For such cases, the heat transfer per unit of mass flow is

$$\left(\frac{\dot{Q}_{cv}}{\dot{m}}\right)_{int\ rev} = \int_1^2 T\ ds \quad\quad (2.28)$$

Combining Equations 2.23b, 2.28, and 2.34b (from Section 2.3.1) gives the work developed per unit of mass flow:

$$\left(\frac{\dot{W}_{cv}}{\dot{m}}\right)_{int\ rev} = -\int_1^2 v\ dp + \tfrac{1}{2}(V_1^2 - V_2^2) + g(z_1 - z_2) \quad (2.29)$$

(heat transfer according to Equation 2.28)

The integrals of Equations 2.28 and 2.29 are performed from inlet to outlet, denoted 1 and 2, and the subscripts int rev stress that the expressions apply only to control volumes in which there are no internal irreversibilities. The integral of Equation 2.29 requires a relationship between pressure and specific volume. An example is the expression $pv^n = $ const, where the value of n is a constant. An internally reversible process described by such an expression is called a *polytropic process* and n is the *polytropic exponent*. When the exponent n is determined for an actual process by fitting pressure-specific volume data, the use of $pv^n = $ const to calculate the integral in Equation 2.29 can yield a plausible approximation of the work per unit of mass flow in the actual process.

Head Loss. Equation 2.29 is a special case of the *generalized Bernoulli equation* [3]:

$$\int_1^2 v\ dp + \tfrac{1}{2}(V_2^2 - V_1^2) + g(z_2 - z_1) + \frac{\dot{W}_{cv}}{\dot{m}} + h_l = 0 \quad (2.30)$$

where $\dot{W}_{cv}/\dot{m}$ denotes the work developed by the control volume per unit of mass flowing, and h_l accounts for the rate at which mechanical energy is irreversibly converted to internal energy. The term h_l is known by various names, including *head loss, friction, friction work,* and *energy dissipation.* In the absence of irreversibilities, such as friction, h_l vanishes and Equation 2.30 reduces to give Equation 2.29.

The term h_l of Equation 2.30 can be obtained experimentally by measuring all the other terms. The accumulated experimental data have been organized to allow estimation of the term for use in designing piping systems. Thus, when h_l is known, Equation 2.30 can be used to find some other quantity, such as the work required for pumping a liquid or the volumetric flow rate when the work is known. The term h_l is conventionally formulated as the sum of two contributions, one associated with wall friction over the length of the conduit carrying the liquid or gas and the other associated with flow through resistances such as valves, elbows, tees, and pipe entrances and exits. Thus, for a piping system consisting of several resistances and several lengths L of pipe each with diameter D we have

$$h_l = \sum_p \underbrace{\left[\left(f \frac{4L}{D} \right) \frac{V^2}{2} \right]_p}_{\text{pipe friction}} + \sum_l \underbrace{\left[K \frac{V^2}{2} \right]_l}_{\text{resistances}} \qquad (2.31)$$

In this expression, f is the *Fanning friction factor,* which is a function of the *relative roughness* k_s/D and the *Reynolds number* $\mathrm{Re}_D = \rho V D/\mu$, where μ denotes viscosity. As discussed further in Section 2.6.2, the function $f(\mathrm{Re}_D, k_s/D)$ is available both graphically and analytically. The *loss coefficients K* for resistances of practical interest are also available from the engineering literature (also see Section 2.6.2).

Momentum Equation. Like mass, energy, and entropy, *linear momentum* is carried into and out of a control volume at the inlet and outlet, respectively. Such transfers are accounted for by expressions of the form

$$\begin{bmatrix} \text{Time rate of linear momentum} \\ \text{transfer into (or out of) a} \\ \text{control volume accompanying} \\ \text{mass flow} \end{bmatrix} = \dot{m}V$$

In this equation, the linear momentum per unit of mass flowing across the boundary of the control volume is given by the velocity vector **V**. In accordance with the one-dimensional flow model, this vector is normal to the inlet or outlet and oriented in the direction of flow. Using the linear momentum transfer concept, an expression of *Newton's second law of motion* is

$$\begin{bmatrix} \text{Time rate of change} \\ \text{of linear momentum} \\ \text{contained within the} \\ \text{control volume} \end{bmatrix} = \begin{bmatrix} \text{resultant external force} \\ \text{acting } on \text{ the} \\ \text{control volume} \end{bmatrix}$$

$$+ \begin{bmatrix} \text{net rate at which linear} \\ \text{momentum is transferred} \\ \text{into the control volume} \\ \text{accompanying mass flow} \end{bmatrix}$$

At steady state, the total amount of linear momentum contained in the control volume is constant with time. Accordingly, when applying Newton's second law of motion of control volumes at steady state, it is necessary to consider only the momentum accompanying the incoming and outgoing streams of matter and the forces acting on the control volume. Newton's law then states that the resultant external force **F** acting *on* the control volume equals the difference between the rates of momentum associated with mass flow exiting and entering the control volume:

$$\mathbf{F} = \dot{m}(\mathbf{V}_2 - \mathbf{V}_1) \tag{2.32}$$

where 1 denotes the inlet and 2 the outlet. The resultant force includes the forces due to pressure acting at the inlet and outlet, forces acting on the portion of the boundary through which there is no mass flow, and the force of gravity. Applications of Equation 2.32 are provided elsewhere [1, 3, 4].

2.3 PROPERTY RELATIONS

To apply the mass, energy, and entropy balances to a system of interest requires knowledge of the properties of the system and how the properties are related. This section reviews property relations for simple compressible systems, which include a wide range of industrially important gases and liquids.

2.3.1 Basic Relations for Pure Substances

The* T dS *Equations. Internal energy, enthalpy, and entropy are calculated using data on properties that are more readily measured. The calculations use the $T\,dS$ *equations* and relations derived from them.

To establish the $T\,dS$ equations consider a closed system undergoing an internally reversible process in the absence of overall system motion and the effect of gravity. An energy balance in differential form is

$$dU = (\delta Q)_{\text{int rev}} - (\delta W)_{\text{int rev}}$$

From Equation 2.11, $(\delta Q)_{\text{int rev}} = T\,dS$, so

$$dU = T \, dS - (\delta W)_{\text{int rev}}$$

A pure, simple compressible system is one in which the only significant energy transfer by work in an internally reversible process is associated with volume change and is given by $(\delta W)_{\text{int rev}} = p \, dV$ [1]. Accordingly, the first $T \, dS$ equation for pure, simple compressible systems takes the form

$$dU = T \, dS - p \, dV \qquad (2.33a)$$

Since $H = U + pV$, $dH = dU + p \, dV + V \, dp$. Combining this with Equation 2.33a the second $T \, dS$ equation follows

$$dH = T \, dS + V \, dp \qquad (2.34a)$$

Using the same approach, a further relation can be derived from the *Gibbs function*, defined as $G = H - TS$. The result is

$$dG = V \, dp - S \, dT \qquad (2.35a)$$

The last three equations can be expressed on a per unit mass (or a per mole) basis:

$$du = T \, ds - p \, dv \qquad (2.33b)$$

$$dh = T \, ds + v \, dp \qquad (2.34b)$$

$$dg = v \, dp - s \, dT \qquad (2.35b)$$

p–v–T *Relations*. Pressure, specific volume, and temperature are relatively easily measured and considerable p–v–T data have been accumulated for industrially important gases and liquids. These data can be represented in the form $p = f(v, T)$, called an *equation of state*. Equations of state can be expressed in tabular, graphical, and analytical forms [1, 5].

The graph of a function $p = f(v, T)$ is a surface in three-dimensional space. Figure 2.2 shows the p–v–T relationship for water, a substance that expands upon freezing. Figure 2.2*b* shows the projection of the surface onto the pressure–temperature plane. The projection onto the p–v plane is shown in Figure 2.2*c*.

Figure 2.2 has three regions labeled solid, liquid, and vapor where the substance exists only in a single phase. Between the single-phase regions lie two-phase regions, where two phases coexist in equilibrium. The line separating a single-phase region from a two-phase region is called a *saturation line*. Any state represented by a point on a saturation line is a *saturation state*. Thus, the line separating the liquid phase and the two-phase liquid–vapor

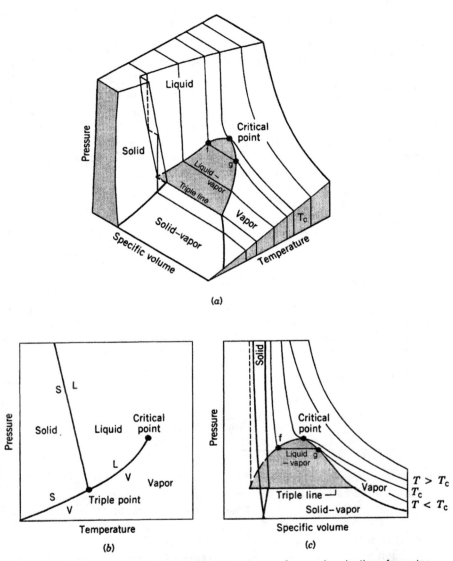

Figure 2.2 Pressure-specific volume–temperature surface and projections for water.

region is the saturated liquid line. The state denoted by f in Figure 2.2c is a saturated liquid state. The saturated vapor line separates the vapor region and the two-phase liquid–vapor region. The state denoted by g in Figure 2.2c is a saturated vapor state. The saturated liquid line and the saturated vapor line meet at a point, called the *critical point*. At the critical point, the pressure is called the *critical pressure* p_c, and the temperature is called the *critical temperature* T_c.

When a phase change occurs during constant pressure heating or cooling, the temperature remains constant as long as both phases are present. Accordingly, in the two-phase liquid–vapor region, a line of constant pressure is also a line of constant temperature. For a specified pressure, the corresponding temperature is called the *saturation temperature.* For a specified temperature, the corresponding pressure is called the *saturation pressure.* The region to the right of the saturated vapor line is often referred to as the *superheated vapor region* because the vapor exists at a temperature greater than the saturation temperature for its pressure. The region to the left of the saturated liquid line is also known as the *compressed liquid region* because the liquid is at a pressure higher than the saturation pressure for its temperature.

When a mixture of liquid and vapor coexist in equilibrium, the liquid phase is at the saturated liquid state and the vapor is at the corresponding saturated vapor state. The total volume of any such mixture is $V = V_f + V_g$, where f denotes saturated liquid and g saturated vapor. This may be written as

$$mv = m_f v_f + m_g v_g$$

where m and v denote mass and specific volume, respectively. Dividing by the total mass of the mixture m and letting the mass fraction of the vapor in the mixture, m_g/m, be symbolized by x, called the *quality,* the apparent specific volume v of the mixture is

$$v = (1 - x)v_f + xv_g$$

$$= v_f + xv_{fg} \tag{2.36}$$

where $v_{fg} = v_g - v_f$. Expressions similar in form to these can be written for internal energy, enthalpy, and entropy.

Property Tables for Gases and Liquids. Values for specific internal energy, enthalpy, and entropy are calculated for gases and liquids using the $T\ dS$ equations and relations developed from them, p–v–T and specific heat data, and other thermodynamic data. The results of such calculations, supplemented by data taken from direct physical measurement or determined from these measurements, are presented in tabular form and, increasingly, as computer software. The form of the tables and the way they are used is assumed to be familiar.

The calculated specific internal energy, enthalpy, and entropy data are determined relative to arbitrary datums and the datums used vary from substance to substance. For example, in Reference 1 the internal energy of saturated liquid water at 0.01°C (32.02°F) is set to zero, but for refrigerants the saturated liquid enthalpy is zero at −40°C (−40°F for tables in English units). When calculations are being performed that involve only differences in a particular specific property, the datum cancels. Care must be taken, however, when there are changes in chemical composition during the process. The approach to be followed when composition changes due to chemical reaction is considered in Section 2.4.

Specific Heats. The properties c_v and c_p are defined as partial derivatives of the functions $u(T, v)$ and $h(T, p)$, respectively

$$c_v = \left.\frac{\partial u}{\partial T}\right)_v \tag{2.37}$$

$$c_p = \left.\frac{\partial h}{\partial T}\right)_p \tag{2.38}$$

where the subscripts v and p denote, respectively, the variables held fixed in the differentiation process. These properties are known commonly as *specific heats*. The property k is simply their ratio

$$k = \frac{c_p}{c_v} \tag{2.39}$$

Incompressible Liquid Model. Reference to property data shows that for the liquid phase of water there are regions where the variation in the specific volume is slight and the dependence of internal energy on pressure (at fixed temperature) is weak. This behavior is also exhibited by other substances. As a model, it is often convenient to assume in such regions that the specific volume (density) is constant and the internal energy depends only on temperature. A liquid modeled in this way is said to be *incompressible*.

Since the incompressible liquid model regards internal energy to depend *only* on temperature, the specific heat c_v is also a function of temperature alone:

$$c_v(T) = \frac{du}{dT}$$

Although specific volume v is constant, enthalpy varies with both temperature and pressure as shown by

$$h(T, p) = u(T) + pv \tag{2.40}$$

Differentiating Equation 2.40 with respect to temperature, holding pressure fixed gives $c_p = c_v$. That is, for an incompressible substance the two specific heats are equal. The common specific heat is commonly shown simply as c.

For an incompressible substance Equation 2.33b reduces to $du = T\,ds$. Then, with $du = c(T)\,dT$, the change in specific entropy is

$$\Delta s = \int_{T_1}^{T_2} \frac{c(T)}{T} \, dT \tag{2.41}$$

As an alternative to the incompressible liquid model when saturated liquid data are available, the following equations can be used to estimate property values at liquid states

$$v(T, p) \approx v_f(T) \tag{2.42a}$$

$$u(T, p) \approx u_f(T) \tag{2.42b}$$

$$h(T, p) \approx h_f(T) + v_f[p - p_{sat}(T)] \tag{2.42c}$$

$$s(T, p) \approx s_f(T) \tag{2.42d}$$

where the subscript f denotes the saturated liquid state at the temperature T and p_{sat} is the corresponding saturation pressure.

Ideal-Gas Model. For vapor and liquid states, the general pattern of the p–v–T relation can be conveniently shown in terms of the compressibility factor Z, defined by $Z = p\bar{v}/\bar{R}T$, where $\bar{R}$ is the universal gas constant (Section 2.1). Using the compressibility factor, generalized charts can be developed from which reasonable approximations to the p–v–T behavior of many substances can be obtained. In one form of generalized charts the compressibility factor Z is plotted versus the reduced pressure p_R and reduced temperature T_R, defined as $p_R = p/p_c$ and $T_R = T/T_c$, where p_c and T_c are, respectively, the critical pressure and critical temperature. Generalized compressibility charts suitable for engineering calculations, together with illustrations of their use, are available from the literature [1, 2, 5].

By inspection of a generalized compressibility chart, it can be concluded that when p_R is small, and for many states when T_R is large, the value of the compressibility factor Z is closely 1. That is, for pressures that are low relative to p_c, and for many states with temperatures high relative to T_c, the compressibility factor approaches a value of 1. Within the indicated limits, it may be assumed with resonable accuracy that $Z = 1$. That is,

$$p\bar{v} = \bar{R}T \tag{2.43}$$

It can be shown that the internal energy and enthalpy for any gas whose equation of state is exactly given by Equation 2.43 depends only on temperature [1].

These considerations lead to the introduction of an *ideal-gas model* for each real gas. For example, the ideal-gas model of nitrogen obeys the equation of state $p\bar{v} = \bar{R}T$, and its internal energy is a function of temperature alone. The ideal-gas model of oxygen also obeys $p\bar{v} = \bar{R}T$, and its internal energy is also a function of temperature alone, but the function differs from that for nitrogen since each gas has a unique internal structure. The real gas approaches the ideal-gas model in the limit of low reduced pressure. At other states the actual behavior may deviate substantially from the predictions of the model.

Dividing the ideal-gas equation of state by the molecular weight M, the equation is placed on a unit mass basis

$$pv = RT$$

where R, defined as $\overline{R}/M$, is the *specific gas constant*. Other forms in common use are

$$pV = n\overline{R}T \quad , \quad pV = mRT$$

Since $p\overline{v} = \overline{R}T$ for an ideal gas, the ideal-gas enthalpy and internal energy are related by $\overline{h}(T) = \overline{u}(T) + \overline{R}T$. Differentiating with respect to temperature

$$\frac{d\overline{h}}{dT} = \frac{d\overline{u}}{dT} + \overline{R}$$

With Equations 2.37 and 2.38, this gives

$$\overline{c}_p(T) = \overline{c}_v(T) + \overline{R} \tag{2.44a}$$

That is, the two ideal-gas specific heats depend on temperature alone, and their difference is the gas constant. Since $k = \overline{c}_p/\overline{c}_v$

$$\overline{c}_p = \frac{k\overline{R}}{k - 1} \quad , \quad \overline{c}_v = \frac{\overline{R}}{k - 1} \tag{2.44b}$$

Specific heat data can be obtained by direct measurement. When extrapolated to zero pressure, ideal-gas model specific heats result. Ideal-gas specific heats also can be obtained from theory based on molecular models of matter using spectroscopic measurements. Ideal-gas specific heat functions in both tabular and equation form are available in the literature for a number of substances [1, 5, 6].

Since $\overline{c}_v = d\overline{u}/dT$ and $\overline{c}_p = d\overline{h}/dT$ for an ideal gas, expressions for internal energy and enthalpy can be obtained by integration:

$$\bar{u}(T) = \int_{T'}^{T} \bar{c}_v(T) \, dT + \bar{u}(T')$$

$$\bar{h}(T) = \int_{T'}^{T} \bar{c}_p(T) \, dT + \bar{h}(T') \qquad (2.45)$$

where T' is an arbitrary reference temperature.

The first $T \, dS$ equation, $T \, d\bar{s} = d\bar{u} + p \, d\bar{v}$, can be used to determine the entropy change of an ideal gas between two states. Using $d\bar{u} = \bar{c}_v \, dT$ and $p/T = \bar{R}/\bar{v}$

$$d\bar{s} = \frac{\bar{c}_v(T) \, dT}{T} + \frac{\bar{R}}{\bar{v}} \, d\bar{v}$$

Integrating between states 1 and 2

$$\bar{s}(T_2, v_2) - \bar{s}(T_1, v_1) = \int_{T_1}^{T_2} \frac{\bar{c}_v(T)}{T} \, dT + \bar{R} \ln \frac{v_2}{v_1} \qquad (2.46)$$

Similarly, from the second $T \, dS$ equation (Equation 2.34b) follows

$$\bar{s}(T_2, p_2) - \bar{s}(T_1, p_1) = \int_{T_1}^{T_2} \frac{\bar{c}_p(T)}{T} \, dT - \bar{R} \ln \frac{p_2}{p_1} \qquad (2.47)$$

Equation 2.47 can be rewritten using $\bar{s}°(T)$ defined by

$$\bar{s}°(T) \equiv \int_{T'}^{T} \frac{\bar{c}_p(T)}{T} \, dT \qquad (2.48)$$

as

$$\bar{s}(T_2, p_2) - \bar{s}(T_1, p_1) = \bar{s}°(T_2) - \bar{s}°(T_1) - \bar{R} \ln \frac{p_2}{p_1} \qquad (2.49)$$

Equations 2.45 and 2.48 provide the basis for a simple tabular display. That is, by specifying the reference temperature T' the quantities $\bar{u}(T)$, $\bar{h}(T)$, $\bar{s}°(T)$ can be tabulated versus temperature. Tabulations for several gases based on the specification $\bar{h} = \bar{u} = \bar{s}° = 0$ at $T' = 0$ K are available in the literature [1]. In Section 2.4, $\bar{s}°$ is identified as the *absolute entropy* at temperature T and a reference pressure p_{ref}.

When the temperature interval is relatively small, the ideal-gas specific heats are nearly constant. It is often convenient in such instances to assume them to be constant, usually their arithmetic average over the interval. The expressions for changes in internal energy, enthalpy, and entropy of an ideal gas then appear as

$$\bar{u}(T_2) - \bar{u}(T_1) = \bar{c}_v(T_2 - T_1) \tag{2.50a}$$

$$\bar{h}(T_2) - \bar{h}(T_1) = \bar{c}_p(T_2 - T_1) \tag{2.50b}$$

$$\bar{s}(T_2, v_2) - \bar{s}(T_1, v_1) = \bar{c}_v \ln \frac{T_2}{T_1} + \bar{R} \ln \frac{v_2}{v_1} \tag{2.50c}$$

$$\bar{s}(T_2, p_2) - \bar{s}(T_1, p_1) = \bar{c}_p \ln \frac{T_2}{T_1} - \bar{R} \ln \frac{p_2}{p_1} \tag{2.50d}$$

Ideal-Gas Mixtures. Consider a phase consisting of a mixture of N gases for which the *Dalton mixture model* applies: Each gas is uninfluenced by the presence of the others, each can be treated as an ideal gas, and each acts as if it exists separately at the volume and temperature of the mixture. Writing the ideal-gas equation of state for the mixture as a whole and for any component k

$$pV = n\bar{R}T, \qquad p_k V = n_k \bar{R}T$$

where p is the mixture pressure and p_k is the partial pressure of component k. The partial pressure p_k is the pressure gas k would exert if n_k moles occupied the mixture volume alone at the mixture temperature.

Forming the ratio of these two equations gives

$$p_k = \frac{n_k}{n} p = x_k p \tag{2.51}$$

where x_k, defined as n_k/n, is the mole fraction of component k. Since $\sum_{k=1}^{N} x_k = 1$, we have $p = \sum_{k=1}^{N} p_k$: The pressure of a mixture of ideal gases is equal to the sum of the partial pressures of the gases. This is known as *Dalton's model of additive partial pressures.*

The internal energy, enthalpy, and entropy of the mixture can be determined as the sum of the respective properties of the component gases, provided that the contribution from each gas is evaluated at the condition at which the gas exists in the mixture. Thus

$$U = \sum_{k=1}^{N} n_k \bar{u}_k \quad \text{or} \quad \bar{u} = \sum_{k=1}^{N} x_k \bar{u}_k \tag{2.52a}$$

$$H = \sum_{k=1}^{N} n_k \bar{h}_k \quad \text{or} \quad \bar{h} = \sum_{k=1}^{N} x_k \bar{h}_k \tag{2.52b}$$

$$S = \sum_{k=1}^{N} n_k \bar{s}_k \quad \text{or} \quad \bar{s} = \sum_{k=1}^{N} x_k \bar{s}_k \tag{2.52c}$$

Since the internal energy and enthalpy of an ideal gas depend only on temperature, the $\bar{u}_k$ and $\bar{h}_k$ terms appearing in these equations are evaluated at the temperature of the mixture. Entropy is a function of two independent properties. Accordingly, the $\bar{s}_k$ terms are evaluated either at the temperature and volume of the mixture or at the mixture temperature and the partial pressure p_k of the component. In the latter case

$$S = \sum_{k=1}^{N} n_k \bar{s}_k(T, p_k)$$

$$= \sum_{k=1}^{N} n_k \bar{s}_k(T, x_k p) \tag{2.53}$$

Differentiation of $\bar{u} = \sum_{k=1}^{N} x_k \bar{u}_k$ and $\bar{h} = \sum_{k=1}^{N} x_k \bar{h}_k$ with respect to temperature results, respectively, in expressions for the two specific heats $\bar{c}_v$ and $\bar{c}_p$ for the mixture in terms of the corresponding specific heats of the components:

$$\bar{c}_v = \sum_{k=1}^{N} x_k \bar{c}_{vk} \tag{2.54a}$$

$$\bar{c}_p = \sum_{k=1}^{N} x_k \bar{c}_{pk} \tag{2.54b}$$

where $\bar{c}_{vk} = d\bar{u}_k/dT$ and $\bar{c}_{pk} = d\bar{h}_k/dT$.

The molecular weight M of the mixture is determined in terms of the molecular weights M_k of the components as

$$M = \sum_{k=1}^{N} x_k M_k \tag{2.55}$$

Inserting the expressions for H and S given by Equations 2.52b and 2.52c into the Gibbs function $G = H - TS$ results in

$$G = \sum_{k=1}^{N} n_k \bar{h}_k(T) - T \sum_{k=1}^{N} n_k \bar{s}_k(T, p_k)$$

$$= \sum_{k=1}^{N} n_k \bar{g}_k(T, p_k) \tag{2.56}$$

where the molar specific Gibbs function of component k is $\bar{g}_k(T, p_k) = \bar{h}_k(T) - T\bar{s}_k(T, p_k)$.

The Gibbs function of component k can be expressed alternatively by integrating Equation 2.35b at fixed temperature from an arbitrarily selected reference pressure p' to pressure $p_k(= x_k p)$ as follows:

$$\bar{g}_k(T, p_k) - \bar{g}_k(T, p') = \int_{p'}^{p_k} \bar{v} \, dp$$

For an ideal gas, $\bar{v} = \bar{R}T/p$; thus

$$\bar{g}_k(T, p_k) = \bar{g}_k(T, p') + \bar{R}T \ln \frac{p_k}{p'}$$

$$= \bar{g}_k(T, p') + \bar{R}T \ln \frac{x_k p}{p'} \tag{2.57}$$

Psychrometrics. In this section we consider some important terms, definitions, and principles of *psychrometrics,* the study of systems consisting of dry air and water. *Moist air* refers to a mixture of dry air and water vapor in which the dry air is treated as if it were a pure component. Ideal-gas mixture principles are assumed to apply to moist air. In particular, the Dalton model is applicable, and so the mixture pressure p is the sum of the partial pressures p_a and p_v of the dry air and water vapor, respectively. *Saturated air* is a mixture of dry air and saturated water vapor. For saturated air, the partial pressure of the water vapor equals the saturation pressure of water p_g corresponding to the mixture temperature T.

The composition of a given moist air sample can be described in terms of the *humidity ratio (specific humidity)* ω, defined as the ratio of the mass of the water vapor to the mass of dry air:

$$\omega = \frac{m_v}{m_a} \tag{2.58a}$$

which can be expressed alternatively as

$$\omega = 0.622 \frac{p_v}{p - p_v} \tag{2.58b}$$

The makeup of moist air also can be described in terms of the *relative humidity* ϕ, defined as the ratio of the water vapor mole fraction in a given moist air sample to the water vapor mole fraction in a saturated moist air sample at the same temperature T and pressure p. The relative humidity can be expressed as

$$\phi = \left. \frac{p_v}{p_g} \right)_{T,p} \tag{2.59}$$

For moist air, the values of H and S can be found by adding the contribution of each component at the condition at which it exists in the mixture. Thus, the enthalpy is

$$H = H_a + H_v = m_a h_a + m_v h_v$$

or

$$\frac{H}{m_a} = h_a + \omega h_v \tag{2.60}$$

where h_a and h_v are each evaluated at the mixture temperature. When using steam table data for the water vapor, h_v is calculated from $h_v = h_g(T)$. To evaluate the entropy of moist air, the contribution of each component in the mixture is determined at the mixture temperature and the partial pressure of the component:

$$\frac{S}{m_a} = s_a(T, p_a) + \omega s_v(T, p_v) \tag{2.61}$$

When using steam table data, s_v is calculated from $s_v(T, p_v) = s_g(T) - R \ln \phi$. This equation is obtained by applying Equation 2.47 on a mass basis between the saturated vapor state at temperature T and the vapor state at T, p_v, and using Equation 2.59.

When a sample of moist air is cooled at constant pressure, the temperature at which the sample becomes saturated is called the *dew point temperature*. Cooling below the dew point temperature results in the condensation of some of the water vapor initially present. When cooled to a final equilibrium state at a temperature below the dew point temperature, the original sample would consist of a gas phase of dry air and saturated water vapor in equilibrium with a liquid water phase. Further discussion including a numerical illustration is provided in Section 2.5.

Several important parameters of moist air are represented graphically by psychrometric charts. For moist air at a pressure of 1 atm, Reference 1 gives detailed charts in SI and English units. Familiarity with such charts is assumed.

2.3.2   Multicomponent Systems

In this section we introduce some general aspects of the properties of multicomponent systems consisting of nonreacting mixtures in a single phase. Elaboration is provided elsewhere [1]. The special case of ideal-gas mixtures is considered in the previous section.

Partial Molal Properties. Any extensive thermodynamic property X of a single-phase, single-component system is a function of two independent intensive properties and the size of the system. Selecting temperature and pressure as the independent properties and the number of moles n as the measure of size, we have $X = X(T, p, n)$. For a single-phase, multicomponent system consisting of N components, the extensive property X must then be a function of temperature, pressure, and the number of moles of each component present in the mixture, $X = X(T, p, n_1, n_2, ..., n_N)$. As X is mathematically *homogeneous of degree 1* in the n's, the function is expressible as

$$X = \sum_{k=1}^{N} n_k \overline{X}_k \tag{2.62}$$

where the *partial molal property* $\overline{X}_k$ is, by definition,

$$\overline{X}_k = \frac{\partial X}{\partial n_k}\bigg)_{T,p,n_l} \tag{2.63}$$

The subscript n_l denotes that all n's except n_k are held fixed during differentiation. The partial molal property $\overline{X}_k$ is a property of the mixture and not simply a property of the kth component. Here, $\overline{X}_k$ depends in general on temperature, pressure, and mixture composition: $\overline{X}_k(T, p, n_1, n_2, ..., n_N)$. Partial molal properties are intensive properties of the mixture.

Selecting the extensive property X in Equation 2.62 to be volume, internal energy, enthalpy, entropy, and the Gibbs function, respectively, gives

$$V = \sum_{k=1}^{N} n_k \overline{V}_k, \qquad U = \sum_{k=1}^{N} n_k \overline{U}_k, \qquad H = \sum_{k=1}^{N} n_k \overline{H}_k$$

$$\tag{2.64}$$

$$S = \sum_{k=1}^{N} n_k \overline{S}_k, \qquad G = \sum_{k=1}^{N} n_k \overline{G}_k$$

where $\overline{V}_k$, $\overline{U}_k$, $\overline{H}_k$, $\overline{S}_k$, and $\overline{G}_k$ denote the partial molal volume, internal energy, enthalpy, entropy, and Gibbs function, respectively.

Chemical Potential. Because of its importance in the study of multicomponent systems, the partial molal Gibbs function of the kth component is given a special name and symbol. It is called the *chemical potential* and symbolized by μ_k:

$$\mu_k = \overline{G}_k = \left.\frac{\partial G}{\partial n_k}\right)_{T,p,n_l} \tag{2.65}$$

Like temperature and pressure, the chemical potential μ_k is an intensive property. The chemical potential is a measure of the *escaping tendency* of a substance: Any substance will try to move from the phase having the higher chemical potential for that substance to the phase having a lower chemical potential. A necessary condition for *phase equilibrium* is that the chemical potential of each component has the same value in every phase.

Using Equation 2.65, together with the expression for the Gibbs function given in Equation 2.64

$$G = \sum_{k=1}^{N} n_k \mu_k \tag{2.66}$$

For a *single-component system,* Equation 2.66 reduces to $G = n\mu$. That is, for a single component the chemical potential equals the molar Gibbs function. Comparing Equations 2.56 and 2.66, we conclude $\mu_k = \overline{g}_k(T, p_k)$. That is, the chemical potential of component k in an ideal-gas mixture is equal to its Gibbs function per mole of gas k, evaluated at the mixture temperature and the partial pressure of the kth gas of the mixture.

With $G = H - TS$ and $H = U + pV$, Equation 2.66 can be expressed as

$$U = TS - pV + \sum_{k=1}^{N} n_k \mu_k \tag{2.67}$$

Using Equation 2.67, the following can be derived:

$$dU = T\,dS - p\,dV + \sum_{k=1}^{N} \mu_k\,dn_k \tag{2.68}$$

When the mixture composition is constant, Equation 2.68 reduces to Equation 2.33a.

2.4   REACTING MIXTURES AND COMBUSTION

The thermodynamic analysis of reactive systems is primarily an extension of principles introduced thus far. The fundamental concepts remain the same: conservation of mass, conservation of energy, and the second law. It is necessary, though, to modify the methods used to evaluate specific enthalpy and entropy. Once appropriate values for these variables are determined, they are used as in earlier discussions in the energy and entropy balances for the system under consideration.

2.4.1   Combustion

When a chemical reaction occurs, the bonds within molecules of the *reactants* are broken, and atoms and electrons rearrange to form *products*. In combustion reactions, rapid oxidation of combustible elements of the fuel results in energy release as combustion products are formed. The three major combustible chemical elements in most common fuels are carbon, hydrogen, and sulfur. Sulfur is usually a relatively unimportant contributor to the energy released, but it can be a significant cause of pollution and corrosion problems. A fuel is said to have burned completely if all of the carbon present in the fuel is burned to carbon dioxide, all of the hydrogen is burned to water, and all of the sulfur is burned to sulfur dioxide. If any of these conditions is not fulfilled, combustion is incomplete.

The minimum amount of air that supplies sufficient oxygen for the complete combustion of all the combustible chemical elements is called the *theoretical,* or *stoichiometric,* amount of air. In practice, the amount of air actually supplied may be greater than or less than the theoretical amount, depending on the application. The amount of air actually supplied is commonly expressed as the *percent of theoretical air* or *percent excess* (or *percent deficiency*) of air. The *air–fuel ratio* and its reciprocal, the *fuel–air ratio,* each of which can be expressed on a mass or molar basis, are other ways to describe fuel–air mixtures. The *equivalence ratio* is the ratio of the actual fuel–air ratio to the fuel–air ratio for complete combustion with the theoretical amount of air.

2.4.2   Enthalpy of Formation

In tables of thermodynamic properties, values for the specific enthalpy and entropy are usually given relative to some arbitrary datum state where the enthalpy (or alternatively the internal energy) and entropy are set to zero. This approach is satisfactory for evaluations involving differences in property values between states of the same composition, because then arbitrary datums cancel. However, when a chemical reaction occurs, reactants disappear and products are formed, so differences cannot be calculated for all substances involved, and it is necessary to evaluate h and s with special care lest ambiguities or inconsistencies arise.

An enthalpy datum for the study of reacting systems can be established by assigning arbitrarily a value of zero to the enthalpy of the stable elements at a state called the *standard reference state* and defined by $T_{ref} = 298.15$ K ($25°C$) and a reference pressure p_{ref}, which may be 1 bar or 1 atm depending on the data source. The term *stable* simply means that the particular element is chemically stable. For example, at the standard state the stable forms of hydrogen, oxygen, and nitrogen are H_2, O_2, and N_2 and not the monatomic H, O, and N.

Using the datum introduced above, enthalpy values can be assigned to compounds. The molar enthalpy of a compound at the standard state equals its *enthalpy of formation*, symbolized by $\overline{h}_f^\circ$. The enthalpy of formation is the energy released or absorbed when the compound is formed from its elements, the compound and elements all being at T_{ref} and p_{ref}. The enthalpy of formation is usually determined by application of procedures from statistical thermodynamics using observed spectroscopic data. The enthalpy of formation also can be found in principle by measuring the heat transfer in a reaction in which the compound is formed from the elements. In this book, the superscript degree symbol is used to denote properties at p_{ref}. For the case of the enthalpy of formation, the reference temperature T_{ref} is also intended by this symbol.

The molar enthalpy of a substance at a state other than the standard state is found by adding the enthalpy change $\Delta \overline{h}$ between the standard state and the state of interest to the molar enthalpy of formation:

$$\overline{h}(T, p) = \overline{h}_f^\circ + [\overline{h}(T, p) - \overline{h}(T_{ref}, p_{ref})] = \overline{h}_f^\circ + \Delta\overline{h} \qquad (2.69)$$

That is, the enthalpy of a substance is composed of $\overline{h}_f^\circ$, associated with the formation of the substance from its elements, and $\Delta\overline{h}$, associated with a change of state at constant composition. An arbitrarily chosen datum can be used to determine $\Delta\overline{h}$, since it is a difference at constant composition. Accordingly, $\Delta\overline{h}$ can be evaluated from tabular sources such as the steam tables, the ideal-gas tables when appropriate, and so on. As a consequence of the enthalpy datum adopted for the stable elements, the specific enthalpy determined from Equation 2.69 is often negative.

The *enthalpy of combustion* ($\overline{h}_{RP}$) is the difference between the enthalpy of the products and the enthalpy of the reactants, each on a per mole of fuel basis, when complete combustion occurs and both reactants and products are at the same temperature and pressure. For hydrocarbon fuels the enthalpy of combustion is negative in value since chemical internal energy is liberated in the reaction. The *heating value* of a fuel is a positive number equal to the magnitude of the enthalpy of combustion. Two heating values are recognized by name, the *higher heating value* (HHV) and the *lower heating value* (LHV). The higher heating value is obtained when all the water formed by combustion is a liquid; the lower heating value is obtained when all the water formed by combustion is a vapor. The higher heating value exceeds the lower heating value by the energy that would be required to vaporize the liquid water formed at the specified temperature. The values of $\overline{HHV}$ and $\overline{LHV}$ are typically reported at a temperature of 25°C (77°F) and a pressure of 1 bar (or 1 atm). These values also depend on whether the fuel is a liquid or a gas.

2.4.3 Absolute Entropy

When reacting systems are under consideration, the same basic problem arises for entropy as for enthalpy: A common datum must be used to assign entropy values for each substance involved in the reaction. This is accomplished using the *third law of thermodynamics,* which deals with the entropy of substances at the absolute zero of temperature. Based on experimental observations, this law states that the entropy of a pure crystalline substance is zero at the absolute zero of temperature, 0 K or 0°R. Substances not having a pure crystalline structure at absolute zero have a nonzero value of entropy at absolute zero. The experimental evidence on which the third law is based is obtained primarily from studies of chemical reactions at low temperatures and specific heat measurements at temperatures approaching absolute zero.

The third law provides a datum relative to which the entropy of each substance participating in a reaction can be evaluated so that no ambiguities or conflicts arise. The entropy relative to this datum is called the *absolute entropy.* The change in entropy of a substance between absolute zero and any given state can be determined from precise measurements of energy transfers and specific heat data or from procedures based on statistical thermodynamics and observed molecular data.

When the absolute entropy is known at the standard state, the specific entropy at any other state can be found by adding the specific entropy change between the two states to the absolute entropy at the standard state. Similarly, when the absolute entropy is known at the pressure p_{ref} and temperature T, the absolute entropy at the same temperature and any pressure p can be found from

$$\bar{s}(T, p) = \bar{s}(T, p_{\text{ref}}) + [\bar{s}(T, p) - \bar{s}(T, p_{\text{ref}})] \tag{2.70}$$

The term in brackets on the right side of Equation 2.70 can be evaluated for an ideal gas by using Equation 2.47, giving

$$\bar{s}(T, p) = \bar{s}°(T) - \overline{R} \ln \frac{p}{p_{\text{ref}}} \quad \text{(ideal gas)} \tag{2.71}$$

In this expression, $\bar{s}°(T)$ denotes the absolute entropy at temperature T and pressure p_{ref}.

The entropy of the kth component of an ideal-gas mixture is evaluated at the mixture temperature T and the partial pressure p_k: $\bar{s}_k(T, p_k)$. The partial pressure is given by $p_k = x_k p$, where x_k is the mole fraction of component k and p is the mixture pressure. Thus, for the kth component of an ideal-gas mixture Equation 2.71 takes the form

$$\bar{s}_k(T, p_k) = \bar{s}_k^\circ(T) - \bar{R} \ln \frac{p_k}{p_{ref}}$$

or

$$\bar{s}_k(T, p_k) = \bar{s}_k^\circ(T) - \bar{R} \ln \frac{x_k p}{p_{ref}} \tag{2.72}$$

(component k of an ideal-gas mixture)

where $\bar{s}_k^\circ(T)$ is the absolute entropy of component k at temperature T and p_{ref}. The molar Gibbs function $\bar{g}$ is

$$\bar{g} = \bar{h} - T\bar{s} \tag{2.73}$$

For reacting systems, values for the Gibbs function are commonly assigned in a way that closely parallels that for enthalpy: A zero value is assigned to the Gibbs function of each stable element at the standard state. The *Gibbs function of formation* of a compound equals the change in the Gibbs function for the reaction in which the compound is formed from its elements. The Gibbs function at a state other than the standard state is found by adding to the Gibbs function of formation the change in the specific Gibbs function $\Delta\bar{g}$ between the standard state and the state of interest:

$$\bar{g}(T, p) = \bar{g}_f^\circ + [\bar{g}(T, p) - \bar{g}(T_{ref}, p_{ref})] = \bar{g}_f^\circ + \Delta\bar{g} \tag{2.74a}$$

With Equation 2.73, $\Delta\bar{g}$ can be written as

$$\Delta\bar{g} = [\bar{h}(T, p) - \bar{h}(T_{ref}, p_{ref})] - [T\bar{s}(T, p) - T_{ref}\bar{s}(T_{ref}, p_{ref})] \tag{2.74b}$$

The Gibbs function of component k in an ideal-gas mixture is evaluated at the partial pressure of the component and the mixture temperature.

2.4.4 Ancillary Concepts

Table Data. Property data suited for the analysis of reactive systems are available in the literature [1, 6]. Tables giving standard state values of the enthalpy of formation, absolute entropy, and Gibbs function of formation in SI and English units are provided in Reference 1 for the specification $p_{ref} =$

1 atm. This source also provides $\bar{s}^\circ$ versus temperature for several gases modeled as ideal gases. Values for specific heat, enthalpy, absolute entropy, and Gibbs function are given in Reference 6 versus temperature for $p_{ref} = 1$ bar. Reference 6 also provides simple analytical representations of these thermochemical functions readily programmable for use with personal computers. See Table C.1 in Appendix C for a sampling; in this table, the behavior of the gases corresponds to the ideal-gas model.

Maximum Work. To illustrate the combustion principles introduced thus far, let us consider the maximum work per mole of fuel that can be developed by the system shown schematically in Figure 2.3. The result, Equation 2.75, is applied in Section 3.4.3. The system considered in this development is similar to such idealized devices as a reversible fuel cell or a van't Hoff equilibrium box. Referring to Figure 2.3, a hydrocarbon fuel C_aH_b and oxygen O_2 enter the system in separate streams; carbon dioxide and water exit separately. All entering and exiting streams are at the same temperature T and pressure p. The reaction is complete:

$$C_aH_b + (a + \tfrac{1}{4}b)O_2 \rightarrow aCO_2 + \tfrac{1}{2}b\ H_2O$$

The derivation of Equation 2.75 parallels that of Equation 2.27: For steady-state operation, the energy rate balance for the system reduces to give on a per mole of fuel basis

$$\frac{\dot{W}_{cv}}{\dot{n}_F} = \frac{\dot{Q}_{cv}}{\dot{n}_F} + \bar{h}_F + (a + \tfrac{1}{4}b)\bar{h}_{O_2} - a\bar{h}_{CO_2} - \tfrac{1}{2}b\ \bar{h}_{H_2O}$$

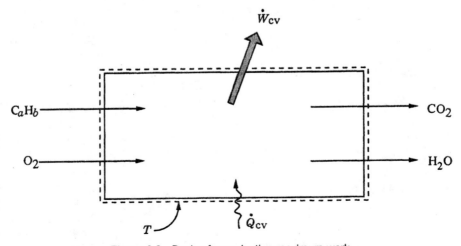

Figure 2.3 Device for evaluating maximum work.

where the subscript F denotes fuel. Kinetic and potential energy effects are regarded as negligible. If heat transfer occurs only at temperature T, an entropy balance for the control volume takes the form

$$0 = \frac{\dot{Q}_{cv}/\dot{n}_F}{T} + \bar{s}_F + (a + \tfrac{1}{4}b)\bar{s}_{O_2} - a\bar{s}_{CO_2} - \tfrac{1}{2}b\,\bar{s}_{H_2O} + \frac{\dot{S}_{gen}}{\dot{n}_F}$$

Eliminating the heat transfer term from these expressions, the work developed per mole of fuel is

$$\frac{\dot{W}_{cv}}{\dot{n}_F} = [\bar{h}_F + (a + \tfrac{1}{4}b)\bar{h}_{O_2} - a\bar{h}_{CO_2} - \tfrac{1}{2}b\,\bar{h}_{H_2O}]$$

$$- T[\bar{s}_F + (a + \tfrac{1}{4}b)\bar{s}_{O_2} - a\bar{s}_{CO_2} - \tfrac{1}{2}b\,\bar{s}_{H_2O}] - T\frac{\dot{S}_{gen}}{\dot{n}_F}$$

In this expression, the enthalpies would be evaluated in terms of enthalpies of formation via Equation 2.69 and the entropies are necessarily absolute entropies. An expression for the maximum value of the work developed per mole of fuel, corresponding to the absence of irreversibilities within the system, is obtained when the entropy generation term is set to zero:

$$\left(\frac{\dot{W}_{cv}}{\dot{n}_F}\right)_{\text{int rev}} = \{[\bar{h}_F + (a + \tfrac{1}{4}b)\bar{h}_{O_2} - a\bar{h}_{CO_2} - \tfrac{1}{2}b\,\bar{h}_{H_2O}](T, p)\}$$

$$- T[\bar{s}_F + (a + \tfrac{1}{4}b)\bar{s}_{O_2} - a\bar{s}_{CO_2} - \tfrac{1}{2}b\,\bar{s}_{H_2O}](T, p) \quad (2.75a)$$

Referring to the discussion of enthalpy of combustion given previously in this section, the term in curly brackets on the right side of Equation 2.75a is recognized as $-\bar{h}_{RP}$. Thus, Equation 2.75a can be written as

$$\left(\frac{\dot{W}_{cv}}{\dot{n}_F}\right)_{\text{int rev}} = -\bar{h}_{RP}(T, p) - T[\bar{s}_F + (a + \tfrac{1}{4}b)\bar{s}_{O_2} - a\bar{s}_{CO_2} - \tfrac{1}{2}b\,\bar{s}_{H_2O}](T, p)$$

$$(2.75b)$$

When the temperature and pressure in this expression correspond, respectively, to 25°C (77°F) and 1 bar (or 1 atm), the term $-\bar{h}_{RP}$ corresponds to the standard heating value: the higher heating value, $\overline{HHV}$, when water exits the system of Figure 2.3 as a liquid, and the lower heating value, $\overline{LHV}$, when water vapor exits.

Alternatively, using the specific Gibbs function $\bar{g} = \bar{h} - T\bar{s}$, Equation 2.75a can be expressed in the form

$$\left(\frac{\dot{W}_{cv}}{\dot{n}_F}\right)_{\text{int rev}} = [\overline{g}_F + (a + \tfrac{1}{4}b)\overline{g}_{O_2} - a\overline{g}_{CO_2} - \tfrac{1}{2}b\,\overline{g}_{H_2O}](T, p) \quad (2.75c)$$

Reaction Equilibrium. Consider a system at equilibrium containing five components, A, B, C, D, and E, at a given temperature and pressure, subject to a chemical reaction of the form

$$\nu_A A + \nu_B B \leftrightarrow \nu_C C + \nu_D D$$

where the ν's are stoichiometric coefficients. Component E is assumed to be inert and thus does not appear in the reaction equation. The form of the equation suggests that at equilibrium the tendency of A and B to form C and D is just balanced by the tendency of C and D to form A and B.

At equilibrium, the temperature and pressure would be uniform throughout the system. Additionally, the following expression, called the *equation of reaction equilibrium,* must be satisfied [1]:

$$\nu_A \mu_A + \nu_B \mu_B = \nu_C \mu_C + \nu_D \mu_D \quad (2.76)$$

where the μ's are the chemical potentials (Section 2.3.2) of A, B, C, and D in the equilibrium mixture. In principle, the composition that would be present at equilibrium for a given temperature and pressure can be determined by solving this equation. The solution procedure is simplified by using the *equilibrium constant $K(T)$,* defined by

$$\ln K = -\frac{\Delta G^\circ}{\overline{R}T} \quad (2.77)$$

where

$$\Delta G^\circ = \nu_C \overline{g}_C + \nu_D \overline{g}_D - \nu_A \overline{g}_A - \nu_B \overline{g}_B \quad (2.78)$$

As before, the superscript degree symbol denotes properties at p_{ref}. Further discussion of the equilibrium constant is provided in Reference 1.

2.5 THERMODYNAMIC MODEL—COGENERATION SYSTEM

Using principles developed thus far, a detailed thermodynamic model is developed and presented in this section for the cogeneration system of Figure 1.7 and Table 1.2. The model is used in subsequent sections of this book, and is illustrated in Examples 2.1 and 2.2, which conclude the present section.

Assumptions. The assumptions underlying the cogeneration system model include the following:

- The cogeneration system operates at steady state.
- Ideal-gas mixture principles apply for the air and the combustion products.
- The fuel (natural gas) is taken as methane modeled as an ideal gas. The fuel is provided to the combustion chamber at the required pressure by throttling from a high-pressure source.
- The combustion in the combustion chamber is complete. N_2 is inert.
- Heat transfer from the combustion chamber is 2% of the fuel lower heating value. All other components operate without heat loss.

Variables. In thermal system design and optimization, it is convenient to identify two types of independent variables: decision variables and parameters. The decision variables may be varied in optimization studies, but the parameters remain fixed in a given application. All other variables are dependent variables. Their values are calculated from the independent variables using the thermodynamic model.

Decision Variables. In this model the compressor pressure ratio p_2/p_1, isentropic compressor efficiency η_{sc}, isentropic turbine efficiency η_{st}, temperature of the air entering the combustion chamber T_3, and temperature of the combustion products entering the turbine T_4 are considered as decision variables. The following nominal values of the decision variables correspond to the case reported in Table 1.2: $p_2/p_1 = 10$, $\eta_{sc} = \eta_{st} = 86\%$, $T_3 = 850$ K, $T_4 = 1520$ K.

Although the decision variables may be varied in optimization studies, each decision variable is normally required to be within a given range. Here we give only the higher limits for the decision variables: To use one of the commercially available gas turbine power plants, the compressor pressure ratio should not exceed 16: $p_2/p_1 \leq 16$. For cost reasons, we also require that the maximum values of the isentropic compressor and isentropic turbine efficiencies are less than 90 and 92%, respectively: $\eta_{sc} \leq 0.90$, $\eta_{st} \leq 0.92$. Because of materials limitations for the gas turbine, the temperature T_4 is also constrained: $T_4 \leq 1550$ K.

Parameters. Parameters are independent variables whose values are specified. They are kept fixed in an optimization study. In this model, the following parameters are identified:

System Products

The net power generated by the system is 30 MW.

Saturated water vapor is supplied by the system at $p_9 = 20$ bars and $\dot{m}_9 = 14$ kg/s.

Air Compressor

$T_1 = 298.15$ K (25°C), $p_1 = 1.013$ bars (1 atm).
Air molar analysis (%): 77.48 N_2, 20.59 O_2, 0.03 CO_2, 1.90 $H_2O(g)$.

Air Preheater

Pressure drops: 3% on the gas side and 5% on the air side.

Heat-Recovery Steam Generator

$T_8 = 298.15$ K, $p_8 = 20$ bars, $p_7 = 1.013$ bars. Pressure drop: 5% on the gas side.

Combustion Chamber

$T_{10} = 298.15$ K. Pressure drop: 5%. Additionally, the pressure p_{10} of the fuel entering the combustion chamber is a parameter. In Table 1.2 we give a nominal value for p_{10}. In the optimization study discussed in Section 9.5, we use $p_{10} = 40$ bars.

Dependent Variables. The dependent variables include the mass flow rates of the air, combustion products, and fuel, the power required by the compressor, the power developed by the turbine, and the following pressures and temperatures:

Air compressor	p_2, T_2
Air preheater	p_3, p_6, T_6
Combustion chamber	p_4
Gas turbine	p_5, T_5
Heat recovery steam generator	T_7

Governing Equations. Using the assumptions listed, a set of governing equations can be developed from the principles introduced thus far in this chapter. This involves consideration of several individual control volumes as follows:

Control Volume Enclosing the Combustion Chamber. Denoting the fuel–air ratio on a molar basis as $\bar{\lambda}$, the molar flow rates of the fuel, air, and combustion products are related by

$$\frac{\dot{n}_F}{\dot{n}_a} = \bar{\lambda}, \qquad \frac{\dot{n}_P}{\dot{n}_a} = 1 + \bar{\lambda} \qquad (2.79a)$$

where the subscripts F, P, and a denote, respectively, fuel, combustion products, and air. For complete combustion of methane the chemical equation takes the form

$$\bar{\lambda}CH_4 + [0.7748N_2 + 0.2059O_2 + 0.0003CO_2 + 0.019H_2O]$$
$$\rightarrow [1 + \bar{\lambda}][x_{N_2}N_2 + x_{O_2}O_2 + x_{CO_2}CO_2 + x_{H_2O}H_2O] \qquad (2.79b)$$

Balancing carbon, hydrogen, oxygen, and nitrogen, the mole fractions of the components of the combustion products are

$$x_{N_2} = \frac{0.7748}{1 + \bar{\lambda}}, \qquad x_{O_2} = \frac{0.2059 - 2\bar{\lambda}}{1 + \bar{\lambda}}$$

$$x_{CO_2} = \frac{0.0003 + \bar{\lambda}}{1 + \bar{\lambda}}, \qquad x_{H_2O} = \frac{0.019 + 2\bar{\lambda}}{1 + \bar{\lambda}} \qquad (2.79c)$$

The molar analysis of the combustion products is fixed once the fuel–air ratio $\bar{\lambda}$ has been determined. The fuel–air ratio can be obtained from an energy rate balance as follows:

$$0 = \dot{Q}_{cv} - \dot{W}_{cv} + \dot{n}_F \bar{h}_F + \dot{n}_a \bar{h}_a - \dot{n}_P \bar{h}_P \qquad (2.79d)$$

As the heat loss is assumed to be 2% of the fuel lower heating value, we have

$$\dot{Q}_{cv} = -0.02\dot{n}_F \overline{LHV} = \dot{n}_a(-0.02\bar{\lambda} \overline{LHV}) \qquad (2.79e)$$

Using data from Table C.1, $\overline{LHV} = 802{,}361$ kJ/kmol. Collecting Equations 2.79d and e, we have

$$0 = -0.02\bar{\lambda} \overline{LHV} + \bar{h}_a + \bar{\lambda}\bar{h}_F - (1 + \bar{\lambda})\bar{h}_P \qquad (2.79f)$$

from which $\bar{\lambda}$ can be evaluated.

With ideal-gas mixture principles, the enthalpies of the air and combustion products are, respectively

$$\bar{h}_a = [0.7748\bar{h}_{N_2} + 0.2059\bar{h}_{O_2} + 0.0003\bar{h}_{CO_2} + 0.019\bar{h}_{H_2O}](T) \qquad (2.79g)$$

$$(1 + \bar{\lambda})\bar{h}_P = [0.7748\bar{h}_{N_2} + (0.2059 - 2\bar{\lambda})\bar{h}_{O_2} + (0.0003 + \bar{\lambda})\bar{h}_{CO_2}$$
$$+ (0.019 + 2\bar{\lambda})\bar{h}_{H_2O}](T) \qquad (2.79h)$$

where the specific enthalpies are evaluated using data from Table C.1. In these equations, h_a is evaluated at temperature T_3 and $\bar{h}_P$ is evaluated at T_4. Thus, the value of λ depends on the decision variables T_3 and T_4.

As the pressure at state 2 is specified, the pressure at the exit of the combustion chamber, p_4, is determined using the assumed pressure drops across the air preheater and the combustion chamber: $p_3 = 0.95p_2$, $p_4 = 0.95p_3$, giving $p_4 = (0.95)^2p_2$.

Since $\dot{n}_F = \lambda\dot{n}_a$, the fuel and air mass flow rates are related by

$$\dot{m}_F = \bar{\lambda}\left(\frac{M_F}{M_a}\right)\dot{m}_a \tag{2.79i}$$

where M_F and M_a denote the molecular weights of the fuel and air, respectively. The molar flow rate $\dot{n}_a$ is evaluated using Equation 2.80a, developed next.

Control Volume Enclosing the Compressor and Turbine. For this control volume the energy rate balance takes the form

$$0 = \dot{Q}_{cv} - \dot{W}_{cv} + \dot{n}_a(\bar{h}_1 - \bar{h}_2) + \dot{n}_P(\bar{h}_4 - \bar{h}_5)$$

where $\dot{W}_{cv}$ is the net power developed ($=30$ MW). With Equation 2.79a

$$0 = -\frac{\dot{W}_{cv}}{\dot{n}_a} + (\bar{h}_1 - \bar{h}_2) + (1 + \bar{\lambda})(\bar{h}_4 - \bar{h}_5) \tag{2.80a}$$

The term $(\bar{h}_1 - \bar{h}_2)$ of Equation 2.80a is evaluated using the isentropic compressor efficiency:

$$\eta_{sc} = \frac{\bar{h}_{2s} - \bar{h}_1}{\bar{h}_2 - \bar{h}_1} \tag{2.80b}$$

where $\bar{h}_{2s}$ denotes the specific enthalpy for an isentropic compression from the inlet state 1 to the specified exit pressure p_2. The state 2s is fixed using $\bar{s}_{2s} - \bar{s}_1 = 0$. The value of $\bar{h}_2$ determined from Equation 2.80b is used to calculate the value of T_2 iteratively.

The term $(\bar{h}_4 - \bar{h}_5)$ of Equation 2.80a is evaluated using the isentropic turbine efficiency:

$$\eta_{st} = \frac{\bar{h}_4 - \bar{h}_5}{\bar{h}_4 - \bar{h}_{5s}} \tag{2.80c}$$

where $\bar{h}_{5s}$ denotes the specific enthalpy for an isentropic expansion from the turbine inlet state 4 to the exit pressure p_5. As the pressure at state 7 is specified, p_5 is determined using the assumed pressure drops across the heat recovery steam generator and the air preheater: $p_6 = p_7/0.95$, $p_5 = p_6/0.97$, giving $p_5 = p_7/(0.95)(0.97)$. The value of $\bar{h}_5$ determined from Equation 2.80c is used to calculate the value of T_5 iteratively.

Using Equations 2.80b and c, Equation 2.80a can be solved for the air molar flow rate $\dot{n}_a$. Then the fuel and product molar flow rates $\dot{n}_F$ and $\dot{n}_P$ can be evaluated from Equation 2.79a. Alternatively, using Equation 2.79i, together with the respective values for the molecular weights, the mass flow rates $\dot{m}_a$, $\dot{m}_F$, and $\dot{m}_P$ can be determined. The mass flow rates depend on the values of all five decision variables.

Control Volume Enclosing the Air Preheater. For this control volume the energy rate balance takes the form

$$0 = \cancel{\dot{Q}}_{cv} - \cancel{\dot{W}}_{cv} + \dot{n}_a(\bar{h}_2 - \bar{h}_3) + \dot{n}_P(\bar{h}_5 - \bar{h}_6) \tag{2.81}$$

After solving Equation 2.81 for $\bar{h}_6$, the temperature T_6 is obtained iteratively.

Control Volume Enclosing the Heat Recovery Steam Generator. For this control volume the energy rate balance takes the form

$$0 = \dot{n}_P(\bar{h}_6 - \bar{h}_7) + \dot{m}_8(h_8 - h_9) \tag{2.82a}$$

After solving Equation 2.82a for $\bar{h}_7$, the temperature T_7 is obtained iteratively.

Owing to the presence of sulfur in natural gas, corrosive sulfuric acid can be formed when the products of combustion are sufficiently cooled. This can be guarded against by maintaining the temperature T_7 above a minimum value, which in the present model is taken as 400 K[1]:

$$T_7 \geq 400 \text{ K} \tag{2.82b}$$

The constraint on T_7, together with the other constraints listed before, assists in identifying workable designs. For example, in an optimization study where the decision variables have been changed from the nominal values of Table 1.2, but Equation 2.82b is not satisfied, the decision variable T_3 would be reduced and/or the decision variable T_4 increased until $T_7 \geq 400$ K. Since T_4 significantly affects the turbine costs (Appendix Table B.2), the preferred approach for satisfying the constraint would be to reduce T_3, however.

As the heat recovery steam generator (HRSG) involves a preheater (economizer) and an evaporator, an additional constraint on workable designs follows: To avoid vaporization in the preheating section, the temperature of the liquid exiting the preheater, T_l, is kept below the saturation temperature: $T_l < T_9$. In the present model we assume $T_l = T_9 - 15$ K $(= 471$ K). The

---

[1]For the case of Table 1.2, we have $T_7 = 427$ K.

temperature of the gas entering the preheater from the evaporator section, T_g, is then obtained by solving the preheater energy rate balance[2]:

$$0 = \dot{n}_P(\bar{h}_g - \bar{h}_7) + \dot{m}_9(h_8 - h_l) \qquad (2.82c)$$

To avoid excessive capital costs, it is necessary for the gas temperature T_g to approach the saturation temperature T_9 no more closely than a specified amount. In the present model we require the following[3]:

$$T_g - T_9 \geq 15 \text{ K} \qquad (2.82d)$$

Accordingly, if Equation 2.82d were not satisfied in an optimization study where the decision variables have been changed from the nominal values of Table 1.2, we would reduce T_3 (and/or increase T_4) as discussed above until $T_g - T_9 \geq 15$ K.

Dew Point Temperature of the Combustion Products. When a mixture consisting of gaseous products of combustion containing water vapor is cooled at constant mixture pressure, the dew point temperature (the saturation temperature corresponding to the partial pressure of the water vapor) marks the onset of condensation of the water vapor. For example, for the products exiting the cogeneration system at state 7, and using values given in Table 1.2, the partial pressure of the water vapor is

$$p_v = x_v p = (0.0807)(1.013 \text{ bars}) = 0.0817 \text{ bar} \qquad (2.83a)$$

The corresponding saturation temperature—the dew point temperature—is closely 42°C.

Cooling at constant mixture pressure below the dew point temperature would result in some condensation of the water vapor. For example, if the mixture were cooled to 25°C at a fixed pressure of 1 atm, some condensation would occur. We model the result at 25°C as a gas phase containing saturated water vapor in equilibrium with a saturated liquid water phase.[4] On the basis of 1 kmol of combustion products formed, the gas phase at 25°C would consist of 0.9193 kmol of dry products (0.7507 N_2, 0.1372 O_2, 0.0314 CO_2) plus n_v kmol of water vapor. The partial pressure of water vapor would be equal to the saturation pressure, p_g (25°C) = 0.0317 bar. The amount of water vapor present can be found from $p_v = x_v p$, which now takes the form

---

[2]For the case of Table 1.2, we have T_g = 526 K.
[3]For the case of Table 1.2, we have $T_g - T_9$ = 40 K.
[4]At normal temperatures and pressures the equilibrium between the liquid water phase and the water vapor is not significantly disturbed by the presence of the dry air. See Reference 1, Sections 12.6.4 and 14.6.1.

$$0.0317 \text{ bar} = \frac{n_v}{0.9193 + n_v} (1.013 \text{ bars}) \qquad (2.83b)$$

Solving Equation 2.83b gives $n_v = 0.0297$ kmol. Thus, for the case of Table 1.2, the composition of the combustion products at 25°C, 1 atm reads

$$\{0.7507 \text{ N}_2, \ 0.1372 \text{ O}_2, \ 0.0314 \text{ CO}_2, \ 0.0297 \text{ H}_2\text{O(g)}, \ 0.0510 \text{ H}_2\text{O(l)}\}$$

where the underlining identifies the gas phase. This result is used in the solution to Example 3.1.

Example 2.1 For the cogeneration system of Figure 1.7 and Table 1.2, determine the fuel–air ratio and the analysis of the combustion products, each on a molar basis.

Solution

MODEL

1. A control volume enclosing the combustion chamber is considered.
2. The control volume operates at steady state.
3. Ideal-gas mixture principles apply for the air and combustion products.
4. Combustion is complete. N_2 is inert.
5. Kinetic and potential energy effects are ignored.
6. Heat transfer from the control volume is 2% of the lower heating value of the fuel.

ANALYSIS. Denoting the fuel–air ratio on a molar basis as $\bar{\lambda}$, the molar flow rates of the fuel, air, and combustion products are related as follows:

$$\frac{\dot{n}_F}{\dot{n}_a} = \bar{\lambda}, \qquad \frac{\dot{n}_P}{\dot{n}_a} = 1 + \bar{\lambda}$$

On a per mole of air basis, the chemical equation then takes the form

$$\bar{\lambda}\text{CH}_4 + [0.7748 \text{ N}_2 + 0.2059 \text{ O}_2 + 0.0003 \text{ CO}_2 + 0.019 \text{ H}_2\text{O}]$$

$$\rightarrow [1 + \bar{\lambda}][x_{\text{N}_2}\text{N}_2 + x_{\text{O}_2}\text{O}_2 + x_{\text{CO}_2}\text{CO}_2 + x_{\text{H}_2\text{O}}\text{H}_2\text{O}]$$

Balancing carbon, hydrogen, oxygen, and nitrogen, the mole fractions of the components of the combustion products are

$$x_{N_2} = \frac{0.7748}{1 + \bar{\lambda}}, \qquad x_{O_2} = \frac{0.2059 - 2\bar{\lambda}}{1 + \bar{\lambda}}$$

$$x_{CO_2} = \frac{0.0003 + \bar{\lambda}}{1 + \bar{\lambda}}, \qquad x_{H_2O} = \frac{0.019 + 2\bar{\lambda}}{1 + \bar{\lambda}}$$

Accordingly, the molar analysis of the products is fixed once $\bar{\lambda}$ has been determined. The fuel–air ratio can be obtained from an energy rate balance as follows:

With indicated idealizations, the energy rate balance takes the form

$$0 = \dot{Q}_{cv} - \dot{W}_{cv} + \dot{n}_F \bar{h}_F + \dot{n}_a \bar{h}_a - \dot{n}_P \bar{h}_P$$

By assumption 6 of the model, the heat transfer rate is

$$\dot{Q}_{cv} = -0.02 \dot{n}_F \, \overline{LHV}$$

$$= \dot{n}_a (-0.02 \bar{\lambda} \, \overline{LHV})$$

Collecting results

$$0 = -0.02 \bar{\lambda} \, \overline{LHV} + \bar{h}_a + \bar{\lambda} \bar{h}_F - (1 + \bar{\lambda}) \bar{h}_P$$

Using ideal-gas mixture principles to determine the enthalpies of the air and combustion products, we have, for $T_3 = 850$ K, $T_4 = 1520$ K,

$$\bar{h}_a = [0.7748 \bar{h}_{N_2} + 0.2059 \bar{h}_{O_2} + 0.0003 \bar{h}_{CO_2} + 0.019 \bar{h}_{H_2O}](T_3)$$

$$(1 + \bar{\lambda}) \bar{h}_P = [0.7748 \bar{h}_{N_2} + (0.2059 - 2\bar{\lambda}) \bar{h}_{O_2} + (0.0003 + \bar{\lambda}) \bar{h}_{CO_2}$$

$$+ (0.019 + 2\bar{\lambda}) \bar{h}_{H_2O}](T_4)$$

Combining the last three equations and solving for $\bar{\lambda}$ yields

$$\bar{\lambda} = \frac{0.7748 \, \Delta\bar{h}_{N_2} + 0.2059 \, \Delta\bar{h}_{O_2} + 0.0003 \, \Delta\bar{h}_{CO_2} + 0.019 \, \Delta\bar{h}_{H_2O}}{\bar{h}_F - 0.02 \, \overline{LHV} - (-2\bar{h}_{O_2} + \bar{h}_{CO_2} + 2\bar{h}_{H_2O})(T_4)}$$

Evaluating enthalpy values in kJ/mol as in Appendix Table C.1, we have

$$\bar{h}_F = -74{,}872 \text{ kJ/kmol}, \qquad \overline{LHV} = 802{,}361 \text{ kJ/kmol}$$

Component	$\bar{h}$ (850 K)	$\bar{h}$ (1520 K)	$\Delta\bar{h}$ (kJ/kmol)
N_2	17,072	39,349	22,277
O_2	17,540	41,138	23,598
CO_2	-367,121	-330,159	36,962
H_2O	-221,321	-192,283	29,038

With these data, $\bar{\lambda} = 0.0321$, and the molar analysis of the products is as follows:

Component	N_2	O_2	CO_2	H_2O
Mole fraction	0.7507	0.1373	0.0314	0.0806

COMMENTS

1. Roundoff accounts for the slight departure from the product molar analysis reported in Table 1.2.
2. Using Equation 2.55, the mixture molecular weight of the combustion products is 28.254.

Example 2.2 For the cogeneration system shown in Figure 1.7 and Table 1.2, determine the mass flow rates of the fuel and the air, each in kilograms per second, for a net power output of 30 MW.

Solution

MODEL

1. A control volume enclosing the compressor and turbine is considered.
2. The control volume operates at steady state.
3. Ideal-gas mixture principles apply for the air and combustion products.
4. Kinetic and potential energy effects are ignored.
5. The compressor and turbine operate adiabatically.

ANALYSIS. The fuel and air molar flow rates are related by the fuel–air ratio: $\dot{n}_F = \bar{\lambda}\dot{n}_a$. Or, expressed alternatively in terms of mass flow rates

$$\dot{m}_F = \overline{\lambda} \frac{M_F}{M_a} \dot{m}_a$$

As $\overline{\lambda}$ is evaluated in Example 2.1, the fuel mass flow rate can be determined once $\dot{m}_a$ is known. The mass flow rate of the air can be obtained from an energy rate balance as follows:

With indicated idealizations, the energy rate balance takes the form

$$0 = \cancel{\dot{Q}}_{cv} - \dot{W}_{cv} + \dot{n}_a(\overline{h}_1 - \overline{h}_2) + \dot{n}_P(\overline{h}_4 - \overline{h}_5)$$

or, in terms of the fuel–air ratio, as

$$0 = -\frac{\dot{W}_{cv}}{\dot{n}_a} + (\overline{h}_1 - \overline{h}_2) + (1 + \overline{\lambda})(\overline{h}_4 - \overline{h}_5)$$

Converting to a mass flow rate basis and solving yields

$$\dot{m}_a = \frac{M_a \dot{W}_{cv}}{(1 + \overline{\lambda})(\overline{h}_4 - \overline{h}_5) + (\overline{h}_1 - \overline{h}_2)}$$

The molar analysis and temperature of the air entering the compressor at state 1 are known. The molar analysis and temperature of the combustion products entering the turbine at state 4 are also known. Thus, the corresponding specific enthalpy values, $\overline{h}_1$ and $\overline{h}_4$, required to evaluate $\dot{m}_a$ can be determined using ideal-gas mixture principles together with enthalpies evaluated as in Appendix Table C.1:

$$\overline{h}_1 = -4713.3 \text{ kJ/kmol}, \qquad \overline{h}_4 = 9304.5 \text{ kJ/kmol}$$

Values for the specific enthalpies $\overline{h}_2$ and $\overline{h}_5$ are also required. They can be evaluated using the compressor and turbine isentropic efficiencies, respectively, as follows: Solving the expression for the compressor isentropic efficiency

$$\eta_{sc} = \frac{\overline{h}_{2s} - \overline{h}_1}{\overline{h}_2 - \overline{h}_1}$$

we have

$$\overline{h}_2 = \overline{h}_1 + \frac{\overline{h}_{2s} - \overline{h}_1}{\eta_{sc}}$$

where $\overline{h}_{2s}$ denotes the specific enthalpy for an isentropic compression from the inlet state 1 to the specified exit pressure p_2. As the air composition is fixed, ideal-gas mixture principles allow the isentropic compression to be described as

$$\bar{s}_{2s} - \bar{s}_1 = 0.7748\left[\bar{s}°(T_{2s}) - \bar{s}°(T_1) - \bar{R}\ln\frac{p_2}{p_1}\right]_{N_2}$$

$$+ 0.2059\left[\bar{s}°(T_{2s}) - \bar{s}°(T_1) - \bar{R}\ln\frac{p_2}{p_1}\right]_{O_2}$$

$$+ 0.0003\left[\bar{s}°(T_{2s}) - \bar{s}°(T_1) - \bar{R}\ln\frac{p_2}{p_1}\right]_{CO_2}$$

$$+ 0.0190\left[\bar{s}°(T_{2s}) - \bar{s}°(T_1) - \bar{R}\ln\frac{p_2}{p_1}\right]_{H_2O}$$

$$= 0$$

Evaluating the specific entropies at $T_1 = 298.15$ K as in Table C.1, this reduces to

$$0.7748\bar{s}°_{N_2}(T_{2s}) + 0.2059\bar{s}°_{O_2}(T_{2s}) + 0.0003\bar{s}°_{CO_2}(T_{2s}) + 0.0190\bar{s}°_{H_2O}(T_{2s})$$

$$= 213.495 \text{ kJ/kmol·K}$$

Inserting the specific entropy expressions for N_2, O_2, CO_2, and H_2O from Table C.1 and solving gives T_{2s} (≈ 563 K) and the corresponding specific enthalpy: $\bar{h}_{2s} = 3293.5$ kJ/kmol. Accordingly,

$$\bar{h}_2 = \bar{h}_1 + \frac{\bar{h}_{2s} - \bar{h}_1}{\eta_{sc}}$$

$$= -4713.3 + \frac{3293.5 + 4713.3}{0.86}$$

$$= 4596.9 \text{ kJ/kmol}$$

Solving the expression for the turbine isentropic efficiency

$$\eta_{st} = \frac{\bar{h}_4 - \bar{h}_5}{\bar{h}_4 - \bar{h}_{5s}}$$

we have

$$\bar{h}_5 = \bar{h}_4 - \eta_{st}(\bar{h}_4 - \bar{h}_{5s})$$

where $\bar{h}_{5s}$ denotes the specific enthalpy for an isentropic expansion from the inlet state 4 to the specified exit pressure p_5. As the product composition is fixed, the isentropic expansion can be described as $\bar{s}_{5s} - \bar{s}_4 = 0$. Then, with ideal-gas mixture principles and specific entropies of the components evaluated at $T_4 = 1520$ K we have

$$0.7507\bar{s}_{N_2}^{\circ}(T_{5s}) + 0.1372\bar{s}_{O_2}^{\circ}(T_{5s}) + 0.0314\bar{s}_{CO_2}^{\circ}(T_{5s}) + 0.0807\bar{s}_{H_2O}^{\circ}(T_{5s})$$

$$= 229.838 \text{ kJ/kmol·K}$$

Inserting the specific entropy expressions for N_2, O_2, CO_2, and H_2O from Table C.1 and solving gives T_{5s} (≈ 920 K) and the corresponding specific enthalpy at state $5s$: $\bar{h}_{5s} = -11{,}792.6$ kJ/kmol. Accordingly,

$$\bar{h}_5 = \bar{h}_4 - \eta_{st}(\bar{h}_4 - \bar{h}_{5s})$$

$$= 9304.9 - 0.86(9304.5 + 11{,}792.6)$$

$$= -8839 \text{ kJ/kmol}$$

With $\bar{h}_1$, $\bar{h}_2$, $\bar{h}_4$, and $\bar{h}_5$ known, the value of $\dot{m}_a$ can be calculated:

$$\dot{m}_a = \frac{(28.649 \text{ kg/kmol})(30{,}000 \text{ kJ/s})}{(1.032)(9304.5 + 8839) - (4713.3 + 4596.9) \text{ kJ/kmol}}$$

$$= 91.3 \text{ kg/s}$$

The mass flow rate of the fuel is then

$$\dot{m}_F = (0.0321)\left(\frac{16.043}{28.649}\right)(91.3 \text{ kg/s})$$

$$= 1.64 \text{ kg/s}$$

COMMENTS

1. Roundoff accounts for the slight departures from the mass flow rates reported in Table 1.2.
2. The temperature of the air exiting the compressor, T_2, can be obtained by inserting the specific enthalpy expressions for N_2, O_2, CO_2, and H_2O, from Table C.1 into the following equation and solving to obtain $T_2 = 603.74$ K:

$$\bar{h}_2 = 0.7748\bar{h}_{N_2}(T_2) + 0.2059\bar{h}_{O_2}(T_2) + 0.0003\bar{h}_{CO_2}(T_2)$$

$$+ 0.0190\bar{h}_{H_2O}(T_2)$$

$$= 4596.9 \text{ kJ/kmol}$$

A similar approach can be used to obtain the temperature at the turbine exit, $T_5 = 1006.16$ K.

2.6 MODELING AND DESIGN OF PIPING SYSTEMS

Industrial plants, and especially chemical-processing plants, frequently exhibit intricate and costly piping configurations. Piping systems are also encountered in many familiar situations, including water supply, fire protection, and district heating applications. Piping systems are commonly composed of single-path elements, as shown in Figure 2.4a, multiple-path elements, as shown in Figure 2.4b, or a combination. The design of piping systems requires the consideration of many issues, some of which conform to the main theme of this book. In this section we provide an overview of the modeling and design of piping systems. For elaboration, readers should consult specialized literature [e.g., 7, 8].

2.6.1 Design Considerations

As pipelines may carry high-temperature, high-pressure, and potentially hazardous fluids, applicable codes and standards must be carefully considered when undertaking the design of such systems. Noteworthy examples are the ASME codes and ANSI standards mentioned in Section 1.4.2. As surveyed next, the design of piping systems also requires the judicious integration of mechanical, thermal-fluid, and economics principles.

General design considerations include a careful assessment of the nature of the fluid being conveyed and, in the event of a leak or general failure, its possible effect on personnel and the environment through toxicity and chemical reaction, including combustion. The impact of a piping element failure on overall plant safety and operation is also an important consideration. Piping system designs must account for both dynamic and static forces owing to hydraulic shock, vibration, wind (exposed piping), earthquakes, and weight. Weight effects include the weight of pipes, valves, insulation, the fluids being conveyed, and, for exposed piping, the weight of ice and snow. To withstand such forces, adequate piping supports must be provided. Appropriate materials of construction must be specified, with attention given to thermal expansion or compression effects. Code requirements for methods of joining (welding, brazing, soldering, etc.) should be met. Installed piping systems should be pressure tested prior to initial operation to assure tightness.

The installed cost of piping systems is another important design consideration. The cost of piping systems varies widely with the complexity of the system and the materials of construction specified. The economics also depend significantly on the piping size and fabrication methods used. Safe-

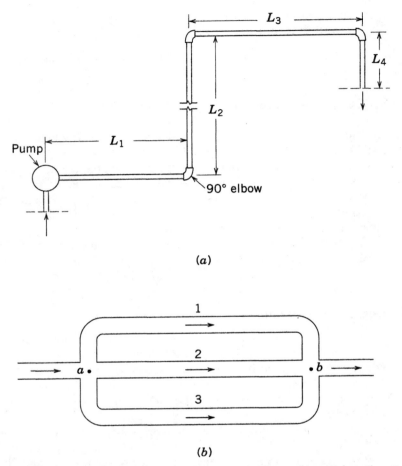

(a)

(b)

Figure 2.4   Elementary piping systems: (a) single-path system; (b) multiple-path system.

guarding measures (thermal insulation, vibration damping, valve bonnet shields, sight glasses, etc.) add to costs as well. A cost-effective practice in the design of piping systems is to specify piping that is commonly available commercially. In this respect, it is noteworthy that pipes are manufactured in a limited number of standard sizes (nominal sizes). Manufacturers' representatives should be contacted for data on sizes, materials, recommended applications, and costs. In Chapter 7, we give further consideration to piping costs.

Additional design considerations are provided in Sections 2.6.3 and 2.6.4 dealing with piping system design analysis and pump selection, respectively. Considered next is the evaluation of head loss, a key design variable.

2.6.2 Estimation of Head Loss

In piping system design, a central role is played by Equation 2.30. If we limit consideration to liquids modeled as incompressible, flowing through and completely filling circular ducts, Equation 2.30 takes the form

$$\frac{p_2 - p_1}{\rho} + \frac{V_2^2 - V_1^2}{2} + g(z_2 - z_1) + \frac{\dot{W}_{cv}}{\dot{m}} + h_l = 0 \qquad (2.84)$$

where the term h_l, the *head loss*, is given by Equation 2.31, repeated here for ease of reference:

$$h_l = \sum_p \left[\left(f \frac{4L}{D} \right) \frac{V^2}{2} \right]_p + \sum_l \left[K \frac{V^2}{2} \right]_l \qquad (2.31)$$

$$\underbrace{\qquad\qquad\qquad}_{\text{pipe friction}} \qquad \underbrace{\qquad\qquad}_{\text{resistances}}$$

With the assumption of one-dimensional flow, the velocities appearing in these expressions are frequently written in terms of the volumetric flow rate Q and cross-sectional area A: $V = Q/A = 4Q/\pi D^2$.

Equation 2.31 represents the head loss as the sum of two contributions called, respectively, the *major* and *minor* losses. The major loss summation term accounts for pipe friction. The minor loss summation term accounts for flow through various resistances to be discussed later.

Referring to the first summation on the right side of Equation 2.31, we see that the major losses are evaluated using the friction factor f for each of the pipes making up the overall system.[5] The friction factor is a function of two dimensionless groups: the relative roughness k_s/D, where k_s denotes the roughness of the pipe's inner wall surface, and the Reynolds number $\rho VD/\mu$, where μ denotes viscosity.[6] The graphical representation of the friction factor function developed from experimental data, called a *Moody diagram,* is presented in Figure 2.5, together with a sampling of values for the roughness.

Three regimes are identified on Figure 2.5: laminar, critical, and turbulent. The laminar flow friction factor is a straight line on the log-log plot and is given by $4f = 64/\text{Re}_D$, an analytical result that is valid to a Reynolds number

---

[5]The Fanning friction factor f is used here and differs from the Darcy friction factor f_D by a factor of 4: $4f = f_D$.

[6]The roughness parameter is denoted by k_s instead of the usual ε to avoid overlap with other uses in this book for the symbol ε.

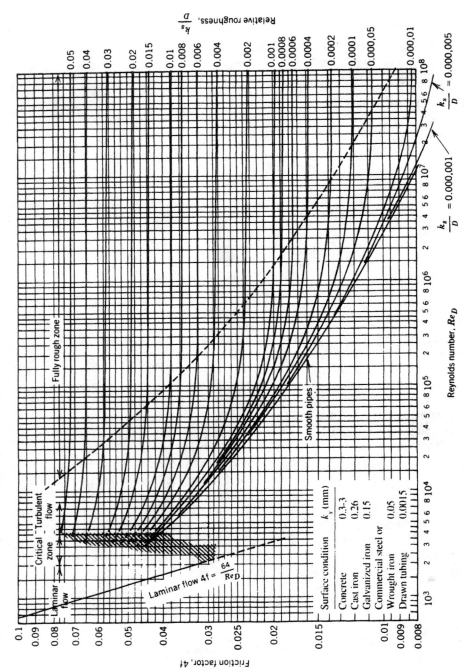

Figure 2.5 Friction factor for fully developed flow in circular pipes. (Data from L. F. Moody, *Trans. ASME*, **66**, 8, 671–684, 1944.)

Surface condition	k_s (mm)
Concrete	0.3–3
Cast iron	0.26
Galvanized iron	0.15
Commercial steel or Wrought iron	0.05
Drawn tubing	0.0015

Reynolds number, Re_D

Friction factor, $4f$

Relative roughness, $\dfrac{k_s}{D}$

Laminar flow $4f = \dfrac{64}{Re_D}$

of about 2000. The Reynolds number in the critical zone, where the flow may be either laminar or turbulent, is about 2000–4000. In the transition from laminar to turbulent flow, the friction factor increases sharply and then decreases gradually for smooth pipes as the Reynolds number increases. For rough pipes, the turbulent friction factor is determined by the relative roughness together with the Reynolds number. At sufficiently high Reynolds numbers, however, the turbulent friction factor is determined by the relative roughness alone: $f \sim f(k_s/D)$. This is indicated on Figure 2.5 as the *fully rough zone*.

To assist computer-aided design and analysis, the friction factor function $f(\text{Re}_D, k_s/D)$ is available in several alternative mathematical forms. The *Colebrook equation* is often cited for turbulent flow:

$$\frac{1}{\sqrt{f}} = -4 \log \left(\frac{k_s/D}{3.7} + \frac{1.256}{\sqrt{f}\ \text{Re}_D} \right) \tag{2.85}$$

Since this expression is implicit in f, iteration is required to obtain the friction factor for a specified Re_D and k_s/D.

The literature includes a number of alternative expressions giving the friction factor explicitly. For example, an expression that represents the friction factor continuously for laminar through turbulent flow is the following [9]:

$$f = 2 \left[\left(\frac{8}{\text{Re}_D} \right)^{12} + \frac{1}{(\alpha + \beta)^{1.5}} \right]^{1/12} \tag{2.86}$$

where

$$\alpha = \left(2.457 \ln \frac{1}{(7/\text{Re}_D)^{0.9} + (0.27 k_s/D)} \right)^{16} \tag{2.87a}$$

and

$$\beta = \left(\frac{37{,}530}{\text{Re}_D} \right)^{16} \tag{2.87b}$$

A plausible empirical approach for evaluating friction factors of noncircular ducts is to use the hydraulic diameter in place of D in the graphical and analytical approaches considered above. The hydraulic diameter is defined by

$$D_h \equiv 4 \frac{\text{cross-sectional area}}{\text{wetted perimeter}} \tag{2.88}$$

Turning now to the second term on the right side of Equation 2.31, note that the minor losses are evaluated using a loss coefficient K for each of the resistances included within the overall system. Typically encountered resistances include pipe inlets and outlets, enlargements and contractions, pipe bends and tees, valves, and other fittings. Experimental loss coefficient data are abundant, but scattered among various sources, including handbooks [7, 8], manufacturers' data books [10, 11], and textbooks [3, 4, 12]. Table 2.1 gives a sampling of loss coefficients for use in Equation 2.31. Except as noted, the velocity downstream of the resistance is used to evaluate the associated loss. As different sources may give somewhat different values for the loss coefficient for the same application, the values of Table 2.1 should be considered as only representative.

2.6.3 Piping System Design and Design Analysis

Single-path and multiple-path piping systems such as pictured in Figure 2.4 can be analyzed (or specified) by using Equation 2.84, together with Equation 2.31 and appropriate friction factor and loss coefficient data. Let us consider this for the case of single-path systems. Although such systems may consist of pipes having different lengths and diameters as well as various fittings, the discussion is considerably simplified if we think only of a single pipe of length L and diameter D without elevation change. Then, Equations 2.84 and 2.31 combine to read

$$\frac{\Delta p}{\rho} = f\left(\frac{4\rho Q}{\pi\mu D}, \frac{k_s}{D}\right)\frac{2L}{D}\left(\frac{4Q}{\pi D^2}\right)^2 \tag{2.89}$$

where we have written the velocity and Reynolds number in terms of the volumetric flow rate: $V = 4Q/\pi D^2$, $Re_D = 4\rho Q/\pi\mu D$, and Δp denotes the pressure drop from inlet to outlet: $\Delta p \equiv p_1 - p_2$.

If the pipe roughness and fluid properties are specified, Equation 2.89 involves four quantities: Δp, Q, L, and D, any one of which may be regarded as the unknown. Depending on the choice for the unknown quantity, the solution may be direct or iterative:

Direct Solutions

  1. Δp unknown; L, Q, and D known
  2. L unknown; Δp, Q, and D known

Iterative Solutions

  3. Q unknown; Δp, L, and D known
  4. D unknown; Δp, L, and Q known

Cases 1 and 3 exhibit an *analysis* character since the geometry (L, D) is known and the corresponding pressure drop or flow rate is to be determined.

Table 2.1 Loss coefficients for use with Equation 2.31[a]

Resistance	K
Changes in Cross-Sectional Area	

Rounded pipe entrance

0.04–0.28

Contraction[b]

0.45(1-AR)

$\theta \leq 60°$
$0.1 \leq AR \leq 0.5$

0.04–0.08

Expansion[b,c]

$$\left(1 - \frac{1}{AR}\right)^2$$

1.0

Valves and Fittings

Gate valve, open	0.2
Globe valve, open	6–10
Check valve (ball), open	70
45° elbow, standard	0.3–0.4
90° elbow, rounded	0.4–0.9

[a]Adapted from References 3, 4, and 7.
[b]Area ratio, AR = (smaller area)/(larger area).
[c]When AR = 0, $K = 1.0$. Velocity used to evaluate the loss is the velocity upstream of the expansion.

Cases 2 and 4 exhibit a *design* character since Q and Δp are known and the appropriate geometry (L or D) is to be determined. Let us consider these four cases in order:

- *Cases 1 and 2.* By inspection of Equation 2.89, it is evident that Δp, or L, can be obtained directly using the respective known quantities. In particular, since Q and D are known in each of these cases, the friction factor required by Equation 2.89 can be obtained readily from the Moody diagram or an expression such as Equations 2.85 and 2.86.

- *Case 3.* The unknown volumetric flow rate Q appears in Equation 2.89 both explicitly as the quadratic term Q^2 and implicitly in the friction factor function via the Reynolds number. Thus an iterative solution is required. Since most practical pipe flow applications involve turbulent flow and the turbulent flow friction factor only weakly depends on the Reynolds number, the first iteration may be made using the fully rough-turbulent friction factor corresponding to the known value for the relative roughness. With this trial value for f, Equation 2.89 is readily solved for the volumetric flow rate. The Reynolds number is computed for this value of Q and a new value for f determined. Equation 2.89 is then solved for a second value for Q. The iterative procedure continues until convergence is attained. Convergence tends to occur quickly, however, and often with as few as two iterations.

- *Case 4.* The unknown diameter D appears in Equation 2.89 both explicitly as D^5 and implicitly in the friction factor function via both the Reynolds number and the relative roughness. Thus an iterative solution is required: Iteration may begin by assuming a first-trial pipe diameter. With this, the Reynolds number and relative roughness are calculated and then used to determine the friction factor. The pressure drop Δp is evaluated next from Equation 2.89 and compared to the known value for Δp. If the calculated value for Δp is smaller than the known value, another iteration would proceed with a smaller assumed diameter. If the calculated value for Δp is greater than the known value, another iteration would proceed with a greater assumed diameter. Iteration would continue until satisfactory convergence is achieved between the calculated and known Δp values.

Fluid mechanics texts typically provide solved examples illustrating these cases [e.g., 4].

The cases we have just considered provide a reasonable basis for the analysis of more complex piping configurations. For example, a system consisting of a series of single-path elements as in Figure 2.4a can be analyzed by applying such pipe flow fundamentals in a systematic, sequential manner from the inlet of the overall system to its outlet. Multiple-path systems such as in Figure 2.4b also can be analyzed with these fundamentals, but we must recognize additionally that (i) the total flow rate is the sum of the individual flow rates in the multiple paths: $Q_{tot} = Q_1 + Q_2 + Q_3$, and (ii) the pressure drop for each of the multiple paths is the same: $p_a - p_b = (\Delta p)_1 = (\Delta p)_2 = (\Delta p)_3$.

Although the pipe flow governing relations have relatively simple mathematical forms, iterative solutions of them can be time consuming, especially for multiple-path systems. Accordingly, computer-aided approaches are desirable. The literature contains many specialized computer programs for this purpose, as for example the programs provided in Reference 12. One particularly systematic computer-aided approach for complex flow networks is based on the highly effective Hardy–Cross method [13]. Several software development companies market generalized Hardy–Cross piping system programs executable on microcomputers and in some versions on programmable hand-held calculators.

2.6.4 Pump Selection

In most cases, sustaining the flow of a fluid through piping or ducting requires a work input to a suitable fluid machine. When the fluid is a liquid or slurry, the machine is called a pump. Gas and vapor handling units are called fans, blowers, or compressors, depending on the pressure rise. In harmony with the presentations of Sections 2.6.2 and 2.6.3, we limit the present discussion to pumps.

Two commonly employed types of pumps are positive-displacement pumps and centrifugal pumps. In positive-displacement machines, volume and pressure changes occur while the liquid is confined within a chamber or passage. Machines that direct the flow with blades or vanes attached to a rotating member (impeller) are called radial-flow or axial-flow turbomachines, depending on whether the flow path is essentially radial or nearly parallel to the machine centerline. Centrifugal pumps are radial-flow turbomachines. As centrifugal pumps are widely used for industrial applications, we further limit the present discussion to this type.

When selecting a pump for a given service, it is necessary to know the nature of the liquid being handled. Corrosive or reactive characteristics of the liquid requiring special materials of construction should be understood. Also, the presence of solids in the liquid may introduce complexities related to erosion and agglomeration. Knowledge of the temperature, pressure, viscosity, and other liquid properties is also generally required. Additionally, the designer must consider the relationship to the flow rate of key parameters such as of the pressure rise (or head), power requirement, and pump efficiency.

Measured performance data for centrifugal pumps are commonly represented compactly on plots called *characteristic curves* giving the variation of total head versus volumetric flow rate. Figure 2.6a shows characteristic curves for centrifugal pumps operating at a fixed speed with different impeller sizes. Figure 2.6b shows characteristic curves at various speeds for a fixed impeller

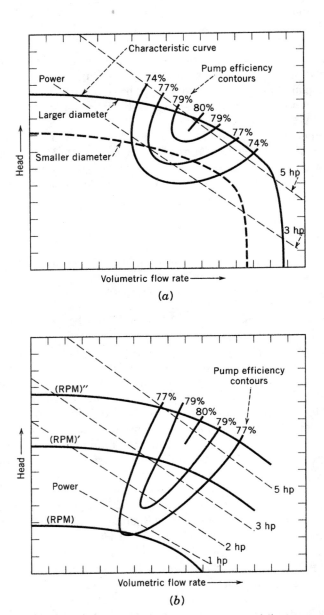

Figure 2.6 Centrifugal pump performance charts: (a) characteristic curves of centrifugal pumps operating at a constant speed for two impeller diameters; (b) characteristic curves of centrifugal pump at various speeds: (RPM) < (RPM)' < (RPM)" for fixed impeller diameter.

size. It is important to note that, at any fixed speed, the pump operates along a particular characteristic curve and at no other points. For this typical machine, the head decreases continuously as the flow rate increases; the power required increases as the flow rate increases; and the pump efficiency increases with capacity until a best efficiency point is reached (80% for the largest diameter and greatest speed cases of Figure 2.6), and then decreases as flow rate increases further. Performance charts such as Figure 2.6 may also contain additional information. An often included parameter is the *net positive suction head* (NPSH), discussed later.

Centrifugal pumps may be combined in parallel to deliver greater flow or in series to provide a greater head. It is common practice to drive pumps with electric motors at nearly constant speed, but variable-speed operation can lead to electricity savings in certain applications. For pumps with variable-speed drives, it is possible to change the characteristic curve, as illustrated in Figure 2.6*b*.

Various operating features influence pump performance. Wear is a noteworthy example. As a pump wears with use, the characteristic curve tends to move downward toward a lower pressure at each flow rate. Other factors that can adversely influence pump performance include pumping hot liquids, two-phase mixtures, or liquids with high viscosities. As the presence of even a small amount of entrained gas can drastically reduce the pump performance, particular attention must be given to preventing air from entering at the suction side of the pump. Another effect that adversely affects the performance is *cavitation*.

Cavitation may arise when the local pressure falls below the vapor pressure of the liquid. If this occurs, liquid may flash to vapor, forming vapor cavities. The growth and collapse of the vapor cavities not only disrupts the flow but also may cause mechanical damage. Since cavitation is detrimental to both pump efficiency and pump life, it must be avoided. Cavitation can be avoided if the pressure everywhere in the pump is kept above the vapor pressure of the liquid. This requires in particular that a pressure in excess of the vapor pressure of the liquid be maintained at the pump inlet (the suction). A measure of the required pressure difference is provided by the NPSH. By locating a pump so that the NPSH is greater than a specified value obtained from manufacturers' data, cavitation can be avoided. As noted before, NPSH data is often shown on pump performance charts.

For further discussion concerning pump selection, including operating data and solved examples, see references such as [7] or contact manufacturers' representatives directly.

2.7 CLOSURE

In this chapter, we have presented some fundamental principles of engineering thermodynamics and illustrated their use for modeling and design analysis in applications involving both engineering thermodynamics and fluid flow. In Chapter 3, the current presentation continues, but with emphasis on the exergy concept. In Chapter 3, we also begin to develop one of the central themes of this book: Greater use should be made in thermal system design of the second law of thermodynamics. Modeling and design analysis considerations related specifically to heat transfer enter the discussion beginning with Chapter 4.

REFERENCES

1. M. J. Moran and H. N. Shapiro, *Fundamentals of Engineering Thermodynamics,* 3rd ed., Wiley, New York, 1995.

2. A. Bejan, *Advanced Engineering Thermodynamics.* Wiley-Interscience, New York, 1988.

3. R. B. Bird, W. E. Stewart, and E. N. Lightfoot, *Transport Phenomena,* Wiley, New York, 1960.

4. R. W. Fox and A. T. McDonald, *Introduction to Fluid Mechanics,* 4th ed., Wiley, New York, 1992.

5. R. C. Reid and T. K. Sherwood, *The Properties of Gases and Liquids,* 2nd ed., McGraw-Hill, New York, 1966.

6. O. Knacke, O. Kubaschewski, and K. Hesselmann, *Thermochemical Properties of Inorganic Substances,* 2nd ed., Springer-Verlag, Berlin, and Verlag Stahleisen, Düsseldorf, 1991.

7. R. H. Perry and D. Green, *Chemical Engineers' Handbook,* 6th ed., McGraw-Hill, New York, 1984.

8. *ASHRAE Handbook 1993 Fundamentals,* American Society of Heating, Refrigerating, and Air Conditioning Engineers, Atlanta, 1993.

9. S. W. Churchill, Friction-factor equation spans all flow regimes, *Chem. Eng.,* 7 November, 1977, pp. 91–92.

10. "Flow of fluids through valves, fittings, and pipes, Crane Company Technical Paper No. 410, New York, 1982.

11. *Engineering Data Book,* Hydraulic Institute, Cleveland, 1979.

12. B. K. Hodge, *Analysis and Design of Energy Systems,* 2nd ed., Prentice-Hall, Englewood Cliffs, NJ, 1990.

13. D. J. Wood and A. G. Rayes, Reliability of algorithms for pipe network analysis, *J. Hydraulic Div., Proc. ASCE,* Vol. 107, No. NY10, 1981, pp. 1145–1161.

PROBLEMS

2.1 Consider an automobile engine as the system. List the principal irreversibilities present during operation. Repeat for an ordinary forced air, natural-gas-fired household furnace.

2.2 At one location the ocean surface temperature is 16°C, while at a depth of 540 m the temperature is 2°C. An inventor claims to have developed a power cycle having a thermal efficiency of 10% that receives and discharges energy by heat transfer at these temperatures, respectively. There are no other heat transfers. Evaluate this claim.

2.3 A patent application describes a device that at steady state generates electricity while heat transfer occurs at temperature T_b only. There are no other energy transfers. Evaluate this device thermodynamically.

2.4 A gas flows through a one-inlet, one-outlet control volume operating at steady state. Heat transfer at the rate $\dot{Q}_{cv}$ occurs only at a location on the boundary where the temperature is T_b. If $\dot{Q}_{cv} \geq 0$, determine whether the outlet specific entropy is greater than, equal to, or less than the inlet specific entropy. What can be said when $\dot{Q}_{cv} < 0$? Discuss.

2.5 Determine the changes in specific enthalpy and specific entropy for each of the following changes of state of water:
(a) $p_1 = 0.15$ MPa, $T_1 = 280°C$; $p_2 = 0.15$ MPa, $v_2 = 0.9$ m³/kg
(b) $T_1 = 320°C$, $v_1 = 0.3$ m³/kg; $p_2 = 2.5$ MPa, $T_2 = 140°C$
(c) $p_1 = 20$ lbf/in.², $T_1 = 500°F$; $p_2 = 20$ lbf/in.², $v_2 = 14$ ft³/lb
(d) $T_1 = 600°F$, $v_1 = 0.5$ ft³/lb; $p_2 = 500$ lbf/in.², $T_2 = 320°F$

2.6 Water vapor at 1.0 MPa, 300°C enters a turbine operating at steady state and expands to 15 kPa. The work developed by the turbine is 630 kJ per kg of steam flowing through the turbine. Ignoring heat transfer with the surroundings and kinetic and potential energy effects, determine (a) the isentropic turbine efficiency, (b) the rate of entropy generation, in kJ/K per kg of steam flowing.

2.7 Air at 40°F, 1 atm enters a compressor operating at steady state and exits at 620°F, 8.6 atm. Ignoring heat transfer with the surroundings and kinetic and potential energy effects, determine (a) the isentropic compressor efficiency and (b) the rate of entropy generation, in Btu/°R per lb of air flowing.

2.8 Liquid water enters a pipe at 50°C, 1 atm and exits at 25°C. Evaluate the change in specific entropy, in kJ/kg·K, using (a) Equation 2.41 and (b) Equation 2.42d. Compare the calculated results, and for operation at steady state interpret the negative sign.

2.9 Derive (a) Equation 2.42c and (b) Equations 2.44b.

2.10 Two approaches are under consideration for the production of hydrogen (H_2) and carbon dioxide (CO_2) by reacting carbon monoxide (CO) with water vapor in an insulated reactor operating at steady state:

(a) The carbon monoxide and water vapor enter the reactor in separate streams, each at 400 K, 1 atm. The products exit as a mixture at 1 atm.

(b) The carbon monoxide and water vapor enter the reactor as a mixture at 400 K, 1 atm. The products exit as a mixture at 1 atm.

In each case determine the rate of entropy generation, in kJ/K per kmol of CO entering. Compare and discuss these results.

2.11 Coal with a mass flow rate of 10 kg/s and mass analysis C, 88%; H, 6%; O, 4%; N, 1%; and S, 1% burns completely with the theoretical amount of air. The combustion products are then supplied at 340 K, 1 atm, to a device for removing the SO_2 formed. Separate streams of SO_2 and the remaining gases exit the device. Each of the two streams exits at 340 K, 1 atm. For isothermal operation at steady state determine the minimum theoretical power required by the device, in kilojoules per kilogram of coal burned.

2.12 Referring to Figure 2.3, determine the maximum theoretical work, in kJ/kmol, at 25°C, 1 bar, if the fuel is (a) C, (b) H_2, (c) CH_4.

2.13 The molar analysis of a gas mixture at 850 K, 9.623 bars is 77.48% N_2, 20.59% O_2, 0.03% CO_2, and 1.90% $H_2O(g)$. For the mixture determine (a) molecular weight, (b) specific enthalpy, in kJ/kmol, (c) specific entropy, in kJ/kmol·K. Obtain enthalpy and entropy data for the components from Table C.1.

2.14 The molar analysis of a gas mixture at 1520 K, 9.142 bars is 75.07% N_2, 13.72% O_2, 3.14% CO_2, and 8.07% $H_2O(g)$. For the mixture determine (a) molecular weight, (b) specific enthalpy, in kJ/kmol, (c) specific entropy, in kJ/kmol·K. Obtain enthalpy and entropy data for the components from Table C.1.

2.15 A final flow sheet specifies a schedule 40 steel pipe with a nominal 3-inch diameter. For such a pipe, what are the actual inner and outer diameters, in inches, and the weight in pounds per linear foot?

2.16 A piping layout carrying liquid water at 70°F at a volumetric flow rate of 0.2 ft³/s is composed of four sections of 4-in. internal diameter steel pipe having a total length of 550 ft, three 90° rounded elbows, and a fully open valve. Evaluate Equation 2.31, in ft²/s², if the valve is (a) a globe valve, (b) a gate valve.

2.17 A desalination plant requires piping for the sea water that has been drawn into the plant. What piping material might you specify for this application? Discuss.

2.18 Design frequently requires choices between different varieties of the same type of device. For what applications might

(a) a positive displacement pump be preferred over a centrifugal pump?

(b) a diesel engine be preferred over a gas turbine?

(c) a fire-tube boiler be preferred over a water-tube boiler?

2.19 A piping system is to conduct water at 20°C at a volumetric flow rate of 0.1 m^3/s from a location where the pressure is 8 bars to a location where the pressure must be at least 5 bars. The total length of piping is 91.4 m, the installation requires four 90° elbows, and there is no significant elevation change. What standard-sized Schedule 40 commercial steel pipe should be installed?

3

EXERGY ANALYSIS

The presentation of thermodynamics, modeling, and design analysis initiated in Chapter 2 continues in the present chapter with emphasis on the exergy concept. Exergy has become an increasingly important tool for the design and analysis of thermal systems. Exergy is also important because it provides the basis for the discussion of thermoeconomics in Chapters 8 and 9. Exergy fundamentals are summarized in Sections 3.1–3.4, and references are provided for further study. Applications are considered in Section 3.5, including an application to the cogeneration system case study. Design guidelines that reflect conclusions evolving from exergy reasoning are listed in Section 3.6.

3.1 EXERGY

In this section we define exergy and represent it in terms of four components: physical, kinetic, potential, and chemical exergy. Additionally, two essential underlying concepts, the environment and the dead state, are discussed.

3.1.1 Preliminaries

The importance of developing thermal systems that effectively use energy resources such as oil, natural gas, and coal is apparent. Effective use is determined with both the first and second laws of thermodynamics. Energy entering a thermal system with fuel, electricity, flowing streams of matter, and so on is accounted for in the products and by-products. Energy cannot be destroyed—a first-law concept. The idea that something can be destroyed is useful in the design and analysis of thermal systems. This idea does not apply to energy, however, but to exergy (availability)—a second-law concept. Moreover, it is exergy and not energy that properly gauges the quality (usefulness) of, say, 1 kJ of electricity generated by a power plant versus 1 kJ of energy in the plant cooling water stream. Electricity clearly has the greater quality and not incidentally, the greater economic value.

The method of *exergy analysis (availability analysis)* presented in this chapter is well suited for furthering the goal of more effective energy resource use, for it enables the location, cause, and true magnitude of waste and loss to be determined. Such information can be used in the design of new energy-efficient systems and for increasing the efficiency of existing systems. Exergy analysis also provides insights that elude a purely first-law approach. Thus, from an energy perspective, the expansion of a gas (or liquid) across a valve without heat transfer (throttling process) occurs without loss. That such an expansion is a site of thermodynamic inefficiency is well known, however, and this can be readily quantified by exergy analysis. From an energy perspective, energy transfers to the environment appear to be the only possible sources of power plant inefficiency. On the basis of first-law reasoning alone, for example, the condenser of a power plant may be mistakenly identified as the component primarily responsible for the plant's seemingly low overall efficiency. An exergy analysis correctly reveals not only that the steam generator is the principal site of thermodynamic inefficiency owing to irreversibilities within it, but also that the condenser is relatively unimportant [1].

3.1.2 Defining Exergy

An opportunity for doing useful work exists whenever two systems at different states are placed in communication, for in principle work can be developed as the two are allowed to come into equilibrium. When one of the two systems is a suitably idealized system called an *environment* and the other is some system of interest, *exergy* is the maximum theoretical useful work (shaft work or electrical work) obtainable as the systems interact to equilibrium, heat transfer occurring with the environment only. Alternatively, exergy is the minimum theoretical useful work required to form a quantity of matter from substances present in the environment and to bring the matter to a specified state.

Exergy is a measure of the *departure* of the state of the system from that of the environment. It is therefore an attribute of the system and environment together. Once the environment is specified, however, a value can be assigned to exergy in terms of property values for the system only, so exergy can be regarded as an extensive property of the system.

Exergy can be destroyed and generally is not conserved. A limiting case is when exergy would be completely destroyed, as would occur if a system were to come into equilibrium with the environment *spontaneously* with no provision to obtain work. The capability to develop work existing initially would be completely wasted in the spontaneous process. Moreover, since no work needs to be done to effect such a spontaneous change, we may conclude that the value of exergy (the maximum theoretical work obtainable) is at least zero and therefore cannot be negative.

As for other extensive properties we have considered—mass, energy, and entropy—exergy can be transferred between systems. In Section 3.3, the exergy, exergy transfer, and exergy destruction concepts are related by the exergy balance for the system under consideration. Additional aspects of the exergy concept are presented in the remainder of the current section and in Section 3.2.

3.1.3 Environment and Dead States

Environment. Any system, whether a component in a larger system such as a steam turbine in a power plant or the larger system (power plant) itself, operates within surroundings of some kind. It is important to distinguish between the environment and the system's surroundings. The term *surroundings* refers to everything not included in the system. The term *environment* applies to some portion of the surroundings, the intensive properties of each phase of which are uniform and do not change significantly as a result of any process under consideration. The environment is regarded as free of irreversibilities. All significant irreversibilities are located within the system and its immediate surroundings. Internal irreversibilities are those located within the system. External irreversibilities reside in the immediate surroundings.

As the physical world is complicated, models with various levels of specificity have been proposed for describing the environment. The environment is normally regarded as composed of common substances existing in abundance within the Earth's atmosphere, oceans, and crust. The substances are in their stable forms as they exist naturally, and there is no possibility of developing work from interactions—physical or chemical—between parts of the environment. Although the intensive properties of the environment are assumed to be unchanging, the extensive properties can change as a result of interactions with other systems. Kinetic and potential energies are evaluated relative to coordinates in the environment, all parts of which are considered to be at rest with respect to one another. Accordingly, a change in the energy of the environment is a change in its internal energy only.

In this book the environment is modeled as a simple compressible system, large in extent, and uniform in temperature T_0 and pressure p_0. In keeping with the idea that the environment has to do with the actual physical world, the values for p_0 and T_0 required for subsequent analyses are taken for simplicity as typical environmental conditions, such as 1 atm and 25°C (77°F). However, for real-world applications the temperature T_0 and pressure p_0 may be specified differently. For example, T_0 and p_0 may be taken as the average ambient temperature and pressure, respectively, for the location at which the system under consideration operates. If the system uses atmospheric air, for example, T_0 would be specified as the average air temperature. If both air and water from the natural surroundings are used, T_0 would be specified as the lower of the average temperatures for air and water. Further discussion of the modeling of the environment is provided in Section 3.4.

Dead States. When the pressure, temperature, composition, velocity, or elevation of a system is different from the environment, there is an opportunity to develop work. As the system changes state toward that of the environment, the opportunity diminishes, ceasing to exist when the two, at rest relative to one another, are in equilibrium. This state of the system is called the *dead state*. At the dead state, the conditions of mechanical, thermal, and chemical equilibrium between the system and the environment are satisfied: The pressure, temperature, and chemical potentials of the system equal those of the environment, respectively. In addition, the system has zero velocity and zero elevation relative to coordinates in the environment. Under these conditions, there is no possibility of a spontaneous change within the system or the environment, nor can there be an interaction between them.

Another type of equilibrium between the system and environment can be identified. This is a restricted form of equilibrium where only the conditions of mechanical and thermal equilibrium must be satisfied. This state of the system is called the *restricted dead state*. At the restricted dead state, the fixed quantity of matter under consideration is imagined to be sealed in an envelope impervious to mass flow, at zero velocity and elevation relative to coordinates in the environment, and at the temperature T_0 and pressure p_0.

3.1.4 Exergy Components

In the absence of nuclear, magnetic, electrical, and surface tension effects, the total exergy of a system E can be divided into four components: *physical exergy E^{PH}, kinetic exergy E^{KN}, potential exergy E^{PT},* and *chemical exergy E^{CH}:*

$$E = E^{PH} + E^{KN} + E^{PT} + E^{CH} \qquad (3.1)$$

The boldface italic distinguishes the total exergy and physical exergy of a *system* from other exergy quantities, including transfers associated with streams of matter as in Equations 3.11–3.13. The sum of the kinetic, potential, and physical exergies is also referred to in the literature as the *thermomechanical exergy* [1, 2].

Although exergy is an extensive property, it is often convenient to work with it on a unit-of-mass or molar basis. The total specific exergy on a mass basis e is given by

$$e = e^{PH} + e^{KN} + e^{PT} + e^{CH} \qquad (3.2a)$$

When evaluated relative to the environment, the kinetic and potential energies of a system are in principle fully convertible to work as the system is brought to rest relative to the environment, and so they correspond to the kinetic and potential exergies, respectively. Accordingly,

$$e^{KN} = \tfrac{1}{2}V^2 \tag{3.2b}$$

$$e^{PT} = gz \tag{3.2c}$$

where V and z denote velocity and elevation relative to coordinates in the environment, respectively. We may then write Equation 3.2a as

$$e = e^{PH} + \tfrac{1}{2}V^2 + gz + e^{CH} \tag{3.2d}$$

Considering a system at rest relative to the environment ($e^{KN} = e^{PT} = 0$), the physical exergy is the maximum theoretical useful work obtainable as the system passes from its initial state where the temperature is T and the pressure is p to the restricted dead state where the temperature is T_0 and the pressure is p_0. The chemical exergy is the maximum theoretical useful work obtainable as the system passes from the restricted dead state to the dead state where it is in complete equilibrium with the environment. (The use of the term *chemical* here does not necessarily imply a chemical reaction, however; see Section 3.4.2 for an illustration.) In each instance heat transfer takes place with the environment only. Physical exergy is considered further in the next section and chemical exergy is the subject of Section 3.4.

3.2 PHYSICAL EXERGY

The physical exergy of a closed system at a specified state is given by the expression

$$E^{PH} = (U - U_0) + p_0(V - V_0) - T_0(S - S_0) \tag{3.3}$$

where U, V, and S denote, respectively, the internal energy, volume, and entropy of the system at the specified state, and U_0, V_0, and S_0 are the values of the same properties when the system is at the restricted dead state.

3.2.1 Derivation

Equation 3.3 for the physical exergy can be derived by applying energy and entropy balances to the combined system shown in Figure 3.1, which consists

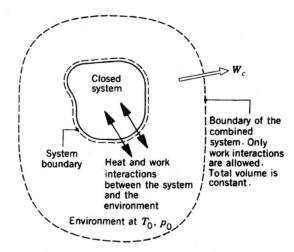

Figure 3.1 Combined system of closed system and environment.

of a closed system and the environment. The system is at rest relative to the environment. As the objective is to evaluate the maximum work that could be developed by the combined system, the boundary of the combined system allows only energy transfers by work across it, ensuring that the work developed is not affected by heat transfers to or from the combined system. And although the volumes of the system and environment may vary, the boundary of the combined system is located so that the total volume remains constant. This ensures that the work developed is useful: fully available for lifting a mass, say, and not expended in merely displacing the surroundings of the combined system.

An energy balance for the combined system reduces to

$$\Delta U_c = \cancel{Q}_c - W_c$$

or

$$W_c = -\Delta U_c$$

where W_c is the work developed by the combined system, and ΔU_c is the internal energy change of the combined system: the sum of the internal energy changes of the closed system and the environment. The internal energy of the closed system initially is denoted by U. At the restricted dead state, the internal energy of the system is denoted by U_0. Accordingly, ΔU_c can be expressed as

$$\Delta U_c = (U_0 - U) + \Delta U^e$$

where ΔU^e denotes the internal energy change of the environment. Since T_0, p_0, and the composition of the environment remain fixed, ΔU^e is related to changes in the entropy S^e and volume V^e of the environment through Equation 2.33a:

$$\Delta U^e = T_0 \, \Delta S^e - p_0 \, \Delta V^e$$

Collecting the last three equations,

$$W_c = (U - U_0) - (T_0 \, \Delta S^e - p_0 \, \Delta V^e)$$

As the total volume of the combined system is constant, the change in volume of the environment is equal in magnitude but opposite in sign to the volume change of the closed system: $\Delta V^e = -(V_0 - V)$. The expression for work then becomes

$$W_c = (U - U_0) + p_0(V - V_0) - T_0 \, \Delta S^e$$

This equation gives the work developed by the combined system as the closed system passes to the restricted dead state while interacting only with the environment. The maximum theoretical value for the work is determined using the entropy balance as follows: Since no heat transfer occurs across its boundary, the entropy balance for the combined system reduces to give

$$\Delta S_c = S_{gen}$$

where S_{gen} accounts for entropy generation within the combined system as the closed system comes into equilibrium with the environment. The entropy change of the combined system, ΔS_c, is the sum of the entropy changes for the closed system and environment, respectively,

$$\Delta S_c = (S_0 - S) + \Delta S^e$$

where S and S_0 denote the entropy of the closed system at the given state and the restricted dead state, respectively. Combining the last two equations, solving for ΔS^e, and inserting the result into the expression for W_c gives

$$W_c = \underline{(U - U_0) + p_0(V - V_0) - T_0(S - S_0)} - T_0 S_{gen}$$

The value of the underlined term is determined by two states of the closed system—the initial state and the restricted dead state—and is independent of the details of the process linking these states. However, the value of S_{gen} depends on the nature of the process as the closed system passes to the restricted dead state. In accordance with the second law, this term is positive when irreversibilities are present and vanishes in the limiting case where there are no irreversibilities; it cannot be negative. Hence, the *maximum theoretical value* for the work of the combined system is obtained by setting S_{gen} to zero, leaving

$$W_{c,max} = (U - U_0) + p_0(V - V_0) - T_0(S - S_0)$$

By definition, the physical exergy, E^{PH}, is this maximum value; and Equation 3.3 is obtained as the appropriate expression for calculating the physical exergy of a system. Various idealized devices can be invoked to visualize the development of work as a system passes from a specified state to the restricted dead state [2, 3].

3.2.2 Discussion

The physical exergy E^{PH} can be expressed on a unit-of-mass or molar basis. On a unit-of-mass basis, we have

$$e^{PH} = (u - u_0) + p_0(v - v_0) - T_0(s - s_0) \tag{3.4}$$

For the special case of an ideal gas with constant specific heat ratio k, Equation 3.4 can be expressed as

$$\frac{e^{PH}}{c_p T_0} = \frac{T}{T_0} - 1 - \ln \frac{T}{T_0} + \frac{k-1}{k}\left[\ln \frac{p}{p_0} + \frac{T}{T_0}\left(\frac{p_0}{p} - 1\right)\right] \tag{3.5}$$

The derivation is left as an exercise. Equation 3.5 is shown graphically in Figure 3.2. That the physical exergy vanishes at the restricted dead state and is positive elsewhere is evidenced by the contours of this figure.

For a wide range of practical applications not involving chemical reaction, mixing, or separation of mixture components, knowledge of the physical, kinetic, and potential exergies at various states of a system suffices. An explicit evaluation of the chemical exergy is not required because the chemical exergy value is the same at all states of interest and thus cancels when differences in exergy values between the states are calculated. This is observed in applications with closed systems and control volumes alike; for illustrations see References 1, 2. In such special applications knowledge of the chemical composition of the environment is not required. Only the pressure p_0 and temperature T_0 have to be specified. Furthermore, the exergy change between two states of a closed system is determined from Equations 3.1 and 3.3 as

$$E_2 - E_1 = (U_2 - U_1) + p_0(V_2 - V_1) - T_0(S_2 - S_1)$$

$$+ (KE_2 - KE_1) + (PE_2 - PE_1) \tag{3.6}$$

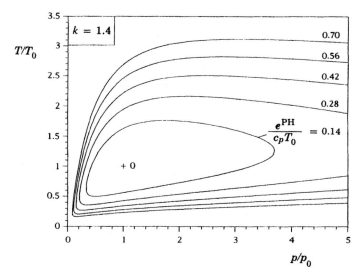

Figure 3.2 Physical exergy of a system consisting of an ideal gas with constant specific heat ratio k, Equation 3.5.

which requires that p_0 and T_0 be specified but not the composition of the environment. For many engineering applications, the kinetic and potential exergy changes are negligible, leaving just the change in physical exergy to determine $E_2 - E_1$ via Equation 3.6.

3.3 EXERGY BALANCE

As for the extensive properties mass, energy, and entropy, exergy balances can be written in alternative forms suitable for particular applications of practical interest. The objective of this section is to present such forms, beginning with the closed system case. The closed system exergy balance is then used as a basis for extending the exergy balance concept to control volumes, which is the case of greater practical utility.

3.3.1 Closed System Exergy Balance

The exergy balance for a closed system is developed by combining the energy and entropy balances:

$$(U_2 - U_1) + (KE_2 - KE_1) + (PE_2 - PE_1) = \int_1^2 \delta Q - W$$

$$S_2 - S_1 = \int_1^2 \left(\frac{\delta Q}{T}\right)_b + S_{gen}$$

where W and Q represent, respectively, transfers of energy by work and heat between the system under study and its surroundings, T_b denotes the temperature on the boundary where energy transfer by heat occurs, and the term S_{gen} accounts for entropy generation owing to internal irreversibilities. Multiplying the entropy balance by the temperature T_0 and subtracting the resulting expression from the energy balance gives

$$(U_2 - U_1) + (KE_2 - KE_1) + (PE_2 - PE_1) - T_0(S_2 - S_1)$$
$$= \int_1^2 \delta Q - T_0 \int_1^2 \left(\frac{\delta Q}{T}\right)_b - W - T_0 S_{gen}$$

Collecting the terms involving δQ and introducing Equation 3.6 on the left side, this expression can be rewritten as

$$(E_2 - E_1) - p_0(V_2 - V_1) = \int_1^2 \left(1 - \frac{T_0}{T_b}\right) \delta Q - W - T_0 S_{gen}$$

Rearranging, the *closed system exergy balance* results:

$$(E_2 - E_1) = \underbrace{\int_1^2 \left(1 - \frac{T_0}{T_b}\right) \delta Q - [W - p_0(V_2 - V_1)]}_{} - \underbrace{T_0 S_{gen}}_{} \qquad (3.7)$$

$$\underbrace{}_{\substack{\text{exergy} \\ \text{change}}} \qquad \underbrace{}_{\substack{\text{exergy} \\ \text{transfers}}} \qquad \underbrace{}_{\substack{\text{exergy} \\ \text{destruction}}}$$

For specified end states, the exergy change on the left side of Equation 3.7 can be evaluated from Equation 3.6 without regard for the nature of the process. The terms on the right side of Equation 3.7 depend explicitly on the process, however. The first term on the right side is associated with heat transfer to or from the system during the process and can be interpreted as the exergy transfer associated with (or accompanying) the transfer of energy by heat:

$$E_q = \int_1^2 \left(1 - \frac{T_0}{T_b}\right) \delta Q \qquad (3.8a)$$

The second term on the right side is associated with the net useful work and can be interpreted as the exergy transfer associated with (or accompanying) the transfer of energy by work:

$$E_w = W - p_0(V_2 - V_1) \qquad (3.8b)$$

The third term on the right side accounts for the destruction of exergy due to irreversibilities within the system. The *exergy destruction* E_D is related to the entropy generation by

$$E_D = T_0 S_{\text{gen}} \qquad (3.8c)$$

In the literature, the exergy destruction is also commonly referred to as the *availability destruction*, the *irreversibility*, and the *lost work*. Equation 3.8c is known also as the *Gouy–Stodola theorem*.

As for the exergy values at the states visited by the system, exergy transfers associated with heat and work are evaluated relative to the environment used to define exergy. Thus, on recognizing the term $(1 - T_0/T_b)$ as the Carnot efficiency (Equation 2.8), the quantity $(1 - T_0/T_b)\, \delta Q$ appearing in Equation 3.8a can be interpreted as the work that could be generated by a reversible power cycle receiving energy by heat transfer δQ at temperature T_b and discharging energy by heat transfer to the environment at temperature T_0 ($<T_b$). It may also be noted that when T_b is less than T_0, the sign of the exergy transfer would be opposite to the sign of the heat transfer, so the heat transfer and the associated exergy transfer would be oppositely directed. Exergy transfers associated with work are also evaluated relative to the environment: The exergy transfer is the work of the system W less the work that would be required to displace the environment whose pressure is uniform at p_0, namely $p_0\, \Delta V$, leaving $W - p_0\, \Delta V$ as shown in Equation 3.8b. Further discussion of these exergy transfers is given in References 1, 2.

The exergy balance can be expressed in various forms that may be more appropriate for particular applications. A convenient form of the exergy balance for closed systems is the rate equation

$$\frac{dE}{dt} = \sum_j \left(1 - \frac{T_0}{T_j}\right) \dot{Q}_j - \left(\dot{W} - p_0 \frac{dV}{dt}\right) - \dot{E}_D \qquad (3.9)$$

where dE/dt is the time rate of change of exergy. The term $(1 - T_0/T_j)\dot{Q}_j$ represents the time rate of exergy transfer associated with heat transfer at the rate $\dot{Q}_j$ occurring at the location on the boundary where the instantaneous temperature is T_j. The term $\dot{W}$ represents the time rate of energy transfer by work, and the associated exergy transfer is given by $\dot{W} - p_0\, dV/dt$, where dV/dt is the time rate of change of system volume. Here, $\dot{E}_D$ accounts for the time rate of exergy destruction due to irreversibilities within the system and is related to the rate of entropy generation within the system by $\dot{E}_D = T_0 \dot{S}_{\text{gen}}$.

3.3.2 Control Volume Exergy Balance

Building on the foregoing, we now introduce forms of the exergy balance applicable to control volumes. To allow for the chemical exergy to play a role, if necessary, when these balances are applied, we must be prepared to specify the chemical makeup of the environment in addition to the temperature T_0 and pressure p_0.

General Form. Like mass, energy, and entropy, exergy is an extensive property, so it too can be transferred into or out of a control volume where streams of matter enter and exit. Accordingly, the counterpart of Equation 3.9 applicable to control volumes requires the addition of terms accounting for such exergy transfers:

$$\underbrace{\frac{dE_{cv}}{dt}}_{\substack{\text{rate of} \\ \text{exergy} \\ \text{change}}} = \underbrace{\sum_j \left(1 - \frac{T_0}{T_j}\right) \dot{Q}_j - \left(\dot{W}_{cv} - p_0 \frac{dV_{cv}}{dt}\right) + \sum_i \dot{m}_i e_i - \sum_e \dot{m}_e e_e}_{\substack{\text{rates of} \\ \text{exergy} \\ \text{transfer}}} - \underbrace{\dot{E}_D}_{\substack{\text{rate of} \\ \text{exergy} \\ \text{destruction}}}$$

$$(3.10a)$$

As for control volume rate balances considered in Chapter 2, the subscripts i and e denote inlets and outlets, respectively.

In Equation 3.10a, the term dE_{cv}/dt represents the time rate of change in the exergy of the control volume. The term $\dot{Q}_j$ represents the time rate of heat transfer at the location on the boundary of the control volume where the instantaneous temperature is T_j, and the associated exergy transfer is given by

$$\dot{E}_{q,j} = \left(1 - \frac{T_0}{T_j}\right) \dot{Q}_j \qquad (3.10b)$$

As in the control volume energy rate balance, $\dot{W}_{cv}$ represents the time rate of energy transfer by work other than flow work. The associated exergy transfer is given by

$$\dot{E}_w = \dot{W}_{cv} - p_0 \frac{dV_{cv}}{dt} \qquad (3.10c)$$

where dV_{cv}/dt is the time rate of change of volume of the control volume itself. The term $\dot{m}_i e_i$ accounts for the time rate of exergy transfer at the inlet i. Similarly, $\dot{m}_e e_e$ accounts for the time rate of exergy transfer at the outlet e. In subsequent discussions, the exergy transfer rates at control volume inlets and outlets are denoted, respectively, as $\dot{E}_i = \dot{m}_i e_i$ and $\dot{E}_e = \dot{m}_e e_e$. Finally, $\dot{E}_D$ accounts for the time rate of exergy destruction due to irreversibilities within the control volume, $\dot{E}_D = T_0 \dot{S}_{gen}$.

Steady-State Form. Since the engineering analyses considered in this book involve control volumes at steady state, it is important to identify the steady-state form of the exergy rate balance. At steady state, $dE_{cv}/dt = 0$ and $dV_{cv}/dt = 0$, so Equation 3.10a reduces to

$$0 = \sum_j \left(1 - \frac{T_0}{T_j}\right) \dot{Q}_j - \dot{W}_{cv} + \sum_i \dot{m}_i e_i - \sum_e \dot{m}_e e_e - \dot{E}_D \qquad (3.11\text{a})$$

This equation states that the rate at which exergy is transferred into the control volume must exceed the rate at which exergy is transferred out; the difference is the rate at which exergy is destroyed within the control volume due to irreversibilities. Expressed in terms of the time rates of exergy transfer and destruction, Equation 3.11a takes the form

$$0 = \sum_j \dot{E}_{q,j} - \dot{W}_{cv} + \sum_i \dot{E}_i - \sum_e \dot{E}_e - \dot{E}_D \qquad (3.11\text{b})$$

where $\dot{E}_i$ and $\dot{E}_e$ are exergy transfer rates at inlets and outlets, respectively, and $\dot{E}_{q,j}$ is given by Equation 3.10b. For the thermodynamic analysis of control volumes at steady state, Equations 3.11 for exergy may be added to the steady-state forms of the mass, energy, and entropy balances given previously as Equations 2.23.

Exergy Transfer at Inlets and Outlets. To complete the introduction of the control volume exergy balance, means are required to evaluate the exergy transfers at inlets and outlets represented by the terms e_i and e_e appearing in Equations 3.10a and 3.11a. As for other exergy transfers, these terms must be evaluated relative to the environment used to define exergy. Accordingly, the exergy associated with a stream of matter entering (or exiting) a control volume is the maximum theoretical work that could be obtained were the stream brought to the dead state, heat transfer occurring with the environment only. This work can be evaluated in two steps as follows:

In the first step the stream is brought to the restricted dead state, and in the second step from the restricted dead state to the dead state. The contribution of the second step to the work developed is evidently the chemical exergy e^{CH}. The contribution of the first step can be obtained in principle with a device of the kind shown in Figure 2.1. In the current application, however, the properties at the inlet of the device are those of the stream under consideration: h, s, V, and z, while at the outlet the corresponding properties are h_0,

s_0, $V_0 = 0$, $z_0 = 0$, where h_0 and s_0 denote, respectively, the specific enthalpy and specific entropy at the restricted dead state. Moreover, as heat transfer occurs with the environment only, the temperature T_b at which heat transfer occurs corresponds to T_0. Thus, using Equation 2.27 the work developed in the first step, per unit of mass flowing, is $(h - h_0) - T_0(s - s_0) + \frac{1}{2}V^2 + gz$. In summary, for the two steps together we have on a unit-of-mass basis the following expression for the total exergy transfer associated with a stream of matter:

$$e = \underline{(h - h_0) - T_0(s - s_0)} + \tfrac{1}{2}V^2 + gz + e^{\mathrm{CH}} \qquad (3.12\mathrm{a})$$

The underlined term in Equation 3.12a is conventionally identified as the physical component, e^{PH}, of the exergy transfer associated with a stream of matter:

$$e^{\mathrm{PH}} = (h - h_0) - T_0(s - s_0) \qquad (3.13)$$

The physical exergy is associated with the temperature and pressure of a stream of matter.

For the special case of an ideal gas with constant specific heat ratio k, Equation 3.13 can be expressed as

$$\frac{e^{\mathrm{PH}}}{c_p T_0} = \left[\frac{T}{T_0} - 1 - \ln \frac{T}{T_0} \right] + \ln \left(\frac{p}{p_0} \right)^{(k-1)/k} \qquad (3.14)$$

The derivation is left as an exercise. Equation 3.14 is given graphically in Figure 3.3. The figure shows that the physical exergy term vanishes at the restricted dead state and can become negative when $p < p_0$. Finally, note that the two terms on the right side of Equation 3.14 depend only on temperature and pressure, respectively. These terms may be referred to, respectively, as the thermal and mechanical components of the exergy associated with the ideal gas stream, each in dimensionless form. In general, however, the physical exergy cannot be represented in terms of these two components.

The character of the exergy transfer term given by Equation 3.12a can be seen from another perspective by expressing it alternatively as

$$e = e + (pv - p_0 v) \qquad (3.12\mathrm{b})$$

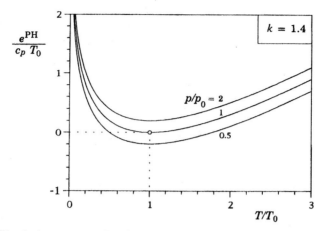

Figure 3.3 Physical exergy associated with a stream consisting of an ideal gas with constant specific heat ratio k, Equation 3.14.

where e is the total specific exergy of the matter entering or exiting a control volume, evaluated from Equations 3.2d and 3.4. The term pv is the *specific flow work* (Section 2.2) at the inlet (or outlet), and $(pv - p_0v)$ accounts for the exergy transfer associated with the flow work [1, 2]. The exergy transfer term e is thus simply the sum of two contributions: the exergy of the flowing matter and the exergy transfer associated with flow work. Although the first term of Equation 3.12b, e, is never negative, the second term is negative when $p < p_0$, giving rise to negative values for e at certain states, as shown in Figure 3.3.

Example 3.1   Using data from Table 1.2 for the cogeneration plant of Figure 1.7, evaluate the physical exergy rate $\dot{E}^{PH}$ $(= \dot{m}e^{PH})$, in megawatts, at states 10, 9, 1, 3, and 4.

Solution

MODEL

1. All data are for operation at steady state.
2. Ideal-gas mixture principles apply for the air and combustion products. Methane is also modeled as an ideal gas.
3. Kinetic and potential energy effects are ignored.
4. For the environment, $T_0 = 298.15$ K (25°C), $p_0 = 1.013$ bars (1 atm).

ANALYSIS. The physical component of the exergy associated with a stream is evaluated via Equation 3.13, where h_0 and s_0 denote the specific enthalpy and entropy, respectively, of the same stream of matter at the state where the temperature is T_0 and the pressure is p_0.

State 10. For methane at state 10,

$$\dot{E}_{10}^{PH} = \dot{m}_{10}[h_{10} - h_0 - T_0(s_{10} - s_0)]$$

Since $T_{10} = T_0$, this reduces with the ideal-gas relations Equations 2.45 and 2.47 to read

$$\dot{E}_{10}^{PH} = \dot{m}_{10}RT_0 \ln \frac{p_{10}}{p_0}$$

$$= (1.6419 \text{ kg/s}) \left(\frac{8.314}{16.043} \text{ kJ/kg·K} \right)$$

$$\times (298.15 \text{ K}) (1 \text{ MW}/10^3 \text{ kJ/s}) \ln \frac{12 \text{ bars}}{1.013 \text{ bars}}$$

$$= 0.63 \text{ MW}$$

State 9. At this state, we have saturated water vapor at 20 bars. Using steam table data, and evaluating h_0 and s_0 via Equations 2.42c and 2.42d, respectively,

$$\dot{E}_9^{PH} = \dot{m}_9[h_9 - h_0 - T_0(s_9 - s_0)]$$
$$= (14.0 \text{ kg/s}) [2799.5 - 104.88 - 298.15(6.3409$$
$$- 0.3674)] \text{ kJ/kg}(1 \text{ MW}/10^3 \text{ kJ/s})$$
$$= 12.79 \text{ MW}$$

The physical exergy rate at state 8, $\dot{E}_8^{PH}$, is calculated similarly.

States 1, 3, and 4. The table below summarizes data for enthalpy (in kJ/kmol) and entropy (in kJ/kmol·K) at states 1, 3, and 4. The $\bar{h}$ and $\bar{s}^\circ$ values are obtained from Table C.1 using the corresponding temperatures of Table 1.2. The tabulated $\bar{s}$ values are evaluated using Equation 2.72, as illustrated by the case of N_2 at state 3 where $x_{N_2} = 0.7748$:

$$\bar{s}_{N_2}(T_3, x_{N_2}p_3) = \bar{s}^{\circ}_{N_2}(T_3) - \bar{R} \ln \frac{x_{N_2}p_3}{p_{\text{ref}}}$$

$$= 223.707 - 8.314 \ln \frac{(0.7748)(9.623 \text{ bars})}{1 \text{ bar}}$$

$$= 207.004 \text{ kJ/kmol·K}$$

The mixture values for $\bar{h}$ and $\bar{s}$ are obtained using Equations 2.52b and 2.52c, as can be verified.

	N_2	O_2	CO_2	$H_2O(g)$	Mixture
$\bar{h}_1$	0	0	−393,521	−241,856	−4713.3
$\bar{h}_3$	17,072	17,540	−367,121	−221,321	12,524.0
$\bar{h}_4$	39,349	41,138	−330,159	−192,283	9304.5
$\bar{s}^{\circ}_1$	191.610	205.146	213.794	188.824	—
$\bar{s}^{\circ}_3$	223.707	238.032	262.522	227.082	—
$\bar{s}^{\circ}_4$	242.978	258.422	294.467	252.114	—
$\bar{s}_1$	193.624	218.178	281.128	221.668	199.24
$\bar{s}_3$	207.004	232.347	311.139	241.209	212.90
$\bar{s}_4$	226.964	256.538	304.843	254.643	235.70

State 1. At this state, $T_1 = T_0$ and $p_1 = p_0$. Accordingly, $\bar{h}_0 = \bar{h}_1$ and $\bar{s}_1 = \bar{s}_0$, and the physical exergy component vanishes: $\dot{E}^{\text{PH}}_1 = 0$.

State 3. Using property values tabulated above, and noting that $\bar{h}_0 = \bar{h}_1$ and $\bar{s}_0 = \bar{s}_1$, we have,

$$\dot{E}^{\text{PH}}_3 = \dot{m}_3 \frac{\bar{h}_3 - \bar{h}_0 - T_0(\bar{s}_3 - \bar{s}_0)}{M}$$

$$= (91.28 \text{ kg/s}) \left[\frac{12524 - (-4713.3) - 298.15(212.90 - 199.24)}{28.649} \right]$$

$$\times (\text{kJ/kg})(1 \text{ MW}/10^3 \text{ kJ/s})$$

$$= 41.94 \text{ MW}$$

The physical exergy rate at state 2, $\dot{E}^{\text{PH}}_2$, is calculated similarly.

State 4. Special considerations apply for the combustion products. When a mixture having the composition at state 4 is brought to p_0, T_0, some condensation would occur: At 25°C, 1 atm, the mixture would consist of N_2, O_2, and CO_2, together with saturated water vapor in equilibrium with saturated liquid.

On the basis of 1 kmol of mixture at 4, the composition at 25°C, 1 atm would be

$$0.7507N_2, \ 0.1372O_2, \ 0.0314CO_2, \ 0.0297H_2O(g), \ 0.0510H_2O(l)$$

as discussed in Section 2.5. The mole fractions of the components of the gas phase, shown underlined, are

$$x'_{N_2} = 0.7910, \qquad x'_{O_2} = 0.1446, \qquad x'_{CO_2} = 0.0331, \qquad x'_{H_2O(g)} = 0.0313$$

Then, with data from Table C.1 the $\bar{h}_0$ and $\bar{s}_0$ values required to evaluate the physical exergy of the combustion products at states 4–7 are, respectively

$$\bar{h}_0 = 0.7507(0) + 0.1372(0) + 0.0314(-393,521) + 0.0297(-241,856)$$

$$+ \ 0.0510(-285,829)$$

$$= -34,117 \text{ kJ/kmol}$$

$$\bar{s}_0 = 0.7507(193.452) + 0.1372(221.115) + 0.0314(242.022)$$

$$+ \ 0.0297(217.530) + 0.0510(69.948)$$

$$= 193.17 \text{ kJ/kmol·K}$$

In these expressions, the enthalpy and entropy of the liquid water phase are evaluated, respectively, as the values listed in Table C.1 at 25°C. The contribution of N_2 to $\bar{s}_0$ is evaluated at T_0 and the partial pressure $x'_{N_2}p_0$:

$$\bar{s}_{N_2}(T_0, x'_{N_2}p_0) = \bar{s}^{\circ}_{N_2}(T_0) - \bar{R} \ln \frac{x'_{N_2}p_0}{p_{ref}}$$

$$= 191.610 - 8.314 \ln \frac{(0.7910)(1.013)}{1} = 193.452 \text{ kJ/kmol·K}$$

A similar procedure is used for the other gases. Accordingly, at state 4

$$\dot{E}^{PH}_4 = \dot{m}_4 \frac{\bar{h}_4 - \bar{h}_0 - T_0(\bar{s}_4 - \bar{s}_0)}{M}$$

$$= (92.92) \left[\frac{9304.5 - (-34,117) - 298.15(235.70 - 193.17)}{28.254} \right]$$

$$= 101.1 \text{ MW}$$

The physical exergy rates at states 5, 6, and 7 ($\dot{E}^{PH}_5, \dot{E}^{PH}_6, \dot{E}^{PH}_7$) are calculated similarly.

COMMENT. For consistency with the property datums for $\bar{h}$ and $\bar{s}$ used for N_2, O_2, and CO_2, all enthalpy and entropy data of water required for calculations at states 1–7 are obtained from Table C.1. Steam table data suffice for the calculations at states 8 and 9.

3.4 CHEMICAL EXERGY

When evaluating chemical exergy (the exergy component associated with the departure of the chemical composition of a system from that of the environment), the substances comprising the system must be referred to the properties of a suitably selected set of environmental substances. To exclude the possibility of developing work from interactions—physical or chemical—between parts of the environment, these reference substances would need to be in mutual equilibrium. Our natural environment is not in equilibrium, however, nor as presumed in the earlier discussion of the environment (Section 3.1.3) are its temperature, pressure, and other intensive properties uniform spatially or with time. Accordingly, it is necessary to compromise significantly between physical reality and the requirements of thermodynamic theory. Such considerations have led to alternative models for evaluating chemical exergy, [1–6], and the terms *exergy reference environment* and *thermodynamic environment* are frequently used to distinguish the thermodynamic concept from the natural environment.

As the modeling of exergy reference environments is discussed in the references cited, no attempt is made to detail this topic here. Rather, for simplicity, the present development features the use of standard chemical exergies determined relative to a standard environment, as considered next.

3.4.1 Standard Chemical Exergy

Standard chemical exergies are based on standard values of the environmental temperature T_0 and pressure p_0, for example, 298.15 K (25°C) and 1 atm, respectively. The standard environment is regarded as consisting of a set of reference substances with standard concentrations reflecting as closely as possible the chemical makeup of the natural environment. The reference substances generally fall into three groups: gaseous components of the atmosphere, solid substances from the lithosphere, and ionic and nonionic substances from the oceans. Two alternative standard exergy reference environments have gained acceptance for engineering evaluations; these are called here model I and model II.

The reference substances for model I, presented in Reference 4, are determined assuming restricted chemical equilibrium for nitric acid and nitrates and unrestricted thermodynamic equilibrium for all other chemical components of the atmosphere, the oceans, and a portion of the lithosphere. A different approach is used in model II, presented in Reference 5: A reference substance is selected for each chemical element from among substances that contain the element being considered and that are abundantly present in the natural environment, even though the substances are not in completely mutual stable equilibrium. An underlying rationale for this approach is that substances found abundantly in nature have little economic value. Model I attempts to satisfy the equilibrium requirement of the thermodynamic theory; also, the chemical composition of the gas phase of this model approximates satisfactorily the composition of the natural atmosphere. On an overall basis, however, the chemical composition of the exergy reference environment of model II is closer to the composition of the natural environment; but the equilibrium requirement is not generally satisfied.

Readers should refer to References 4 and 5 for additional details about the specific choices of the reference substances, the methods used to calculate the standard chemical exergies, tables of standard chemical exergies, and comparisons of the two approaches. Table C.2 in Appendix C gives the standard chemical exergies for selected substances obtained using models I and II. Throughout a given application, only chemical exergy values corresponding to the same model should be used.

The use of a table of standard chemical exergies greatly facilitates the application of exergy principles. The term *standard* is somewhat misleading, however, for there is no one specification of the environment that suffices for all applications. Still, chemical exergies calculated relative to alternative specifications of the environment are generally in good agreement. For a broad range of engineering applications the simplicity and ease of use of standard chemical exergies generally outweighs any slight lack of accuracy that might result. In particular, the effect of slight variations in the values of T_0 and p_0 about the values used to calculate the standard chemical exergies reported in Table C.2 can be neglected.

3.4.2 Standard Chemical Exergy of Gases and Gas Mixtures

A common feature of standard exergy reference environments is a gas phase, intended to represent air, that includes N_2, O_2, CO_2, $H_2O(g)$, and other gases. The kth gas present in this gas phase is at temperature T_0 and the partial pressure $p_k^e = x_k^e p_0$, where the superscript e denotes the environment and x_k^e is the mole fraction of gas k in the environmental gas phase. Referring to the device at steady state shown in Figure 3.4 (a special case of Figure 2.1), the evaluation of the standard chemical exergy for a gas included in the environmental gas phase can be accomplished as follows: The kth gas enters at temperature T_0 and pressure p_0, expands isothermally with heat transfer only with the environment, and exits to the environment at temperature T_0 and the partial pressure $x_k^e p_0$. The maximum theoretical work per mole of gas k would be developed when the expansion occurs without irreversibilities. Accord-

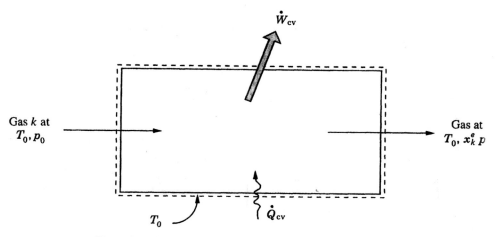

Figure 3.4 Device for evaluating the chemical exergy of a gas.

ingly, with Equation 2.27 and ideal-gas relations for enthalpy and entropy, the chemical exergy per mole of gas k is

$$\bar{e}_k^{CH} = -\bar{R}T_0 \ln \frac{x_k^e p_0}{p_0}$$

$$= -\bar{R}T_0 \ln x_k^e \qquad (3.15)$$

The details are left as an exercise.

As an application of Equation 3.15, consider CO_2 with $x_{CO_2}^e = 0.00328$ and $T_0 = 298.15$ K, as in model I. Inserting values into Equation 3.15 gives $\bar{e}_{CO_2}^{CH} = 14,179$ kJ/kmol, which, allowing for roundoff, is the value for CO_2 listed in Table C.2 for model I. The mole fraction of CO_2 in model II is $x_{CO_2}^e = 0.00033$, and thus a different value is obtained for the chemical exergy of CO_2 when model II is used: $\bar{e}_{CO_2}^{CH} = 19,871$ kJ/kmol.

The chemical exergy of a mixture of N gases, each of which is present in the environmental gas phase, can be obtained similarly. In such a case we may think of a set of N devices like that shown in Figure 3.4, one for each gas in the mixture. Gas k, whose mole fraction in the gas mixture at T_0, p_0 is x_k, enters at T_0 and the partial pressure $x_k p_0$ and exits at T_0 and the partial pressure $x_k^e p_0$. Paralleling the previous development, the work per mole of k is $-\bar{R}T_0 \ln(x_k^e/x_k)$. Summing over all components, the chemical exergy per mole of mixture is

$$\bar{e}^{CH} = -\bar{R}T_0 \sum x_k \ln \frac{x_k^e}{x_k}$$

Expressing the natural logarithm term as $x_k(\ln x_k^e - \ln x_k)$ and introducing Equation 3.15, this can be written alternatively as

$$\bar{e}^{CH} = \sum x_k \bar{e}_k^{CH} + \bar{R}T_0 \sum x_k \ln x_k \qquad (3.16)$$

Equation 3.16 is valid for mixtures containing gases other than those present in the reference environment, for example, gaseous fuels. This equation also can be extended to mixtures (and solutions) that do not adhere to the ideal gas model [3]. In all such applications, the terms $\bar{e}_k^{CH}$ are selected from a table of standard chemical exergies.

3.4.3 Standard Chemical Exergy of Fuels

In principle, the standard chemical exergy of a substance not present in the environment can be evaluated by considering an idealized reaction of the substance with other substances (usually reference substances) for which the chemical exergies are known. To illustrate this for the case of a pure hydrocarbon fuel C_aH_b at T_0, p_0 refer to the system shown in Figure 3.5 (a special case of Figure 2.3), where the fuel reacts with oxygen to form carbon dioxide

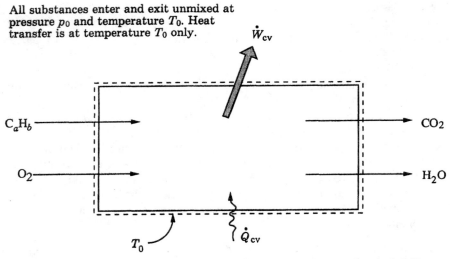

Figure 3.5 Device for evaluating the chemical exergy of a hydrocarbon fuel, C_aH_b.

and liquid water. All substances are assumed to enter and exit unmixed at T_0, p_0 and heat transfer occurs only at temperature T_0.

Assuming no irreversibilities, an exergy balance for the system reads

$$0 = \sum_j \left(1 - \frac{T_0}{T_j}\right)\frac{\dot{Q}_j}{\dot{n}_F} - \left(\frac{\dot{W}_{cv}}{\dot{n}_F}\right)_{int\ rev} + \bar{e}_F^{CH}$$
$$+ \left(a + \frac{b}{4}\right)\bar{e}_{O_2}^{CH} - a\bar{e}_{CO_2}^{CH} - \frac{b}{2}\bar{e}_{H_2O(l)}^{CH} - \dot{E}_D$$

where as before the subscript F denotes the fuel. Solving for the fuel chemical exergy

$$\bar{e}_F^{CH} = \left(\frac{\dot{W}_{cv}}{\dot{n}_F}\right)_{int\ rev} + a\bar{e}_{CO_2}^{CH} + \frac{b}{2}\bar{e}_{H_2O(l)}^{CH} - \left(a + \frac{b}{4}\right)\bar{e}_{O_2}^{CH}$$

Using Equation 2.75b and recognizing that $-\bar{h}_{RP}$ is the molar higher heating value $\overline{HHV}$, the work term in the previous expression can be eliminated, giving

$$\bar{e}_F^{CH} = \overline{HHV}(T_0, p_0) - T_0\left[\bar{s}_F + \left(a + \frac{b}{4}\right)\bar{s}_{O_2} - a\bar{s}_{CO_2} - \frac{b}{2}\bar{s}_{H_2O(l)}\right](T_0, p_0)$$
$$+ \left\{a\bar{e}_{CO_2}^{CH} + \frac{b}{2}\bar{e}_{H_2O(l)}^{CH} - \left(a + \frac{b}{4}\right)\bar{e}_{O_2}^{CH}\right\} \tag{3.17a}$$

Similarly, with Equation 2.75c the following alternative expression is obtained:

$$\bar{e}_F^{CH} = \left[\bar{g}_F + \left(a + \frac{b}{4}\right)\bar{g}_{O_2} - a\bar{g}_{CO_2} - \frac{b}{2}\bar{g}_{H_2O(l)}\right](T_0, p_0)$$
$$+ \left\{a\bar{e}_{CO_2}^{CH} + \frac{b}{2}\bar{e}_{H_2O(l)}^{CH} - \left(a + \frac{b}{4}\right)\bar{e}_{O_2}^{CH}\right\} \tag{3.17b}$$

The first term of this expression involving the Gibbs functions can be written more compactly as $-\Delta G$: the negative of the change in Gibbs function for the reaction. To illustrate the use of Equation 3.17b, consider the case of methane, CH_4. With $a = 1$, $b = 4$, Gibbs function data,[1] and standard chemical exergies for CO_2, $H_2O(l)$, and O_2, from Table C.2 (model I), Equation 3.17b gives 824,720 kJ/kmol. This agrees closely with the value listed for methane in Table C.2 for this model.

With Equations 3.17, the standard chemical exergy of the hydrocarbon C_aH_b can be calculated using the standard chemical exergies of O_2, CO_2, and $H_2O(l)$, together with selected property data: the fuel higher heating value and absolute entropies, or Gibbs functions. Note that only the terms in curly brackets of Equations 3.17 are affected by the choice of the model used for the exergy reference environment. When the necessary fuel data are lacking, as for example in the cases of coal, char, and fuel oil, the approach of Equation 3.17a can be invoked using a measured or estimated fuel heating value and an estimated value of the fuel absolute entropy determined with procedures discussed in the literature. See Section 3.5.4 for an illustration.

By paralleling the development given above for hydrocarbon fuels leading to Equation 3.17b, we can in principle determine the standard chemical exergy of any substance not present in the environment. With such a substance playing the role of the fuel in the previous development, we consider a reaction of the substance involving other substances (usually reference substances) for which the standard chemical exergies are known, and write

$$\bar{e}^{CH} = -\Delta G + \left\{ \sum_P n\bar{e}^{CH} - \sum_R n\bar{e}^{CH} \right\} \tag{3.18}$$

where ΔG is the change in Gibbs function for the reaction, regarding each substance as separate at temperature T_0 and pressure p_0. The term in curly brackets corresponds to the terms in curly brackets of Equations 3.17, and is evaluated using the known standard chemical exergies, together with the n's giving the moles of these reactants and products per mole of the substance whose chemical exergy is being evaluated.

To illustrate the use of Equation 3.18 consider the case of ammonia, NH_3. Letting NH_3 play the role of hydrocarbon in the development leading to Equations 3.17, we may consider any reaction of NH_3 involving other substances for which the standard chemical exergies are known. As an example, consider

---

[1]Values obtained from Table C.1 at 25°C, 1 bar. The very small corrections to the Gibbs functions, calculable via Equations 2.74, owing to the difference between 1 bar and the pressure of Table C.2 (model I): 1.019 atm, are ignored.

$$NH_3 + \tfrac{3}{4}O_2 \rightarrow \tfrac{1}{2}N_2 + \tfrac{3}{2}H_2O(l)$$

for which Equation 3.18 takes the form

$$\overline{e}_{NH_3}^{CH} = \left[\overline{g}_{NH_3} + \tfrac{3}{4}\overline{g}_{O_2} - \tfrac{1}{2}\overline{g}_{N_2} - \tfrac{3}{2}\overline{g}_{H_2O(l)} \right](T_0, p_0) + \tfrac{1}{2}\overline{e}_{N_2}^{CH} + \tfrac{3}{2}\overline{e}_{H_2O(l)}^{CH} - \tfrac{3}{4}\overline{e}_{O_2}^{CH}$$

Using Gibbs function data from Table C.1, and standard chemical exergies for O_2, N_2, and $H_2O(l)$ from Table C.2 (model I), we have $\overline{e}_{NH_3}^{CH} = 336{,}650$ kJ/kmol. This agrees closely with the value for ammonia listed in Table C.2 for this model.

Example 3.2 Using data from Table 1.2 for the cogeneration system of Figure 1.7, evaluate the chemical exergy rate $\dot{E}^{CH}$ ($= \dot{m}e^{CH}$), in megawatts, at states 10, 9, 4, 1, 2, and 3.

Solution

MODEL

1. Standard chemical exergies are from Table C.2 (model I). The small difference between $p_0 = 1$ atm and the pressure used in calculating the table values is ignored.
2. The model of Example 3.1 is also applicable.

ANALYSIS. The standard chemical exergy of CH_4 obtained from Table C.2 is 824,348 kJ/kmol. Thus, for state 10

$$\dot{E}_{10}^{CH} = (1.6419 \text{ kg/s})\left(\frac{824{,}348}{16.043} \text{ kJ/kg} \right)\left(\frac{1 \text{ MW}}{10^3 \text{ kJ/s}} \right) = 84.3668 \text{ MW}$$

From the same source, the standard chemical exergy of liquid water is 45 kJ/kmol. Accordingly, for state 9

$$\dot{E}_9^{CH} = (14 \text{ kg/s})\left(\frac{45}{18.015} \text{ kJ/kg} \right)\left(\frac{1 \text{ MW}}{10^3 \text{ kJ/s}} \right) = 0.035 \text{ MW}$$

As noted in Example 3.1, the restricted dead state corresponding to the mixture at state 4 consists of a liquid water phase and a gas phase with mole fractions x'_{N_2}, x'_{O_2}, x'_{CO_2}, and $x'_{H_2O(g)}$. The contribution of the liquid water to the chemical exergy is determined as for state 9: 45 kJ per kmol of liquid water. The contribution of the gas phase to the chemical exergy is evaluated via Equation 3.16. That is, with standard chemical exergies from Table C.2,

$$\Sigma \, x_k' \, \overline{e}_k^{CH} + \overline{R}T_0 \, \Sigma \, x_k' \ln x_k' = 0.7910(639) + 0.1446(3951)$$
$$+ \ 0.0331(14176) + 0.0313(8636)$$
$$+ \ (8.314)(298.15)[0.7910 \ln 0.7910$$
$$+ \ 0.1446 \ln 0.1446 + 0.0331 \ln(0.0331)$$
$$+ \ 0.0313 \ln(0.0313)]$$
$$= \ 115 \text{ kJ/kmol (gas)}$$

On the basis of 1 kmol of mixture at state 4, we have 0.949 kmol as a gas phase and 0.051 kmol as liquid water; thus $\overline{e}_4^{CH} = 0.949(115) + 0.051(45) = 111.43$ kJ/kmol. Finally, the chemical exergy rate of the combustion products is

$$\dot{E}_4^{CH} = (92.9176 \text{ kg/s})\left(\frac{111.43}{28.254} \text{ kJ/kg}\right)\left(\frac{1 \text{ MW}}{10^3 \text{ kJ/s}}\right)$$
$$= \ 0.3665 \text{ MW}$$

At states 1, 2, and 3 the air composition closely corresponds to that of model I and so the chemical exergy is taken as zero in value: $\dot{E}_1^{CH} = \dot{E}_2^{CH} = \dot{E}_3^{CH} = 0$.

COMMENT. As noted above, the composition of the air at states 1, 2, and 3 only closely corresponds to the air composition of model I. This fact, together with the manner in which the reference environment of model I has been formulated, leads to a value for the chemical exergy at these states on the order of -0.04 MW, which has been ignored in the present analysis. This omission causes a negligible error in subsequent evaluations.

3.5 APPLICATIONS

Thus far in this chapter the emphasis has been on exergy fundamentals. In the present section we turn to important exergy applications, several of which are essential for progress with the presentation of thermoeconomics in Chapters 8 and 9. We begin by discussing the exergy analysis of the cogeneration system of Figure 1.7.

3.5.1 Cogeneration System Exergy Analysis

Thermal systems are typically supplied with exergy inputs associated directly or indirectly with fossil fuels or other energy resources. Accordingly, destructions and losses of exergy represent the waste of these energy resources. The method of *exergy analysis* aims at the quantitative evaluation of the exergy destructions and losses associated with a system.

As an illustration of exergy analysis, refer to Tables 3.1 and 3.2, which give data for the case study cogeneration system shown in Figure 1.7. Sample calculations of the physical and chemical exergies given in Table 3.1 are provided in Examples 3.1 and 3.2, respectively. Table 3.2 provides a rank-ordered listing of the exergy destructions within the principal components. Example 3.3 details the exergy destruction calculations for three representative components.

An *exergetic efficiency* for the cogeneration system can be calculated as the percentage of the exergy supplied to the system that is recovered in the product of the system. Identifying the product of the cogeneration system as the sum of the net power generated and the net increase of the exergy of the feedwater, we have

$$\varepsilon = \frac{\dot{W}_{net} + (\dot{E}_9 - \dot{E}_8)}{\dot{E}_{10} + \dot{E}_1}$$

$$= \frac{(30 + 12.7486)\ \text{MW}}{84.9939\ \text{MW}} = 0.503\ (50.3\%)$$

where the exergy values are obtained from Table 3.1. The exergy carried out

Table 3.1 Exergy data for the cogeneration system of Figure 1.7[a]

State	Substance	$\dot{E}^{PH}$	+	$\dot{E}^{CH}$	=	$\dot{E}$
1	Air	0.0000		0.0000		0.0000
2	Air	27.5382		0.0000		27.5382
3	Air	41.9384		0.0000		41.9384
4	Combustion products	101.0873		0.3665		101.4538
5	Combustion products	38.4158		0.3665		38.7823
6	Combustion products	21.3851		0.3665		21.7516
7	Combustion products	2.4061		0.3665		2.7726
8	Water	0.0266		0.0350		0.0616
9	Water	12.7752		0.0350		12.8102
10	Methane	0.6271		84.3668		84.9939

Exergy Rates (MW)

[a]See Examples 3.1 and 3.2 for sample calculations. Roundoff accounts for slight differences between the table values and sample calculations.

Table 3.2 Exergy destruction data for the cogeneration system of Figure 1.7[a]

Component	Exergy Destruction		
	Rate (MW)	Percentage[b]	Percentage[c]
Combustion chamber[d]	25.48	64.56	29.98
Heat-recovery steam generator	6.23	15.78	7.33
Gas turbine	3.01	7.63	3.54
Air preheater	2.63	6.66	3.09
Air compressor	2.12	5.37	2.49
Overall plant	39.47	100.00	46.43[e]

[a]See Example 3.3 for discussion.
[b]Exergy destruction rate within a component as a percentage of the total exergy destruction rate within the cogeneration system (Equation 3.26).
[c]Exergy destruction rate within a component as a percentage of the exergy rate entering the cogeneration system with the fuel (Equation 3.25).
[d]Includes the exergy loss accompanying heat transfer from the combustion chamber. Value determined using the nominal pressure at state 10 given in Table 1.2.
[e]An additional 3.26% of the fuel exergy is carried out of the system at state 7 and is charged as an exergy loss.

of the system at state 7, amounting to 3.26% of the fuel exergy, is regarded as a loss. Further discussion of exergetic efficiencies is provided in Section 3.5.3.

The exergy destruction and loss data summarized in Table 3.2 clearly identify the combustion chamber as the major site of thermodynamic inefficiency. The next most prominent site is the heat-recovery steam generator (HRSG). Roughly equal contributions to inefficiency are made by the gas turbine, air preheater, and the loss associated with stream 7. The air compressor is an only slightly smaller contributor.

The exergy destructions in these components stem from one or more of three principal irreversibilities associated, respectively, with chemical reaction, heat transfer, and friction. All three irreversibilities are present in the combustion chamber, where chemical reaction is the most significant source of exergy destruction. For the HRSG and air preheater, heat transfer and friction are the sources of exergy destruction, with the most significant irreversibility being related to the stream-to-stream heat transfer. Exergy destruction in the adiabatic gas turbine and air compressor is caused primarily by friction.

Combustion is intrinsically a very significant source of irreversibility, and a dramatic reduction in its effect on exergy destruction by conventional means cannot be expected. Still, it is known that the inefficiency of combustion can be reduced by preheating the combustion air and reducing the air–fuel ratio. Exergy destruction associated with heat transfer decreases as the temperature difference between the streams is reduced. This is achievable through the specification of the heat exchanger, though there is normally an accompanying increase in exergy destruction by friction. The exergy destruction within the gas turbine and the air compressor decreases as friction is reduced. The significance of the exergy loss associated with stream 7 reduces as the temperature T_7 is reduced.

The foregoing considerations provide a basis for implementing practical engineering measures aimed at improving the thermodynamic performance of the cogeneration system. As discussed in Section 1.2, such measures have to be applied judiciously, however. Measures that improve the thermodynamic performance of one component might adversely affect another, leading to no net overall improvement. (Additional discussion of the interdependency of exergy destructions occurring in different components is found in Section 8.2.1.) Moreover, measures to improve thermodynamic performance invariably have economic consequences. The objective in thermal system design normally is to identify the cost-optimal configuration and the cost-optimal values of the decision variables. This requires consideration of both thermodynamics and economics, as discussed in Chapters 8 and 9.

Example 3.3 Using data from Table 3.1 for the cogeneration system of Figure 1.7, evaluate the rate of exergy destruction, in megawatts, for (a) the combustion chamber, (b) the air preheater, and (c) the compressor. Express each as a percentage of the total exergy destruction within the system and of the exergy entering the system with the fuel.

Solution

MODEL

1. The control volume enclosing the combustion chamber is selected to encompass the combustion chamber and enough of its nearby surroundings so that the heat loss occurs at the ambient temperature, 25°C.
2. The models of Examples 3.1 and 3.2 are also applicable.

ANALYSIS. For each component, the value the exergy destruction rate is obtained from the exergy rate balance, Equation 3.11b.

(a) For the combustion chamber

$$\dot{E}_D = \sum_j \left(1 - \frac{T_0}{T_j}\right)\dot{Q}_j - \dot{W}_{cv} + \dot{E}_3 + \dot{E}_{10} - \dot{E}_4$$

where the heat transfer term vanishes because heat transfer is assumed to occur at T_0. With data from Table 3.1

$$\dot{E}_D = (41.9384 + 84.9939 - 101.4538) \text{ MW} = 25.48 \text{ MW}$$

Expressed as a percentage, the contribution of the combustion chamber to the total exergy destruction is $(25.48/39.47)(100) = 64.56\%$. From Table 3.1, the rate exergy enters with the fuel is closely 84.99 MW. Of this $(25.48/84.99)(100) = 29.98\%$ is destroyed by irreversibilities within the combustion chamber.

(b) For the air preheater, an exergy rate balance and data from Table 3.1 give

$$\dot{E}_D = \sum_j \left(1 - \frac{T_0}{T_j}\right)\dot{Q}_j - \dot{W}_{cv} + \dot{E}_2 + \dot{E}_5 - \dot{E}_3 - \dot{E}_6$$

$$= (27.5382 + 38.7823 - 41.9384 - 21.7516) \text{ MW} = 2.63 \text{ MW}$$

The two exergy destruction ratios expressed as percentages are, respectively,

$$\frac{2.63}{39.47}(100) = 6.66\%, \qquad \frac{2.63}{84.99}(100) = 3.09\% \; .$$

(c) Using $\bar{h}_1$ and $\bar{h}_2$ from Example 2.2, an energy rate balance for the compressor gives

$$\dot{W}_{cv} = \dot{m}_1(h_1 - h_2) = \dot{m}_1 \frac{\bar{h}_1 - \bar{h}_2}{M_a}$$

$$= (91.2757 \text{ kg/s}) \left[\frac{(-4713.3) - 4596.9}{28.649} \text{ kJ/kg}\right] \left(\frac{1 \text{ MW}}{10^3 \text{ kJ/s}}\right)$$

$$= -29.6623 \text{ MW}$$

Then, applying an exergy rate balance

$$\dot{E}_D = \sum_j \left(1 - \frac{T_0}{T_j}\right)\dot{Q}_j - \dot{W}_{cv} + \dot{E}_1 - \dot{E}_2$$

$$= 0 - (-29.6623 \text{ MW}) + 0 - (27.5382 \text{ MW}) = 2.12 \text{ MW}$$

The two exergy destruction ratios expressed as percentages are, respectively,

$$\frac{2.12}{39.47}(100) = 5.37\%, \qquad \frac{2.12}{84.99}(100) = 2.49\%$$

COMMENT. The exergy loss associated with heat transfer from the combustion chamber is counted here as an exergy destruction for an enlarged control volume that encompasses the combustion chamber and a portion of its nearby surroundings. See Section 3.5.2 for further discussion.

3.5.2 Exergy Destruction and Exergy Loss

The cogeneration system exergy analysis just considered illustrates several aspects of exergy analysis, including the roles of exergy destruction and exergy loss in determining thermodynamic performance. Exergy analysis additionally often involves the calculation of measures of performance: *exergy destruction ratios*, *exergy loss ratios*, and *exergetic efficiencies*. In this section and the next, such measures of performance are considered.

System Boundary. The first step in any thermodynamic analysis is to define the system or control volume. System definition is especially important in exergy analysis because the choice of boundary can determine whether the effects of heat transfer should be charged as an exergy destruction or an exergy loss.

To illustrate how boundary selection influences an exergy analysis, refer to Figure 3.6 depicting a gas (or liquid) flowing through a pipe section at steady state. An energy transfer by heat from the gas occurs at the rate $\dot{Q}$, where $\dot{Q}$ is taken as positive in the direction of the arrow. Two alternative boundaries, denoted I and II, are indicated. Boundary I includes just the cylindrical volume through which the gas flows. In this case, heat transfer occurs at temperatures that generally vary in the direction of flow. Boundary II is located outside the pipe where the temperature corresponds to the ambient temperature, taken here as the temperature of the exergy reference environment T_0.

Exergy enters the control volume at inlet i at the rate $\dot{E}_i = \dot{m}e_i$ and leaves

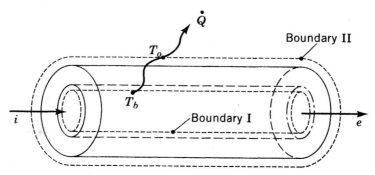

Figure 3.6 Flow of a gas (or liquid) through a pipe at steady state with heat transfer to the surroundings.

at outlet e at a lesser rate $\dot{E}_e = \dot{m}e_e$, where $\dot{m}$ denotes the mass flow rate and the terms e_i, e_e are evaluated from Equation 3.12a at the inlet and outlet, respectively. Owing to exergy destruction and exergy loss, the exergy rate at the outlet is less than the exergy rate at the inlet. These exergy quantities are related by the exergy rate balance, which at steady state can be expressed as

$$\dot{E}_i = \dot{E}_e + \dot{E}_D + \dot{E}_L \qquad (3.19)$$

where $\dot{E}_D$ and $\dot{E}_L$ represent rates of exergy destruction and loss, respectively. By inspecting Equation 3.19, it is apparent that the sum of exergy destruction and exergy loss, $\dot{E}_D + \dot{E}_L$, remains constant for given inlet and outlet states, regardless of which boundary is chosen, I or II. The individual values of $\dot{E}_D$ and $\dot{E}_L$ depend explicitly on the boundary choice, however, as will now be considered.

With boundary I the rate of exergy loss $\dot{E}_L$ equals the rate of exergy transfer associated with heat transfer, and is thus given by

$$\dot{E}_q = \int_i^e \left(1 - \frac{T_0}{T_b}\right) q' \, dL \qquad (3.20)$$

where q' is the heat transfer rate per unit of length L. Exergy is also destroyed within this control volume owing to the effect of friction.

With boundary II, heat transfer occurs at the temperature T_0, and thus there is no associated exergy transfer: $\dot{E}_q = 0$. Accordingly, the exergy loss term of Equation 3.19 vanishes. In this case, however, the exergy destruction term $\dot{E}_D$ accounts for the total exergy destruction within the enlarged control volume: exergy destruction owing to friction *and* the irreversibility of heat transfer within the control volume.

To evaluate the exergy transfer via Equation 3.20 requires information about the rate of heat transfer q' across each portion of the boundary and the temperature T_b on the boundary where heat transfer occurs. In many practical applications this exergy transfer is not amenable to evaluation, however, because the required information is unknown or mathematical complexities intrude. In such applications, exemplified by the combustion chamber analysis of Example 3.3, the use of boundary II provides an alternative approach that often suffices. In special cases, the evaluation can be conducted using the thermodynamic average temperature considered next.

Thermodynamic Average Temperature. In the absence of internal irreversibilities, the heat transfer experienced by a one-inlet, one-outlet control volume at steady state is given by Equation 2.28. Such heat transfer can be regarded for simplicity to occur at a *thermodynamic average temperature* T_a: $(\dot{Q}_{cv}/\dot{m})_{\text{int rev}} = T_a(s_e - s_i)$, where T_a is defined by

$$T_a = \frac{\displaystyle\int_i^e T \, ds}{s_e - s_i} \qquad (3.21a)$$

or with Equation 2.34b

$$T_a = \frac{(h_e - h_i) - \int_i^e v \, dp}{s_e - s_i} \tag{3.21b}$$

For applications involving heat exchangers and certain other elementary components, heat transfer occurs ideally at constant pressure. Equation 3.21b then reduces to the special form

$$T_a = \frac{h_e - h_i}{s_e - s_i} \qquad \text{(constant pressure)} \tag{3.22}$$

In terms of the thermodynamic average temperature, the exergy transfer rate associated with the heat transfer rate $\dot{Q}$ is simply

$$\dot{E}_q = \left(1 - \frac{T_0}{T_a}\right)\dot{Q} \tag{3.23}$$

Although strictly adhering to cases where the idealizations used to develop them are satisfied, Equations 3.21 and 3.22 provide plausible means for approximating T_a for use in Equation 3.23 even when nonidealities such as pressure drops owing to friction occur. As discussed in Section 8.1.1 (Equation 8.7), this approach allows for an estimate of the monetary loss associated with an exergy loss due to heat transfer.

Exergy Destruction Through Heat Transfer and Friction. The thermodynamic average temperature concept can be used to develop an understanding of the effects of heat transfer and friction on exergy destruction. Referring to the counterflow heat exchanger of Figure 3.7, let T_{ha} and T_{ca} denote thermodynamic average temperatures for the hot and cold streams, respectively. Two subsystems, labeled A and B, are shown on the figure.

Subsystem A is a closed system consisting of the wall separating the hot and cold streams. Energy flows by heat transfer through the wall at the rate $\dot{Q}$. The exergy destruction in subsystem A is caused by heat transfer only. At steady state the closed system exergy rate equation, Equation 3.9, reads

$$0 = \left(1 - \frac{T_0}{T_{ha}}\right)\dot{Q} - \left(1 - \frac{T_0}{T_{ca}}\right)\dot{Q} - \dot{E}_D$$

where $\dot{E}_D$ accounts for the exergy destruction in subsystem A due to heat transfer. Reducing this expression gives

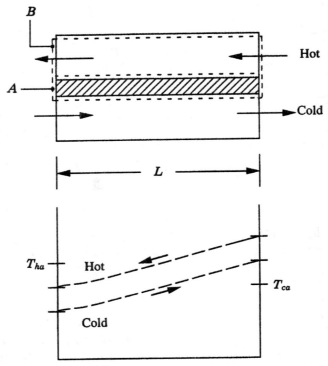

Figure 3.7 Counterflow heat exchanger used to discuss the effects of heat transfer and friction on exergy destruction.

$$\dot{E}_D = T_0\dot{Q}\,\frac{T_{ha} - T_{ca}}{T_{ha}T_{ca}}$$

Since the heat transfer rate is proportional to the temperature difference between the streams: $\dot{Q} \propto (T_{ha} - T_{ca})$, we have, qualitatively,

$$\dot{E}_D \propto \frac{T_0(T_{ha} - T_{ca})^2}{T_{ha}T_{ca}}$$

That is, the rate of exergy destruction associated with heat transfer varies quadratically with the stream-to-stream temperature difference and inversely with the product of the temperature levels, as gauged by T_{ha} and T_{ca}.

Subsystem B is a control volume encompassing the channel through which the hot stream flows. Friction is the only irreversibility of this subsystem. At steady state the control volume exergy rate equation, Equation 3.11a, reduces with Equation 3.12a to give

$$\dot{E}_D = \left(1 - \frac{T_0}{T_{ha}}\right)\dot{Q} + \dot{m}[(h_i - h_e) - T_0(s_i - s_e)]$$

where $\dot{E}_D$ accounts for the exergy destruction in subsystem B owing to friction. In writing this, the chemical exergy at the inlet and outlet cancels, and the kinetic and potential exergy terms are ignored. From an energy rate balance, $\dot{Q} = \dot{m}(h_e - h_i)$, and so the expression for the exergy destruction rate becomes

$$\dot{E}_D = T_0\dot{m}\left[(s_e - s_i) - \frac{h_e - h_i}{T_{ha}}\right]$$

As T_{ha} represents the thermodynamic average temperature, the last equation can be rewritten with Equation 3.21b as

$$\dot{E}_D = \frac{-T_0\dot{m}\int_i^e v\, dp}{T_{ha}}$$

With Equation 2.30 we have

$$\dot{E}_D = \frac{T_0\dot{m}h_l}{T_{ha}} \tag{3.24a}$$

and with Equation 2.31 this becomes

$$\dot{E}_D = \frac{T_0\dot{m}(4L/D)(V^2/2)f}{T_{ha}} \tag{3.24b}$$

where h_l is the head loss and f is the friction factor. That is, the rate of exergy destruction associated with friction varies directly with the head loss and inversely with the temperature level, as gauged by T_{ha}.

Equations 3.24 have been developed for the special case of Figure 3.7. Still, these equations can be used as points of departure for introducing considerations that apply more generally. Thus, we take this opportunity to introduce without additional development some conclusions about the relationship of exergy destruction to head loss and friction. To begin, note that head loss and exergy destruction are not synonymous. Head loss accounts for the effect of friction and other resistances in converting mechanical energy to internal energy (Section 2.2.3). Energy is conserved in all such conversions. Exergy destruction, on the other hand, accounts for an irrevocable reduction in the magnitude of total exergy owing to friction and other resistances. That the two concepts differ is brought out clearly by Equation 3.24a, which shows that exergy destruction and head loss differ by the temperature ratio T_0/T_{ha}.

Although the friction factor f varies generally with the relative roughness and Reynolds number (Figure 2.5), f is determined mainly by pipe roughness in most applications. Accordingly, Equation 3.24b suggests that the exergy destruction rate varies directly with the roughness level. Moreover, for a fixed friction level f, the exergy destruction rate varies directly with the mass flow rate $\dot{m}$, and inversely with the temperature level, as gauged by T_{ha}. Thus, the effect of friction is more significant at higher mass flow rates and/or lower temperature levels. Let us consider the role of temperature in more detail:

- Equation 3.24a suggests that for components operating at temperatures above T_0 the magnitude of the exergy destruction is less than that of the head loss. For components operating below T_0, however, the magnitude of the exergy destruction is greater than the head loss. Thus, the effect of friction is especially significant for systems operating at temperatures below that of the environment—cryogenic systems, for example.
- Equation 3.24a also suggests that the lower the working fluid temperature the higher the exergy destruction associated with a given value of the head loss. Relatively greater attention should be paid, therefore, to the design of the lower temperature stages of turbines and compressors— the last stages of turbines and the first stages of compressors—than to the remaining stages of these devices.
- Although invariably contributing to exergy destruction, the full effect of friction depends on the process being considered: Referring to Figure

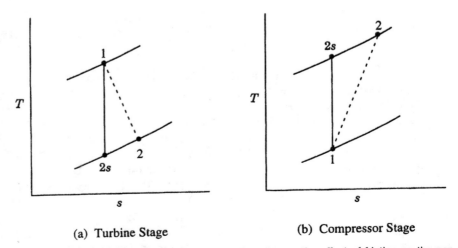

(a) Turbine Stage   (b) Compressor Stage

Figure 3.8 Temperature–entropy diagrams used to discuss the effect of friction on the performance of turbine and compressor stages.

3.8a, which shows the adiabatic expansion of a gas across a turbine stage, note that the exit temperature for the actual expansion with friction, T_2, exceeds the exit temperature for an ideal expansion without friction, T_{2s}. The higher temperature of the actual expansion can be traced to the conversion of mechanical energy into internal energy via friction. Since the temperature at state 2 is higher than at state 2s, the specific exergy at state 2 exceeds that at state 2s, and consequently the capacity to develop work in the next turbine state is enhanced. Turning next to Figure 3.8b, we see that due to friction the exit temperature for a stage of compression is higher than for an ideal compression: $T_2 > T_{2s}$. However, in this case the higher temperature T_2 is unfavorable because the higher temperature requires the next compressor stage to accommodate a greater volumetric flow rate.

Exergy Destruction and Exergy Loss Ratios. The values of the rates of exergy destruction $\dot{E}_D$ and exergy loss $\dot{E}_L$ provide thermodynamic measures of the system inefficiencies. Related to these measures, respectively, are the exergy destruction ratios y_D and y_D^* and the exergy loss ratio y_L.

The rate of exergy destruction in a system component can be compared to the exergy rate of the fuel provided to the overall system, $\dot{E}_{F,\text{tot}}$, giving the exergy destruction ratio

$$y_D = \frac{\dot{E}_D}{\dot{E}_{F,\text{tot}}} \tag{3.25}$$

Alternatively, the component exergy destruction rate can be compared to the total exergy destruction rate within the system, $\dot{E}_{D,\text{tot}}$, giving the ratio

$$y_D^* = \frac{\dot{E}_D}{\dot{E}_{D,\text{tot}}} \tag{3.26}$$

The two exergy destruction ratios are useful for comparisons among various components of the same system. Referring to Table 3.2, for example, the columns identified by footnotes b and c give y_D^* and y_D, respectively, for the cogeneration system. The exergy destruction ratio y_D can also be invoked for comparisons among similar components of different systems using the same, or closely similar, fuels.

The exergy loss ratio is defined similarly to Equation 3.25, by comparing the exergy loss to the exergy of the fuel provided to the overall system:

$$y_L = \frac{\dot{E}_L}{\dot{E}_{F,\text{tot}}} \tag{3.27}$$

Referring again to Table 3.2 for the cogeneration system, footnote e gives the exergy loss ratio associated with stream 7. As illustrated by the data of Section 3.5.1, Equations 3.25 and 3.27 express the reduction in overall efficiency caused by exergy destruction and loss, respectively: $\varepsilon = 1 - \Sigma y_D - \Sigma y_L = 1 - 0.4643 - 0.0326 = 0.503$.

3.5.3 Exergetic Efficiency

The exergy analysis of the cogeneration system considered in Section 3.5.1 introduces the exergetic efficiency as a parameter for evaluating thermodynamic performance. The exergetic efficiency (second-law efficiency, effectiveness, or rational efficiency) provides a true measure of the performance of an energy system from the thermodynamic viewpoint. In this section we consider the concept in detail.

In defining the exergetic efficiency it is necessary to identify both a *product* and a *fuel* for the thermodynamic system being analyzed. The product represents the desired result produced by the system. Accordingly, the definition of the product must be consistent with the purpose of purchasing and using the system. The fuel represents the resources expended to generate the product, and is not necessarily restricted to being an actual fuel such as natural gas, oil, or coal. Both the product and the fuel are expressed in terms of exergy.

As an illustration, let us consider a system at steady state where, in terms of exergy, the rates at which the fuel is supplied and the product is generated are $\dot{E}_F$ and $\dot{E}_P$, respectively. An exergy rate balance for the system reads

$$\dot{E}_F = \dot{E}_P + \dot{E}_D + \dot{E}_L \tag{3.28}$$

where as before $\dot{E}_D$ and $\dot{E}_L$ denote the rates of exergy destruction and exergy loss, respectively. The exergetic efficiency ε is the ratio between product and fuel:

$$\varepsilon = \frac{\dot{E}_P}{\dot{E}_F} = 1 - \frac{\dot{E}_D + \dot{E}_L}{\dot{E}_F} \tag{3.29}$$

The exergetic efficiency shows the percentage of the fuel exergy provided to a system that is found in the product exergy. Moreover, the difference between 100% and the actual value of the exergetic efficiency, expressed as a percent, is the percentage of the fuel exergy wasted in this system as exergy destruction and exergy loss.

An important use of exergetic efficiencies is to assess the thermodynamic performance of a component, plant, or industry relative to the performance of *similar* components, plants, or industries. By this means the performance of a gas turbine, for instance, can be gauged relative to the typical present-day performance level of gas turbines. A comparison of exergetic efficiencies for *dissimilar* devices—gas turbines and heat exchangers, for example—is generally not meaningful, however.

When Equation 3.29 is applied to a system, decisions must be made concerning what is to be counted as the fuel and as the product of the system. Table 3.3 provides illustrations for several common components. Let us consider these beginning with the first entry: a compressor, pump, or fan.

Compressor, Pump, or Fan. In a compressor, pump, or fan, a gas or liquid is caused to flow in the direction of increasing pressure and/or elevation by means of a mechanical or electrical power input. As the exergy of the stream increases, we consider the product to be the exergy increase between inlet and outlet ($\dot{E}_2 - \dot{E}_1$). In this case we consider the fuel as the power input $\dot{W}$. The exergetic efficiency is then

$$\varepsilon = \frac{\dot{E}_2 - \dot{E}_1}{\dot{W}} \quad \text{(compressor, pump, or fan)} \qquad (3.30)$$

where $\dot{W}$ is taken as positive in the direction of the arrow shown in the schematic of Table 3.3. Applying Equation 3.30 to the gas compressor of the cogeneration system of Table 3.1, $\varepsilon = 92.8\%$.

Turbine or Expander. The purpose of purchasing and using a turbine or expander is to generate power from the expansion of a gas or liquid. Thus, for a turbine without extraction we consider the power developed $\dot{W}$ as the product and the decrease in the exergy of the expanding gas or liquid from inlet to outlet ($\dot{E}_1 - \dot{E}_2$) as the fuel. The exergetic efficiency is then

$$\varepsilon = \frac{\dot{W}}{\dot{E}_1 - \dot{E}_2} \quad \text{(turbine or expander)} \qquad (3.31a)$$

Applying Equation 3.31a to the gas turbine of the cogeneration system of Table 3.1, $\varepsilon = 95.2\%$. Table 3.3 gives the definitions of product and fuel for a turbine with one extraction, and the exergetic efficiency is

Table 3.3 Exergy rates associated with fuel and product for selected components at steady-state

Component	Compressor, Pump, or Fan	Turbine or Expander	Heat Exchanger[a]	Mixing Unit	Gasifier or Combustion Chamber	Boiler
Schematic	(schematic)	(schematic)	(schematic)	(schematic)	(schematic)	(schematic)
Exergy rate of product, $\dot{E}_P$	$\dot{E}_2 - \dot{E}_1$	$\dot{W}$	$\dot{E}_2 - \dot{E}_1$	$\dot{E}_3$	$\dot{E}_3$	$(\dot{E}_6 - \dot{E}_5) + (\dot{E}_8 - \dot{E}_7)$
Exergy rate of fuel, $\dot{E}_F$	$\dot{W}$	$\dot{E}_1 - \dot{E}_2 - \dot{E}_3$	$\dot{E}_3 - \dot{E}_4$	$\dot{E}_1 + \dot{E}_2$	$\dot{E}_1 + \dot{E}_2$	$(\dot{E}_1 + \dot{E}_2) - (\dot{E}_3 + \dot{E}_4)$

[a]These definitions assume that the purpose of the heat exchanger is to heat the cold steam ($T_1 \geq T_0$). If the purpose of the heat exchanger is to provide cooling($T_3 \leq T_0$), then the following relations should be used: $\dot{E}_P = \dot{E}_4 - \dot{E}_3$ and $\dot{E}_F = \dot{E}_1 - \dot{E}_2$. For simple coolers or condensers see discussion in text.

$$\varepsilon = \frac{\dot{W}}{\dot{E}_1 - \dot{E}_2 - \dot{E}_3} \qquad (3.31b)$$

Heat Exchangers. As heat exchangers can have different purposes, alternative choices for the fuel and product lead to alternative expressions for the exergetic efficiency:

1. For the case listed in Table 3.3, we assume that all heat transfers occur at or above T_0 and that the purpose of the heat exchanger is to increase the exergy of the cold stream (from state 1 to state 2) at the expense of the exergy of the hot stream (from state 3 to state 4). Accordingly, the product is $(\dot{E}_2 - \dot{E}_1)$ and the fuel is $(\dot{E}_3 - \dot{E}_4)$, and

$$\varepsilon = \frac{\dot{E}_2 - \dot{E}_1}{\dot{E}_3 - \dot{E}_4} \quad \text{(heat exchanger; } T_1 \geq T_0) \qquad (3.32)$$

With this approach, the exergetic efficiencies of the heat recovery steam generator and air preheater of the cogeneration system of Table 3.1 are 67.2 and 84.6%, respectively.

2. In refrigeration applications, where the hot steam is cooled by the cold stream and all heat transfers occur at or below T_0 ($T_3 \leq T_0$), exergy is transferred from the cold stream to the hot stream (exergy is transferred in the direction opposite to the direction of heat transfer). Thus, the product is $(\dot{E}_4 - \dot{E}_3)$, the fuel is $(\dot{E}_1 - \dot{E}_2)$ and

$$\varepsilon = \frac{\dot{E}_4 - \dot{E}_3}{\dot{E}_1 - \dot{E}_2} \quad \text{(heat exchanger; } T_3 \leq T_0) \qquad (3.33)$$

3. In the special case of a heat exchanger providing cooling for a cold chamber ($T < T_0$), the exergy product equals the exergy rate associated with the energy removed by heat transfer from the chamber.

4. No meaningful product, and thus no meaningful exergetic efficiency, can be defined for a heat exchanger that allows heat transfer across the temperature of the environment T_0—that is, heat transfer between two streams at $T > T_0$ and $T < T_0$, respectively. Cost considerations might sanction the use of such a heat exchanger, however, even though this would appear to be a design error from an exergy perspective.

Mixing Unit, Gasifier, Combustion Chamber. The definitions of fuel and product given in Table 3.3 for a mixing unit, a combustion chamber, or a gasification reactor are not the only ones that might be used and are provided mainly for illustrative purposes; for further discussion, see Reference 7. With the approach of Table 3.3, the exergetic efficiency of the combustion chamber of the cogeneration system of Table 3.1 is 79.9%.

Boiler. The purpose of the boiler shown in Table 3.3 is to increase the exergy rate between inlet and outlet for the boiler feedwater (from state 5 to state 6) and the reheat stream (from state 7 to state 8). Thus the product is taken as the sum $(\dot{E}_6 - \dot{E}_5) + (\dot{E}_8 - \dot{E}_7)$. The exergies of the flue-gas stream 4 and the ash stream 3 might be treated as losses, but usually it is not appropriate to do so. Accordingly, as indicated by Table 3.3, in determining the fuel for the boiler the exergy rates of streams 3 and 4 are subtracted from the sum of exergy rates associated with the coal and the air supplied to the boiler to give $(\dot{E}_1 + \dot{E}_2) - (\dot{E}_3 + \dot{E}_4)$. Thus, the exergetic efficiency reads

$$\varepsilon = \frac{(\dot{E}_6 - \dot{E}_5) + (\dot{E}_8 - \dot{E}_7)}{(\dot{E}_1 + \dot{E}_2) - (\dot{E}_3 + \dot{E}_4)} \qquad (3.34)$$

Guidelines for Defining Exergetic Efficiencies. The cases of Table 3.3 provide illustrations of how the fuel and product might be defined for several common components of thermal systems. For other cases the following guidelines are useful when identifying the fuel and product:

- The definition of exergetic efficiency should be meaningful from both the thermodynamic and the economic viewpoints. The purpose of owning and operating a component determines the product of a thermal system.
- Identifying the fuel as the sum of all exergy inputs and the product as the sum of all exergy outputs can result in misleading conclusions for single plant components.
- When a stream crosses the boundary of a system twice with no change in chemical composition, generally only the difference in the exergy values of the stream should be considered in the calculation of the fuel or product. That is, the *net* exergy supplied *by* such a stream would be identified with the fuel and the *net* exergy supplied *to* such a stream would be identified with the product.
- Exergy losses associated with material streams should be included among the losses when the *overall* system efficiency is evaluated, but not when evaluating the efficiencies of the system *components* that the streams *last* exit. In a component efficiency, such streams would be included (with a minus sign) in the fuel term, as for $\dot{E}_3$ and $\dot{E}_4$ in Equation 3.34.

When defining exergetic efficiencies, special considerations can apply. For example, a throttling valve is a component for which a product is not readily defined when the valve is considered singly. Throttling valves typically serve other components. Accordingly, when formulating an exergetic efficiency, the throttling valve and the components it serves should be considered together. Similar considerations apply to heat exchangers (coolers) that achieve the cooling of a stream by the heating of another stream, the exergy gain of which is discarded. Such a case is considered in the discussion of Figure 3.9.

When the purpose of owning and operating a plant component also involves other components, an exergetic efficiency generally should be defined for an enlarged system consisting of the component and the other components it directly affects. This can be illustrated by referring to Figure 3.9 depicting four components: two compressors, a cooler, and a feedwater heater. Two enlarged systems are shown on the figure: System A comprises the cooler and one of the compressors. System B comprises system A and the feedwater heater. Exergetic efficiencies for systems A and B can be defined as follows:

Referring first to system A, note that the cooler serves the compressor by cooling the working fluid (from state 3 to state 4), reducing thereby both the power required by the compressor and the compressor size. Identifying the product of system A as the increase in exergy of the working fluid from state 3 to state 5 and the fuel as the power input, the exergetic efficiency is

$$\varepsilon_A = \frac{\dot{E}_5 - \dot{E}_3}{\dot{W}_{II}}$$

In writing this we have assumed that the exergy gain of the cooling water, $\dot{E}_9 - \dot{E}_8$, is discarded. It is counted as an exergy loss. Although the cooler contributes to both the exergy destruction and exergy loss of system A, it plays an overall positive role by reducing the costs associated with the compressor: a lower operating cost owing to a reduced power requirement and a lower investment cost owing to a reduction in its size.

System B involves similar considerations. But note that the feedwater heater provides preheated feedwater for use elsewhere (heating from state 6 to state 7) and also assists in reducing the power required for compressor II by cooling the working fluid (cooling from state 2 to state 3). Accordingly,

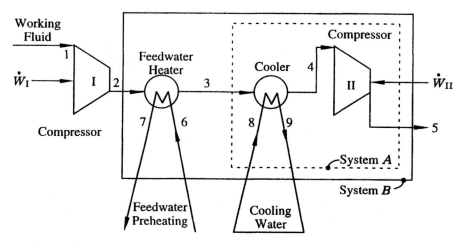

Figure 3.9 Schematic representation of a compression process with intercooling.

the product of system B is the sum of the net exergy increases of the streams flowing from state 2 to state 5 and from state 6 to state 7: $(\dot{E}_5 - \dot{E}_2) + (\dot{E}_7 - \dot{E}_6)$. As before, the fuel is the power input. The corresponding exergetic efficiency reads

$$\varepsilon_B = \frac{(\dot{E}_5 - \dot{E}_2) + (\dot{E}_7 - \dot{E}_6)}{\dot{W}_{II}}$$

Concluding Comments. The exergetic efficiency defined by Equation 3.29 is generally more meaningful, objective, and useful than any other efficiency based on the first or second law of thermodynamics, including the thermal efficiency of a power plant, the isentropic efficiency of a compressor or turbine, and the effectiveness of a heat exchanger. The thermal efficiency of a cogeneration system, for instance, is misleading because it treats both work and heat transfer as having equal thermodynamic value. The isentropic turbine efficiency, which compares the actual process with an isentropic process, does not consider that the working fluid at the outlet of the turbine has a higher temperature (and consequently a higher exergy that may be used in the next component) in the actual process than in the isentropic process. The heat exchanger effectiveness fails, for example, to identify the exergy waste associated with the pressure drops of the heat exchanger working fluids.

3.5.4 Chemical Exergy of Coal, Char, and Fuel Oil

We conclude this section considering exergy applications by evaluating the chemical exergy of the coal whose composition is given in Table 3.4. The presentation follows that of Reference 8. For additional discussion concerning the estimation of the chemical exergy of fuels, see References 2, 8.

The procedure for determining the chemical exergy of pure hydrocarbon fuels developed in Section 3.4.3 can be adapted to the cases of coal, char,

Table 3.4 Coal composition[a]

Constituent	As Received Mass Fraction (%)	Dry and Ash Free (DAF)	
		Mass Fraction (%)[b]	Kmol/kg (DAF)[c]
C	63.98	80.80	0.0673
H	4.51	5.70	0.0565
O	6.91	8.73	0.0055
N	1.26	1.59	0.0011
S	2.52	3.18	0.0010
Ash	9.70		
H_2O	11.12		

[a]Illinois No. 6 bituminous coal.
[b]Denoted C, H, O, N, S for constituent C, H, O, N, and S, respectively.
[c]Denoted c, h, o, n, s for constituent C, H, O, N, and S, respectively.

and fuel oil. In particular, the approach of Equation 3.17a can be invoked using a measured or estimated fuel heating value and an estimated value for the fuel absolute entropy. This will now be illustrated for the coal of Table 3.4, with $T_0 = 25°C$ and $p_0 = 1$ atm.

For this case, the counterpart of Figure 3.5 involves dry and ash free (DAF) coal entering the control volume at T_0, p_0 and reacting completely with oxygen entering separately at T_0, p_0 to form CO_2, SO_2, and $H_2O(l)$, which exit separately at T_0, p_0. The nitrogen contained in the coal also exits separately at T_0, p_0 as N_2. All heat transfer occurs at temperature T_0. On the basis of 1 kg of DAF coal entering the control volume, the combustion reaction is described by

$$(cC + hH + oO + nN + sS) + \nu_{O_2} O_2 \rightarrow$$
$$\nu_{CO_2} CO_2 + \nu_{H_2O} H_2O(l) + \nu_{SO_2} SO_2 + \nu_{N_2} N_2$$

$$(3.35a)$$

where c, h, o, n, and s, each in kmol/kg (DAF), are provided in Table 3.4. Balancing the reaction equation, we have

$$\nu_{CO_2} = c, \qquad \nu_{H_2O} = \tfrac{1}{2}h, \qquad \nu_{SO_2} = s, \qquad \nu_{N_2} = \tfrac{1}{2}n$$
$$\nu_{O_2} = c + \tfrac{1}{4}h + s - \tfrac{1}{2}o \qquad\qquad (3.35b)$$

For this case, the counterpart of Equation 3.17a takes the form

$$e^{CH}_{DAF} = (HHV)_{DAF} - T_0[s_{DAF} + \nu_{O_2}\bar{s}_{O_2} - \nu_{CO_2}\bar{s}_{CO_2} - \nu_{H_2O}\bar{s}_{H_2O} - \nu_{SO_2}\bar{s}_{SO_2}$$
$$- \nu_{N_2}\bar{s}_{N_2}] + [\nu_{CO_2}\bar{e}^{CH}_{CO_2} + \nu_{H_2O}\bar{e}^{CH}_{H_2O} + \nu_{SO_2}\bar{e}^{CH}_{SO_2} + \nu_{N_2}\bar{e}^{CH}_{N_2} - \nu_{O_2}\bar{e}^{CH}_{O_2}]$$

$$(3.36a)$$

Ignoring the slight difference between p_0 (1 atm) and p_{ref} (1 bar), Table C.1 gives

$$\bar{s}_{O_2} = 205.15 \text{ kJ/kmol·K}, \qquad \bar{s}_{CO_2} = 213.79 \text{ kJ/kmol·K}$$
$$\bar{s}_{H_2O} = 69.95 \text{ kJ/kmol·K}, \qquad \bar{s}_{SO_2} = 248.09 \text{ kJ/kmol·K} \qquad (3.37a)$$
$$\bar{s}_{N_2} = 191.61 \text{ kJ/kmol·K}$$

The contribution of Equations 3.37a to the value of the first bracketed term of Equation 3.36a is -0.367 kJ/kg(DAF)·K, as can be readily verified.

With data from model I of Table C.2,

$$\overline{e}_{O_2}^{CH} = 3.951 \text{ MJ/kmol}, \qquad \overline{e}_{CO_2}^{CH} = 14.176 \text{ MJ/kmol}$$
$$\overline{e}_{H_2O}^{CH} = 0.045 \text{ MJ/kmol}, \qquad \overline{e}_{SO_2}^{CH} = 301.939 \text{ MJ/kmol}$$
$$\overline{e}_{N_2}^{CH} = 0.639 \text{ MJ/kmol} \tag{3.37b}$$

Thus, the second bracketed term of Equation 3.36a has the value 0.94 MJ/kg(DAF), as can be readily verified.

To complete the calculation of Equation 3.36a, values are required for the higher heating value of the dry and ash free coal $(HHV)_{DAF}$ and the absolute entropy s_{DAF}, each at T_0, p_0. In the absence of a measured value, the higher heating value can be estimated, in MJ/kg, as follows, [9]:

$$(HHV)_{DAF} = [152.19H + 98.767][(C/3) + H - (O - S)/8] \tag{3.38a}$$

where the mass fractions H, C, O, and S are provided in Table 3.4. That is,

$$(HHV)_{DAF} = [(152.19)(0.057) + 98.767][(0.808/3)$$
$$+ 0.057 - (0.0873 - 0.0318)/8]$$
$$= 34.32 \text{ MJ/kg(DAF)} \tag{3.38b}$$

The absolute entropy for the dry and ash free coal can be estimated, in kJ/kg·K, as follows, [9]:

$$s_{DAF} = c\left[37.1653 - 31.4767 \exp\left(-0.564682 \frac{h}{c+n}\right) + 20.1145 \frac{o}{c+n} \right.$$
$$\left. + 54.3111 \frac{n}{c+n} + 44.6712 \frac{s}{c+n} \right] \tag{3.39a}$$

where c, h, o, n, and s are provided in Table 3.4. That is,

$$s_{DAF} = 0.0673\left[37.1653 - 31.4767 \exp\left(-0.564682 \frac{0.0565}{0.0684}\right) \right.$$
$$\left. + 20.1145 \frac{0.0055}{0.0684} + 54.311 \frac{0.0011}{0.0684} + 44.6712 \frac{0.0010}{0.0684} \right]$$
$$= 1.384 \text{ kJ/kg(DAF)·K} \tag{3.39b}$$

Collecting results, Equation 3.36a gives the specific chemical exergy of the dry and ash free coal

$$e_{DAF}^{CH} = 34.32 \text{ MJ/kg(DAF)} - [(298.15 \text{ K})(1.384$$
$$- 0.367) \text{ kJ/kg(DAF)·K}](1 \text{ MJ}/10^3 \text{ kJ})$$
$$+ 0.94 \text{ MJ/kg(DAF)}$$
$$= 34.96 \text{ MJ/kg(DAF)} \tag{3.36b}$$

The value for the chemical exergy in this case is determined mainly by the first term: the higher heating value. The use of the higher heating value of a fuel to approximate the fuel chemical exergy is frequently observed in the technical literature.

Referring again to Table 3.4, the specific chemical exergy of the coal (as received) is then

$$e^{CH} = 0.7918e_{DAF}^{CH} + \frac{0.1112}{18.015}\bar{e}_{H_2O(l)}^{CH}$$
$$= (0.7918 \text{ kg(DAF)/kg})(34.96 \text{ MJ/kg(DAF)})$$
$$+ \left(\frac{0.1112}{18.015} \text{ kmol/kg}\right)(0.045 \text{ MJ/kmol})$$
$$= 27.68 \text{ MJ/kg} \tag{3.40}$$

The contribution of the water is negligible in this calculation; the chemical exergy of the ash has been ignored.

3.6   GUIDELINES FOR EVALUATING AND IMPROVING THERMODYNAMIC EFFECTIVENESS

Several design guidelines are listed in Table 1.1 to assist the concept development stage of the design process. In the present section additional guidelines are introduced. Most of these stem from experience with the evaluation and improvement of the thermodynamic effectiveness of thermal systems, and most reflect conclusions evolving from exergy analysis. The guidelines provided here also extend and amplify guidelines presented in Reference 10.

An important guideline for achieving cost-effective thermal systems is to maximize the use of cogeneration when it is feasible. Cogeneration achieves economies by combining the production of power together with process steam generation or heating. This is exemplified by the gas turbine cogeneration system we have considered several times, beginning with Chapter 1. The technical literature is abundant with discussions of various cogeneration alternatives and practical applications [11, 12], assuring that cogeneration is firmly founded in current practice and will continue to be so in the future.

At some point in the design process it is necessary to deal directly with the issue of thermodynamic inefficiencies. We have previously noted that sources of inefficiency are related to exergy destruction and exergy loss. The primary contributors to exergy destruction have been identified as chemical reaction, heat transfer, mixing, and friction, including unrestrained expansions of gases and liquids. To increase the thermodynamic effectiveness of a particular system, the principal contributors to its inefficiencies not only should be understood qualitatively but also determined quantitatively. Although the relative magnitude of the exergy destructions and losses must be evaluated, extreme precision in such evaluations is seldom required to guide decisions directed at reducing inefficiencies. When evaluating the relative magnitudes of the exergy destruction and loss, therefore, engineers should not hesitate to:

- Make reasonable simplifying assumptions
- Use simplified exergy calculations

Examples of simplified calculations include the appropriate use of ideal-gas property relations with constant specific heats, the incompressible liquid model, and an approximate thermodynamic average temperature.

An important issue encountered in the discussion of Figure 3.9 can be framed as a question: Does the exergy destruction and loss associated with a component reduce capital investment in the overall system and/or fuel costs in another component? If not, it is generally good practice to eliminate or reduce this source of inefficiency.

Design changes to improve efficiency must be done judiciously, however, for as shown in Chapter 8 the unit costs associated with different sources of inefficiency can be different. For example, the unit cost of the electrical or mechanical power required to feed the fuel exergy destroyed owing to a pressure drop is generally higher than the unit cost of the fuel (e.g., natural gas, oil, or coal) required to feed the exergy destruction caused by combustion or heat transfer. As the unit cost attributed to exergy destruction depends on the nature of the source of inefficiency and on the position of the component in the system (Section 8.2.1), steps taken to reduce inefficiencies should take these aspects into account.

Since chemical reaction is a significant source of thermodynamic inefficiency, it is generally good practice to minimize the use of combustion. In many applications the use of combustion equipment such as boilers is unavoidable, however. In these cases a significant reduction in the combustion irreversibility by conventional means simply cannot be expected, for the major part of the exergy destruction introduced by combustion is an inevitable consequence of incorporating such equipment. Still, the exergy destruction in practical combustion systems can be reduced by the following:

- Minimize the use of excess air
- Preheat the reactants

In most cases only a small part of the exergy destruction in a combustion chamber can be avoided by such means. Thus after considering options such as these for reducing the exergy destruction related to combustion, efforts to improve the thermodynamic performance should be centered on components of the overall system that are more amenable to betterment by cost-effective conventional means. This illustrates another guideline:

- Some exergy destructions and exergy losses can be avoided, others cannot. Efforts should be centered on those that can be avoided.

We have noted previously that nonidealities associated with heat transfer typically contribute heavily to inefficiency. An important general guideline, therefore, is that unnecessary or cost-ineffective heat transfer must be avoided. Other guidelines specific to heat transfer include the following:

- The higher the temperature T at which a heat transfer occurs in cases where $T > T_0$, the more valuable the heat transfer and, consequently, the greater the need to avoid direct heat transfer to the ambient, to cooling water, or to a refrigerated stream. Avoid heat transfer across T_0.
- The lower the temperature T at which a heat transfer occurs in cases where $T < T_0$, the more valuable the heat transfer and, consequently, the greater the need to avoid direct heat transfer with the ambient or a heated stream. Avoid heat transfer across T_0.
- The lower the temperature level, the greater the need to minimize the stream-to-stream temperature difference.
- Avoid the use of intermediate heat transfer fluids when exchanging energy by transfer between two streams.
- In the design of heat exchanger networks, consider these additional guidelines:
  (a) Try to match streams where the final temperature of one stream is close to the initial temperature of the other.
  (b) If there is a significant difference in the *heat capacity rates* (product of the mass flow rate and specific heat c_p) of two streams exchanging energy by heat transfer, consider splitting the stream with the larger heat capacity rate.
  (c) Use the *pinch method* (Section 9.3).

Irreversibilities related to friction, unrestrained expansion, and mixing are generally secondary in importance to those of combustion and heat transfer. Still, they are not to be overlooked, and the following guidelines apply:

- Relatively more attention should be paid to the design of the lower temperature stages of turbines and compressors (the last stages of turbines and the first stages of compressors) than to the remaining stages of these devices.

- For turbines, compressors, and motors consider the most thermodynamically efficient options.
- Minimize the use of throttling; check whether power recovery expanders are a cost-effective alternative for pressure reduction.
- Avoid processes using excessively large thermodynamic driving forces (differences in temperature, pressure, and chemical composition). In particular, minimize the mixing of streams differing significantly in temperature, pressure, or chemical composition.
- The greater the mass rate of flow, the greater the need to use the exergy of the stream effectively.
- The lower the temperature level, the greater the need to minimize friction.

These guidelines aim to improve the use of energy resources in thermal systems by reducing the sources of thermodynamic inefficiency. We should always keep in mind, however, that the objectives of thermal design normally include development of the most effective system from the cost viewpoint. Still, in the cost optimization process, particularly of complex energy systems, it is often expedient to begin by identifying a design that is nearly optimal thermodynamically; such a design can then be used as the starting solution for cost optimization. Further elaboration of this is provided in Chapter 9.

3.7 CLOSURE

In Chapters 2 and 3 we have presented principles of thermodynamics and fluid flow, together with some related modeling and design analysis considerations. These principles can play important roles in thermal system design and optimization to be sure. Still, the principles do not suffice, and we must augment them by bringing in fundamentals from other fields. Heat transfer and engineering economics are two important examples. Heat transfer fundamentals enter the presentation in Chapter 4 to follow and are considered further in the applications of Chapters 5 and 6. Economic analysis enters in Chapter 7 and underpins the discussions of Chapters 8 and 9.

REFERENCES

1. M. J. Moran and H. N. Shapiro, *Fundamentals of Engineering Thermodynamics,* 3rd ed., Wiley, New York, 1995.
2. M. J. Moran, *Availability Analysis: A Guide to Efficient Energy Use,* ASME Press, New York, 1989.

3. T. J. Kotas, *The Exergy Method of Thermal Plant Analysis,* Krieger, Melbourne, Fl, 1995.

4. J. Ahrendts, Reference states, *Energy—Int. J.,* Vol. 5, 1980, pp. 667–677.

5. J. Szargut, D. R. Morris, and F. R. Steward, *Exergy Analysis of Thermal, Chemical, and Metallurgical Processes,* Hemisphere, New York, 1988.

6. R. A. Gaggioli and P. J. Petit, Use the second law, first, *Chemtech,* Vol. 7, 1977, pp. 496–506.

7. G. Tsatsaronis, Thermoeconomic analysis and optimization of energy systems, *Prog. Energy Combustion Sci.,* Vol. 19, 1993, pp. 227–257.

8. G. Tsatsaronis and T. A. Tawfik, *Application of the Exergy Concept to the Analysis of a Gasifier,* Project Report for the U.S. Department of Energy, Project DE-FC21-89MC26019, Center for Electric Power, Tennessee Technological University, Cookeville, TN, 1989.

9. W. Eiserman, P. Johnson, and W. L. Conger, Estimating thermodynamic properties of coal, char, tar and ash, *Fuel Proc. Tech.,* Vol. 3, 1980, pp. 39–53.

10. D. A. Sama, The use of the second law of thermodynamics in the design of heat exchangers, heat exchanger networks and processes, J. Szargut, Z. Kolenda, G. Tsatsaronis, and A. Ziebik, eds. in *Proc. Int. Conf. Energy Syst. Ecol.,* Cracow, Poland, July 5–9, 1993, pp. 53–76.

11. W. F. Kenney, *Energy Conservation in the Process Industries,* Academic Press, Orlando, FL, 1984.

12. H. G. Stoll, *Least-Cost Electric Utility Planning,* Wiley-Interscience, New York, 1989.

PROBLEMS

3.1 For discussion:

    (a) Is it possible for exergy to be negative? Discuss.

    (b) Consider an evacuated space with volume V as the system. Evaluate its exergy and discuss.

    (c) Is it possible for the specific physical exergy e^{PH} associated with a stream of matter to be negative? Discuss.

3.2 Showing all essential steps, derive the following equations:

    (a) Equation 3.5   **(f)** Equation 3.15

    (b) Equation 3.6   **(g)** Equation 3.16

    (c) Equation 3.7   **(h)** Equation 3.17a, b

    (d) Equation 3.12a, b   **(i)** Equation 3.24a, b

    (e) Equation 3.14

3.3 Evaluate the specific physical exergy, e^{PH}, in kJ/kg, of a stream of (a) saturated water vapor at 100°C, (b) saturated liquid water at 100°C, (c) Refrigerant 134a at 0.2 MPa, 40°C, and (d) dry air at 0.3 MPa, 400 K.

Let $T_0 = 20°C$, $p_0 = 0.1$ MPa. Neglect the effects of motion and gravity.

3.4 Using the approach of Equations 3.17b and 3.18, and model I, evaluate the chemical exergy, in kJ/kmol, for (a) CH_4, (b) SO_2, (c) H_2S.

3.5 The following flow rates, in pounds per hour, are reported for the product SNG (substitute natural gas) stream in a process for producing SNG from bituminous coal: CH_4, 429,684; CO_2, 9,093; N_2, 3,741; H_2, 576; CO, 204; H_2O, 60. If SNG exits at 77°F, 1 atm, determine the rate exergy exits, in Btu/h.

3.6 Products of combustion at 480 K, 1 atm, with the molar analysis N_2, 76.9%; O_2, 11.98%; CO_2, 6.66%; H_2O, 4.15%; and SO_2, 0.31% exit a steam generator at a mass flow rate of 17.323 kg per kg of fuel consumed. Evaluate the rate exergy exits, in MW, if the fuel flow rate is 5487 kg/h.

3.7 Verify the exergetic efficiency values provided in Section 3.5.3 for the cogeneration system of Table 3.1.

3.8 A gas enters a turbine operating at steady state and expands adiabatically to a lower pressure. When would the value of the turbine exergetic efficiency, Equation 3.31a, be greater than the value of the turbine isentropic efficiency, Equation 2.25? Discuss.

3.9 Methane (CH_4) at 77°F, 1 atm, enters an adiabatic combustion chamber operating at steady state and reacts completely with 140% of the theoretical amount of air entering separately at 77°F, 1 atm. The products of combustion exit as a mixture at 1 atm. Calculate an exergetic efficiency for this combustion chamber.

3.10 Consider a coal gasification reactor making use of the carbon steam process in which carbon at 25°C, 1 bar, and water vapor at 316°C, 1 bar, enter the reactor and the product gas mixture exits at 927°C, 1 bar. The overall reaction is

$$C + 1.25H_2O \rightarrow CO + H_2 + 0.25H_2O$$

The energy required for this endothermic reaction is provided by an electric resistance heating unit. For operation at steady state, determine, in kJ per kmol of carbon (a) the electricity requirement, (b) the exergy entering with the carbon, (c) the exergy entering with the steam, (d) the exergy exiting with the product gas, (e) the exergy destroyed, and (f) the exergetic efficiency.

3.11 Methane (CH_4) at 77°F, 1 atm, enters the steam generator of a simple vapor plant operating at steady state and burns completely with 200% theoretical air also entering at 77°F, 1 atm. Steam exits the steam generator at 900°F, 500 lbf/in.^{2}, expands through a turbine, and exits to the condenser at 1 lbf/in.^{2} and a quality of 97%. Cooling water passing through the condenser enters at 77°F and exits at 90°F. The combustion products exit the plant at 500°F, 1 atm. Pump work can be ignored. Evaluate the following as a percent of the exergy entering the plant as fuel: (a) the power generated by the turbine, (b) the exergy destroyed in the steam generator.

3.12 For a counterflow heat exchanger, use sketches of the hot and cold stream temperatures (as in Figure 3.7) drawn relative to T_0 to discuss Equations 3.32, 3.33, and the case 4 which follows.

3.13 Define an exergetic efficiency for the heat-recovery steam generator of the case study cogeneration system. Using data from Table 3.1, evaluate the exergetic efficiency. Discuss.

4

HEAT TRANSFER, MODELING, AND DESIGN ANALYSIS

In Chapters 2 and 3 we presented fundamental concepts and definitions of engineering thermodynamics and fluid flow and illustrated their use for design analysis and thermal modeling. In the current chapter, we present the fundamental concepts, definitions, and relations of engineering heat transfer needed to support more wide-ranging design analysis and thermal modeling activities. We assume the reader has had an introduction to heat transfer. Accordingly, heat transfer principles are introduced in only enough detail to refresh understanding. If further elaboration is required, readers should consult References 1–3 or other standard references.

Also in this chapter we begin to illustrate effective assumption making in adapting heat transfer fundamentals to practical problems. This objective is also pursued in Chapters 5 and 6. Throughout, we want to encourage readers to make bold but informed choices in their thermal modeling efforts.

4.1 THE OBJECTIVE OF HEAT TRANSFER

In Chapters 2 and 3 we have been careful to emphasize that Q and $\dot{Q}$ account for transfers of *energy* and not transfers of heat: *Heat* is the energy transfer triggered by a difference between the temperatures of the interacting systems, the transfer of energy being in the direction of the lower temperature (Section 2.1.2). However, to achieve economy of expression in subsequent discussions Q and $\dot{Q}$ are simply referred to as heat transfer and heat transfer rate, respectively.

The objective of the discipline of *heat transfer* is to describe precisely the way in which the difference between the two temperatures (say, T_A and T_B)

167

governs the magnitude of the heat transfer *rate* between the two systems (A and B). This description can be complicated because, in general, the magnitude of the heat transfer rate is influenced not only by the fact that T_A and T_B are different but also by the physical configuration of the heat-exchanging entities:

$$\dot{Q} = \text{function } (T_A, T_B, \text{ time and, for both } A \text{ and } B,$$

$$\text{thermophysical properties, geometry, flow)} \qquad (4.1)$$

A more common alternative to Equation 4.1 that shows the temperature difference explicitly is

$$\dot{Q} = (T_A - T_B) \cdot \text{function } (T_A, T_B,$$

$$\text{time and, for both } A \text{ and } B, \qquad (4.2)$$

$$\text{thermophysical properties, geometry, flow)}$$

The heat transfer results reviewed in this chapter illustrate the many forms that can be assumed by the function shown on the right-hand side of Equation 4.2. This function is often represented in nondimensional terms using one or more of the dimensionless groups listed in Table 4.1. Knowing how to identify the proper form of Equation 4.2 is an important prerequisite for thermal design work.

Although the engineering problems we encounter are extremely diverse, they can be grouped loosely into three categories that illustrate the generality of Equation 4.1 or 4.2:

Thermal Insulation. In this class the temperature extremes (T_A, T_B) are usually fixed. The unknown is the heat transfer rate $\dot{Q}$, which is also named *heat loss* or, in cryogenics, *heat leak*. Among others, the thermal design objective might be to select the insulation (its material, size, shape, orientation, structure) so that $\dot{Q}$ is minimized cost effectively while T_A and T_B remain fixed.

Heat Transfer Enhancement (Augmentation). In heat exchanger design, for example, the total heat transfer rate between the two streams is usually a prescribed quantity. Extremely desirable from a thermodynamic standpoint is the transfer of $\dot{Q}$ across a *minimum* temperature difference $T_A - T_B$, because in this way the rate of entropy generation (or, proportionally, the destruction of exergy) is reduced. The unknown in Equation 4.2 is the temperature difference: The design objective is to improve the thermal contact between the heat-exchanging entities, that is, to minimize the temperature difference $T_A - T_B$. This can be done by changing the flow patterns of the

Table 4.1   Dimensionless groups used in heat transfer[a]

Biot number	$\text{Bi} = hL/k_s$
Boussinesq number	$\text{Bo} = g\beta\,\Delta\text{T}\,H^3/\alpha^2$
Eckert number	$\text{Ec} = \text{U}^2/c_p\,\Delta T$
Fourier number	$\text{Fo} = \alpha t/L^2$
Graetz number	$\text{Gz} = D^2\text{U}/\alpha x = \text{Re}_D\text{Pr}\,\dfrac{D}{x}$
Grashof number	$\text{Gr} = g\beta\,\Delta T\,H^3/\nu^2$
Lewis number[b]	$\text{Le} = \alpha/D = \text{Sc}/\text{Pr}$
Mass transfer Stanton number	$\text{St}_m = \dfrac{h_m}{\text{U}}$
Mass transfer Rayleigh number[b]	$\text{Ra}_m = g\beta_c\,\Delta\rho_i\,H^3/\nu D$
Nusselt number	$\text{Nu} = hL/k_f$
Péclet number	$\text{Pe} = UL/\alpha = \text{Re Pr}$
Prandtl number	$\text{Pr} = \nu/\alpha = \text{Sc}/\text{Le}$
Pressure difference number	$\text{II} = \Delta p\,L^2/\mu\alpha$
Rayleigh number	$\text{Ra} = g\beta\,\Delta T\,H^3/\alpha\nu$
Rayleigh number based on heat flux	$\text{Ra}^* = g\beta\,q''\,H^4/\alpha\nu k$
Reynolds number	$\text{Re} = UL/\nu$
Schmidt number[b]	$\text{Sc} = \nu/D = \text{Le Pr}$
Sherwood number[b]	$\text{Sh} = h_m L/D$
Stanton number	$\text{St} = h/\rho c_p\text{U} = \text{Nu}/\text{Re Pr}$
Stefan number	$\text{Ste} = c_f\,\Delta T/h_{sf}$

[a]Subscripts: s = solid, f = fluid.
[b]In this definition D is the mass diffusivity.

two streams and the shapes of the solid surfaces bathed by the fluid streams (by using fins, for example).

Temperature Control. There are many applications in which the main concern is the overheating of the warm surface (T_A) that produces the heat transfer rate $\dot{Q}$. In a tightly packaged set of electronic circuits, for example, $\dot{Q}$ is generated by *Joule heating,* while the lower temperature T_B is provided by the ambient (e.g., a stream of atmospheric air). The temperature of the electrical conductors (T_A) should not rise above a maximum temperature because high temperatures threaten the operation of the electrical circuitry. In this design problem, the heat transfer rate must be maximized, and the flow configuration must vary in such a way that T_A does not exceed a certain ceiling value.

Examples of temperature control applications include also the cooling of nuclear reactor cores, and the cooling of the outer skin of space vehicles during reentry. In these examples the high-temperature T_A must be kept below the temperature at which the mechanical strength characteristics of the hot surface begin to deteriorate.

4.2 CONDUCTION

4.2.1 Steady Conduction

Conduction or pure thermal diffusion is the mode of heat transfer in which energy transfer from the region of high temperature to the region of low temperature is due to a temperature gradient between the regions. The simplest example of this type is the plane wall at steady state of thickness L and area $A = HW$ illustrated in Table 4.2. The two faces of the wall are maintained at different temperatures T_1 and T_2. The heat transfer rate from T_1 to T_2 is

$$\dot{Q} = k A \frac{T_1 - T_2}{L} \tag{4.3}$$

in which k is the *thermal conductivity* of the material. Units for the thermal conductivity are W/(m·K) and Btu/(h·ft·°R). In Equation 4.3 and the other heat transfer rate results reviewed in this chapter, the thermal conductivity is regarded as a constant, that is, a number independent of temperature, the position between T_1 and T_2, and the direction of $\dot{Q}$.

Thermal Resistance. An alternative to Equation 4.3 or any other proportionality between heat transfer rate and temperature difference is to define the *thermal resistance R_t*:

$$R_t = \frac{T_1 - T_2}{\dot{Q}} \tag{4.4}$$

The flow of $\dot{Q}$ down the temperature step $T_1 - T_2$ is completely analogous to the flow of an electrical current through a resistance with a potential difference across its ends. Units of R_t are K/W and °R/(Btu/h). According to Equation 4.3, the thermal resistance of the plane wall is $R_t = L/k A$.

Shape Factor. Another alternative to expressing the proportionality between heat transfer rate and temperature difference during steady conduction is by using the *shape factor S*:

$$S = \frac{\dot{Q}}{k(T_1 - T_2)} = \frac{1}{k R_t} \tag{4.5}$$

The units of S are those of length. The shape factors of the plane wall and several other geometric configurations are summarized in Table 4.2.

Table 4.2 The shape factors (S) of several configurations involving conduction heat transfer between isothermal surfaces [1], $\dot{Q} = Sk\,(T_1 - T_2)$

Plane Walls

Plane wall with isothermal surfaces (T_1, T_2)

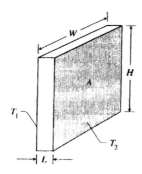

$$S = \frac{A}{L} = \frac{WH}{L}$$

$$\left(H > \frac{L}{5}, W > \frac{L}{5}\right)$$

Edge prism, joining two plane walls with isothermal surfaces:

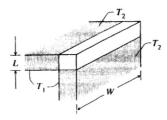

$$S = 0.54W$$
$$\left(W > \frac{L}{5}\right)$$

T_1 = temperature of internal surfaces of plane walls

T_2 = temperature of external surfaces of plane walls

T_2 = temperature of two exposed faces of the prism

Corner cube, joining three plane walls. The temperature difference $T_1 - T_2$ is maintained between the inner and outer surfaces of the three-wall structure.

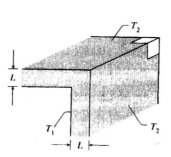

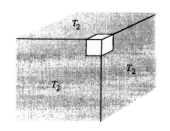

$$S = 0.15L$$

(continued

Table 4.2 (*Continued*)

Cylindrical Surfaces		
Bar with square cross section, isothermal outer surface, and cylindrical hole through the center	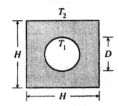	$$S = \frac{2\pi L}{\ln(1.08 H/D)}$$ (L = length normal to the plane of the figure)
Infinite slab with isothermal surfaces (T_2) and thickness $2H$, with an isothermal cylindrical hole (T_1) positioned midway between the surfaces	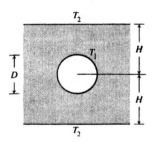	$$S = \frac{2\pi L}{\ln(8H/\pi D)}$$ ($L \gg D$, $H > D/2$, L = length of cylindrical hole)
Cylinder of length L and diameter D, with cylindrical hole positioned eccentrically	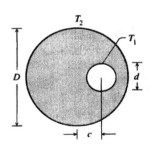	$$S = \frac{2\pi L}{\cosh^{-1}\left(\dfrac{D^2 + d^2 - 4c^2}{2Dd}\right)}$$ ($L \gg D$)
Semi-infinite medium with isothermal surface (T_2) and isothermal cylindrical hole (T_1) of length L, parallel to the surface (i.e., normal to the plane of the figure)	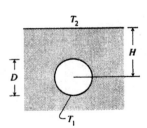	$$S = \frac{2\pi L}{\cosh^{-1}(2H/D)}$$ ($L \gg D$) $$S \cong \frac{2\pi L}{\ln(4H/D)} \quad \text{if } H/D > 3/2$$

Table 4.2 (Continued)

Infinite medium with two parallel isothermal cylindrical holes of length L	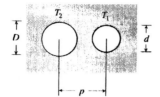	$S = \dfrac{2\pi L}{\cosh^{-1}\left(\dfrac{4p^2 - D^2 - d^2}{2Dd}\right)}$ $(L >> D, d, p)$
Semi-infinite medium with isothermal surface (T_2) and with a cylindrical hole (T_1) drilled to a depth H normal to the surface. The hole is open. The medium temperature far from the hole is T_2	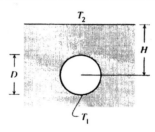	$S = \dfrac{2\pi H}{\ln(4H/D)}$ $(H >> D)$

Spherical Surfaces

Semi-infinite medium with isothermal surface (T_2) and isothermal spherical cavity	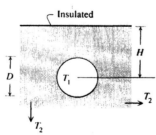	$S = \dfrac{2\pi D}{1 - (D/4H)}$ $(H > D/2)$ Spherical cavity in an infinite medium: $S \cong 2\pi D \quad$ if $H/D > 3$
Semi-infinite medium with insulated surface and far-field temperature T_2 containing an isothermal spherical cavity (T_1)	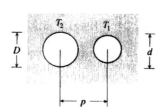	$S = \dfrac{2\pi D}{1 + (D/4H)}$ $(H > D/2)$
Infinite medium with two isothermal spherical cavities		$S = \dfrac{4\pi}{\dfrac{d}{D}\left[1 - \dfrac{(D/2p)^4}{1 - (d/2p)^2}\right] - \dfrac{d}{p}}$ $(p/D > 3)$

Table 4.2 (*Continued*)

Hemispherical isothermal dimple (T_1) into the insulated surface of a semi-infinite medium with far-field temperature T_2	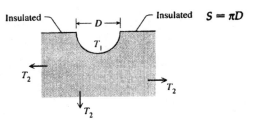	$S = \pi D$

Thin Discs and Plates

Semi-infinite medium with isothermal surface (T_2) and isothermal disc (T_1) parallel to the surface	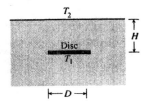	$S = \dfrac{2\pi D}{(\pi/D) - \tan^{-1}(D/4H)}$ $(H/D > 1)$ Disc attached to the surface: $S = 2D \quad (H = 0)$
Semi-infinite medium with insulated surface and far-field temperature T_2, containing an isothermal disc (T_1) parallel to the surface	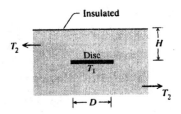	$S = \dfrac{2\pi D}{(\pi/2) + \tan^{-1}(D/4H)}$ $(H/D > 1)$
Infinite medium with two parallel, coaxial and isothermal discs	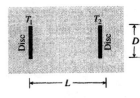	$S = \dfrac{2\pi D}{(\pi/2) - \tan^{-1}(D/2L)}$ $(L/D > 2)$
Semi-infinite medium with isothermal surface (T_2) and isothermal plate (T_1) parallel to the surface (W is the width of the plate, in the direction normal to the plane of the figure)	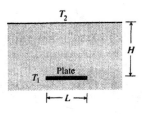	$S = \dfrac{2\pi L}{\ln(4L/W)} \quad (H \gg L)$ Plate attached to the surface: $S = \dfrac{\pi L}{\ln(4L/W)} \quad (H = 0)$

Cylindrical Shells. The thermal resistance of a long cylindrical shell with isothermal inner and other surfaces is

$$R_t = \frac{\ln (r_o/r_i)}{2 \pi k L} \tag{4.6}$$

The outer and inner radii are r_o and r_i, respectively, while L is the shell axial length ($L \gg r_0$). If the outer surface exchanges heat with a fluid through a constant heat transfer coefficient h (Section 4.3), the total thermal resistance between the inner surface and the fluid is

$$R_t = \frac{\ln (r_o/r_i)}{2 \pi k L} + \frac{1}{2 \pi r_o L h} \tag{4.7}$$

The *critical insulation radius* of a cylindrical shell is k/h. This means that if $r_i > k/h$, the wrapping of a material with thermal conductivity k on a cylinder of radius r_i will always have an insulating effect—that is, will induce an increase in the thermal resistance between the cylinder and the surrounding fluid.

Spherical Shells. The thermal resistance of a spherical shell with isothermal surfaces is

$$R_t = \frac{1}{4 \pi k} \left(\frac{1}{r_i} - \frac{1}{r_o} \right) \tag{4.8}$$

in which r_o and r_i are the outer and inner radii. The critical insulation radius is $2k/h$, such that if $r_i > 2k/h$, the wrapping of a material with thermal conductivity k on the surface of radius r_i leads to an increase in the resistance between that surface and the surrounding fluid.

Fins. The use of extended surfaces or fins is one of the most common techniques of augmenting the heat transfer, or enhancing the thermal contact between a solid "base" surface (at temperature T_b) and a fluid (at temperature T_∞). If the geometry of the fin is such that the cross-sectional area (A_c), the wetted perimeter of the cross section (p), and the heat transfer coefficient (h) are *constant* (e.g., Reference 1, p. 54), the heat transfer rate $\dot{Q}_b$ through the base of the fin is approximated well by

$$\dot{Q}_b \cong (T_b - T_\infty)(kA_c hp)^{1/2} \tanh\left[\left(\frac{hp}{kA_c} \right)^{1/2} \left(L + \frac{A_c}{p} \right) \right] \tag{4.9}$$

In this expression L is the fin length, measured from the tip of the fin to the base surface, and k is the thermal conductivity of the fin material. The heat transfer rate through a fin with variable cross-sectional area and perimeter can be calculated by writing

$$\dot{Q}_b = (T_b - T_\infty)h\, A_{\text{exp}}\, \eta \qquad\qquad (4.10)$$

where A_{exp} is the total exposed (wetted) area of the fin and η is the *fin efficiency,* a dimensionless number between 0 and 1, which can be found in heat transfer texts and handbooks [1–5].

Equations 4.9 and 4.10 are based on the very important assumption that the conduction through the fin is essentially *unidirectional* and oriented along the fin. This assumption is valid when $(ht/k)^{1/2} < 1$, where t is the thickness of the fin, that is, the dimension perpendicular to the conduction heat current [1].

4.2.2 Unsteady Conduction

Immersion Cooling or Heating of a Conducting Body. The temperature history inside a conducting body immersed suddenly (at time $t = 0$) in a bath of fluid at a different temperature (T_∞) is important in many applications, for example, the heat treating of special alloys. Figure 4.1 shows the evolution of the center temperature T_c in three geometries: a plate of thickness $2L$, a long cylinder of radius r_o, and a sphere of radius r_o. The thermal conductivity, thermal diffusivity, and initial temperature of each body are k, α, and T_i, respectively. The constant heat transfer coefficient between the body surface and the fluid at T_∞ is h.

Figure 4.1 shows only the early part of the time-dependent conduction process: $0.4 < (T_c - T_\infty)/(T_i - T_\infty)$, because for larger times the curves can be extended as straight lines. The dimensionless groups used in these graphs are the *Fourier numbers,* or the dimensionless time, $\text{Fo} = \alpha t/L^2$ and $\text{Fo} = \alpha t/r_o^2$, and the *Biot numbers,* $\text{Bi} = hL/k$ and $\text{Bi} = hr_o/k$. The temperature distributions inside simple three-dimensional bodies can be determined based on the information of Figure 4.1 and a multiplication rule presented in heat transfer textbooks [1, 2].

When the heat transfer coefficient is large enough that $\text{Bi} > 1$, the surface temperature of the immersed body is almost steady and equal to the fluid temperature T_∞. In this limit the temperature within spheres, long cylinders and flat plates varies as shown in Figure 4.2, where $\overline{T}$ is the instantaneous temperature averaged over the body volume V. Listed on the ordinate is $(1 - Q/Q_i)$, where Q is the total heat transfer from the body to the fluid during the time interval 0 to t, namely $Q = \rho V c(T_i - \overline{T})$, and Q_i is the maximum (long time) value of this quantity, $Q_i = \rho V c(T_i - T_\infty)$, where ρ and c denote the density and specific heat of the body. For larger times the three curves can be extended as straight lines.

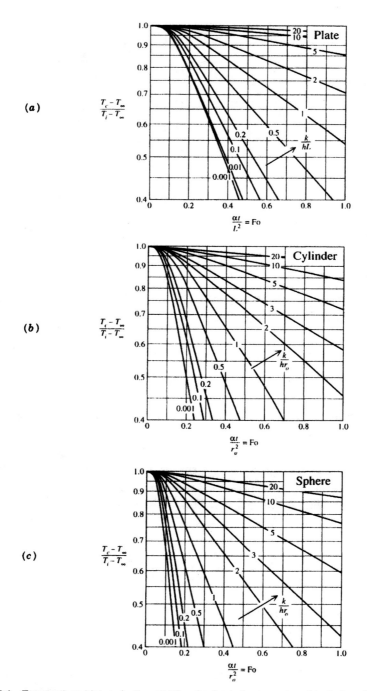

Figure 4.1 Temperature history in the middle of a body immersed suddenly in a fluid at a different temperature: (a) L = plate half-thickness; (b) r_o = cylinder radius; (c) r_o = sphere radius [1].

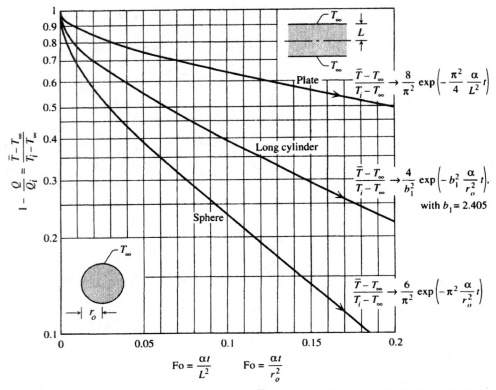

Figure 4.2 The volume-averaged temperature $\overline{T}$ and the total heat transfer Q from a body with the surface temperature fixed at T_∞ [1].

In cases where the heat transfer coefficient is small: Bi $<$ 1, and the elapsed time is large: Fo $>$ 1, the temperature gradients inside the body have decayed, and the instantaneous (time-dependent) body temperature is approximated well by a single value, $T(t)$. This temperature approaches the fluid bath temperature exponentially:

$$\frac{T - T_\infty}{T_1 - T_\infty} = \exp\left[-\frac{hA}{\rho Vc}(t - t_1)\right] \tag{4.11}$$

where A is the wetted body surface and $T_1 = T(t_1)$. The group ρVc is the *lumped capacitance* of the body, and Equation 4.11 is known as the *lumped capacitance model* for the temperature of the immersed body.

Semi-Infinite Solid in Contact with Fluid Flow. When the elapsed time is so short that Fo $<$ 1, any conducting body that is immersed in a flow is penetrated by thermal diffusion only a small distance. Relative to this skin region of time-dependent conduction, the body looks semi-infinite and the wetted surface looks plane. This means that the body may be modeled as semi-infinite. The temperature history in a semi-infinite body whose surface ($x = 0$) is placed at $t = 0$ in contact with a fluid (h, T_∞) is given as

$$\frac{T(x, t) - T_\infty}{T_i - T_\infty} = \text{erf}\left[\frac{x}{2(\alpha t)^{1/2}}\right] + \exp\left(\frac{hx}{k} + \frac{h^2 \alpha t}{k^2}\right) \text{erfc}\left[\frac{x}{2(\alpha t)^{1/2}} + \frac{h}{k}(\alpha t)^{1/2}\right]$$

(4.12)

As before, T_i, k, and α are the initial temperature, thermal conductivity, and thermal diffusivity of the body. The distance x is measured from the surface into the body.

The temperature distribution in a semi-infinite solid whose surface temperature is raised instantly from T_i to T_∞ is a special case of Equation 4.12 as $h \to \infty$:

$$\frac{T(x, t) - T_\infty}{T_i - T_\infty} = \text{erf}\left[\frac{x}{2(\alpha t)^{1/2}}\right]$$

(4.13)

Equation 4.13 approximates Equation 4.12 well when $(h/k)(\alpha t)^{1/2} \gtrsim 10$. The temperature field under the surface of a semi-infinite solid that, starting with the time $t = 0$, is exposed to a constant heat flux q'' (heat transfer rate per unit area) is

$$T(x, t) - T_i = 2\frac{q''}{k}\left(\frac{\alpha t}{\pi}\right)^{1/2} \exp\left(-\frac{x^2}{4\alpha t}\right) - \frac{q''}{k} x \, \text{erfc}\left[\frac{x}{2(\alpha t)^{1/2}}\right]$$

(4.14)

Concentrated Sources and Sinks. Common phenomena that can be described in terms of concentrated heat sources are underground fissures filled with geothermal steam, underground explosions, canisters of nuclear and chemical waste, and buried electrical cables. It is important to distinguish between instantaneous heat sources and continuous heat sources. Consider first the class of *instantaneous heat sources* released at $t = 0$ in a conducting medium with constant properties (ρ, c, k, α) and uniform initial temperature T_i. The temperature distribution in the vicinity of the source depends on the shape of the source:

$$T(x, t) - T_i = \frac{Q''}{2\rho c(\pi \alpha t)^{1/2}} \exp\left(-\frac{x^2}{4\alpha t}\right)$$

[instantaneous plane source, strength Q'' (J/m^2 or Btu/ft^2) at $x = 0$]

(4.15)

$$T(r, t) - T_i = \frac{Q'}{4\rho c\, \pi \alpha t}\, \exp\left(-\frac{r^2}{4\alpha t}\right)$$

[instantaneous line source, strength
Q' (J/m or Btu/ft) at $r = 0$]

(4.16)

$$T(r, t) - T_i = \frac{Q}{8\rho c(\pi \alpha t)^{3/2}}\, \exp\left(-\frac{r^2}{4\alpha t}\right)$$

[instantaneous point source, strength
Q (J or Btu) at $r = 0$]

(4.17)

The corresponding results for the temperature near *continuous heat sources,*
which release heat at constant rate when $t > 0$, are

$$T(x, t) - T_i = \frac{q''}{\rho c}\left(\frac{t}{\pi \alpha}\right)^{1/2} \exp\left(-\frac{x^2}{4\alpha t}\right) - \frac{q''|x|}{2k}\, \mathrm{erfc}\left[\frac{|x|}{2(\alpha t)^{1/2}}\right]$$

[continuous plane source, strength
q'' (W/m² or Btu/h·ft²) at $x = 0$]

(4.18)

$$T(r, t) - T_i = \frac{q'}{4\pi k}\int_{r^2/4\alpha t}^{\infty} \frac{e^{-u}}{u}\, du$$

$$\cong \frac{q'}{4\pi k}\left[\ln\left(\frac{4\alpha t}{r^2}\right) - 0.5772\right] \quad \text{if } \frac{r^2}{4\alpha t} < 1$$

[continuous line source, strength
q' (W/m or Btu/h·ft) at $r = 0$]

(4.19)

$$T(r, t) - T_i = \frac{\dot{Q}}{4\pi kr}\, \mathrm{erfc}\left[\frac{r}{2(\alpha t)^{1/2}}\right]$$

$$\cong \frac{\dot{Q}}{4\pi kr} \quad \text{if } \frac{r}{2(\alpha t)^{1/2}} < 1$$

[continuous point source, strength
$\dot{Q}$ (W or Btu/h) at $r = 0$]

(4.20)

The temperature within wakes behind *moving sources* are described by the
following expressions, in which U is the constant speed of the source and x
is measured away from the source in the downstream direction:

$$T(x, y) - T_i = \frac{q'/\rho c}{(4 \pi \, \mathrm{U} \, \alpha x)^{1/2}} \exp\left(-\frac{\mathrm{U} \, y^2}{4\alpha x}\right)$$

[moving continuous line source, strength
q' (W/m or Btu/h·ft) at $x = 0$]   (4.21)

$$T(x, r) - T_i = \frac{\dot{Q}/\rho c}{4 \pi \alpha x} \exp\left(-\frac{\mathrm{U} \, r^2}{4\alpha x}\right)$$

[moving continuous point source, strength
$\dot{Q}$ (W or Btu/h) at $x = 0$]   (4.22)

In Equations 4.21 and 4.22 the y and r coordinates are perpendicular to the x direction. These expressions are valid provided the wake is slender, in other words, when the Peclet number is large, $Ux/\alpha > 1$.

Equations 4.15–4.22 apply unchanged to the temperature fields near concentrated *heat sinks*. In such cases the numerical values of the source strengths (Q'', Q', Q, q'', q', $\dot{Q}$) are negative.

Melting and Solidification. A semi-infinite solid that is isothermal and at the melting point (T_m) melts if its surface is raised to another temperature (T_0). In the absence of the effect of convection, the liquid layer thickness δ increases in time according to

$$\delta(t) \cong \left[2 \, \frac{kt}{\rho h_{sf}} (T_0 - T_m) \right]^{1/2}$$ (4.23)

where ρ and k are the density and thermal conductivity of the liquid, and h_{sf} is the latent heat of melting. Equation 4.23 is valid provided $c(T_0 - T_m)/h_{sf} < 1$, where c is the specific heat of the liquid.

The solidification of a motionless pool of liquid is described by Equation 4.23, in which $T_0 - T_m$ is replaced by $T_m - T_0$ because the liquid is saturated at T_m, and the surface temperature is lowered to T_0. In the resulting expression δ is the thickness of the solid layer, and k, c, and α are properties of the solid layer.

Example 4.1   The shape in which a potato is cut has an important effect on how fast each piece is cooked. Three different shapes have been proposed, each containing 5 g of potato matter: (a) sphere, (b) cylinder with a length of 6 cm, (c) thin disk with a diameter of 4 cm. Each piece is initially at the temperature $T_i = 30°C$. At time $t = 0$, each piece is placed in boiling water at the temperature $T_\infty = 100°C$. The heat transfer coefficient is constant, $h = 2 \times 10^4$ W/m²·K, and the properties of potato matter are approximately $\rho =$

0.9 g/cm^3, $k = 0.6$ W/m·K, and $\alpha = 0.0017$ cm^2/s. For each shape, calculate the time t until the volume-averaged temperature rises to 65°C. Comment on the best shape for the shortest cooking time.

Solution

MODEL

1. The potato matter is homogeneous and isotropic with constant properties.
2. The heat transfer coefficient at the wetted surface is assumed constant, that is, independent of time.
3. The transfer of water through the potato surface and any swelling of the potato are neglected.

ANALYSIS. As shown below, the requirement of fixed mass, $m = 5$ g, determines the size of each piece. For example, the radius of the sphere is calculated as 1.1 cm. Calculating the Biot number for the sphere, we have

$$\text{Bi} = \frac{hr_o}{k} \sim (2 \times 10^4 \text{ W/m}^2\text{·K})(0.011 \text{ m})/(0.6 \text{ W/m·K})$$
$$= 367$$

With similar calculations, we find that in each case $\text{Bi} \gg 1$, and so we can rely on Figure 4.2. Accordingly, using the given temperatures, the ordinate reads (regardless of shape)

$$\frac{\overline{T} - T_\infty}{T_i - T_\infty} = \frac{(65 - 100)°C}{(30 - 100)°C} = 0.5$$

The three curves indicate the following readings on the abscissa:

$$\text{Fo}_{\text{sphere}} \cong 0.031 = \alpha t/r_o^2 \qquad \text{(a)}$$

$$\text{Fo}_{\text{cylinder}} \cong 0.064 = \alpha t/r_o^2 \qquad \text{(b)}$$

$$\text{Fo}_{\text{plate}} \cong 0.196 = \alpha t/L^2 \qquad \text{(c)}$$

(a) In the case of the spherical shape, the radius is fixed by the mass $m = 5$ g,

$$m = \rho \frac{4\pi}{3} r_o^3$$

$$r_o^3 = \frac{5 \text{ g}}{0.9 \text{ g/cm}^3} \frac{3}{4\pi} = 1.33 \text{ cm}^3$$

$$r_o = 1.1 \text{ cm}$$

and Equation (a) pinpoints the warming time,

$$t = 0.031 \frac{(1.1)^2 \text{ cm}^2}{0.0017 \text{ cm}^2/\text{s}} \cong 22 \text{ s}$$

(b) For a cylinder of length $Z = 6$ cm, the radius is

$$m = \rho Z \pi r_o^2$$

$$r_o^2 = \frac{5 \text{ g}}{0.9 \text{ g/cm}^3} \frac{1}{6\pi \text{ cm}} = 0.29 \text{ cm}^2$$

$$r_o = 0.54 \text{ cm}$$

and Equation (b) gives

$$t = 0.064 \frac{(0.54)^2 \text{ cm}^2}{0.0017 \text{ cm}^2/\text{s}} \cong 11 \text{ s}$$

(c) Finally, in the case of a disk of diameter $D = 4$ cm, the disk half-thickness L is obtained as follows:

$$m = \rho \frac{\pi D^2}{4} 2L$$

$$L = \frac{5 \text{ g}}{0.9 \text{ g/cm}^3} \frac{2}{16\pi \text{ cm}^2} = 0.22 \text{ cm}$$

with the corresponding warming time from Equation (c),

$$t = 0.196 \frac{(0.22)^2 \text{ cm}^2}{0.0017 \text{ cm}^2/\text{s}} \cong 6 \text{ s}$$

COMMENT. Comparing these three time intervals (22, 11, and 6 s), we see that in accord with intuition the disk shape (the potato slice) promises to cook much faster than the other two shapes.

4.3   CONVECTION

Convection is the heat transfer process in which a flowing material (gas, fluid, solid) acts as a conveyor for the energy that it draws from (or delivers to) a solid wall, and, as a consequence, the heat transfer rate is affected greatly by the characteristics of the flow (e.g., velocity distribution, turbulence). To know the flow distribution and the regime (laminar vs. turbulent) is an important prerequisite for calculating convection heat transfer rates. Also, the nature of the boundary layers (hydrodynamic and thermal) plays an important role in evaluating convection.

In this section we review the most important results of convection heat transfer together with the corresponding fluid mechanics results. It is assumed that the fluid is Newtonian, homogeneous, and isotropic and has a nearly constant density.

4.3.1   External Forced Convection

Convection is said to be *external* when a much larger space filled with flowing fluid (the *free stream*) exchanges heat with a body immersed in the fluid. According to Equation 4.2, the objective is to determine the relation between the heat transfer rate (or the heat flux through a spot on the wall, q''), and the wall-fluid temperature difference ($T_w - T_\infty$). The alternative is to determine the *convective heat transfer coefficient h*, which in an external flow is defined by

$$h = \frac{q''}{T_w - T_\infty} \qquad (4.24)$$

where q'' is the heat flux (heat transfer rate per unit area). Units for the convective heat transfer coefficient are $W/m^2 \cdot K$ and $Btu/h \cdot ft^2 \cdot °R$. Figure 4.3 shows the order of magnitude of h in various cases.

Boundary Layer over a Plane Wall. When the fluid velocity U_∞ is uniform and parallel to a wall of length L, the hydrodynamic boundary layer along the wall is *laminar* over L if $Re_L \leqslant 5 \times 10^5$, where the *Reynolds number* is defined by $Re_L = U_\infty L/\nu$ and ν is the kinematic viscosity. The leading edge of the wall is perpendicular to the direction of the free stream (U_∞). The wall shear stress in laminar flow averaged over the length L (L averaged) is

$$\bar{\tau} = 0.664 \rho U_\infty^2 \, Re_L^{-1/2} \qquad (Re_L \leqslant 5 \times 10^5) \qquad (4.25)$$

so that the total tangential force experienced by a plate of width W and length

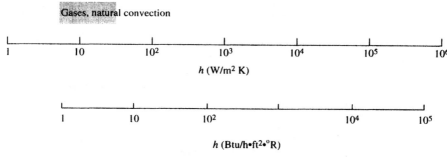

Figure 4.3 Effect of flow configuration and fluid type on the convective heat transfer coefficient [1].

L is $F = \overline{\tau}LW$. The length L is measured in the flow direction. The thickness of the hydrodynamic boundary layer at the trailing edge of the plate is of the order of $L\,\mathrm{Re}_L^{-1/2}$.

If the wall is *isothermal* at T_w, the heat transfer coefficient $\overline{h}$ averaged over the flow length L is ($\mathrm{Re}_L \lesssim 5 \times 10^5$):

$$\frac{\overline{h}L}{k} = \begin{cases} 0.664\ \mathrm{Pr}^{1/3}\mathrm{Re}_L^{1/2} & (\mathrm{Pr} \gtrsim 0.5) \\ 1.128\ \mathrm{Pr}^{1/2}\mathrm{Re}_L^{1/2} & (\mathrm{Pr} \lesssim 0.5) \end{cases} \qquad (4.26)$$

In these expressions k and Pr are the fluid thermal conductivity and the Prandtl number $Pr = \nu/\alpha$. The free stream is isothermal at T_∞. The total heat transfer rate through the wall of area LW is $\dot{Q} = \bar{h}LW(T_w - T_\infty)$.

When the wall heat flux, q'', is uniform, the wall temperature T_w increases away from the leading edge $(x = 0)$ as $x^{1/2}$:

$$T_w(x) - T_\infty = \frac{2.21q''x}{k\ Pr^{1/3}Re_x^{1/2}} \qquad (Pr \geqslant 0.5, \quad Re_x \leqslant 5 \times 10^5) \quad (4.27)$$

where the Reynolds number is based on the distance from the leading edge, $Re_x = U_\infty x/\nu$. The wall temperature averaged over the flow length L, $\bar{T}_w$, is obtained by substituting, respectively, $\bar{T}_w$, Re_L, and 1.47 in place of $T_w(x)$, Re_x, and 2.21 in Equation 4.27.

At Reynolds numbers Re_L greater than approximately 5×10^5, the boundary layer begins with a laminar section that is followed by a turbulent section. For $5 \times 10^5 < Re_L < 10^8$ and $0.6 < Pr < 60$, the average heat transfer coefficient $\bar{h}$ and wall shear stress $\bar{\tau}$ are [2]

$$\frac{\bar{h}L}{k} = 0.037\ Pr^{1/3}\ (Re_L^{4/5} - 23,550) \qquad (4.28)$$

$$\frac{\bar{\tau}}{\rho U_\infty^2} = \frac{0.037}{Re_L^{1/5}} - \frac{871}{Re_L} \qquad (4.29)$$

Equation 4.28 is sufficiently accurate for isothermal walls (T_w) as well as for uniform-heat flux walls. The total heat transfer rate from an isothermal wall is $\dot{Q} = \bar{h}LW(T_w - T_\infty)$, and the average temperature of the uniform-flux wall is $\bar{T}_w = T_\infty + q''/\bar{h}$. The total tangential force experienced by the wall is $F = \bar{\tau}LW$, where W is the wall width.

In most of the wall friction and heat transfer formulas reviewed in this section the fluid properties (k, ν, μ, α) should be evaluated at the *film temperature* $\frac{1}{2}(T_w + T_\infty)$. However, there are special correlations (e.g., Equation 4.31) in which the effect of temperature-dependent properties is taken into account by means of explicit correction factors.

Single Cylinder in Cross Flow. When the free stream is uniform (U_∞, T_∞) and perpendicular to the cylinder axis, the heat transfer coefficient $\bar{h}$ averaged over the cylinder perimeter is [6]

$$\frac{\bar{h}D}{k} = 0.3 + \frac{0.62\ Re_D^{1/2}Pr^{1/3}}{[1 + (0.4/Pr)^{2/3}]^{1/4}}\left[1 + \left(\frac{Re_D}{282,000}\right)^{5/8}\right]^{4/5} \qquad (4.30)$$

Here D is the cylinder diameter and $Re_D = U_\infty D/\nu$. Equation 4.30 is sufficiently accurate for all values of Re_D and Pr, provided the Peclet number $Pe_D = Re_D Pr$ is greater than 0.2. In the intermediate Re_D range 7×10^4 to

4×10^5, Equation 4.30 yields $\bar{h}$ values that can be 20% smaller than those obtained by direct measurement. The total heat transfer rate per unit of cylinder axial length is $q' = \bar{h}\,\pi D(T_w - T_\infty)$ when the cylinder surface is isothermal at T_w. If the heat flux q'' is uniform around the perimeter, the surface temperature averaged over the cylinder perimeter is $\bar{T}_w = T_\infty + q''/\bar{h}$.

The total drag force F_D experienced by the cylinder can be calculated by using the drag coefficient C_D defined and presented in Figure 4.4. The wake behind the cylinder buckles (meanders) when $\mathrm{Re}_D \gtrsim 40$, and the frequency with which the large vortices shed from the cylinder is approximately $0.21 U_\infty/D$ [7].

Sphere. The recommended correlation for the average heat transfer coefficient between a sphere of diameter D and an isothermal uniform free stream (U_∞, T_∞) is [8]

$$\frac{\bar{h}D}{k} = 2 + (0.4\,\mathrm{Re}_D^{1/2} + 0.06\,\mathrm{Re}_D^{2/3})\mathrm{Pr}^{0.4}\left(\frac{\mu_\infty}{\mu_w}\right)^{1/4} \tag{4.31}$$

This correlation has been tested for $0.71 < \mathrm{Pr} < 380$, $3.5 < \mathrm{Re}_D < 7.6 \times 10^4$, and $1 < \mu_\infty/\mu_w < 3.2$. All the physical properties are evaluated at the free-stream temperature, except the viscosity μ_w, which is determined by the surface temperature. The total heat transfer rate from an isothermal sphere at T_w is $\dot{Q} = \bar{h}\,\pi D^2(T_w - T_\infty)$, while the average surface temperature of a sphere with uniform flux is $\bar{T}_w = T_\infty + q''/\bar{h}$. The drag coefficient for calculating the drag force on the sphere is presented in Figure 4.4.

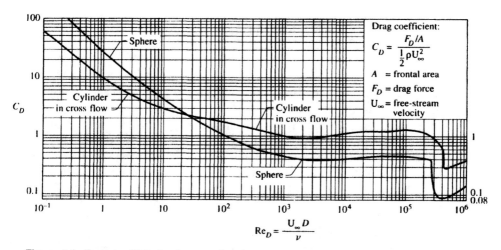

Figure 4.4 Drag coefficients of a smooth sphere and a single smooth cylinder in cross flow. (Drawn with permission, after the data compiled in H. Schlichting, *Boundary Layer Theory*, 4th ed., McGraw-Hill, New York, 1960.)

Arrays of Cylinders in Cross Flow. Figure 4.5 shows that for cylinders in cross flow the array geometry is characterized by the array type (aligned vs. staggered), the cylinder diameter D, the longitudinal pitch X_l, and the transversal pitch X_t. The average heat transfer coefficient $\bar{h}$ in *aligned arrays* is [9]

$$\frac{\bar{h}D}{k} = \begin{cases} 0.9C_n \text{Re}_D^{0.4}\text{Pr}^{0.36}\left(\dfrac{\text{Pr}}{\text{Pr}_w}\right)^{1/4}, & \text{Re}_D = 1\text{--}10^2 \\[2mm] 0.52C_n \text{Re}_D^{0.5}\text{Pr}^{0.36}\left(\dfrac{\text{Pr}}{\text{Pr}_w}\right)^{1/4}, & \text{Re}_D = 10^2\text{--}10^3 \\[2mm] 0.27C_n \text{Re}_D^{0.63}\text{Pr}^{0.36}\left(\dfrac{\text{Pr}}{\text{Pr}_w}\right)^{1/4}, & \text{Re}_D = 10^3\text{--}2 \times 10^5 \\[2mm] 0.033C_n \text{Re}_D^{0.8}\text{Pr}^{0.4}\left(\dfrac{\text{Pr}}{\text{Pr}_w}\right)^{1/4}, & \text{Re}_D = 2 \times 10^5\text{--}2 \times 10^6 \end{cases} \tag{4.32}$$

where C_n is a function of the total number of rows (n) in the array, Figure 4.6. The recommended correlation for *staggered arrays* is [9]

$$\frac{\bar{h}D}{k} = \begin{cases} 1.04C_n \text{Re}_D^{0.4}\text{Pr}^{0.36}\left(\dfrac{\text{Pr}}{\text{Pr}_w}\right)^{1/4}, & \text{Re}_D = 1\text{--}500 \\[2mm] 0.71C_n \text{Re}_D^{0.5}\text{Pr}^{0.36}\left(\dfrac{\text{Pr}}{\text{Pr}_w}\right)^{1/4}, & \text{Re}_D = 500\text{--}10^3 \\[2mm] 0.35C_n \text{Re}_D^{0.6}\text{Pr}^{0.36}\left(\dfrac{\text{Pr}}{\text{Pr}_w}\right)^{1/4}\left(\dfrac{X_t}{X_l}\right)^{0.2}, & \text{Re}_D = 10^3\text{--}2 \times 10^5 \\[2mm] 0.031C_n \text{Re}_D^{0.8}\text{Pr}^{0.36}\left(\dfrac{\text{Pr}}{\text{Pr}_w}\right)^{1/4}\left(\dfrac{X_t}{X_l}\right)^{0.2}, & \text{Re}_D = 2 \times 10^5\text{--}2 \times 10^6 \end{cases}$$

$$(4.33)$$

The $\bar{h}$ values calculated with Equations 4.32 and 4.33 have been averaged over the entire surface of the array.

The Reynolds number Re_D is based on the maximum average velocity, which occurs in the narrowest cross section formed by the array, $\text{Re}_D = V_{max}D/\nu$. For example, in the case of aligned cylinders the narrowest flow cross section forms in the plane that contains the centers of all the cylinders of one row. Conservation of mass requires that $V_{max} = U_\infty X_t/(X_t - D)$, where U_∞ is the uniform velocity of the stream that approaches the array. All the physical properties except $\text{Pr}_w = \text{Pr}(T_w)$ should be evaluated at the mean temperature of the fluid that flows through the spaces formed between the cylinders. The total heat transfer area of the array is $nm\pi DL$, where m is the number of cylinders in each row and L is the length of the cylinder.

The pressure drop experienced by the cross flow is proportional to the number of rows counted in the flow direction, n:

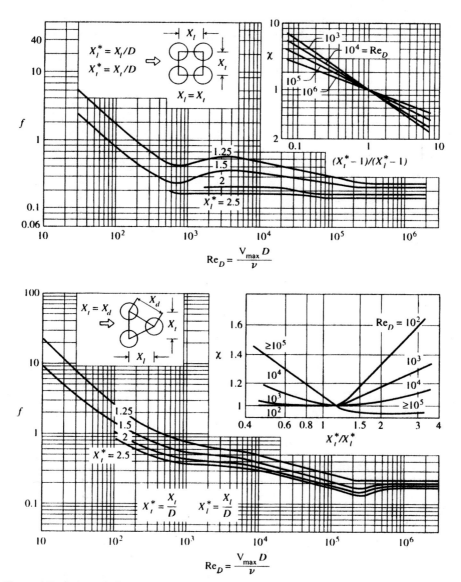

Figure 4.5 Arrays of aligned cylinders (top) and staggered cylinders (bottom): the coefficients f and χ for the array pressure drop formula, Equation 4.34. (From Ref. 9, with permission from John Wiley & Sons, Inc.)

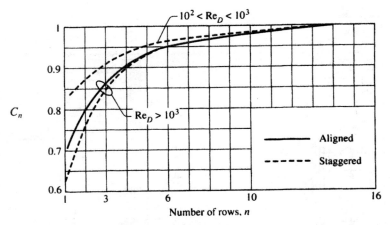

Figure 4.6 The effect of the number of rows on the array-averaged heat transfer coefficient for banks of cylinders in cross flow. (Drawn after Ref. 9.) [1].

$$\Delta p = n f \chi (\tfrac{1}{2}\rho V_{max}^2) \qquad (4.34)$$

The nondimensional factors f and χ are presented in Figure 4.5, which is valid for $n > 9$. Equations 4.32–4.34 and Figures 4.5 and 4.6 were constructed based on extensive experiments in which the test fluids were air, water, and several oils [9].

4.3.2 Internal Forced Convection

In flows through ducts, the heat transfer surface surrounds and guides the stream, and the convection process is said to be *internal*. For internal flow, Equation 4.2 can be reduced to obtain the *heat transfer coefficient*

$$h = \frac{q''}{T_w - T_m} \qquad (4.35)$$

In this definition q'' is the heat flux (heat transfer rate per unit area) through the wall where the temperature is T_w, and T_m is the *mean (bulk) temperature* of the stream:

$$T_m = \frac{1}{\text{U}A} \int_A \text{u}T \, dA \qquad (4.36)$$

The mean temperature is a weighted average of the local fluid temperature T over the duct cross section A. The role of weighting factor is played by the longitudinal fluid velocity u, which is zero at the wall and large in the center of the duct cross section. The mean velocity U is defined by

$$\text{U} = \frac{1}{A} \int_A \text{u} \, dA \qquad (4.37)$$

The mean velocity symbol U, which is standard in heat transfer, should not be confused with the symbols used for overall heat transfer coefficient and internal energy.

Laminar Flow Through a Duct. The velocity distribution in a duct has two distinct regions: first, the entrance region, where the walls are lined by growing boundary layers, and, farther downstream, the fully developed region, where the longitudinal velocity is independent of the position along the duct. It is assumed that the duct geometry (cross section A, internal wetted perimeter p) does not change with the longitudinal position. A measure of the duct cross section is the *hydraulic diameter*

$$D_h = \frac{4A}{p} \qquad (4.38)$$

The hydrodynamic entrance length X for laminar flow can be calculated with the formula

$$\frac{X}{D_h} \sim 0.05 \, \text{Re}_{D_h} \qquad (4.39)$$

where the Reynolds number is based on mean velocity and hydraulic diameter, $\text{Re}_{D_h} = \text{U}D_h/\nu$. The pressure drop experienced by a stream that flows through a duct of length L is given by

$$\Delta p = f \frac{4L}{D_h} \frac{\rho \text{U}^2}{2} \qquad (4.40)$$

where f is the *Fanning friction factor* introduced in Section 2.2.3. Equation 4.40 is a special case of Equations 2.30 and 2.31. Equation 4.40 applies equally to laminar and turbulent flow.

When the pipe is much longer than the flow entrance length, $L >> X$, the flow is fully developed along most of the length L, and the friction factor is independent of L. Table 4.3 shows that in *fully developed laminar flow* the friction factor behaves as $f = C/\mathrm{Re}_{D_h}$, where the value of the constant C is reported in the table. Two special cases of Table 4.3 and Equation 4.40 for fully developed laminar flow are

$$\Delta p = 32 \frac{\mu L U}{D^2} \quad \text{(round pipe diameter } D; \; D_h = D\text{)} \tag{4.41}$$

$$\Delta p = 12 \frac{\mu L U}{a^2} \quad \text{(parallel plate channel of spacing } a; \; D_h = 2a\text{)} \tag{4.42}$$

Table 4.3 Friction factors (f) and heat transfer coefficients (h) for hydrodynamically fully developed laminar flows through ducts [3]

Cross Section Shape	$C = f\,\mathrm{Re}_{D_h}$	hD_h/k	
		Uniform q''	Uniform T_w
60° triangle	13.3	3	2.35
square	14.2	3.63	2.89
circle	16	4.364	3.66
4a rectangle (a)	18.3	5.35	4.65
parallel plates	24	8.235	7.54
parallel plates, One side insulated	24	5.385	4.86

The heat transfer coefficient in fully developed flow (laminar or turbulent) is constant, that is, independent of longitudinal position. Table 4.3 lists the h values for fully developed laminar flow for two heating models: duct with uniform heat flux (q'') and duct with isothermal wall (T_w). Using these h values is appropriate when the duct length L is considerably greater than the thermal entrance X_T over which the temperature distribution is developing (i.e., changing) from one longitudinal position to the next. The thermal entrance length can be estimated using Equation 4.43 for the entire range of Prandtl number Pr [3]

$$\frac{X_T}{D_h} \sim 0.05 \ \mathrm{Pr} \ \mathrm{Re}_{D_h} \qquad (4.43)$$

Means for calculating the heat transfer coefficient for laminar duct flows in which X_T is not much smaller than L can be found in heat transfer textbooks [1–3].

Turbulent Flow Through a Duct. The flow through a straight duct ceases to be laminar when Re_{D_h} exceeds approximately 2300. It is also observed that the turbulent flow becomes fully developed hydrodynamically *and* thermally after a relatively short entrance distance:

$$X \sim X_T \sim 10 D_h \qquad (4.44)$$

In fully developed turbulent flow ($L \gg X$) the friction factor f is independent of L, as shown by the family of curves drawn for turbulent flow on the Moody chart (Figure 2.5) [10]. The chart is for a pipe with the internal diameter D. The turbulent flow curves can be used for ducts with other cross-sectional shapes, provided D is replaced by the appropriate hydraulic diameter of the duct, D_h. The pressure drop is evaluated using Equation 4.40. Two closed-form expressions for the friction factor in the smooth-wall limit are worth recording:

$$f \cong \begin{cases} 0.079 \ \mathrm{Re}_{D_h}^{-1/4}, & 3 \times 10^3 < \mathrm{Re}_{D_h} < 2 \times 10^4 \\ 0.046 \ \mathrm{Re}_{D_h}^{-1/5}, & 2 \times 10^4 < \mathrm{Re}_{D_h} < 10^6 \end{cases} \qquad (4.45)$$

The heat transfer coefficient h is constant in fully developed turbulent flow and can be estimated based on the *Colburn analogy* [11] between heat transfer and momentum transfer

$$\frac{h}{\rho c_p \mathrm{U}} = \frac{\frac{1}{2}f}{\mathrm{Pr}^{2/3}} \qquad (4.46)$$

This formula holds for Pr $\geqslant$ 0.5 and is to be used in conjunction with Figure 2.5, which supplies the f value. Equation 4.46 applies to ducts of various cross-sectional shapes, with wall surfaces having uniform temperature or uniform heat flux, and various degrees of roughness. A more accurate alternative to Equation 4.46 is given by the correlation [12]

$$\frac{hD_h}{k} = \frac{(f/2)\,(\mathrm{Re}_{D_h} - 10^3)\mathrm{Pr}}{1 + 12.7(f/2)^{1/2}\,(\mathrm{Pr}^{2/3} - 1)} \tag{4.47}$$

for which f is again supplied by Figure 2.5. Equation 4.47 is accurate within $\pm 10\%$ in the range $0.5 < \mathrm{Pr} < 10^6$ and $2300 < \mathrm{Re}_{D_h} < 5 \times 10^6$. Two simpler versions of Equation 4.47 for a smooth duct are [12]

$$\frac{hD_h}{k} = 0.0214(\mathrm{Re}_{D_h}^{0.8} - 100)\,\mathrm{Pr}^{0.4}$$

$$(0.5 < \mathrm{Pr} < 1.5,\ 10^4 < \mathrm{Re}_{D_h} < 5 \times 10^6) \tag{4.48}$$

$$\frac{hD_h}{k} = 0.012(\mathrm{Re}_{D_h}^{0.87} - 280)\,\mathrm{Pr}^{0.4}$$

$$(1.5 < \mathrm{Pr} < 500,\ 3 \times 10^3 < \mathrm{Re}_{D_h} < 10^6) \tag{4.49}$$

The Total Heat Transfer Rate. Referring to Figure 4.7, an energy balance can be applied to an elemental duct length dx to obtain

$$\frac{dT_m}{dx} = \frac{p}{A}\,\frac{q''}{\rho c_p U} \tag{4.50}$$

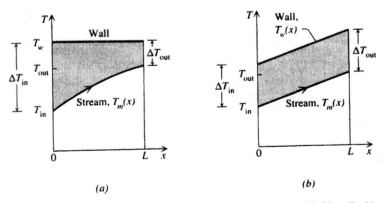

Figure 4.7 Distribution of temperature along a duct: (a) isothermal wall; (b) wall with uniform heat flux [1].

where p is the wetted internal perimeter of the duct and A is the cross-sectional area. Equation 4.50 holds for both laminar and turbulent flow. It can be combined with Equation 4.35 and the h value (furnished by Table 4.3 or Equations 4.46–4.49) to determine the longitudinal variation of the mean temperature of the stream, $T_m(x)$: When the duct wall is *isothermal* at T_w (Figure 4.7a), the total rate of heat transfer between the wall and a stream with the mass flow rate $\dot{m}$ is

$$\dot{Q} = \dot{m}c_p\, \Delta T_{in} \left[1 - \exp\left(-\frac{hA_w}{\dot{m}c_p} \right) \right] \tag{4.51}$$

where A_w is the wall surface, $A_w = pL$.

A more general alternative to Equation 4.51 is

$$\dot{Q} = hA_w\, \Delta T_{lm} \tag{4.52}$$

where ΔT_{lm} is the *log-mean temperature difference*

$$\Delta T_{lm} = \frac{\Delta T_{in} - \Delta T_{out}}{\ln\, (\Delta T_{in}/\Delta T_{out})} \tag{4.53}$$

Equations 4.52 and 4.53 are more general than Equation 4.51 because they apply when T_w is not constant, for example, when $T_w(x)$ is the temperature of a second stream in counterflow with the stream whose mass flow rate is $\dot{m}$. Equation 4.51 can be deduced from Equations 4.52 and 4.53 by writing $T_w = \text{const}$, and $\Delta T_{in} = T_w - T_{in}$ and $\Delta T_{out} = T_w - T_{out}$. When the wall *heat flux* is *uniform* (Figure 4.7b), the local temperature difference between the wall and the stream does not vary with the longitudinal position x: $T_w(x) - T_m(x) = \Delta T$ (constant). In particular, $\Delta T_{in} = \Delta T_{out} = \Delta T$, and Equation 4.53 yields $\Delta T_{lm} = \Delta T$. Equation 4.52 reduces in this case to $\dot{Q} = hA_w\, \Delta T$.

4.3.3 Natural Convection

In natural convection, or free convection, the fluid flow is driven by the effect of buoyancy. This effect is distributed throughout the fluid and is associated with the tendency of most fluids to expand when heated. The heated fluid becomes less dense and flows upward, while packets of cooled fluid become more dense and sink.

Vertical Plane Wall. Consider first the natural convection boundary layer formed along an *isothermal* vertical wall of temperature T_w, height H, and width W, which is in contact with a fluid whose temperature away from the wall is T_∞. Experiments show that the transition from the laminar section to the turbulent section of the boundary layer occurs at the altitude y (between the leading edge, $y = 0$, and the trailing edge, $y = H$), where [13]

$$\mathrm{Ra}_y \sim 10^9 \mathrm{Pr} \qquad (10^{-3} < \mathrm{Pr} < 10^3) \tag{4.54}$$

In this expression, Ra_y is the *Rayleigh number based on temperature difference* defined by

$$\mathrm{Ra}_y = \frac{g\beta(T_w - T_\infty)y^3}{\alpha\nu} \tag{4.55}$$

where β is the coefficient of volumetric thermal expansion, $\beta = (-1/\rho)(\partial\rho/\partial T)_p$. If the fluid behaves as an ideal gas, β equals $1/T$, where T is expressed in K (or °R). The heat transfer results assembled in this section are valid when $|\beta(T_w - T_\infty)| \ll 1$. The boundary layer remains laminar over its entire height H when $\mathrm{Ra}_H < 10^9 \mathrm{Pr}$.

An empirical correlation giving the average heat transfer coefficient for laminar and turbulent flow is [14]

$$\frac{\bar{h}H}{k} = \left\{ 0.825 + \frac{0.387\,\mathrm{Ra}_H^{1/6}}{[1 + (0.492/\mathrm{Pr})^{9/16}]^{8/27}} \right\}^2 \tag{4.56}$$

This correlation is recommended for $10^{-1} < \mathrm{Ra}_H < 10^{12}$ and all Prandtl numbers. For gases with $\mathrm{Pr} = 0.72$, Equation 4.56 reduces to

$$\frac{\bar{h}H}{k} = (0.825 + 0.325\,\mathrm{Ra}_H^{1/6})^2 \qquad (\mathrm{Pr} = 0.72) \tag{4.57}$$

A vertical wall that releases the *uniform heat flux q''* into the fluid has a temperature that increases with altitude, $T_w(y)$. The relation between q'' and the wall–fluid temperature difference $(\bar{T}_w - T_\infty)$ is represented adequately by Equations 4.56 and 4.57, provided $\bar{h}$ is replaced by $q''(\bar{T}_w - T_\infty)$, and $\mathrm{Ra}_H = g\beta(\bar{T}_w - T_\infty)H^3/\alpha\nu$. Note that the temperature difference $(\bar{T}_w - T_\infty)$ is averaged over the wall height H.

Inclined Walls. Assume that a plane wall makes an angle ϕ with the vertical direction, and that this angle is restricted to the range $-60° < \phi < 60°$. In the laminar regime, Equations 4.56 and 4.57 continue to hold for both isothermal and uniform-flux walls provided g is replaced by $g \cos \phi$ in the Rayleigh number definition: $\text{Ra}_H = g \cos \phi\ \beta(T_w - T_\infty)H^3/\alpha\nu$. In the turbulent regime the heat transfer measurements are correlated sufficiently well by Equations 4.56 and 4.57 with Ra_H based on g, as in Equation 4.55, $\text{Ra}_H = g\beta(T_w - T_\infty)H^3/\alpha\nu$.

Horizontal Walls. Measurements for the average heat transfer coefficient have been correlated using the *characteristic length* of the horizontal wall: $L = A/p$, where A is the wall area and p is the perimeter of A. The following formulas are valid for $\text{Pr} > 0.5$, for both isothermal surfaces, $\text{Ra}_L = g\beta(T_w - T_\infty)L^3/\alpha\nu$, and uniform-flux surfaces, $\text{Ra}_L = g\beta(\bar{T}_w - T_\infty)L^3/\alpha\nu$:

$$\frac{\bar{h}L}{k} = \begin{cases} 0.54\ \text{Ra}_L^{1/4} & (10^4 < \text{Ra}_L < 10^7) \\ 0.15\ \text{Ra}_L^{1/3} & (10^7 < \text{Ra}_L < 10^9) \end{cases}$$

(hot surface facing upward,
or cold surface facing downward [15]) (4.58)

$$\frac{\bar{h}L}{k} = 0.27\ \text{Ra}_L^{1/4} \qquad (10^5 < \text{Ra}_L < 10^{10})$$

(hot surface facing downward,
or cold surface facing upward [2]) (4.59)

In the case of a wall with uniform flux q'', the wall temperature $\bar{T}_w$ can be calculated by substituting $\bar{h} = q''/(\bar{T}_w - T_\infty)$ into Equations 4.58 and 4.59.

Horizontal Cylinder. The recommended correlation for the average heat transfer coefficient for a single horizontal cylinder (diameter D, length L) immersed in a fluid is [14]

$$\frac{\bar{h}D}{k} = \left\{ 0.6 + \frac{0.387\ \text{Re}_D^{1.6}}{[1 + (0.559/\text{Pr})^{9/16}]^{8/27}} \right\}^2 \qquad (4.60)$$

This is valid for $10^{-5} < \text{Ra}_D < 10^{12}$ and the entire Pr range. When the cylinder surface is isothermal (T_w), $\text{Ra}_D = g\beta(T_w - T_\infty)D^3/\alpha\nu$, and the total heat transfer rate is $\dot{Q} = \bar{h}\pi DL(T_w - T_\infty)$. In the case of uniform heat flux (q''), Ra_D is based on the average temperature difference, $\text{Ra}_D = g\beta(\bar{T}_w - T_\infty)D^3/\alpha\nu$, and $\bar{h} = q''/(\bar{T}_w - T_\infty)$.

Sphere. The average heat transfer coefficient for a sphere of diameter D immersed in a fluid is given by [16]

$$\frac{\bar{h}D}{k} = 2 + \frac{0.589 \, \text{Ra}_D^{1/4}}{[1 + (0.469/\text{Pr})^{9/16}]^{4/9}} \tag{4.61}$$

in the range $\text{Pr} > 0.7$ and $\text{Ra}_D < 10^{11}$. Additional relations for an isothermal (T_w) sphere are $\text{Ra}_D = g\beta(T_w - T_\infty)D^3/\alpha\nu$ and $\dot{Q} = \bar{h}\pi D^2(T_w - T_\infty)$, and, for a sphere with uniform heat flux (q''), $\text{Ra}_D = g\beta(\overline{T}_w - \overline{T}_\infty)D^3/\alpha\nu$ and $\bar{h} = q''/(\overline{T}_w - T_\infty)$.

Vertical Channels. The following results apply to a fully developed laminar flow driven by buoyancy through a vertical duct of hydraulic diameter D_h, height H, and inner surface temperature T_w. The top and bottom ends of the duct are open to a fluid at a temperature T_∞. Following Reference 3, it is assumed that the duct is sufficiently slender so that $\text{Ra}_{D_h} < H/D_h$, where $\text{Ra}_{D_h} = g\beta(T_w - T_\infty)D_h^3/\alpha\nu$. The average heat transfer coefficient depends on the shape of the duct cross section:

$$\frac{\bar{h}H/k}{\text{Ra}_{D_h}} = \begin{cases} \dfrac{1}{192} & \text{(parallel plates)} \\[6pt] \dfrac{1}{128} & \text{(round)} \\[6pt] \dfrac{1}{113.6} & \text{(square)} \\[6pt] \dfrac{1}{106.4} & \text{(equilateral triangle)} \end{cases} \tag{4.62}$$

The total heat transfer rate between the duct and the stream is approximately $\dot{Q} \cong \dot{m}c_p(T_w - T_\infty)$ when the group $\bar{h}A_w/\dot{m}c_p$ is greater than 1 (see Equation 4.51), where the mass flow rate is $\dot{m} = \rho AU$, and the duct cross-sectional area is A. The mean velocity U can be estimated using Equation 4.40 in which $\Delta p/L$ is now replaced by $\rho g\beta(T_w - T_\infty)$. Many other results for laminar and turbulent flow and for mixed (forced and natural) convection in vertical channels are reviewed in Reference 17.

Enclosures Heated from the Side. Consider a two-dimensional vertical rectangular enclosure of height H and horizontal spacing L, which is filled with a fluid. The vertical walls are maintained at different temperatures, T_h and T_c, while the top and bottom walls are assumed insulated. If W is the enclosure width (perpendicular to the $H \times L$ cross section), the total rate of heat transfer $\dot{Q}$ from T_h to T_c is given by the correlations [1]

$$\frac{\dot{Q}/W}{k(T_h - T_c)} = \begin{cases} 0.22 \left(\dfrac{\text{Pr}}{0.2 + \text{Pr}}\, \text{Ra}_H\right)^{0.28} \left(\dfrac{L}{H}\right)^{0.09} \\[2mm] \left(2 < \dfrac{H}{L} < 10,\ \text{Pr} < 10^5,\ \text{Ra}_H < 10^{13}\right) \\[4mm] 0.18 \left(\dfrac{\text{Pr}}{0.2 + \text{Pr}}\, \text{Ra}_H\right)^{0.29} \left(\dfrac{L}{H}\right)^{-0.13} \\[2mm] \left(1 < \dfrac{H}{L} < 2,\ \text{Pr} < 10^5,\ \dfrac{\text{Pr}}{0.2 + \text{Pr}}\, \text{Ra}_H \left(\dfrac{L}{H}\right)^3 > 10^3\right) \end{cases}$$

$$(4.63)$$

where the Rayleigh number is based on the height of the enclosure, $\text{Ra}_H = g\beta(T_h - T_c)H^3/\alpha\nu$. Additional results for shallow enclosures ($H/L < 1$), inclined enclosures, concentric cylinders and spheres, and enclosures with uniform flux on the heated and cooled walls have been compiled in References 1 and 3.

Enclosures Heated from Below. In enclosures heated from the side natural convection is present as soon as a very small temperature difference is imposed between the two side walls. By contrast, in enclosures heated from below the imposed temperature difference must exceed a critical value before the first signs of buoyant flow are detected. Consider an enclosure formed between two large horizontal walls with area A and vertical spacing H and filled with a fluid that expands upon heating. Assume the bottom wall (at T_H) is heated and the top wall (at T_L) is cooled. Natural convection occurs if $\text{Ra}_H > 1708$, where $\text{Ra}_H = g\beta(T_h - T_c)H^3/\alpha\nu$. At high Rayleigh numbers the total heat transfer rate $\dot{Q}$ between the two walls can be estimated by using the empirical correlation [18]

$$\frac{\dot{Q}/A}{k(T_h - T_c)/H} = 0.069\, \text{Ra}_H^{1/3}\text{Pr}^{0.074} \qquad (4.64)$$

which is valid in the range $3 \times 10^5 < \text{Ra}_H < 7 \times 10^9$. When natural convection is absent, $\text{Ra}_H < 1708$, the transfer of heat is by pure conduction, and $\dot{Q}/A = k(T_h - T_c)/H$.

4.3.4 Condensation

In this section we review several results for heat transfer in circumstances where a gas is cooled sufficiently that it condenses to a liquid. Condensation is an example of *convection with change of phase*. The following results are based on the assumptions that the condensate flows as a *continuous film* over a surface at T_w, and the gas is at the saturation temperature T_{sat} of the condensing substance.

Vertical Wall. The total rate of condensation on a vertical wall of height L can be estimated using Figure 4.8, which was constructed [1] based on a correlation presented in Reference 19 for laminar as well as turbulent film condensation. The rate at which the condensate leaves the bottom edge of the wall, per unit width, is $\Gamma(L)$. The calculation begins with finding the *condensation driving parameter B*

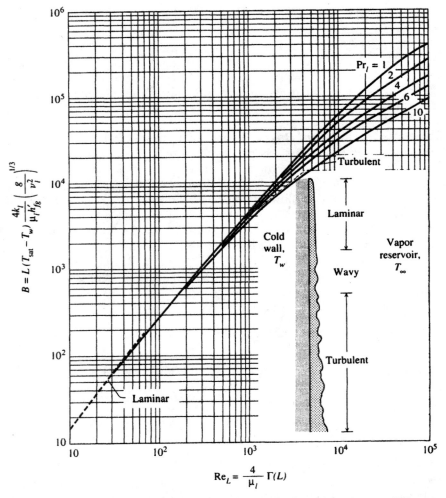

Figure 4.8 Film condensation on a vertical surface: the total condensation rate $\Gamma(L)$ versus the condensation driving parameter B [1].

$$B = L(T_{sat} - T_w) \frac{4k_l}{\mu_l h'_{fg}} \left(\frac{g}{\nu_l^2}\right)^{1/3} \qquad (4.65)$$

in which the subscript l indicates liquid properties, and $h'_{fg} = h_{fg} + 0.68c_{p,l}(T_{sat} - T_w)$, where h_{fg} is the enthalpy of vaporization at T_{sat}. Figure 4.8 provides the Reynolds number of the liquid film at the bottom of the wall, $\mathrm{Re}_L = 4\Gamma(L)/\mu_l$, which is the nondimensional counterpart of the unknown of the problem, the total flow rate $\Gamma(L)$. The flow is laminar at the bottom edge of the wall if $\mathrm{Re}_L < 30$, wavy if $30 < \mathrm{Re}_L < 1800$, and turbulent if $\mathrm{Re}_L > 1800$. From an energy balance, the total heat transfer rate per unit of length q' absorbed by the vertical wall is

$$q' = h'_{fg} \Gamma(L) \qquad (4.66)$$

Horizontal Cylinders. The total rate of condensate accumulated on n horizontal cylinders (temperature T_w, diameter D) aligned in a vertical column can be calculated in two steps. First, if the film of condensate is laminar on all the cylinders, the heat transfer coefficient $\bar{h}$ averaged over all the cylindrical surfaces in the vertical column is given by [20]

$$\frac{\bar{h}D}{k_l} = 0.729 \left[\frac{D^3 h'_{fg} g (\rho_l - \rho_v)}{n k_l \nu_l (T_{sat} - T_w)}\right]^{1/4} \qquad (4.67)$$

The subscripts l and v indicate liquid and vapor properties. Second, the total heat transfer rate per unit of cylinder length is $q' = \bar{h}n \pi D(T_{sat} - T_w)$ and using an energy balance as in Equation 4.66, the condensate mass flow rate per unit of cylinder length is q'/h'_{fg}. The case of the single horizontal cylinder is represented by $n = 1$ in Equation 4.67.

Sphere. The condensation rate on a sphere of diameter D and at temperature T_w in the laminar film regime can be estimated using the average heat transfer coefficient calculated from the following expression [21]

$$\frac{\bar{h}D}{k_l} = 0.815 \left[\frac{D^3 h'_{fg} g(\rho_l - \rho_v)}{k_l \nu_l (T_{sat} - T_w)}\right]^{1/4} \qquad (4.68)$$

The total heat transfer rate into the spherical wall is $\dot{Q} = \bar{h} \pi D^2(T_{sat} - T_w)$, and the condensate mass flow rate is $\dot{Q}/h'_{fg}$. This calculation is valid for condensation on the outside or inside of the spherical surface, provided the film thickness is small relative to the sphere diameter.

Horizontal Flat Wall. For a horizontal surface A and perimeter p we can define a characteristic length as $L = A/p$. If this surface has the temperature T_w and faces upward toward saturated vapor at T_{sat}, the average heat transfer coefficient for laminar film condensation is [1]

$$\frac{\overline{h}L}{k_l} \cong 0.8 \left[\frac{L^3 \, h'_{fg} \, g \, (\rho_l - \rho_v)}{k_l \, \nu_l \, (T_{sat} - T_w)} \right]^{1/5} \tag{4.69}$$

The total heat transfer rate and condensation mass flow rate are, respectively, $\dot{Q} = \overline{h}A(T_{sat} - T_w)$ and $\dot{Q}/h'_{fg}$.

Drop Condensation. The heat transfer coefficient during dropwise condensation is roughly 10 times larger than during film condensation under the same conditions. Since surface coatings used to promote drop formation are gradually removed by the scraping action of drop movement, it is a good idea to base the calculation of $\overline{h}$ on the worst-case assumption that the condensate flows as a film.

4.3.5 Boiling

Heat transfer by boiling occurs when the surface temperature (T_w) is sufficiently higher than the saturation temperature (T_{sat}) of the liquid with which it comes in contact. In this section we review the main results for *pool boiling,* where the hot surface is immersed in a pool of initially stagnant saturated liquid. Figure 4.9 shows qualitatively the relationship between the surface heat flux q'' and the excess temperature $T_w - T_{sat}$. The various pool-boiling regimes are related to the way in which the newly generated vapor bubbles interact in the vicinity of the hot surface.

Nucleate Boiling. Experimental excess temperature data agree within $\pm 25\%$ with $T_w - T_{sat}$ based on the correlation [22]

$$T_w - T_{sat} = \frac{h_{fg}}{c_{p,l}}(\mathrm{Pr}_l)^s \, C_{sf} \left[\frac{q''}{\mu_l \, h_{fg}} \left(\frac{\sigma}{g(\rho_l - \rho_v)} \right)^{1/2} \right]^{1/3} \tag{4.70}$$

This correlation applies to clean surfaces and depends on two empirical constants, C_{sf} and s, which are listed in Table 4.4. The subscripts l and v indicate

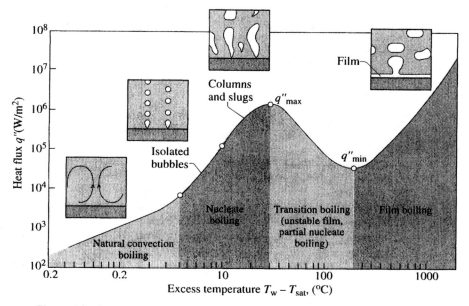

Figure 4.9 The four regimes of pool boiling in water at atmospheric pressure [1].

saturated liquid and saturated vapor, and σ is the surface tension of the liquid in contact with its own vapor. In the reverse case, where the excess temperature is specified, the heat flux calculated from Equation 4.70 agrees within a factor of 2 with actual heat flux measurements. As an engineering approximation, Equation 4.70 can be applied without regard for the shape and orientation of the hot surface.

Peak Heat Flux for Nucleate Boiling. The recommended correlation for the peak heat flux for nucleate boiling on a large horizontal surface is [24]

$$q''_{max} = 0.149 h_{fg} \rho_v^{1/2} [\sigma g(\rho_l - \rho_v)]^{1/4} \tag{4.71}$$

This formula is independent of the surface material. It applies to a surface that is considerably larger than the vapor bubble. Equation 4.71 can be used also for a horizontal cylinder by replacing the 0.149 factor with 0.116 [25].

Minimum Heat Flux for Film Boiling. For a horizontal plane surface the minimum heat flux (Figure 4.9) can be estimated using [26]

$$q''_{min} = 0.09 h_{fg} \rho_v \left[\frac{\sigma g(\rho_l - \rho_v)}{(\rho_l + \rho_v)^2} \right]^{1/4} \tag{4.72}$$

Table 4.4 Constants for Rohsenow's nucleate pool-boiling correlation [22, 23]

Liquid–Surface Combination	C_{sf}	s
Water–copper		
Polished	0.013	1.0
Scored	0.068	1.0
Emery polished, paraffin treated	0.015	1.0
Water–stainless steel		
Ground and polished	0.008	1.0
Chemically etched	0.013	1.0
Mechanically polished	0.013	1.0
Teflon pitted	0.0058	1.0
Water–brass	0.006	1.0
Water–nickel	0.006	1.0
Water–platinum	0.013	1.0
CCl_4–copper	0.013	1.7
Benzene–chromium	0.010	1.7
n-Pentane–chromium	0.015	1.7
n-Pentane–copper		
Emery polished	0.0154	1.7
Emery rubbed	0.0074	1.7
Lapped	0.0049	1.7
n-Pentane–nickel		
Emery polished	0.013	1.7
Ethyl alcohol–chromium	0.0027	1.7
Isopropyl alcohol–copper	0.0025	1.7
35% K_2CO_3–copper	0.0054	1.7
50% K_2CO_3–copper	0.0027	1.7
n-Butyl alcohol–copper	0.0030	1.7

This correlation does not depend on the excess temperature. The calculated minimum heat flux agrees within 50% with laboratory measurements at low and moderate pressures, and the accuracy deteriorates as the pressure increases. The surface roughness has a negligible effect on q''_{min} because the asperities are covered by the vapor film.

Film Boiling. The correlations developed for film boiling have the same analytical form as the formulas for film condensation and laminar boundary layer natural convection. For example, the average heat transfer coefficient on an immersed horizontal cylinder, $\bar{h}_D$, is given by [27]

$$\frac{\bar{h}_D D}{k_v} = 0.62 \left[\frac{D^3 \, h'_{fg} \, g(\rho_l - \rho_v)}{k_v \, \nu_v (T_w - T_{sat})} \right]^{1/4} \tag{4.73}$$

where $h'_{fg} = h_{fg} + 0.4c_{p,v}(T_w - T_{sat})$. The formula for film boiling on an immersed sphere is the same as Equation 4.73 except that the 0.62 factor is replaced by 0.67 [24]. The average heat flux is $q'' = \overline{h}_D (T_w - T_{sat})$.

As the surface temperature increases, the effect of *thermal radiation* (Section 4.4) across the vapor film contributes more and more to the overall heat transfer rate. The thermal radiation effect can be incorporated into an effective heat transfer coefficient $\overline{h}$ [27]:

$$\overline{h} = \overline{h}_D \left(\frac{\overline{h}_D}{\overline{h}}\right)^{1/3} + \overline{h}_{rad} \qquad (4.74)$$

for which $\overline{h}_D$ is furnished by Equation 4.73, and $\overline{h}_{rad}$ is the radiation coefficient

$$\overline{h}_{rad} = \frac{\sigma\varepsilon_w(T_w^4 - T_{sat}^4)}{T_w - T_{sat}} \qquad (4.75)$$

where σ is the *Stefan–Boltzmann constant*, $\sigma = 5.669 \times 10^{-8}$ W/m^2·K^4, ε_w is the *emissivity* of the heater surface, and the temperatures (T_w, T_{sat}) must be expressed in K. The heat flux is $q'' = \overline{h}(T_w - T_{sat})$. In water, the thermal radiation effect begins to be felt as $T_w - T_{sat}$ increases above 550–660°C [28].

Example 4.2 It has been proposed to reduce the drag experienced by a submarine by heating its outer surface electrically to a high enough temperature that the viscosity of the water in the adjacent hydrodynamic boundary layer decreases. Evaluate the merit of this proposal assuming that the surface temperature is raised to 90°C, the water free-stream temperature is 10°C, and the vessel's speed is 10 m/s.

Solution

MODEL

1. The outer surface is a smooth, plane wall of length L.
2. The hydrodynamic boundary layer is turbulent, and the length of the laminar leading section is negligible.
3. The turbulence of the free stream can be neglected.
4. The variation of temperature across the boundary layer is sufficiently small so that the film temperature can be used to evaluate the water properties.

ANALYSIS. The work rate concept of mechanics allows the power (Section 2.2) expended, per unit of width, on moving the flat surface through the fluid to be evaluated as the product of force and velocity: $\dot{W}' = F'U$, where F' is the total tangential force, per unit of width, experienced by the wall and U is the wall speed. Here, F' can be evaluated in terms of the average shear stress: $F' = \bar{\tau}L$. Then, with Equation 4.29, which at high Reynolds numbers reduces to $\bar{\tau} = 0.037\rho U^2 \, \mathrm{Re}_L^{-4/5}$, the power required per unit of width is

$$\dot{W}' = F'U = 0.037\rho U^3 L \left(\frac{\nu}{UL}\right)^{1/5} \tag{1}$$

Using subscripts c and h to represent the cold-wall and hot-wall conditions, respectively, the power requirement changes according to the ratio

$$\frac{\dot{W}'_h}{\dot{W}'_c} = \frac{\rho_h}{\rho_c} \left(\frac{\nu_h}{\nu_c}\right)^{1/5}$$

The properties of cold water (10°C) are $\rho_c \cong 1 \text{ g/cm}^3$, $\mu_c = 0.013 \text{ g/cm·s}$. When the surface is heated, the water film temperature is (90°C + 10°C)/2 = 50°C, with the corresponding properties, $\rho_h \cong 1 \text{ g/cm}^3$, $\mu_h = 0.00548 \text{ g/cm·s}$. The ratio is then

$$\frac{\dot{W}'_h}{\dot{W}'_c} = \frac{1}{1} \left(\frac{0.00548}{0.013}\right)^{1/5} = 0.84$$

which shows that by heating the surface to 90°C we can reduce the drag power by roughly 16%.

The power that has been saved by heating the water boundary layer can be calculated by using Equation 1 to estimate $\dot{W}'_h - \dot{W}'_c$:

$$\dot{W}'_c - \dot{W}'_h = 0.037\rho U^{14/5} L^{4/5} \nu_c^{1/5} \left[1 - \left(\frac{\nu_h}{\nu_c}\right)^{1/5}\right] \tag{2}$$

The electric power per unit of width for heating the surface is $\bar{q}''L = \bar{h}\,\Delta T\,L$, where $\bar{h}$ is given by Equation 4.28 simplified for high Reynolds numbers:

$$\frac{\bar{h}L}{k_h} = 0.037 \, \mathrm{Pr}_h^{1/3} \left(\frac{UL}{\nu_h}\right)^{4/5} \tag{3}$$

The result is

$$\bar{q}''L = 0.037 k_h \, \Delta T \, \text{Pr}_h^{1/3} \left(\frac{UL}{\nu_h} \right)^{4/5} \tag{4}$$

where k_h and Pr_h are also evaluated at the 50°C film temperature,

$$k_h = 0.64 \text{ W/m·K}, \qquad \text{Pr}_h = 3.57$$

By dividing Equations 2 and 4 we obtain a dimensionless measure of how effectively the surface heating has been converted into a power savings:

$$\frac{\dot{W}'_c - \dot{W}'_h}{\bar{q}''L} = \frac{U^2}{c \, \Delta T} \text{Pr}_h^{2/3} \left(\frac{\nu_c}{\nu_h} \right)^{1/5} \left[1 - \left(\frac{\nu_h}{\nu_c} \right)^{1/5} \right]$$

In this expression, $c = 4.18$ kJ/kg·K, $\Delta T = 90°C - 10°C = 80°C$, $(\nu_h/\nu_c)^{1/5} = 0.84$, and $U = 10$ m/s, giving

$$\frac{\dot{W}'_c - \dot{W}'_h}{\bar{q}''L} = 1.3 \times 10^{-4}$$

COMMENT. This example illustrates how an apparently complex suggestion originating, for example, in a brainstorming session (Section 1.5.1) might be evaluated using elementary modeling. The analysis shows that the power savings are much smaller than the electrical heating requirement when the speed is 10 m/s.

4.4 RADIATION

Unlike conduction and convection, radiation is the heat transfer mechanism by which bodies can exchange energy without making direct contact. Net heat transfer by radiation can occur even when the space between two bodies at different temperatures (T_1, T_2) is completely evacuated. The two bodies emit their respective streams of thermal radiation in all directions. Only a fraction of the energy stream emitted by body (T_1) is intercepted and possibly absorbed by body (T_2). This fraction depends not only on the shapes and sizes of the two bodies but also on their relative position, on the condition (e.g., smoothness, cleanliness) of their surfaces, and on the nature of their surroundings. Similarly, only a fraction of the radiation emitted by body (T_2) is intercepted and possibly absorbed by body (T_1).

4.4.1 Blackbody Radiation

The surfaces of solids can be classified according to their ability to absorb the radiation energy streams that strike them. Figure 4.10 shows the *total irradiation G*, or the total radiation heat flux that impinges on the elemental area *dA*. A portion of the total irradiation (αG) can be absorbed, another portion (ρG) can be reflected back, and the remainder (τG) can pass through the body and exit through the other side. Conservation of energy for a system that encloses *dA* requires $\alpha G + \rho G + \tau G = G$, or

$$\alpha + \rho + \tau = 1 \qquad (4.76)$$

where α is the *total absorptivity*, ρ is the *total reflectivity*, and τ is the *total transmissivity* of the material within the system. A material is *opaque* when its total transmissivity is zero. If, in addition, the total reflectivity is zero, the surface is said to be *black*. The body defined by a black surface is referred to as a *blackbody*, and is represented by

$$\alpha = 1 \qquad (\text{black, } \rho = \tau = 0) \qquad (4.77)$$

The *monochromatic hemispherical emissive power* of a black surface, $E_{b,\lambda}$, represents the energy emitted by the surface in all the directions of the hemisphere per unit time, wavelength λ, and surface area. The parameter $E_{b,\lambda}$ has the units W/m³, or Btu/h·ft³, and is given by

$$E_{b,\lambda} = \frac{C_1 \lambda^{-5}}{\exp(C_2/\lambda T) - 1} \qquad (4.78)$$

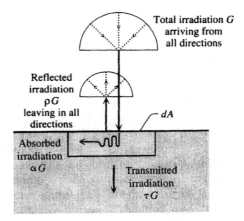

Figure 4.10 The definition of total absorptivity, reflectivity, and transmissivity [1].

where $C_1 = 3.742 \times 10^{-16}$ W·m^2 = 1.375×10^{-14} Btu ft^2/h, and $C_2 = 1.439 \times 10^{-2}$ m·K = 9.44×10^{-3} ft·°R. The *total hemispherical emissive power* E_b is the energy emitted per unit time and surface area,

$$E_b = \int_0^\infty E_{b,\lambda} \, d\lambda = \sigma T^4 \qquad (4.79)$$

The product σT^4 is commonly called the *potential* of the black surface of absolute temperature T.

4.4.2 Geometric View Factors

Assume that A_1 and A_2 are the areas of the surfaces of bodies (T_1) and (T_2). The geometric view factor F_{12} is defined as the ratio

$$F_{12} = \frac{\text{radiation leaving } A_1 \text{ and being intercepted by } A_2}{\text{radiation leaving } A_1 \text{ in all directions}} \qquad (4.80)$$

Accordingly, the value of a geometric view factor will fall between 0 and 1. Figure 4.11 and Table 4.5 present the view factors of some of the most common two-surface configurations encountered in radiation heat transfer calculations. Several other configurations are discussed in the literature [29, 30].

In some instances it is possible to deduce the F_{12} value by manipulating the known view factors of one or more related configurations. Essential to this approach are the relationships expressed by Equations 4.81–4.83, also known as *view factor algebra*. These relationships are purely geometric, that is, independent of surface description (black versus nonblack).

Reciprocity. If we happen to know F_{21} from a chart, formula, or table, we can calculate F_{12} if we also know the areas involved in the configuration (A_1, A_2),

$$A_1 F_{12} = A_2 F_{21} \qquad (4.81)$$

Additivity. Let us assume that the surface A_2 is made up of n smaller pieces, $A_2 = A_{2_1} + A_{2_2} + \cdots + A_{2_n}$. The view factor from A_1 to A_2 is the sum of the individual view factors from A_1 to each component of A_2,

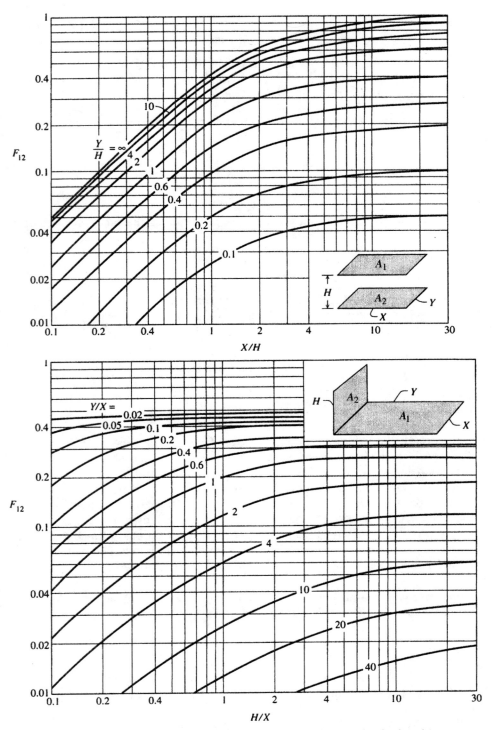

Figure 4.11 The geometric view factor between two parallel rectangles (top) and two perpendicular rectangles with a common edge (bottom) [1].

Table 4.5 Sampling of geometric view factors [1, 29, 30]

Configuration	Geometric View Factor
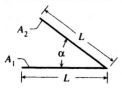	Two infinitely long plates of width L, joined along one of the long edges: $$F_{12} = F_{21} = 1 - \sin\frac{\alpha}{2}$$
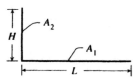	Two infinitely long plates of different widths (H, L), joined along one of the long edges and with a 90° angle between them: $$F_{12} = \tfrac{1}{2}[1 + x - (1 + x^2)^{1/2}]$$ where $x = H/L$
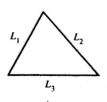	Triangular cross section enclosure formed by three infinitely long plates of different widths (L_1, L_2, L_3): $$F_{12} = \frac{L_1 + L_2 - L_3}{2L_1}$$
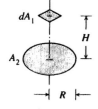	Disc and parallel infinitesimal area positioned on the disc centerline: $$F_{12} = \frac{R^2}{H^2 + R^2}$$
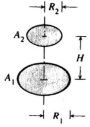	Parallel discs positioned on the same centerline: $$F_{12} = \frac{1}{2}\left\{ X - \left[X^2 - 4\left(\frac{x_2}{x_1}\right)^2 \right]^{1/2} \right\}$$ where $x_1 = \dfrac{R_1}{H}$, $x_2 = \dfrac{R_2}{H}$, and $X = 1 + \dfrac{1 + x_2^2}{x_1^2}$
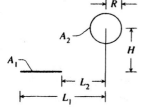	Infinite cylinder parallel to an infinite plate of finite width $(L_1 - L_2)$: $$F_{12} = \frac{R}{L_1 - L_2}\left(\tan^{-1}\frac{L_1}{H} - \tan^{-1}\frac{L_2}{H} \right)$$

Table 4.5   *(Continued)*

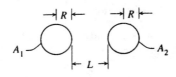

Two parallel and infinite cylinders:

$$F_{12} = F_{21} = \frac{1}{\pi}\left[\left(X^2 - 1\right)^{1/2} + \sin^{-1}\left(\frac{1}{X}\right) - X\right]$$

where $X = 1 + \dfrac{L}{2R}$

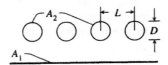

Row of equidistant infinite cylinders parallel to an infinite plate:

$$F_{12} = 1 - (1 - x^2)^{1/2} + x \tan^{-1}\left(\frac{1 - x^2}{x^2}\right)^{1/2}$$

where $x = \dfrac{D}{L}$

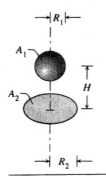

Sphere and disc positioned on the same centerline:

$$F_{12} = \tfrac{1}{2}[1 - (1 + x^2)^{-1/2}]$$

where $x = \dfrac{R_2}{H}$

$$F_{12} = F_{12_1} + F_{12_2} + \cdots + F_{12_n} \qquad (4.82)$$

Enclosure. Consider a configuration in which the surface of interest (A_1) and all the surfaces that surround it $(A_2, A_3, \ldots, A_n)$ form a complete enclosure. In this case, the sum of the view factors is unity:

$$F_{11} + F_{12} + \cdots + F_{1n} = 1 \qquad (4.83)$$

The view factor F_{11} is zero if the surface A_1 is plane or convex and positive if A_1 is concave. Equation 4.83 can be written for each of the surfaces that participates in the enclosure (i.e., not just for A_1), to yield a system of n equations.

4.4.3 Diffuse-Gray Surface Model

The monochromatic hemispherical emissive power of a real surface, E_λ, is smaller than that of a black surface at the same temperature and wavelength, $E_{b,\lambda}$. The ratio $E_\lambda/E_{b,\lambda}$ is the *monochromatic hemispherical emissivity* of the real surface:

$$E_\lambda(\lambda, T) = \frac{E_\lambda}{E_{b,\lambda}} \leq 1 \qquad (4.84)$$

which generally depends on wavelength and temperature. A surface is *gray* when its monochromatic hemispherical emissivity is not a function of wavelength:

$$E_\lambda(\lambda, T) = E_\lambda(T) \quad \text{(gray)} \qquad (4.85)$$

Similarly, the total hemispherical emissive power of a real surface, E, is smaller than that of a black surface, E_b:

$$\varepsilon(T) = \frac{E}{E_b} \leq 1 \qquad (4.86)$$

and their ratio $\varepsilon(T)$ is known as the *total hemispherical emissivity*. A gray surface is characterized by $\varepsilon(T) = \varepsilon_\lambda(T)$.

A real surface may emit radiation in certain directions, as shown in Figure 4.12a. A *diffuse emitter* is the surface that emits uniformly in all the directions of the hemisphere centered on dA. Similarly, a real surface may reflect an incident ray of radiation in one or several directions (Figure 4.12c). When every incident ray is reflected uniformly in all directions, the surface is said to be a *diffuse reflector*. Figures 4.12e,f show a similar classification with regard to the absorptive properties of a real surface. The large hemisphere drawn above dA indicates that the surface is irradiated uniformly from all directions. Directional absorbers are surfaces that absorb a greater fraction of the radiation arriving from certain directions. A *diffuse absorber* absorbs the same fraction of the incident radiation regardless of its direction.

The simplest and most commonly used surface model to estimate the radiation heat transfer between generally nonblack surfaces is the *diffuse-gray surface model*. A surface is diffuse-gray if it is (i) opaque, (ii) a diffuse emitter, (iii) a diffuse absorber, (iv) a diffuse reflector, and (v) gray. Tables 4.6 and 4.7 show a compilation of $\varepsilon(T)$ values of real surfaces that have been modeled as diffuse-gray.

Kirchhoff's law for a diffuse-gray surface states that the total hemispherical absorptivity $\alpha(T)$ is equal to the total hemispherical emissivity $\varepsilon(T)$ of the same surface. The statement $\alpha(T) = \varepsilon(T)$ means that α can be estimated

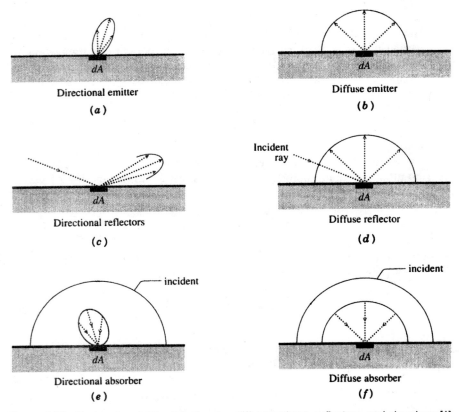

Figure 4.12 Real surfaces: Directional versus diffuse emitters, reflectors, and absorbers [1].

based on the emissivity data of Tables 4.6 and 4.7, provided that the incident radiation has the same temperature as the surface, T. In applications in which the temperature of the incident radiation does not differ substantially from that of the target surface, the requirement of Kirchoff's law can be overlooked so that $\alpha(T)$ can be approximated by using the $\varepsilon(T)$ value found in the tables. This approximation is not appropriate when the incident radiation and the target surface have vastly different temperatures.

4.4.4 Two-Surface Enclosures

Consider now the heat transfer rate between two diffuse-gray surfaces that, together, form a complete enclosure. The space is evacuated or filled with a medium that is perfectly transparent to the radiation that passes through it (such a medium is called *nonparticipating*, and its total transmissivity is $\tau = 1$). The areas (A_1, A_2), *absolute* temperatures (T_1, T_2) and total hemispherical emissivities (ε_1, ε_2) are specified. The *net* heat transfer rate from A_1 to A_2 is

Table 4.6 Metallic surfaces: representative values of the total hemispherical emissivity [1, 31, 32]

Material		$\varepsilon(T)$
Aluminum,	crude	0.07–0.08 (0°C–200°C)
	foil, bright	0.01 (−9°C), 0.04 (1°C), 0.087 (100°C)
	highly polished	0.04–0.05 (1°C)
	ordinarily rolled	0.035 (100°C), 0.05 (500°C)
	oxidized	0.11 (200°C), 0.19 (600°C)
	roughed with abrasives	0.044–0.066 (40°C)
	unoxidized	0.022 (25°C), 0.06 (500°C)
Bismuth,	unoxidized	0.048 (25°C), 0.061 (100°C)
Brass,	after rolling	0.06 (30°C)
	browned	0.5 (20°C–300°C)
	polished	0.03 (300°C)
Chromium,	polished	0.07 (150°C)
	unoxidized	0.08 (100°C)
Cobalt,	unoxidized	0.13 (500°C), 0.23 (1000°C)
Copper,	black oxidized	0.78 (40°C)
	highly polished	0.03 (1°C)
	liquid	0.15
	matte	0.22 (40°C)
	new, very bright	0.07 (40°C–100°C)
	oxidized	0.56 (40°C–200°C), 0.61 (260°C), 0.88 (540°C)
	polished	0.04 (40°C), 0.05 (260°C), 0.17 (1100°C), 0.26 (2800°C)
	rolled	0.64 (40°C)
Gold,	polished or electrolytically deposited	0.02 (40°C), 0.03 (1100°C)
Inconel,	sandblasted	0.79 (800°C), 0.91 (1150°C)
	stably oxidized	0.69 (300°C), 0.82 (1000°C)
	untreated	0.3 (40°C–260°C)
	rolled	0.69 (800°C), 0.88 (1150°C)
Inconel X		0.74–0.81 (100°C–440°C)
	stably oxidized	0.89 (300°C), 0.93 (1100°C)
Iron (see also Steel)		
	cast	0.21 (40°C)
	cast, freshly turned	0.44 (40°C), 0.7 (1100°C)
	galvanized	0.22–0.28 (0°C–200°C)
	molten	0.02–0.05 (1100°C)
	plate, rusted red	0.61 (40°C)
	pure polished	0.06 (40°C), 0.13 (540°C), 0.35 (2800°C)
	red iron oxide	0.96 (40°C), 0.67 (540°C), 0.59 (2800°C)
	rough ingot	0.95 (1100°C)
	smooth sheet	0.6 (1100°C)
	wrought, polished	0.28 (40°C–260°C)
Lead,	oxidized	0.28 (0°C–200°C)
	unoxidized	0.05 (100°C)
Magnesium		0.13 (260°C), 0.18 (310°C)
Mercury		0.09 (0°C), 0.12 (100°C)
Molybdenum		0.071 (100°C), 0.13 (1000°C), 0.19 (1500°C)
	oxidized	0.78–0.81 (300°C–540°C)

Table 4.6   *(Continued)*

Material		$\varepsilon\,(T)$
Monel,	oxidized	0.43 (20°C)
	polished	0.09 (20°C)
Nichrome,	rolled	0.36 (800°C), 0.8 (1150°C)
	sandblasted	0.81 (800°C), 0.87 (1150°C)
Nickel,	electrolytic	0.04 (40°C), 0.1 (540°C), 0.28 (2800°C)
	oxidized	0.31–0.39 (40°C), 0.67 (540°C)
	wire	0.1 (260°C), 0.19 (1100°C)
Platinum,	oxidized	0.07 (260°C), 0.11 (540°C)
	unoxidized	0.04 (25°C), 0.05 (100°C), 0.15 (1000°C)
Silver,	polished	0.01 (40°C), 0.02 (260°C), 0.03 (540°C)
Steel,	calorized	0.5–0.56 (40°C–540°C)
	cold rolled	0.08 (100°C)
	ground sheet	0.61 (1100°C)
	oxidized	0.79 (260°C–540°C)
	plate, rough	0.94–0.97 (40°C–540°C)
	polished	0.07 (40°C), 0.1 (260°C), 0.14 (540°C), 0.23 (1100°C), 0.37 (2800°C)
	rolled sheet	0.66 (40°C)
	type 347, oxidized	0.87–0.91 (300°C–1100°C)
	type AISI 303, oxidized	0.74–0.87 (300°C–1100°C)
	type 310, oxidized and	
	rolled	0.56 (800°C), 0.81 (1150°C)
	sandblasted	0.82 (800°C), 0.93 (1150°C)
Stellite		0.18 (20°C)
Tantalum		0.19 (1300°C), 0.3 (2500°C)
Tin,	unoxidized	0.04–0.05 (25°C–100°C)
Tungsten,	filament	0.03 (40°C), 0.11 (540°C), 0.39 (2800°C)
Zinc,	oxidized	0.11 (260°C)
	polished	0.02 (40°C), 0.03 (260°C)

$$\dot{Q}_{1-2} = \frac{\sigma(T_1^4 - T_2^4)}{(1 - \varepsilon_1)/\varepsilon_1 A_1 + 1/A_1 F_{12} + (1 - \varepsilon_2)/\varepsilon_2 A_2} \qquad (4.87)$$

in which the denominator on the right side represents the total radiation thermal resistance from A_1 (or the potential σT_1^4) to the colder surface A_2 (potential σT_2^4). The term $(1 - \varepsilon)/\varepsilon A$ represents the *internal resistance* of the particular diffuse-gray surface. The internal resistance is zero if the surface is black ($\varepsilon = 1$).

Table 4.7 Nonmetallic surfaces: representative values of the total hemispherical emissivity [1, 31, 32]

Material	$\varepsilon (T)$
Bricks	
chrome refractory	0.94 (540°C), 0.98 (1100°C)
fire clay	0.75 (1400°C)
light buff	0.8 (540°C), 0.53 (1100°C)
magnesite refractory	0.38 (1000°C)
sand lime red	0.59 (1400°C)
silica	0.84 (1400°C)
various refractories	0.71–0.88 (1100°C)
white refractory	0.89 (260°C), 0.68 (540°C)
Building materials	
asbestos, board	0.96 (40°C)
asphalt pavement	0.85–0.93 (40°C)
clay	0.39 (20°C)
concrete, rough	0.94 (0°C–100°C)
granite	0.44 (40°C)
gravel	0.28 (40°C)
gypsum	0.9 (40°C)
marble, polished	0.93 (40°C)
mica	0.75 (40°C)
plaster	0.93 (40°C)
quartz	0.89 (40°C), 0.58 (540°C)
sand	0.76 (40°C)
sandstone	0.83 (40°C)
slate	0.67 (40°C–260°C)
Carbon	
baked	0.52–0.79 (1000°C–2400°C)
filament	0.95 (260°C)
graphitized	0.76–0.71 (100°C–500°C)
rough	0.77 (100°C–320°C)
soot (candle)	0.95 (120°C)
soot (coal)	0.95 (20°C)
unoxidized	0.8 (25°C–500°C)
Ceramics	
coatings	
alumina on inconel	0.65 (430°C), 0.45 (1100°C)
zirconia on inconel	0.62 (430°C), 0.45 (1100°C)
earthenware, glazed	0.9 (1°C)
matte	0.93 (1°C)
porcelain	0.92 (40°C)
refractory, black	0.94 (100°C)
light buff	0.92 (100°C)
white Al_2O_3	0.9 (100°C)
Cloth	
cotton	0.77 (20°C)
silk	0.78 (20°C)
Glass	
Convex D	0.8–0.76 (100°C–500°C)
fused quartz	0.75–0.8 (100°C–500°C)
Nonex	0.82–0.78 (100°C–500°C)
Pyrex	0.8–0.9 (40°C)
smooth	0.92–0.95 (0°C–200°C)
waterglass	0.96 (20°C)

Table 4.7 *(Continued)*

Material	$\varepsilon(T)$
Ice, smooth	0.92 (0°C)
Oxides	
Al_2O_3	0.35–0.54 (850°C–1300°C)
C_2O	0.27 (850°C–1300°C)
Cr_2O_3	0.73–0.95 (850°C–1300°C)
Fe_2O_3	0.57–0.78 (850°C–1300°C)
MgO	0.29–0.5 (850°C–1300°C)
NiO	0.52–0.86 (500°C–1200°C)
ZnO	0.3–0.65 (850°C–1300°C)
Paints	
aluminum	0.27–0.7 (1°C–100°C)
enamel, snow white	0.91 (40°C)
lacquer	0.85–0.93 (40°C)
lampblack	0.94–0.97 (40°C)
oil	0.89–0.97 (0°C–200°C)
white	0.89–0.97 (40°C)
Paper, white	0.95 (40°C), 0.82 (540°C)
Roofing materials	
aluminum surfaces	0.22 (40°C)
asbestos cement	0.65 (1400°C)
bituminous felt	0.89 (1400°C–2800°C)
enameled steel, white	0.65 (1400°C)
galvanized iron, dirty	0.90 (1400°C–2800°C)
new	0.42 (1400°C)
roofing sheet, brown	0.8 (1400°C)
green	0.87 (1400°C)
tiles, uncolored	0.63 (1400°C–2800°C)
brown	0.87 (1400°C)
black	0.94 (1400°C)
asbestos cement	0.66 (1400°C–2800°C)
weathered asphalt	0.88 (1400°C–2800°C)
Rubber	
hard, black, glossy	
surface	0.95 (40°C)
soft, gray	0.86 (40°C)
Snow	
fine	0.82 (−10°C)
frost	0.98 (0°C)
granular	0.89 (−10°C)
Soils	0.92–0.96 (0°C–20°C)
black loam	0.66 (20°C)
plowed field	0.38 (20°C)
Water	0.92–0.96 (0°C–40°C)
Wood	
beech	0.91 (70°C)
oak, planed	0.91 (40°C)
sawdust	0.75 (40°C)
spruce, sanded	0.82 (100°C)

Two Infinite Parallel Plates. Figure 4.13a shows the case of two infinite parallel plates. In this case we recognize that $F_{12} = 1$ and $A_1 = A_2 = A$, and Equation 4.87 reduces to

$$\dot{Q}_{1-2} = \frac{\sigma A(T_1^4 - T_2^4)}{1/\varepsilon_1 + 1/\varepsilon_2 - 1} \tag{4.88}$$

This result is valid not only for infinite parallel plates but also for the space (e.g., vacuum jacket) between two concentric surfaces provided that the spacing is small.

Space Between Two Infinite Cylinders or Two Spheres. Consider next Figures 4.13b and 4.13c showing two infinite cylinders and two spheres, respectively. Assume that A_1 is positioned inside A_2. However, the cylinders or spheres do not have to be positioned concentrically. The net heat transfer rate from A_1 to A_2 is (note that $F_{12} = 1$)

$$\dot{Q}_{1-2} = \frac{\sigma A_1(T_1^4 - T_2^4)}{1/\varepsilon_1 + (A_1/A_2)\left(\dfrac{1}{\varepsilon_2} - 1\right)} \tag{4.89}$$

Extremely Large Surface Surrounding a Convex Surface. Consider next the case of Figure 4.13b (or Figure 4.13c) in the limit $A_1 \ll A_2$. Note that

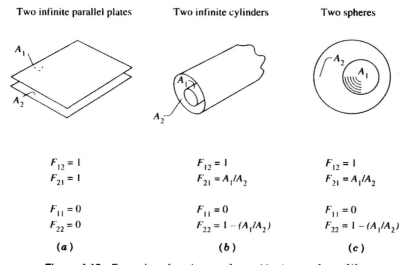

Two infinite parallel plates   Two infinite cylinders   Two spheres

$F_{12} = 1$ $F_{12} = 1$ $F_{12} = 1$

$F_{21} = 1$ $F_{21} = A_1/A_2$ $F_{21} = A_1/A_2$

$F_{11} = 0$ $F_{11} = 0$ $F_{11} = 0$

$F_{22} = 0$ $F_{22} = 1 - (A_1/A_2)$ $F_{22} = 1 - (A_1/A_2)$

       (a) (b) (c)

Figure 4.13 Examples of enclosures formed by two surfaces [1].

A_2 is the extremely large outer surface and A_1 is the smaller internal surface (convex, $F_{11} = 0$). The net heat transfer rate from A_1 to A_2 is

$$\dot{Q}_{1-2} = \sigma A_1 \varepsilon_1 (T_1^4 - T_2^4) \tag{4.90}$$

4.4.5 Enclosures with More Than Two Surfaces

As a generalization of the two-surface enclosure results reviewed in the preceding section, consider the enclosure of Figure 4.14, in which all n surfaces are diffuse-gray. The number of view factors that can be specified independently is $n(n - 1)/2$. It has been shown that because the surface is diffuse-gray the net heat current that must be supplied through the back of surface A_1, *toward* the enclosure, is [1]

$$\dot{Q}_{1-2} = \frac{\varepsilon_1 A_1}{1 - \varepsilon_1} (\sigma T_1^4 - J_1) \tag{4.91}$$

where J_1 denotes the *radiosity*. The radiosity is the total heat flux that leaves a unit area of A_1, that is, the radiation emitted by A_1 plus the irradiation reflected by A_1. Equation 4.91 shows that what drives the energy current $\dot{Q}_{1-2}$ out of the surface is the potential difference $\sigma T_1^4 - J_1$, or $E_{b,1} - J_1$. This flow of energy is impeded by an *internal resistance* $(1 - \varepsilon_1)/\varepsilon_1 A_1$, which is illustrated by the resistance drawn between nodes σT_i^4 and J_i in the network of Figure 4.14.

Relations equivalent to Equation 4.91 can be written for the remaining $n - 1$ surfaces of the enclosure,

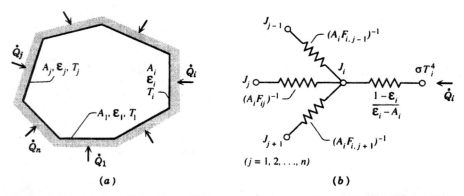

(a)   (b)

Figure 4.14 Enclosure formed by n diffuse-gray surfaces and the resistance network portion associated with the surface A_i [1].

$$\dot{Q}_i = \frac{\varepsilon_i A_i}{1 - \varepsilon_i} (\sigma T_i^4 - J_i) \qquad (i = 1, 2, \ldots, n) \qquad (4.92)$$

In general, an observer stationed on A_i (at node J_i) may see the radiosities of all n surfaces of the enclosure. The point of view of this observer is illustrated in Figure 4.14b. The net radiation between J_i and another surface, J_j, is impeded by the resistance $(A_i F_{ij})^{-1}$, or $(A_j F_{ji})^{-1}$, which is due to the geometric configuration of the enclosure. Figure 4.14b shows the radiation resistances associated with surface A_i. The radiation network for the complete enclosure consists of drawing Figure 4.14b for each of the n surfaces of Figure 4.14a. These drawings will be connected at the n nodes $J_1, J_2, \ldots, J_n$, which represent the surfaces of the enclosure. Such a construction is illustrated in Example 4.3. The analysis presented in Reference 1 shows that the n radiosities J_i can be calculated by solving the system

$$J_i = (1 - \alpha_i) \sum_{j=1}^{n} J_j F_{ij} + \varepsilon_i \sigma T_i^4 \qquad (i = 1, 2, \ldots, n) \qquad (4.93)$$

When all the surface absolute temperatures T_i are specified, Equations 4.93 can be solved to determine all the radiosities J_i, leaving the heat currents $\dot{Q}_i$ to be calculated from Equations 4.92. When only some of T_i and $\dot{Q}_i$ are specified, the two systems of Equations 4.92 and 4.93 must be solved simultaneously. If a surface (A_k) is insulated with respect to the exterior of the enclosure, its net (through-the-back) heat current is known because $\dot{Q}_k = 0$. Such a surface is called *reradiating,* or adiabatic.

4.4.6 Gray Medium Surrounded by Two Diffuse-Gray Surfaces

The simplest model of a two-surface enclosure filled with a participating (absorbing, emitting) medium is presented in Figure 4.15. Each surface is diffuse-gray and isothermal, and the medium (e.g., gas) is isothermal and *gray.* This means that the gas is represented by the unique absolute temperature T_g and that the gas emissivity is equal to the gas absorptivity regardless of the source of incident radiation. Means for estimating ε_g and α_g for participating gases are presented in heat transfer textbooks [1, 2].

In the resistance network of Figure 4.15, the node that represents the gas volume floats (i.e., settles at an intermediate temperature level) when the gas is not heated electrically or by a chemical reaction. The network is constructed by noting that there are three participants in the enclosure, the two surfaces (T_1, T_2) and the gas (T_g). Each surface contributes to the network the resistances identified in Figure 4.14b, namely, one internal resistance and two external resistances. The external resistance between J_1 and J_2 is $(A_1 F_{12} \tau_g)^{-1}$, where τ_g is the gas transmissivity, $\tau_g = 1 - \alpha_g$. The internal resistance associated with the gas node is not visible (does not play a role) because the

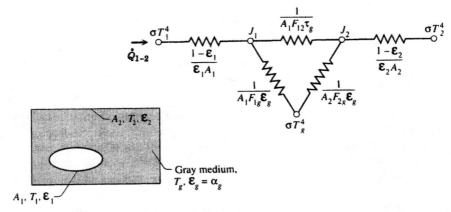

Figure 4.15 Gray medium enclosed by two diffuse-gray surfaces and the radiation network for the case in which the temperature of the medium is floating [1].

net heat transfer out of the gas is zero, and the σT_g^4 and J_g nodes coincide. The step-by-step construction of the network of Figure 4.15 is given in Reference 1. The net heat transfer rate from A_1 to A_2 is

$$\dot{Q}_{1-2} = \frac{\sigma(T_1^4 - T_2^4)}{(1 - \varepsilon_1)/\varepsilon_1 A_1 + R_\Delta + (1 - \varepsilon_2)/\varepsilon_2 A_2} \quad (4.94)$$

where R_Δ is the equivalent series resistance of the triangular loop of the network:

$$R_\Delta = \frac{(A_1 F_{12} \tau_g)^{-1}[(A_1 F_{1g} \varepsilon_g)^{-1} + (A_2 F_{2g} \varepsilon_g)^{-1}]}{(A_1 F_{12} \tau_g)^{-1} + (A_1 F_{1g} \varepsilon_g)^{-1} + (A_2 F_{2g} \varepsilon_g)^{-1}} \quad (4.95)$$

Example 4.3   Hot ash from the boiler of a cogeneration plant is stored temporarily in a hopper located in a 5-m-deep pit in the floor of the plant. The hopper is long in the direction perpendicular to the plane of the accompanying figure. The distance from the side wall of the hopper to the vertical wall of the pit is 2.5 m. The temperature of the outer surface of the hopper is 300°C. This raises concerns about the safety of workers who may have to descend into the pit to unclog the flow of ash through the bottom of the hopper. A key safety parameter is the temperature of the pit wall and floor. Estimate this temperature by assuming that the hopper surface is diffuse-gray with $\varepsilon_1 = 0.6$ and that the pit surface is reradiating (adiabatic). The ambient at temperature 40°C completes the enclosure.

Solution

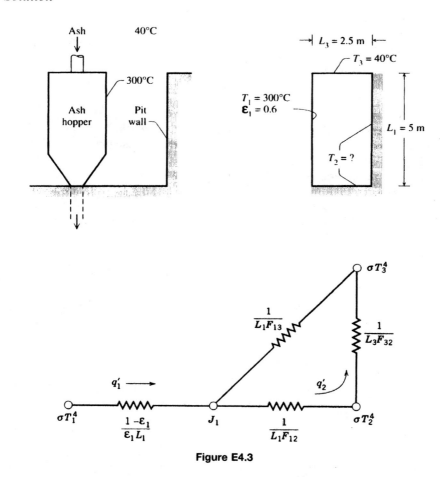

Figure E4.3

MODEL

1. The geometry is two dimensional, that is, long in the direction perpendicular to the figure.
2. The side wall and the floor of the pit are represented by one temperature, T_2.
3. In the enclosure model the ambient is represented by a black surface, T_3.
4. All the heat transfer is due to radiation, Convection effects are negligible.

5. The pit surface is reradiating, that is, its back is adiabatic.

ANALYSIS. The side of the ash hopper (T_1), the pit wall (T_2), and the ambient (T_3) form the two-dimensional rectangular enclosure illustrated in the figure. We know the following:

$$T_1 = 300°C, \qquad \varepsilon_1 = 0.6, \qquad T_3 = 40°C$$
$$L_1 = 5 \text{ m}, \qquad L_2 = 7.5 \text{ m}, \qquad L_3 = 2.5 \text{ m}$$

The unknown is T_2. To construct the radiation network, we note that there are three surfaces (T_1, T_2, T_3). We draw an internal resistance only for T_1. The internal resistance of the top surface is zero because that surface (the opening to the ambient) is modeled as black. The internal resistance of the adiabatic surface is not visible: The nodes J_2 and σT_2^4 coincide because the net heat transfer through the surface is zero. With reference to the heat currents q_1' and q_2' indicated on the network, we write the relations between heat current and drop in potential,

$$q_1' = \frac{\varepsilon_1 L_1}{1 - \varepsilon_1} (\sigma T_1^4 - J_1) \tag{1}$$

$$q_1' = L_1 F_{12}(J_1 - \sigma T_2^4) + L_1 F_{13}(J_1 - \sigma T_3^4) \tag{2}$$

$$q_2' = L_1 F_{12}(J_1 - \sigma T_2^4) \tag{3}$$

$$q_2' = L_3 F_{32}(\sigma T_2^4 - \sigma T_3^4) \tag{4}$$

Eliminating q_2' between the last two equations,

$$J_1 = \sigma T_2^4 + \frac{L_3}{L_1} \frac{F_{32}}{F_{12}} \sigma(T_2^4 - T_3^4) \tag{5}$$

for which the view factor can be calculated sequentially, by starting with the second formula listed in Table 4.5:

$$F_{13} = \tfrac{1}{2}[1 + x - (1 + x^2)^{1/2}] = 0.191 \qquad \left(x = \frac{2.5}{5} = 0.5\right)$$

$$F_{12} = 1 - F_{13} = 0.809$$

$$L_1 F_{13} = L_3 F_{31}$$

$$F_{31} = F_{13} \frac{L_1}{L_3} = 0.191 \frac{5}{2.5} = 0.382$$

$$F_{32} = 1 - F_{31} = 0.618$$

Equation 5 becomes

$$\frac{J_1}{\sigma} = 1.382T_2^4 - 0.382T_3^4 \qquad (6)$$

A second relationship between T_2^4 and J_1/σ is obtained by eliminating q_1' between Equations 1 and 2:

$$\frac{J_1}{\sigma} = 0.6T_1^4 + 0.323T_2^4 + 0.0764T_3^4 \qquad (7)$$

Finally, by eliminating J_1/σ between Equations 6 and 7, we arrive at

$$T_2^4 = 0.567T_1^4 + 0.433T_3^4$$

which yields

$$T_2 = [0.567(300 + 273.15)^4 + 0.433(40 + 273.15)^4]^{1/4} \text{ K}$$

$$= 506 \text{ K} = 232°C$$

COMMENT. This estimate shows, as might have been anticipated, that the temperature of the pit wall would endanger the safety of those who would work in the space between the ash hopper and the pit wall. The ash pit should be made safer, for example, by inserting a radiation shield over the hopper surface and circulating ambient air through the space.

4.5 CLOSURE

Design analysis and modeling activities involving heat transfer are seldom compartmentalized into problems dealing solely with conduction, convection, or radiation but generally require the consideration of simultaneous effects—convection *and* radiation, for example. Furthermore, thermal modeling often requires heat transfer principles to be complemented by principles of engineering thermodynamics and fluid flow. The presentation of this book continues, therefore, with a discussion of applications with heat transfer and fluid flow in Chapter 5. In Chapter 6 we consider applications with thermodynamics, heat transfer, and fluid flow.

REFERENCES

1. A. Bejan, *Heat Transfer*, Wiley, New York, 1993.
2. F. P. Incropera and D. P. DeWitt, *Fundamentals of Heat and Mass Transfer*, 3rd ed., Wiley, New York, 1990.

3. A. Bejan, *Convection Heat Transfer,* 2nd ed., Wiley, New York, 1995.

4. W. M. Rohsenow, J. P. Hartnett, and E. N. Ganic, *Handbook of Heat Transfer Applications,* 2nd ed., McGraw-Hill, New York, 1985.

5. A. P. Fraas, *Heat Exchanger Design,* 2nd ed., Wiley, New York, 1989.

6. S. W. Churchill and M. Bernstein, A correlating equation for forced convection from gases and liquids to a circular cylinder in crossflow, *J. Heat Transfer,* Vol. 99, 1977, pp. 300–306.

7. J. H. Lienhard, *A Heat Transfer Textbook,* 2nd ed., Prentice-Hall, Englewood Cliffs, NJ, 1987, p. 346.

8. S. Whitaker, Forced convection heat transfer correlations for flow in pipes, past flat plates, single cylinders, single spheres, and flow in packed beds and tube bundles, *AIChE J.,* Vol. 18, 1972, pp. 361–371.

9. A. A. Zukauskas, Convective heat transfer in cross flow, S. Kakac, R. K. Shah, and W. Aung, eds., *Handbook of Single-Phase Convective Heat Transfer,* Wiley, New York, 1987, Chapter 6.

10. L. F. Moody, Friction factors for pipe flows, *Trans. ASME,* Vol. 66, 1944, pp. 671–684.

11. A. P. Colburn, A method for correlating forced convection heat transfer data and a comparison with fluid friction, *Trans. Am. Inst. Chem. Eng.,* Vol. 29, 1933, pp. 174–210.

12. V. Gnielinski, New equations for heat and mass transfer in turbulent pipe and channel flow, *Int. Chem. Eng.,* Vol. 16, 1976, pp. 359–368.

13. A. Bejan and J. L. Lage, The Prandtl number effect on the transition in natural convection along a vertical surface, *J. Heat Transfer,* Vol. 112, 1990, pp. 787–790.

14. S. W. Churchill and H. H. S. Chu, Correlating equations for laminar and turbulent free convection from a vertical plate, *Int. J. Heat Mass Transfer,* Vol. 18, 1975, pp. 1323–1329.

15. J. R. Lloyd and W. R. Moran, Natural convection adjacent to horizontal surfaces of various planforms, ASME Paper 74-WA/HT-66, 1974.

16. S. W. Churchill, Free convection around immersed bodies, E. U. Schlünder, ed., *Heat Exchanger Design Handbook,* Section 2.5.7, Hemisphere, New York, 1983.

17. W. Aung, Mixed convection in internal flow, S. Kakac, R. K. Shah, and W. Aung, eds., *Handbook of Single-Phase Convective Heat Transfer,* Wiley, New York, 1987, Chapter 15.

18. S. Globe and D. Dropkin, Natural convection heat transfer in liquids confined by two horizontal plates and heated from below, *J. Heat Transfer,* Vol. 81, 1959, pp. 24–28.

19. S. L. Chen, F. M. Gerner, and C. L. Tien, General film condensation correlations, *Exper. Heat Transfer,* Vol. 1, 1987, pp. 93–107.

20. M. M. Chen, An analytical study of laminar film condensation: Part 2—single and multiple horizontal tubes, *J. Heat Transfer,* Vol. 83, 1961, pp. 55–60.

21. V. K. Dhir and J. H. Lienhard, Laminar film condensation on plane and axisymetric bodies in non-uniform gravity, *J. Heat Transfer,* Vol. 93, 1971, pp. 97–100.

22. W. M. Rohsenow, A method for correlating heat transfer data for surface boiling of liquids, *Trans. ASME,* Vol. 74, 1952, pp. 969–976.

23. R. I. Vachon, G. H. Nix, and G. E. Tanger, Evaluation of constants for the Roh-senow pool-boiling correlation, *J. Heat Transfer,* Vol. 90, 1968, pp. 239–247.

24. J. H. Lienhard and V. K. Dhir, Extended hydrodynamic theory of the peak and minimum pool boiling heat fluxes, NASA CR-2270, July 1973.

25. K. H. Sun and J. H. Lienhard, The peak pool boiling heat flux on horizontal cylinders, *Int. J. Heat Mass Transfer,* Vol. 13, 1970, pp. 1425–1439.

26. P. J. Berenson, Film boiling heat transfer for a horizontal surface, *J. Heat Transfer,* Vol. 83, 1961, pp. 351–358.

27. A. L. Bromley, Heat transfer in stable film boiling, *Chem. Eng. Progress,* Vol. 46, 1950, pp. 221–227.

28. M. R. Duignan, G. A. Greene, and T. F. Irvine, Jr., Film boiling heat transfer to large superheats from a horizontal flat surface, *J. Heat Transfer,* Vol. 113, 1991, pp. 266–268.

29. J. R. Howell, *A Catalog of Radiation Configuration Factors,* McGraw-Hill, New York, 1982.

30. R. Siegel and J. R. Howell, *Thermal Radiation Heat Transfer,* McGraw-Hill, New York, 1972.

31. G. G. Gubareff, J. E. Janssen, and R. H. Torborg, *Thermal Radiation Properties Survey,* 2nd ed., Honeywell Research Center, Minneapolis, MN, 1960.

32. Y. S. Touloukian and C. Y. Ho, eds., *Thermophysical Properties of Matter,* Vols. 7–9, Plenum, New York, 1972.

33. J. S. Lim, A. Bejan, and J. H. Kim, The optimal thickness of a wall with con-vection on one side, *Int. J. Heat Mass Transfer,* Vol. 35, 1992, pp. 1673–1679.

PROBLEMS

4.1 As shown in Figure P4.1, the pipe that supplies drinking water to a community in a cold region is buried at a depth $H = 3$ m below the ground surface. The external surface of the pipe can be modeled as an isothermal cylinder of diameter $D = 0.5$ m and temperature 4°C. The ground surface is at 0°C, and the thermal conductivity of the soil is $k = 1$ W/m · K. Calculate the heat transfer rate per unit of length from the pipe to the surrounding soil, q', in W/m. During a very harsh winter, the top layer of the soil freezes to a depth of 1 m. This new position of the freezing front can be modeled as an isothermal plane of temperature 0°C situated at $H = 2$ m above the pipe centerline. Calculate again the heat transfer rate q' and the increase in q' that is due to the 1-m advancement of the freezing front.

4.2 The cooking of a beef steak can be modeled as time-dependent con-duction in a slab. The thickness of the steak is 4 cm, its initial tem-perature is 25°C, and the properties of beef are $k = 0.4$ W/m · K and $\alpha = 1.25 \times 10^{-7}$ m²/s. The steak is cooked with both sides exposed in a convection oven of temperature 150°C. The heat transfer coeffi-cient between meat and oven air is 60 W/m² · K.

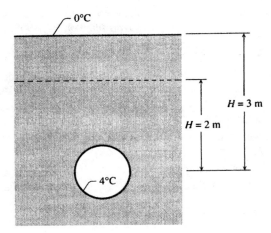

Figure P4.1

(a) If the steak is cooked well done when its centerplane temperature reaches 80°C, calculate the required cooking time.

(b) It is well known that meat shrinks during cooking. For example, the thickness of a steak cooked well done drops to about 75% of the original thickness. To see the extent to which the shrinking of meat shortens the cooking time, repeat part (a) by assuming that the steak thickness is equal to 3 cm throughout the cooking process. Is the actual cooking time longer or shorter than the estimate obtained in part (b)? Discuss.

4.3   The design of the cooling jacket for the apparatus shown in Figure P4.3 calls for the use of a copper tube of inner diameter $D = 0.5$ cm

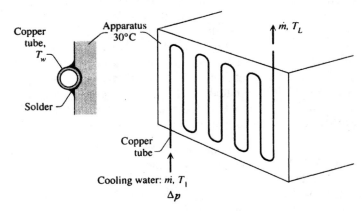

Figure P4.3

and total length $L = 4$ m. Cooling water flows through this tube with the mean velocity $U = 10$ cm/s. The tube is soldered to the side of the apparatus, the temperature of which is 30°C. Model the tube wall as isothermal with the temperature $T_1 = 20$°C. The physical properties of water can be evaluated at 25°C. Neglect the effect of bends.

(a) Verify that the flow is hydrodynamically fully developed over most of the tube length. Calculate the pressure drop along the tube.

(b) Verify also that the flow is thermally fully developed along most of the tube. Calculate the mean outlet temperature of the water stream and the total heat transfer rate extracted by the stream from the apparatus to which the tube is attached.

4.4 An immersion heater for water consists of a thin vertical plate shaped as a rectangle of height 8 cm and length 15 cm. The plate is heated electrically and maintained at 55°C, while the average temperature in the surrounding water tank is 15°C. The plate is bathed by water on both sides.

(a) Show that the natural convection boundary layer that rises along the plate is laminar.

(b) Calculate the total heat transfer rate released by the heater into the water pool.

4.5 Consider the question of how to shape a bar that must be stiff in tension and a good thermal insulator. As shown in Figure P4.5, a low-temperature apparatus (T_0) is supported from a wall at T_L through a vertical bar in tension. The lateral surface of the bar is well insulated so that convection and radiation effects are negligible. The bar must be strong

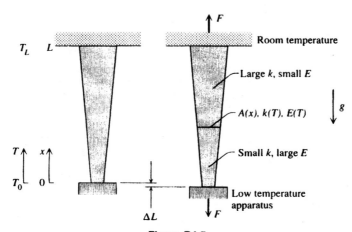

Figure P4.5

enough to carry the weight of the apparatus (F), long and thin enough to allow the smallest heat transfer rate $\dot{Q}$ to flow from T_L to T_0, and rigid enough to fix the position of the apparatus on the vertical line. The rigidity requirement means that the total elastic elongation ΔL caused by the longitudinal force F must be as small as possible.

The thermal conductivity and modulus of elasticity of structural materials vary roughly as $k(T) \cong \overline{k}[1 + b(2\theta - 1)]$ and $E(T) \cong \overline{E}[1 - c(2\theta - 1)]$, where $\overline{k}$ and $\overline{E}$ are temperature averaged values, b and c are positive numbers less than 1, proportional to dk/dT and $-dE/dT$, and $\theta = (T - T_0)/(T_L - T_0)$.

The material model described above accounts for the fact that k decreases and E increases toward lower temperatures. It is reasonable to think that the overall elongation is reduced when more material is positioned near the warm end, where E is smaller than near the cold end. These characteristics of structural materials give a designer the idea that a tapered bar ($0 < \tilde{a} < 1$)

$$A(x) = \overline{A}\left[1 + \tilde{a}\left(2\frac{x}{L} - 1\right)\right] \qquad (0 < \tilde{a} < 1)$$

$$\overline{A} = \frac{1}{L}\int_0^L A(x)\, dx \qquad \text{(constant)}$$

might exhibit a smaller elongation in tension than the bar with x-independent geometry ($\tilde{a} = 0$). Evaluate the merit of this proposal by determining the effect of the taper parameter $\tilde{a}$ not only on the total elongation ΔL but also on the heat transfer rate. Comment on how you might design a rigid support in tension so that it can serve as a thermal insulator at the same time.

4.6   Determine the best shape of a bar that must be both stiff in bending and a good thermal insulator. The horizontal beam with variable thickness $H(x)$ shown in Figure P4.6 connects a body of weight F and temperature T_0 to a support of temperature T_L. The beam geometry is slender and two-dimensional, with the width B measured in the direction perpendicular to the figure. The distance L and the amount of beam material are fixed. The thermal conductivity k and modulus of elasticity E of this material are known constants.

The function of the beam is to support the weight F as rigidly as possible, while impeding the transfer of heat from T_L to T_0. It is, simultaneously, a mechanical support and a thermal insulation. Arrange the beam material [the thickness, $H(x)$] in such a way that the tip deflection y_0 and the end-to-end heat transfer rate $\dot{Q}$ are minimized. *One* approach: Shape the beam cross section such that the thickness H increases as x^n as x increases, where the exponent n is a number between 0 and 1. Determine analytically y_0 and $\dot{Q}$ as functions of n and

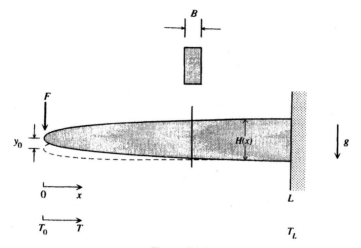

Figure P4.6

comment on how the shape of the beam cross section (n) affects the success of the design.

4.7   As shown in Figure P4.7, hot ash with the temperature $T_i = 900°C$ accumulates at the bottom of a fluidized-bed combustor at the rate $\dot{m} = 500$ kg/h. The ash must be discharged and stored temporarily in a hopper for the purpose of lowering its temperature below 550°C, before it can be shipped out of the plant. The cooling effect is due to combined convection and radiation at the upper surface of the ash pile through an effective heat transfer coefficient $h = 30$ W/m²·K. The atmospheric

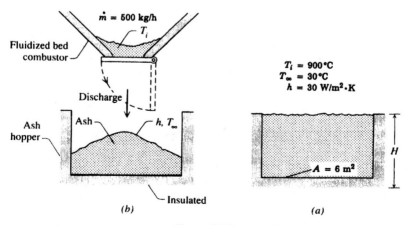

Figure P4.7

temperature is $T_\infty = 30°C$. The thermal properties of this particular ash are $k = 1$ W/m·K, $c = 1$ kJ/kg·K, and $\rho = 1900$ kg/m³. The hopper has the depth $H = 1$ m and horizontal cross-sectional area $A = 6$ m². Its bottom and side walls are insulated. The hopper is to be filled intermittently in a sequence of equal discharges that are spaced equally in time. As the engineer responsible for safe operation of the ash hopper installation, you must estimate the number of discharges so that the temperature at the bottom of the ash pile settles to a value below 550°C. For simplicity, assume that the top surface of the ash pile is always flat, as shown in Figure 4.7b, and that conduction proceeds in the vertical direction only.

4.8   The critical insulation radii mentioned in connection with Equations 4.7 and 4.8 are derived based on the assumption that the heat transfer coefficient at the outer surface of the insulation is independent of the insulation outer diameter D. In reality h depends on D. For example, over sufficiently narrow ranges of Re_D values, the correlations giving h for forced convection from a cylinder in cross flow and a sphere are approximated by power laws of the type

$$\frac{hD}{k_{\text{fluid}}} = C \left(\frac{U_\infty D}{\nu} \right)^n$$

where C is a constant and n is a number between 0 and 1 (usually about 0.5). This power-law approximation shows that h varies at D^{n-1}. Using this result:

(a) Determine the critical outer radius of the insulation ($r_{o,c}$) for a cylinder of radius r_i, which experiences forced convection in cross flow.

(b) Derive an expression for the corresponding critical insulation radius for a sphere with forced convection. The thermal conductivity of the insulation material is k.

4.9   One method of extracting the exergy contained in a geothermal reservoir underground consists of using a downhole coaxial heat exchanger as shown in Figure P4.9. After reaching the bottom of the well, the heated stream returns to the surface by flowing through the inner pipe. An effective layer of insulation is built into the wall of diameter D_i, which separates the downflowing cold stream from the upflowing hot stream. This wall can be regarded as thin.

A fundamental design question concerns the selection of the diameter of the inner pipe, D_i. If D_i is much smaller than D_o, or if D_i is nearly the same as D_o, the flow is impeded. In both extremes, the overall presure drop that must be overcome by the pump is excessive. When D_o is fixed, an optimal inner diameter D_i (or an optimal ratio

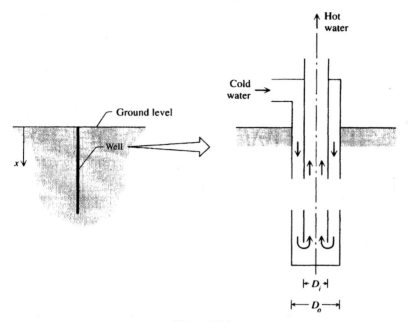

Figure P4.9

D_i/D_o) exists such that the total pressure drop experienced by the stream is minimum.

(a) Determine the optimal ratio D_i/D_o in the large Reynolds number limit of the turbulent regime, where the friction factors for the annular space (f_a) and for the upflow through the inner pipe (f_i) are each constant. For simplicity, assume that $f_a = f_i$.

(b) Consider next the regime in which the flow is laminar both through the annular space and through the inner pipe. Assume that the friction factor for the annular space is approximately equal to the friction factor for flow between two parallel plates positioned $\frac{1}{2}(D_o - D_i)$ apart. Calculate the ratio D_i/D_o for minimum total pressure drop, and show that this result is almost the same as the result obtained in part (a).

4.10 It is proposed to install tapered window glass as an energy saving feature in buildings, that is, as a means of minimizing the heat leak from a room to the cold outside air. This proposal was stimulated by the thought that since the room-side heat transfer coefficient is higher near the top of the window (at the start of the descending boundary layer), it is there that a thicker glass layer can have the greatest effect on reducing the local heat flux [33]. This suggests the tapered glass design illustrated on the left side of Figure P4.10. The thickness of the

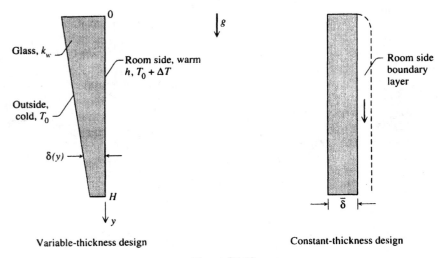

Variable-thickness design   Constant-thickness design

Figure P4.10

glass layer has been exaggerated in order to make the notation clearer.

One way to evaluate the merit of the proposed design is to compare the total heat leak per unit length through the window q' with the corresponding heat leak through a constant-thickness glass window. The amount of glass used in the variable-thickness design is the same as in the constant-thickness (or reference) design. For the tapered-glass window assume that the glass thickness decreases linearly in the downward direction:

$$\delta = \overline{\delta} + b(\tfrac{1}{2} - \xi)$$

In this expression $\xi = y/H$, and the taper parameter $b = -d\delta/d\xi$ is a design variable that must be determined optimally. The thickness $\overline{\delta}$ averaged over the height H is fixed because H and the glass volume are fixed.

For the heat transfer coefficient on the room-air side assume a y dependence consistent with that found in laminar natural convection boundary layers: $h = h_{\min}\xi^{-1/4}$, where $h_{\min}$ corresponds to the bottom of the window where the room-side boundary layer is the thickest. The heat transfer coefficient on the outside of the glass layer is sufficiently large, so that the temperature of that surface is equal to the atmospheric temperature.

Determine the ratio q'/q'_{ref} as a function of two dimensionless groups, the taper parameter

$$S = \frac{H}{\bar{\delta}}\left(-\frac{d\delta}{dy}\right)$$

and the bottom-end Biot number Bi $= h_{min}\delta/k_w$, where k_w is the thermal conductivity of glass. Using this result, determine the best taper parameter S for the smallest q'/q'_{ref} ratio for a fixed Bi. Determine the Bi range in which a common window is likely to operate. Comment on the practicality of the heat leak reduction promised by the tapered-glass design.

4.11 Consider the thermal insulation shown in Figure P4.11a consisting of a horizontal layer of fluid of thickness H and bottom-to-top temperature difference $T_h - T_c = \Delta T$. These two parameters, H and ΔT, are large enough that convection currents form in the fluid. To suppress the formation of these currents, it is proposed to install a horizontal partition at some level between the bottom wall and the top wall, as shown in Figure P4.11b. Assume that the partition can be modeled as an isothermal wall with a temperature between the bottom wall temperature and the top wall temperature. Assume further that convection currents are absent above and below the partition. Find the partition level that maximizes the overall temperature difference ΔT, while preserving this state of pure conduction.

4.12 The heat transfer through a layer of fibrous insulation is a complex mix of fiber-to-fiber radiation and parallel conduction through the air trapped between the fibers. This is the case when the spaces between fibers are small enough so that the air motion is slow and the convection effect negligible. A characteristic of such layers is that their thermal insulation effect is the greatest when the packing (layer porosity) has a certain value [1]. The porosity is defined as the ratio $\varphi =$ (air space)/[total (air + fiber) space]. Animal fur is a good example of a

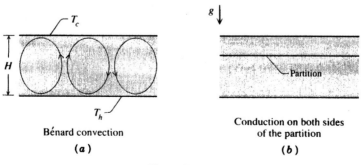

Bénard convection

(a)

Conduction on both sides of the partition

(b)

Figure P4.11

fibrous layer permeated by air. A characteristic of animal fur is that the observed porosity falls in the narrow range $\varphi \cong 0.95$ to 0.99, that is, in a way that seems to be independent of the animal body size or hair strand thickness.

To determine the optimal fiber packing analytically, consider the simple two-dimensional model shown in Figure P4.12. The fibers are modeled as a stack of n equidistant shields of thickness equal to the fiber thickness D. The layer thickness L and the two temperatures T_0 and T_L separated by the insulating layer are given. The overall temperature difference is considerably smaller than the local absolute temperature, $T_0 - T_L << T_0$. Each surface is diffuse-gray, having the same total hemispherical emissivity ε.

(a) Derive an expression for the heat transfer rate across the layer, and show that it is minimized when the fiber-to-fiber spacing reaches a certain optimal value.

(b) Consider the fiber thickness D in the range 10–100 μm, which is the range of the observed thickness of hair strands of animals of various body sizes (20 cm–3 m). Show that the optimal porosity that corresponds to the optimal spacing agrees approximately with the observed porosities of animal fur.

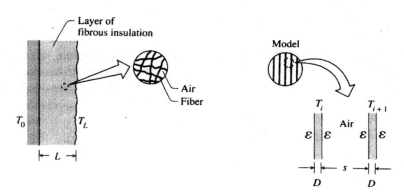

Figure P4.12

5

APPLICATIONS WITH HEAT
AND FLUID FLOW

This chapter deals with some fundamental design aspects: Thermal devices must be put together (e.g., sized, assembled) to perform specified functions, and the functioning of the devices typically involves more than one of the basic phenomena encountered in introductory engineering courses. In this chapter, we also stress first-level design notions such as (a) one design may perform better than another, (b) choices can be made: there are degrees of freedom, (c) design constraints generally have to be taken into account, and (d) mathematical optimization methods may have to be invoked. Applications considered in this chapter include thermal insulation, fins, and electronic packages. We begin with a case involving only heat transfer.

5.1 THERMAL INSULATION

In this section we consider the question of how a given amount of insulation should be arranged (distributed) over a wall with nonuniform temperature so that the heat loss from the wall to the ambient is minimized [1]. This question is important because many industrial applications require the use of insulation on nonisothermal walls. Examples include the outer wall of a long reheating oven through which steel laminates ride slowly on a conveyor belt, and the outer wall of an energy storage tank filled with a thermally stratified liquid.

Next to the minimization of the heat loss to the ambient, the judicious use of insulation material is an important design consideration. The purchase, installation, and maintenance of insulation can be expensive. In some applications even the size (volume, weight) of the material used must not exceed a certain limit. Examples include aircraft applications and installations where the integrity of the support structure is threatened by the weight of the insulation (e.g., an insulated suspended pipe). In this section we have our first encounter with the task of optimizing a design *subject to a constraint*. The underlying idea is that a given amount of insulation can be distributed optimally so that the overall insulation effect is maximized.

Consider for this purpose the plane wall shown in Figure 5.1a, in which the wall temperature varies in the x direction, $T(x)$. The wall length is L, and the width perpendicular to the plane of Figure 5.1a is W. The wall is separated from the ambient at temperature T_0 by a layer of insulation of thermal conductivity k and unspecified thickness $t(x)$. We know only that the total volume of insulation material is given:

$$V = \int_0^L t(x)W \, dt \tag{5.1}$$

or that the average thickness of the insulation is fixed:

$$t_{avg} = \frac{1}{L} \int_0^L t(x) \, dx = \frac{V}{LW} \tag{5.2}$$

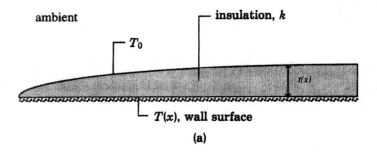

(a)

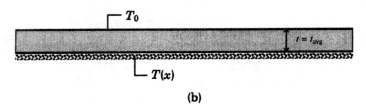

(b)

Figure 5.1 The distribution of a fixed amount of insulation on a nonisothermal wall: (a) insulation with variable thickness, and (b) insulation with uniform thickness.

We assume that the temperature of the outer surface of the insulation is closely equal to T_0 and that the insulation layer is sufficiently thin so that the direction of heat transfer is perpendicular to the wall.

The general problem in which both the wall temperature $T(x)$ and the insulation thickness $t(x)$ are not specified consists of minimizing the heat loss integral

$$\dot{Q} = \int_0^L kW \frac{T(x) - T_0}{t(x)} \, dx \tag{5.3}$$

subject to the volume integral constraint (Equation 5.1). The objective is to find the optimal distribution of insulation, $t_{opt}(x)$, that minimizes the integral of Equation 5.3. The solution to this type of optimization problem is based on the method of *variational calculus*. The method and its general result (the Euler equation) are summarized in Appendix A. Here we note only that the problem stated in the preceding paragraph is analogous to minimizing the integral

$$\Phi = \int_0^L F(x) \, dx$$
$$= \int_0^L \left[kW \frac{T(x) - T_0}{t(x)} + \lambda Wt(x) \right] dx \tag{5.4}$$

in which λ is a *Lagrange multiplier.* Note also that the integrand $F(x)$ is a linear combination of the integrands of the integrals appearing in Equations 5.1 and 5.3.

In the present problem, Equation A.2 of Appendix A (the Euler equation) reduces to solving $\partial F / \partial t = 0$. The result is

$$t_{opt}(x) = K[T(x) - T_0]^{1/2} \tag{5.5}$$

in which K denotes $(k/\lambda)^{1/2}$. By substituting Equation 5.5 into Equation 5.2, the constant K, and thus the Lagrange multiplier λ, can be eliminated, giving

$$t_{opt}(x) = \frac{t_{avg}L}{\int_0^L [T(x) - T_0]^{1/2} \, dx} [T(x) - T_0]^{1/2} \tag{5.6}$$

Equation 5.6 shows that for maximum insulation effect the insulation thickness must be proportional to the square root of the local temperature difference across the insulation. The corresponding minimum heat transfer rate through the area $L\,W$ is

$$\dot{Q}_{min} = \frac{kW}{t_{avg}L} \left\{ \int_0^L [T(x) - T_0]^{1/2} \, dx \right\}^2 \tag{5.7}$$

It is worth noting that the same $t_{opt}(x)$ result is obtained when the amount of insulation material is minimized subject to a fixed rate of heat loss to the ambient. In other words, variational calculus leads again to Equation 5.6 when the volume integral (Equation 5.1) is minimized while holding the heat loss integral (Equation 5.3) fixed.

Example 5.1   Referring to Figure 5.1*a*, if the wall temperature increases linearly in the *x* direction, calculate the total heat transfer rate through the insulation in two competing designs: (a) The insulation material is distributed optimally according to Equation 5.6, and (b) the insulation material is spread evenly over the wall surface of length *L*, as shown in Figure 5.1*b*. Compare the two heat transfer rates.

Solution

MODEL

1. The wall temperature variation is $T(x) = T_0 + x(T_L - T_0)/L$.
2. The heat transfer through the insulation is by conduction in the direction perpendicular to the wall.
3. The temperature on the outer surface of the insulation is T_0, the ambient temperature.

ANALYSIS   By substituting the given wall temperature distribution in Equation 5.6, we obtain

$$t_{opt}(x) = \frac{3}{2} t_{avg} \left(\frac{x}{L}\right)^{1/2}$$

which satisfies also the volume constraint of Equation 5.1. This optimal thickness distribution resembles the one sketched in Figure 5.1*a*. The corresponding heat transfer rate is

$$\dot{Q}_{min} = \frac{4}{9} kWL \frac{T_L - T_0}{t_{avg}}$$

When the same amount of insulation is spread evenly over the wall, $t(x) = t_{avg}$, Equation 5.3 shows that the heat transfer rate to the ambient is

$$\dot{Q}_c = \frac{1}{2} kWL \frac{T_L - T_0}{t_{avg}}$$

If we compare $\dot{Q}_c$ with $\dot{Q}_{min}$, we find that in the constant-thickness case the heat loss is 12.5% larger than in the optimal design.

COMMENT   For further discussion of this class of problems, see Reference 1.

5.2   FINS

The sizing of a fin generally requires the selection of more than one physical dimension. In the plate fin of Figure 5.2, for example, there are three such dimensions, the length L, the thickness t, and the width W of the fin. One of these dimensions can be selected optimally when the total volume of fin material is fixed. Such a constraint is often justified by the cost of the high-thermal-conductivity metals employed in the manufacture of finned surfaces (e.g., copper, aluminum) and weight of the fin in aircraft applications, say.

To select one of the two dimensions of a plate fin optimally means to maximize $\dot{Q}_b$, the total heat transfer rate through the fin root, when the fin volume, V, is fixed [2]:

$$V = LtW \qquad \text{(fixed)} \tag{5.8}$$

5.2.1   Known Fin Width

Let us consider the problem posed above when the fin width is assumed known. We will see in the next section (Equation 5.15) that this assumption also justifies the assumption that the heat transfer coefficient is constant in the longitudinal direction of the plate fin.

The heat transfer rate $\dot{Q}_b$ is given by Equation 4.9 with $p = 2W$, $A_c = Wt$. If we assume further that the fin is sufficiently slender ($t \ll L$) that $L + A_c/p = L + t/2 \approx L$, then L can be eliminated between Equations 4.9 and 5.8 to give the heat transfer rate as a function of the thickness t as the lone design variable:

$$\dot{Q}_b = at^{1/2}\tanh(bt^{-3/2}) \tag{5.9}$$

where the constants a and b are given by

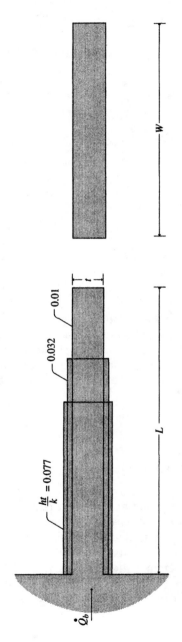

Figure 5.2 Scale drawing of the optimal profile of a plate fin with fixed volume [2].

$$a = (T_b - T_\infty)W(2kh)^{1/2}, \qquad b = \frac{V}{W}\left(\frac{2h}{k}\right)^{1/2} \tag{5.10}$$

The maximum of $\dot{Q}_b$ with respect to the plate fin thickness t is found next by solving $\partial \dot{Q}_b / \partial t = 0$, which yields $bt^{-3/2} \cong 1.42$. Rearranging this result, we obtain the optimal plate fin thickness,

$$t \cong \left(\frac{V}{W}\right)^{2/3}\left(\frac{h}{k}\right)^{1/3} \tag{5.11}$$

and, in view of the volume constraint (Equation 5.8), the corresponding length of the plate fin,

$$L \cong \left(\frac{V}{W}\right)^{1/3}\left(\frac{h}{k}\right)^{-1/3} \tag{5.12}$$

Equations 5.11 and 5.12 can be used to size the plate fin. Thus, after some manipulation, the slenderness ratio of the rectangular profile of the plate fin, t/L, is obtained as follows:

$$\frac{t}{L} \cong \left(\frac{ht}{k}\right)^{1/2} \tag{5.13}$$

It is worth recalling that Equation 4.9 is based on the assumption that the heat flux lines are oriented longitudinally through the fin. It has been shown [2] that this assumption is valid when the Biot number ht/k is smaller than 1. Consequently, the slenderness ratio given by Equation 5.13 must also be smaller than 1, as assumed at the outset. Figure 5.2 shows a sequence of optimal slenderness ratios calculated with Equation 5.13 and drawn to scale. The optimal rectangular profile of the plate fin becomes more slender as the Biot number ht/k decreases. Throughout the sequence the fin volume (or profile area $L\,t$) remains constant.

Finally, the corresponding extremum of $\dot{Q}_b$ is

$$\dot{Q}_{b,\max} \cong 1.258\, kW(T_b - T_\infty)\left(\frac{ht}{k}\right)^{1/2} \tag{5.14}$$

where we note again the requirement $ht/k < 1$.

5.2.2 Known Fin Thickness

Consider now the optimization of the plate fin [3] when the heat transfer coefficient depends on the width W swept by the uniform flow (Figure 5.3). Assume that the fin is shaped as a knife blade ($t << W << L$) and that the boundary layers aligned with W are laminar. In this case the plate thickness t is fixed, while W and L may vary so that the volume constraint $V = LtW$ is satisfied. We are interested in determining the optimal swept width, or the optimal size of the $L \times W$ surface that maximizes the heat transfer rate through the fin, $\dot{Q}_b$.

The expression for $\dot{Q}_b$ is furnished by Equation 4.9 with $p = 2W$, $A_c = Wt$, and with L in place of $L + A_c/p$. The heat transfer coefficient averaged over W is given by Equation 4.26:

$$\bar{h} = 0.664 \frac{k_f}{W} \Pr^{1/3} \left(\frac{U_\infty W}{\nu} \right)^{1/2} \tag{5.15}$$

where k_f and ν denote, respectively, the thermal conductivity and kinematic viscosity of the fluid.

There is only one degree of freedom in this design problem, W or L. Selecting W and eliminating L between Equation 4.9 and $V = LtW$, the resulting heat transfer expression in dimensionless form is

$$\frac{\dot{Q}_b}{kt(T_b - T_\infty)} = (m^{3/4} C^{1/2}) \left(\frac{W}{mt} \right)^{3/4} \tanh \left[\left(\frac{W}{mt} \right)^{-5/4} \right] \tag{5.16}$$

where the constants m and C are given by

$$m = \left(\frac{V}{t^3} C^{1/2} \right)^{4/5}, \qquad C = 1.328 \, \Pr^{1/3} \frac{k_f}{k} \left(\frac{U_\infty t}{\nu} \right)^{1/2} \tag{5.17}$$

We conclude that the fin width W influences $\dot{Q}_b$ in the same way that ξ influences the function $f(\xi) = \xi^{3/4} \tanh(\xi^{-5/4})$, where $\xi = W/mt$. Since

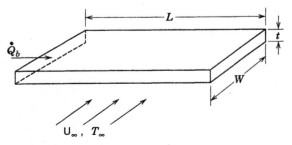

Figure 5.3 Plate fin with laminar boundary layers over the swept width W.

$f \rightarrow 0$ as $\xi \rightarrow 0$ and $\xi \rightarrow \infty$, the function $f(\xi)$ must have a maximum at an intermediate value of ξ. This value can be found numerically as $\xi = 1.07$. The corresponding width is

$$W = 1.07mt \qquad (5.18)$$

or, with Equations 5.17,

$$\frac{W}{t} = 1.12 \left(\frac{V}{t^3}\right)^{4/5} \left(\mathrm{Pr}^{1/3}\, \frac{k_f}{k}\right)^{2/5} \left(\frac{U_\infty t}{\nu}\right)^{1/5} \qquad (5.19)$$

The two fin optimization methods outlined in Sections 5.2.1 and 5.2.2 are based on a very simple fin model. Essential features of this model are (a) the conduction heat transfer through the fin is unidirectional, (b) the temperature T_b is uniform over the root of the fin, and (c) the heat transfer coefficient is independent of position in the direction of heat transfer along the fin. When these assumptions break down, the same optimization ideas can be pursued numerically by modeling the conduction as two dimensional or three dimensional. Additional ways of optimizing the geometry of a fin are presented in the literature [4–6].

Example 5.2   An electronic package includes several parallel plate fins of width $W = 2.2$ cm swept by forced air with $U_\infty = 1.75$ m/s and $T_\infty = 20°C$. The fin material is aluminum. Weight limitations on the overall package permit the use of only 1 g of aluminum for each fin. Determine the plate fin thickness (t) and length (L) that maximizes the heat transfer rate extracted per fin, and an expression for the corresponding heat transfer rate $\dot{Q}_b$.

Solution

MODEL

1. The conduction through the fin is mainly in the L direction.
2. The heat transfer coefficient does not vary in the L direction.
3. The temperature is uniform over the root of the fin.

ANALYSIS   Since the swept length W is specified, this fin optimization problem is of the same type as in Section 5.2.1. We calculate, in order,

$$Re_W = \frac{U_\infty W}{\nu} = \frac{(1.75 \text{ m/s})(0.022 \text{ m})}{0.19 \times 10^{-4} \text{ m}^2/\text{s}} = 2026$$

(laminar boundary layer flow)

$$\bar{h} = 0.664 \frac{k_{\text{air}}}{W} Pr^{1/3} Re_W^{1/2} \quad \text{(Equation 4.26)}$$

$$= 0.664 \frac{0.028 \text{ W/m·K}}{0.022 \text{ m}} (0.72)^{1/3} (2026)^{1/2}$$

$$= 34.1 \text{ W/m}^2\text{·K}$$

$$V = \frac{m}{\rho} = \frac{1 \text{ g}}{2.707 \text{ g/cm}^3} = 0.369 \text{ cm}^3$$

$$t = \left(\frac{V}{W}\right)^{2/3} \left(\frac{\bar{h}}{k_{\text{Al}}}\right)^{1/3} \quad \text{(Equation 5.11)}$$

$$= \left(\frac{0.369 \text{ cm}^3}{2.2 \text{ cm}}\right)^{2/3} \left(\frac{34.1 \text{ W/m}^2\text{·K}}{204 \text{ W/m·K}}\right)^{1/3} = 0.036 \text{ cm}$$

$$L = \frac{V}{tW} = \frac{0.369 \text{ cm}^3}{(0.036 \text{ cm})(2.2 \text{ cm})} = 4.66 \text{ cm}$$

We note at this point that the Biot number is much smaller than 1:

$$\frac{\bar{h}t}{k_{\text{Al}}} = 6 \times 10^{-5} \ll 1$$

which means that assumption 1 is justified. Assumption 2 is also justified because $\bar{h}$ varies as $W^{-1/2}$ (Equation 4.26), and the swept length W is independent of the distance away from the fin base.

The corresponding heat transfer rate is found using Equation 4.10, where

$$m = \left(\frac{2\bar{h}}{k_{\text{Al}}t}\right)^{1/2} = 0.305 \text{ cm}^{-1}, \qquad L_c = L + \frac{t}{2} = 4.68 \text{ cm},$$

$$\eta = \frac{\tanh(mL_c)}{mL_c} = 0.624 \text{ [2, p. 60]}$$

Substituting into Equation 4.10, we have

$$\frac{\dot{Q}_b}{T_b - T_\infty} = 2\bar{h}WL\eta$$

$$= 2(34.1 \text{ W/m}^2\text{·K})(0.022 \text{ m})(0.0466 \text{ m})(0.624) = 0.0436 \text{ W/K}$$

5.3 ELECTRONIC PACKAGES

In the design of packages of electronic components, there are strong incentives to mount as much circuitry as possible in a given space. This can be achieved by judiciously selecting the *geometry* of the package: the way in which the components are arranged relative to the coolant and to each other in the fixed space. An important constraint is that the highest temperature (the "hot spot") in the package must not exceed a specified value. If the temperature rises above the allowable limit, the operation of the electronic circuit is threatened. Since each component in the package generates heat, the design objective is to maximize the total rate of heat transfer from the finite space occupied by the package to the coolant that flows through the package.

As there is a great diversity of components, packages, and cooling techniques, each optimal cooling arrangement that emerges from a design process tends to be specific to the application at hand, and lacks general applicability. Still, in the topics featured in this section we identify several fundamental ways in which the geometric features of entire classes of electronic packages can be optimized, and similar reasoning may be applicable to other cases of practical interest.

5.3.1 Natural Convection Cooling

Figure 5.4 shows a large number of parallel, equidistant electronic boards cooled by natural convection in the space of height H, stack thickness L, and width W. The width is perpendicular to the plane of the figure, and the board thickness is negligible. Cold air at temperature T_∞ enters through the bottom of the package, rises through the board-to-board channels, and exits through the upper opening. The lateral walls of area $H W$ that confine the package are insulated. The maximum allowed board temperature T_{max} is set by electronic operational constraints.

The objective of the design is to maximize the total heat transfer rate $\dot{Q}$ removed by the coolant from the $H \times L \times W$ volume. The only variable is the number of channels

$$n = \frac{L}{D} \tag{5.20}$$

or the board-to-board spacing D. To determine the spacing D that maximizes $\dot{Q}$ can be a laborious task. We will show, however, that with a little ingenuity the problem can be solved easily. The method, outlined first in Reference 3, begins with the assumptions that (a) the flow is laminar, (b) the board surfaces are sufficiently smooth to justify the use of heat transfer results for natural convection over vertical smooth walls, and (c) the maximum temperature T_{max} is closely representative of the temperature at every point on the board surface. The method consists of two steps. In the first, we identify two extremes:

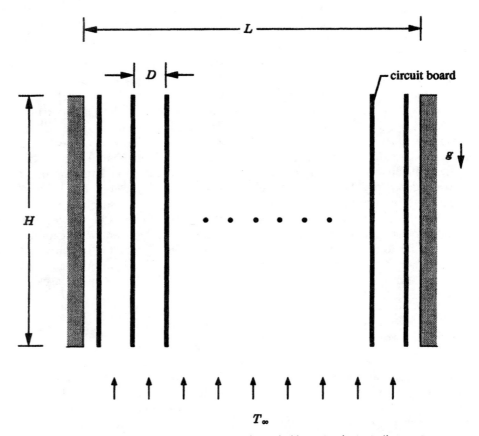

Figure 5.4 Electronic circuit boards cooled by natural convection.

the small-D limit and the large-D limit. In the second step, the two extreme regimes are intersected for the purpose of locating the D value that maximizes $\dot{Q}$.

The Small-D Limit. When D becomes sufficiently small, the channel formed between two boards becomes narrow enough for the flow and heat transfer to be fully developed. Accordingly, from the discussion of vertical channels in Section 4.3.3, we have $\dot{Q}_1 = \dot{m}_1 c_p (T_{max} - T_\infty)$ for the heat transfer rate extracted by the coolant from one of the channels of spacing D, where T_∞ is the inlet temperature and T_{max} is the outlet temperature. The mass flow rate is $\dot{m}_1 = \rho(DW)U$, where the mean velocity U can be estimated by replacing $\Delta p / L$ with $\rho g \beta (T_{max} - T_\infty)$ in Equation 4.40:

$$\rho g \beta (T_{max} - T_\infty) = f \frac{4}{D_h} \frac{1}{2} \rho U^2 \tag{5.21}$$

Substituting $f = 24/(U D_h / \nu)$ from Table 4.3 and $D_h = 2D$, we have

$$U = \frac{g \beta (T_{max} - T_\infty) D^2}{12 \nu} \tag{5.22}$$

Accordingly,

$$\dot{m}_1 = \rho DWU = \frac{\rho W g \beta (T_{max} - T_\infty) D^3}{12 \nu} \tag{5.23}$$

The total rate at which heat is removed from the package is $\dot{Q} = n \dot{Q}_1$, or

$$\dot{Q} = \rho c_p WL \frac{g \beta (T_{max} - T_\infty)^2 D^2}{12 \nu} \tag{5.24}$$

In conclusion, in the $D \to 0$ limit the total heat transfer rate varies as D^2. This trend is indicated by the small-D asymptote plotted in Figure 5.5.

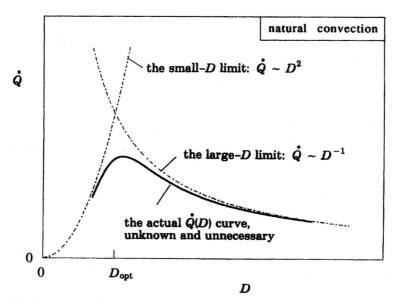

Figure 5.5 The maximization of the total heat transfer rate removed by natural convection from the stack of Figure 5.4.

The Large-D Limit. Consider next the limit in which D is large enough that it exceeds the thickness of the thermal boundary layer that forms on each vertical surface: $\delta_T \sim H \, \text{Ra}_H^{-1/4}$, where $\text{Ra}_H = g\beta H^3(T_{max} - T_\infty)/\alpha\nu$. In other words, the spacing is considered large when

$$D > H \left[\frac{g\beta H^3(T_{max} - T_\infty)}{\alpha\nu} \right]^{-1/4} \tag{5.25}$$

This inequality is valid when $\text{Pr} \geq 0.5$ [3]. In this limit the boundary layers are distinct (thin compared with D), and the center region of the board-to-board spacing is occupied by fluid of temperature T_∞. The number of distinct boundary layers is $2n = 2L/D$ because there are two for each D spacing. The heat transfer rate through one boundary layer is $\bar{h}HW(T_{max} - T_\infty)$ for which $\bar{h}$ is furnished by Equation 4.56 or for laminar flow only by [2]

$$\frac{\bar{h}H}{k} = 0.517 \, \text{Ra}_H^{1/4} \tag{5.26}$$

where $\text{Ra}_H = g\beta(T_{max} - T_\infty)H^3/\alpha\nu$. The total rate of heat transfer extracted from the entire package is $2n$ times larger than $\bar{h}HW(T_{max} - T_\infty)$:

$$\dot{Q} = 2 \frac{L}{D} HW(T_{max} - T_\infty) \frac{k}{H} 0.517 \, \text{Ra}_H^{1/4} \tag{5.27}$$

Equation 5.27 shows that in the large-D limit the total heat transfer rate varies as D^{-1} as the board-to-board spacing changes. This second asymptote also has been plotted in Figure 5.5.

The Optimal Board-to-Board Spacing. What we have determined so far are the two asymptotes of the actual (unknown) curve of $\dot{Q}$ versus D. Figure 5.5 shows that the asymptotes intersect above what would be the peak of the actual $\dot{Q}(D)$ curve. It is unnecessary to determine the actual $\dot{Q}(D)$ relation, however. The optimal spacing D_{opt} for maximum $\dot{Q}$ can be estimated as the D value where Equations 5.24 and 5.27 intersect [3]:

$$\frac{D_{opt}}{H} \cong 2.3 \left[\frac{g\beta(T_{max} - T_\infty)H^3}{\alpha\nu} \right]^{-1/4} \tag{5.28}$$

This D_{opt} estimate is within 20% of the optimal spacing deduced based on much lengthier methods, such as the maximization of the $\dot{Q}(D)$ relation [7] and the finite-difference simulations of the complete flow and temperature fields in the package [8].

An estimate of the maximum heat transfer rate can be obtained by substituting D_{opt} into Equation 5.27 (or Equation 5.24),

$$\dot{Q}_{\text{max}} \approx 0.45k(T_{\text{max}} - T_{\infty}) \frac{LW}{H} \text{Ra}_H^{1/2} \qquad (5.29)$$

The approximation sign is a reminder that the peak of the actual $\dot{Q}(D)$ curve falls under the intersection of the two asymptotes (Figure 5.5). This result also can be expressed as the maximum volumetric rate of heat generation in the $H \times L \times W$ volume,

$$\frac{\dot{Q}_{\text{max}}}{HLW} \approx 0.45 \frac{k}{H^2} (T_{\text{max}} - T_{\infty}) \text{Ra}_H^{1/2} \qquad (5.30)$$

In conclusion, if the heat transfer mechanism is natural convection, the maximum volumetric rate of heat generation is proportional to $(T_{\text{max}} - T_{\infty})^{3/2}$, $H^{-1/2}$, and the property group $k(g\beta/\alpha\nu)^{1/2}$.

5.3.2 Forced Convection Cooling

Consider now the problem of installing the optimal number of heat generating boards in a space cooled by forced convection [9]. As shown in Figure 5.6, the swept length of each board is L, while the transverse dimension of the entire package is H. The width of the stack, W, is perpendicular to the plane of the figure. We retain the assumptions (a)–(c) listed under Equation 5.20. The thickness of the individual board is again negligible relative to the board-to-board spacing D, so that the number of channels is

$$n = \frac{H}{D} \qquad (5.31)$$

The pressure difference across the package, Δp, is assumed constant and known. This is a good model for electronic systems in which several packages and other features (e.g., channels) receive their coolant *in parallel,* from the same plenum. The plenum pressure is maintained by a fan, which may be located upstream or downstream of the package.

The Small-D Limit. When D becomes sufficiently small, the channel formed between two boards becomes narrow enough for the flow and heat transfer to be fully developed. In this limit, the mean outlet temperature of the fluid approaches the board temperature T_{max}. The total rate of heat transfer from

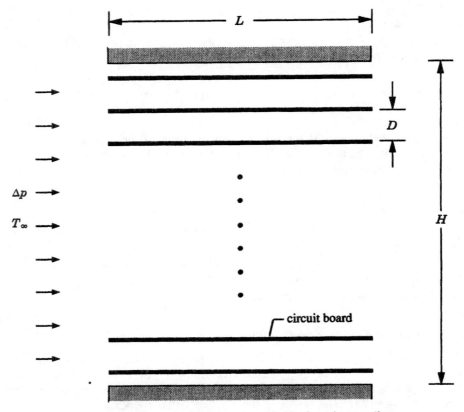

Figure 5.6 Electronic circuit boards cooled by forced convection.

the $H \times L \times W$ volume is $\dot{Q} = \dot{m}c_p(T_{max} - T_\infty)$, where $\dot{m} = \rho HWU$. The mean velocity through the channel, U, can be evaluated from Equation 4.42 with $a = D$:

$$U = \frac{D^2}{12\mu} \frac{\Delta p}{L} \qquad (5.32)$$

Accordingly,

$$\dot{Q} = \rho HW \frac{D^2}{12\mu} \frac{\Delta p}{L} c_p(T_{max} - T_\infty) \qquad (5.33)$$

In this way we conclude that the total heat transfer rate varies as D^2. This trend is illustrated by the small-D asymptote in Figure 5.7.

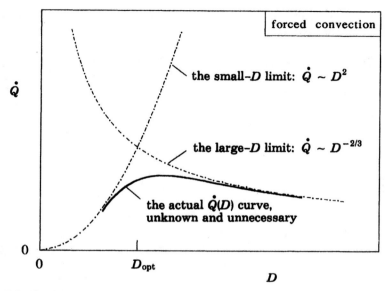

Figure 5.7 The maximization of the total heat transfer rate removed by forced convection from the stack of Figure 5.6.

The Large-D Limit. Consider next the limit in which D is large enough that it exceeds the thickness of the thermal boundary layer that forms on each horizontal surface. In this case it is necessary to determine the free-stream velocity U_∞ that sweeps these boundary layers. Since the pressure drop Δp is fixed, a force balance on the $H \times L \times W$ control volume reads

$$\Delta p H W = \bar{\tau}(2n)LW \tag{5.34}$$

where $\bar{\tau}$ is the wall shear stress averaged over L given by Equation 4.25. Thus, Equations 5.34 and 4.25 yield

$$U_\infty = \left(\frac{\Delta p H}{1.328 n L^{1/2}\rho\nu^{1/2}}\right)^{2/3} \tag{5.35}$$

The heat transfer rate through a single board surface is $\dot{Q}_1 = \bar{h}LW(T_{max} - T_\infty)$, where $\bar{h}$ (the heat transfer coefficient averaged over L) is provided by Equation 4.26 when $\mathrm{Pr} \gtrsim 0.5$. The total heat transfer rate from the entire package is $\dot{Q} = 2n\dot{Q}_1$ or, after using Equation 4.26 and the expression for U_∞ listed above,

$$\dot{Q} = 1.21 k H W(T_{max} - T_\infty)\left(\frac{\mathrm{Pr}L\Delta p}{\rho\nu^2 D^2}\right)^{1/3} \tag{5.36}$$

In conclusion, in the large-D limit the total heat transfer rate varies as $D^{-2/3}$ as the board-to-board spacing changes. This trend is shown in Figure 5.7.

The Optimal Board-to-Board Spacing. The intersection of the two $\dot{Q}(D)$ asymptotes, Equations 5.33 and 5.36, yields an estimate for the board-to-board spacing for maximum heat transfer rate:

$$\frac{D_{\text{opt}}}{L} \cong 2.7 \left(\frac{\Delta p L^2}{\mu \alpha}\right)^{-1/4} \tag{5.37}$$

This spacing increases as $L^{1/2}$ and decreases as $\Delta p^{-1/4}$ with increasing L and Δp, respectively. The D_{opt} estimate of Equation 5.37 underestimates by 12% the more exact value obtained by locating the maximum of the actual $\dot{Q}(D)$ curve [9] and is adequate when the board surface is modeled either as uniform flux or isothermal. It has been shown [10] that Equation 5.37 holds even when the board thickness is not negligible relative to the board-to-board spacing.

At this stage it is useful to introduce the dimensionless group called the *pressure drop number* [2], which was used extensively in Reference 11 for correlating electronics cooling data:

$$\Pi_L = \frac{\Delta p L^2}{\mu \alpha} \tag{5.38}$$

The manner in which the design parameters influence the maximum rate of heat removal from the package can be expressed as

$$\dot{Q}_{\text{max}} \approx 0.6 k (T_{\text{max}} - T_\infty) \frac{HW}{L} \Pi_L^{1/2} \tag{5.39}$$

which is obtained by setting $D = D_{\text{opt}}$ in Equation 5.33 or Equation 5.36. Once again, the approximation sign is a reminder that the acual $\dot{Q}_{\text{max}}$ is as much as 19% smaller because the peak of the $\dot{Q}(D)$ curve is situated under the point where the two asymptotes cross in Figure 5.7. The maximum volumetric rate of heat generation in the $H \times L \times W$ volume is

$$\frac{\dot{Q}_{\text{max}}}{HLW} \approx 0.6 \frac{k}{L^2} (T_{\text{max}} - T_\infty) \Pi_L^{1/2} \tag{5.40}$$

The similarity between the forced convection results (Equations 5.39 and 5.40) and the corresponding results for natural convection cooling (Equations 5.29 and 5.30) is worth noting. The role played by the Rayleigh number Ra_H in the free convection case is played in forced convection by the pressure drop number Π_L.

Example 5.3   Figure E5.3 shows the plate fins of an electronic package that is cooled by forced air. Seen from above, the package is shaped as a square with the side $L = 2.2$ cm. The height of each plate fin is 1.2 cm, the fin thickness is $t = 1$ mm, and the air stream is characterized by $U_\infty = 1.75$ m/s and $T_\infty = 20°C$. The package temperature is 100°C. The fin material (aluminum) is highly conductive such that the fin efficiency η is 0.98. Estimate the number of fins that should be installed on top of the package to maximize the rate of heat transfer.

Solution

MODEL

1. The fin efficiency $\eta = 0.98$ means that the plate fins are essentially isothermal in the direction perpendicular to the package: Their temperature is closely the same as the package temperature.
2. The cooling of the $L \times L$ stack of plate fins is analogous to the cooling of the $L \times H$ stack of hot plates shown in Figure 5.6. In the latter, the stack width H would be equal to the swept length L.

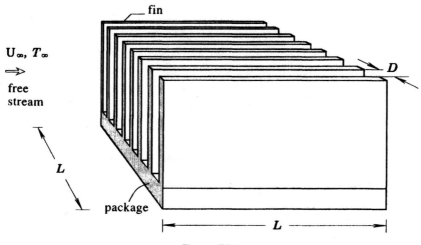

Figure E5.3

3. The pressure difference that drives the air through the plate fins is related to the free-stream velocity U_∞ by

$$\Delta p \sim \tfrac{1}{2}\rho U_\infty^2$$

4. The heat transfer through the unfinned portions of the package surface is negligible.

ANALYSIS The analogy between the finned heat sink and the stack of parallel plates of Figure 5.6 means that we can use Equation 5.37 to estimate the optimal fin-to-fin spacing [11]. Using $\Delta p = \tfrac{1}{2}\rho U_\infty^2$ to eliminate Δp from Equation 5.37 we obtain

$$\frac{D_{\text{opt}}}{L} \sim 3.2 \ Pr^{-1/4} Re_L^{-1/2}$$

where $Re_L = U_\infty L/\nu$. In the specified design, the coolant is air with $Pr = 0.72$ and $\nu = 0.19 \ cm^2/s$ at the film temperature 60°C, and the flow is laminar:

$$Re_L = \frac{U_\infty L}{\nu} = \frac{(1.75 \ \text{m/s})(0.022 \ \text{m})}{0.19 \times 10^{-4} \ \text{m}^2/\text{s}} = 2026$$

The optimal spacing given by the D_{opt} formula shown above is

$$D_{\text{opt}} \sim 3.2(0.72)^{-1/4}(2026)^{-1/2}(2.2 \ \text{cm}) = 0.17 \ \text{cm}$$

which means that the optimal number of plate fins is closely

$$n_{\text{opt}} \sim \frac{L}{D_{\text{opt}} + t} = \frac{2.2 \ \text{cm}}{(0.17 + 0.1) \ \text{cm}} \sim 8$$

COMMENT Nakayama et al. [12] optimized the same finned heat sink numerically and experimentally and found that the optimal number of fins is 8 when $Re_L = 2000$. Also, heat sinks with 7 or 9 plate fins were found to perform nearly the same as the sink with 8 fins. In other words, the peak of the overall thermal conductance curve is sufficiently flat that a nearly optimal solution (Section 1.2) is obtained for $n_{\text{opt}} \sim 7$ to 9.

5.3.3 Cooling of a Heat-Generating Board Inside a Parallel-Plate Channel

In certain types of electronic packages, the coolant flows through a set of two-dimensional parallel channels formed by a row of printed circuit boards plugged into a common board. Each board may be surrounded by a metal or metal-coated plastic case whose function is to shield the electronic circuitry from external electromagnetic noise. It is important to know the optimal geometry of each cassette (i.e., the board and its parallel-plate casing) so that the board operating temperature is minimum. Optimization of the geometry of the cassette requires not only finding the best position for the board inside the channel but also the best spacing of the channel in which the heat-generating board is encased.

Board with Large Thermal Conductance in the Transverse Direction. Consider the problem of cooling a board of length L by positioning it in a stream of coolant that flows through an insulated parallel-plate channel of the same length. The channel spacing D is fixed. The geometry sketched in Figure 5.8 is two dimensional, as the board and the channel are sufficiently wide (the width is W) in the direction perpendicular to the figure, $W > L$.

We assume that the pressure difference across the channel, Δp, is fixed. The flow is driven by a fan with diameter considerably greater than the channel spacing D. In an actual application, the fan would blow air through a stack of 10 or more cassettes like the one shown in Figure 5.8.

The total rate of heat transfer $\dot{Q}$ from the heated plate to the fluid, through *both* sides of the plate, is fixed by the electric circuit design. The plate thickness is negligible with respect to the channel spacing D. The only degree of freedom in choosing the best cooling arrangement is the position of the heated plate inside the channel. This position is defined by the subchannel spacings above and below the heated plate, D_1 and D_2, such that $D_1 + D_2 = D$.

In the present development, we assume that (a) the heated plate is isothermal at T_w, (b) the flow is fully developed and laminar on both sides of the plate, (c) the surfaces of the plate and the channel walls are smooth, and (d) $T_{\text{out,1}} \approx T_w$ and $T_{\text{out,2}} \approx T_w$, where as shown in Figure 5.8, $T_{\text{out,1}}$ and $T_{\text{out,2}}$ are the outlet bulk temperatures, above and below the heated plate. The

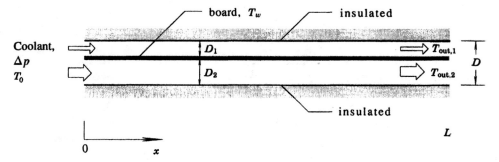

Figure 5.8 Heat-generating board cooled by a stream in an insulated parallel-plate channel [13].

objective is to determine the ratio D_1/D so that the thermal conductance $\dot{Q}/(T_w - T_0)$ is the greatest.

If $\dot{Q}_1$ and $\dot{Q}_2$ are the heat transfers rates through the upper side and the lower side of the heated plate, respectively, then assumption (d) permits us to write

$$\dot{Q}_1 = \dot{m}_1 c_p(T_w - T_0), \qquad \dot{Q}_2 = \dot{m}_2 c_p(T_w - T_0) \qquad (5.41)$$

where, for fully developed laminar flow (Equation 4.42),

$$\dot{m}_1 = \frac{\rho W}{12\mu} \frac{\Delta p}{L} D_1^3, \qquad \dot{m}_2 = \frac{\rho W}{12\mu} \frac{\Delta p}{L} D_2^3 \qquad (5.42)$$

Next, we write y and $1 - y$ for the dimensionless spacings of the upper and lower subchannels:

$$D_1 = yD, \qquad D_2 = (1 - y)D \qquad (5.43)$$

and calculate the total heat transfer rate $\dot{Q} = \dot{Q}_1 + \dot{Q}_2$, using Equations 5.41–5.42. The result is

$$\frac{\dot{Q}}{T_w - T_0} \frac{12\mu L}{\rho c_p W D^3 \, \Delta p} = y^3 + (1 - y)^3 \qquad (5.44a)$$

As the maximum value of the right side of this expression is 1, corresponding to $y = 0$ and $y = 1$, the equation can be rewritten as

$$\frac{\dot{Q}/(T_w - T_0)}{[\dot{Q}/(T_w - T_0)]_{max}} = y^3 + (1 - y)^3 \qquad (5.44b)$$

where

$$\left(\frac{\dot{Q}}{T_w - T_0}\right)_{max} = \frac{\rho c_p W D^3 \, \Delta p}{12\mu L} \qquad (5.44c)$$

Equation 5.44b is shown graphically in Figure 5.9, where it can be noted that the minimum value for the ratio is $\frac{1}{4}$ when $y = \frac{1}{2}$. This is the worst cooling arrangement. We may conclude that the best arrangement is when the board is attached to one of the insulated walls of the channel, even though in that case the entire heat transfer rate $\dot{Q}$ must leave the plate through only one of its side surfaces. When the board is attached to one of the walls, the thermal conductance $\dot{Q}/(T_w - T_0)$ is four times greater than when the board is positioned in the center of the channel.

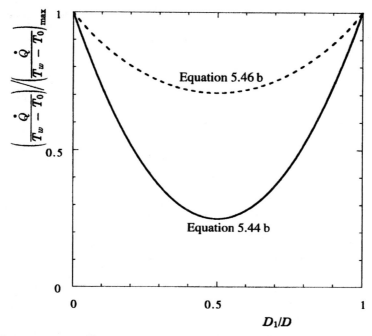

Figure 5.9 The effect of the board position on the overall thermal conductance: (———) laminar, fully developed; (-------) turbulent, fully rough, fully developed [13].

The conclusion that the worst cooling position is $y = \frac{1}{2}$ remains valid even when some of the simplifying assumptions (a)–(d) are relaxed. For example, let us discard (b) and (c) and assume instead that the flow is turbulent fully developed and the board surfaces are very rough. This is a good model for cassettes with L/D ratios greater than 10, so that the entrance region is relatively small, and with boards densely covered with chips and circuitry that rise as large-scale asperities above the surface. Under these circumstances (i.e., the "fully rough" limit of turbulent duct flow) the friction factor is practically independent of the Reynolds number (Figure 2.5) and Equations 5.42 are replaced by

$$\dot{m}_1 = \left(\frac{\rho \, \Delta p}{f_1 L}\right)^{1/2} W D_1^{3/2}, \qquad \dot{m}_2 = \left(\frac{\rho \, \Delta p}{f_2 L}\right)^{1/2} W D_2^{3/2} \qquad (5.45)$$

The constant friction factors f_1 and f_2 depend on the dimensions of the roughness elements (assumed the same for both board surfaces) and on the respective subchannel spacings (D_1, D_2). When the board is placed in the stream at y values comparable with $\frac{1}{2}$, the spacings D_1 and D_2 are also comparable and, as a first approximation, f_1 and f_2 may be taken as equal to the same constant f. This is a conservative approximation to which we return shortly. In the end, Equation 5.44 is replaced by

$$\frac{\dot{Q}}{T_w - T_0} \left(\frac{fL}{\rho \, \Delta p} \right)^{1/2} \frac{1}{c_p W D^{3/2}} = y^{3/2} + (1 - y)^{3/2} \qquad (5.46a)$$

or expressed alternatively as

$$\frac{\dot{Q}/(T_w - T_0)}{[\dot{Q}/(T_w - T_0)]_{max}} = y^{3/2} + (1 - y)^{3/2} \qquad (5.46b)$$

where

$$\left(\frac{\dot{Q}}{T_w - T_0} \right)_{max} = c_p W D^{3/2} \left(\frac{\rho \, \Delta p}{fL} \right)^{1/2} \qquad (5.46c)$$

Equation 5.46b is shown graphically in Figure 5.9, where it can be noted once again that the minimum value for the ratio occurs when the board is placed in the middle of the channel. The thermal conductance minimum is not as sharp as for fully developed laminar flow, however, suggesting that the optimal positioning of the board is not as critical in the fully rough limit. If the analysis is repeated by taking into account the difference between f_1 and f_2 as the board is positioned close to one of the walls, a curve that falls under the dashed curve in Figure 5.9 would be obtained.

Board with Finite Thermal Conductance in the Transverse Direction.

Consider now a more realistic model in which the board of Figure 5.8 (the substrate of an electronic circuit board) has a finite thermal conductivity k_w and thickness t. The thickness continues to be small relative to D, however. The two surfaces of the board are loaded equally and uniformly with electronics. The constant heat generation rate per unit board surface is q''. It is important to note that the heat fluxes removed by the two streams are generally not equal because of the conduction heat transfer across the board. The temperatures of the two board surfaces (T_1, T_2) increase in the downstream direction, and reach their highest levels at the trailing edge, $x = L$. The objective is to minimize the larger of these two trailing-edge temperatures, by choosing the board position y.

We obtain the temperature distributions $T_1(x)$ and $T_2(x)$ on the two surfaces by recognizing that the temperature increase along each surface (for example, $T_1(L) - T_1(0)$) is considerably greater than the local temperature difference between the surface and the corresponding stream. Accordingly, the local temperature of each stream is close to the temperature of the neighboring spot on the board surface bathed by the stream. Energy balances for elemental slices dx of the D_1 and D_2 subchannels read

$$\dot{m}_1 c_p \, dT_1 = q_1'' W \, dx, \qquad \dot{m}_2 c_p \, dT_2 = q_2'' W \, dx \qquad (5.47)$$

The mass flow rates $\dot{m}_1$ and $\dot{m}_2$ are given by Equations 5.42, as each sub-channel stream is assumed laminar and fully developed. Although the heat flux q'' generated by the electronics mounted on each surface of the board is uniform, the heat fluxes removed by the two streams are influenced by the conduction across the board: $(k_w/t)(T_2 - T_1)$. Thus, we have

$$q_1'' = q'' + \frac{k_w}{t}(T_2 - T_1), \qquad q_2'' = q'' - \frac{k_w}{t}(T_2 - T_1) \qquad (5.48)$$

Equations 5.42, 5.47, and 5.48 suffice for determining the surface temperatures $T_1(x)$ and $T_2(x)$ subject to the entrance condition $T_1(0) = T_2(0) = T_0$. By introducing the dimensionless variables $D_1/D = y$, $D_2/D = 1 - y$, $\xi = x/L$, and

$$\begin{cases} \theta_1 = (T_1 - T_0) \dfrac{\rho c_p \, \Delta p D^3}{12 \mu L^2 q''} \\[2mm] \theta_2 = (T_2 - T_0) \dfrac{\rho c_p \, \Delta p D^3}{12 \mu L^2 q''} \\[2mm] B = 12 \dfrac{k_w}{k} \dfrac{\mu \alpha L^2}{\Delta p \, D^3 t} \end{cases} \qquad (5.49)$$

the problem reduces to integrating the two equations

$$y^3 \frac{d\theta_1}{d\xi} = 1 + B(\theta_2 - \theta_1) \qquad (5.50)$$

$$(1 - y)^3 \frac{d\theta_2}{d\xi} = 1 - B(\theta_2 - \theta_1) \qquad (5.51)$$

subject to $\theta_1(0) = \theta_2(0) = 0$. The solution for the temperature of the board surface facing the D_2 subchannel is

$$\theta_2(\xi) = \left(\frac{a}{r} - \frac{b}{r^2} \right) [1 - \exp(-r\xi)] + \frac{b}{r} \xi \qquad (5.52)$$

where

$$r = B \left[\frac{1}{(1 - y)^3} + \frac{1}{y^3} \right], \qquad a = \frac{1}{(1 - y)^3}, \qquad b = \frac{2B}{(1 - y)^3 y^3} \qquad (5.53)$$

The highest temperature occurs at the trailing edge, $\theta_{h,2} = \theta_2(1)$, namely

$$\theta_{h,2} = \left(\frac{a}{r} - \frac{b}{r^2} \right) [1 - \exp(-r)] + \frac{b}{r} \qquad (5.54)$$

This temperature is a function of the board position y and the board conductance parameter B, as shown by the solid curves in the lower frame of Figure 5.10. The highest temperature of the board surface facing the D_1 subchannel, $\theta_{h,1} = \theta_1(1)$, is obtained by switching y and $1 - y$ in the $\theta_{h,2}$ solution of Equation 5.54. The resulting family of curves $\theta_{h,1}(y, B)$ is superimposed with dashed lines on Figure 5.10. On this composite graph, we seek the board position y_{min} that gives the lowest board temperatures when B is specified. The best board position y_{min} depends on B, that is, on the degree to which the board substrate is a good thermal conductor:

(i) When $B >> 1$, the $\theta_{h,1}$ and $\theta_{h,2}$ curves are bell shaped and fall on top of each other. The lowest temperatures are at $y_{min} = 0$ and $y_{min} = 1$, that is, when the board is positioned close to one of the insulated walls of the channel. The worst position (called y_{max}) is in the middle of the channel, $y_{max} = \frac{1}{2}$, where the highest temperature rise ($\theta_{h,1}$ or $\theta_{h,2}$) is about four times greater than when the board is mounted close to one of the insulated walls. These conclusions agree with those for the isothermal board model (i.e., $B \rightarrow \infty$) considered in the first part of this section.

(ii) When the board is a poor thermal conductor, $B << 1$, the $\theta_{h,1}$ and $\theta_{h,2}$ curves intersect at $y = \frac{1}{2}$. That intersection corresponds to the lowest ($\theta_{h,1} = \theta_{h,2}$) values, indicating that the best position for the board is along the midplane of the channel. The worst position, y_{max}, approaches 0.8 and 0.2 as B decreases.

These conclusions are summarized in Figure 5.11. The transition from conducting boards to poorly conducting boards occurs when B drops below 0.166. This transition is also illustrated in Figure 5.10. The best location for poorly conducting boards, $y_{min} = \frac{1}{2}$, happens to be exactly the same as the worst location for highly conducting boards. This observation stresses the important role of the dimensionless number B. This number should be calculated early, in order to determine the problem type: (i) or (ii), as discussed above.

The trailing-edge temperature that corresponds to the location y_{min} is presented as θ_{min} versus B in Figure 5.12. The same figure shows the uppermost

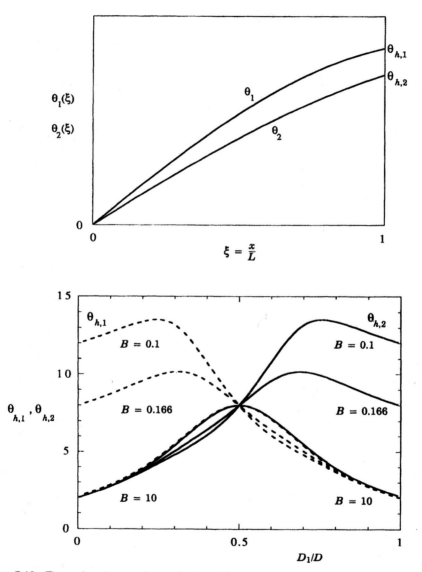

Figure 5.10 The surface temperatures of a board with finite transverse thermal conductance (top) and the trailing-edge temperatures (bottom) [13].

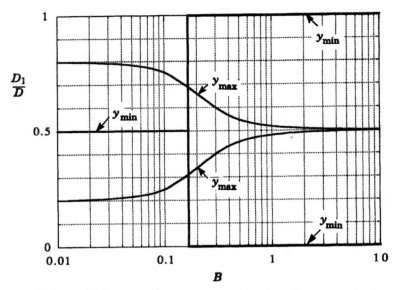

Figure 5.11 The best position (y_{min}) and worst position (y_{max}) of a heat-generating board with finite transverse thermal conductance [13].

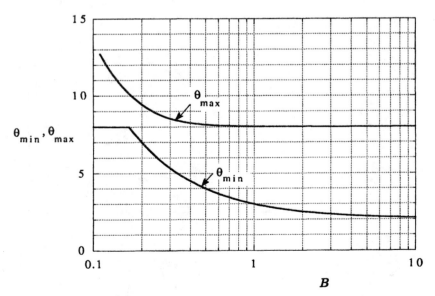

Figure 5.12 The lowest temperature (θ_{min}) and uppermost temperature (θ_{max}) that correspond, respectively, to the best position (y_{min}) and the worst position (y_{max}) of Figure 5.11 [13].

trailing-edge temperature that corresponds to the worst position, y_{max}. The lowest temperature (θ_{min}) is considerably smaller than the highest temperature (θ_{max}), regardless of the B value. This shows the importance of knowing not only the best design (y_{min}) but also the worst design (y_{max}).

5.4 CLOSURE

The introduction of thermal design and optimization principles in this book progresses gradually from the simple to the complex. No attempt is made to cover exhaustively the broad field of thermal systems design, however. Our aim is to foster the use in design of effective reasoning, judicious modeling, and a systematic approach. In this chapter, we consider elementary problems that are representative of those encountered in thermal design. These problems require only heat transfer and fluid flow fundamentals. In the next chapter we consider design problems that require the use of thermodynamics in addition to heat transfer and fluid mechanics. In Chapters 7–9 engineering economics is integrated.

REFERENCES

1. A. Bejan, How to distribute a finite amount of insulation on a wall with nonuniform temperature, *Int. J. Heat Mass Transfer,* Vol. 36, 1993, pp. 49–56.

2. A. Bejan, *Heat Transfer,* Wiley, New York, 1993.

3. A. Bejan, *Convection Heat Transfer,* Wiley, New York, 1984.

4. A. D. Snider and A. D. Kraus, The quest for the optimum longitudinal fin profile, *ASME HTD,* Vol. 64, 1986, pp. 43–48.

5. D. Poulikakos and A. Bejan, Fin geometry for minimum entropy generation in forced convection, *J. Heat Transfer,* Vol. 104, 1982, pp. 616–623.

6. P. Jany and A. Bejan, Ernst Schmidt's approach to fin optimization: an extension to fins with variable conductivity and the design of ducts for fluid flow, *Int. J. Heat Mass Transfer,* Vol. 31, 1988, pp. 1635–1644.

7. A. Bar-Cohen and W. M. Rohsenow, Thermally optimum spacing of vertical, natural convection cooled, parallel plates, *J. Heat Transfer,* Vol. 106, 1984, pp. 116–123.

8. N. K. Anand, S. H. Kim, and L. S. Fletcher, The effect of plate spacing on free convection between heated parallel plates, *J. Heat Transfer,* Vol. 114, 1992, pp. 515–518.

9. A. Bejan and E. Sciubba, The optimal spacing of parallel plates cooled by forced convection, *Int. J. Heat Mass Transfer,* Vol. 35, 1992, pp. 3259–3264.

10. S. Mereu, E. Sciubba, and A. Bejan, The optimal cooling of a stack of heat generating boards with fixed pressure drop, flow rate or pumping power, *Int. J. Heat Mass Transfer,* Vol. 36, 1993, pp. 3677–3686.

11. Al. M. Morega and A. Bejan, The optimal spacing of parallel boards with discrete heat sources cooled by laminar forced convection, *Numerical Heat Transfer, A. Appl.*, Vol. 25, 1994, pp. 373–392.

12. W. Nakayama, H. Matsushima, and P. Goel, Forced convective heat transfer from arrays of finned packages, in W. Aung, ed., *Cooling Technology for Electronic Equipment*, Hemisphere, New York, 1988, pp. 195–210.

13. A. Bejan, Al. M. Morega, S. W. Lee, and S. J. Kim, The cooling of a heat generating board inside a parallel-plate channel, *Int. J. Heat Fluid Flow*, Vol. 14, 1993, pp. 170–176.

14. D. B. Tuckerman and R. F. W. Pease, High-performance heat sinking for VLSI, *IEEE Electron Device Letters*, Vol. EDL-2, 1981, pp. 126–129.

15. R. W. Knight, J. S. Goodling, and D. J. Hall, Optimal thermal design of forced convection heat sinks—analytical, *J. Electronic Packaging*, Vol. 113, 1991, pp. 313–321.

16. A. Bar-Cohen and A. D. Kraus, eds., *Advances in Thermal Modeling of Electronic Components and Systems*, Vol. 2, ASME Press, New York, 1990.

PROBLEMS

5.1 Consider the problem of how to distribute a finite amount of insulation on a nonisothermal wall assuming that the wall temperature varies exponentially:

$$T(x) = T_0 + (T_L - T_0) \frac{\exp[n(x/L)] - 1}{\exp(n) - 1}$$

In this expression the nondimensional parameter n has the same sign as the curvature of the wall temperature distribution (d^2T/dx^2). Repeat the analysis of Section 5.1 and determine the heat loss reduction due to using an insulation with optimal thickness, relative to using an insulation with uniform thickness. Show that the relative heat loss reduction increases when the curvature d^2T/dx^2 increases.

5.2 As shown in Figure P5.2, a fixed amount of insulation material of volume V must be distributed optimally over the outer surface of a cylinder of radius r. The cylinder wall temperature varies as $T(x) = T_0 + (x/L)(T_L - T_0)$. The thermal conductivity of the insulation material (k) and the length of the wall (L) are known. The outer surface of the insulation is at the ambient temperature (T_0). Follow the steps outlined in Section 5.1 and determine $t_{opt}(x)$ and the corresponding minimum heat transfer rate to the ambient, $\dot{Q}_{min}$. Show that the heat loss reduction due to using an insulation with optimal thickness on a cylindrical wall is smaller than on the corresponding plane wall.

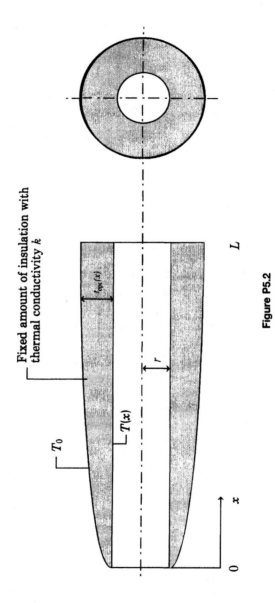

Figure P5.2

5.3   As shown in Figure P5.3, a hot stream originally at the temperature T_h flows through an insulated pipe suspended in a space at temperature T_0. The stream temperature $T(x)$ decreases in the longitudinal direction because of the heat transfer from $T(x)$ to T_0 that takes place everywhere along the pipe. The function of the pipe is to deliver the stream at a temperature (T_{out}) as close as possible to the original temperature, T_h. If the amount of insulation is fixed, determine the best way of distributing it over the pipe.

5.4   The volume of a cylindrical pin fin of diameter D and length L is fixed. Determine the diameter and length for which the heat transfer rate through the root of the fin is maximum. Assume that the heat transfer coefficient is constant along the fin, but varies with diameter according to $\bar{h} = \bar{h}_0 (D/D_0)^{-n}$, in which $\bar{h}_0$ is the heat transfer coefficient for a cylinder of diameter D_0, and n is a number between $\frac{1}{2}$ and 0.

5.5   The temperature distribution along the two-dimensional fin with sharp tip shown in Figure P5.5 is linear, $\theta(x) = (x/L)\theta_b$, where the local temperature difference is $\theta(x) = T(x) - T_\infty$. The fin temperature is $T(x)$, and the fluid temperature is T_∞. The fin width is W. The tip is at the temperature of the surrounding fluid: $\theta(0) = 0$.

(a)   Show that the profile thickness δ must be parabolic in x: $\delta = (h/k)x^2$.

(b)   Drive an expression for the total heat transfer rate through the base of this fin, $\dot{Q}_b$.

(c)   Show that if the total volume of the parabolic profile fin, V, is the same as the volume of the plate fin optimized in Section 5.2.1, then $\dot{Q}_b$ is 14.8% larger than the maximum heat transfer rate that can be accommodated by the plate fin, Equation 5.14.

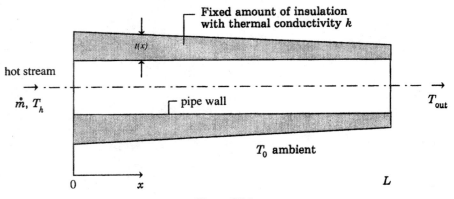

Figure P5.3

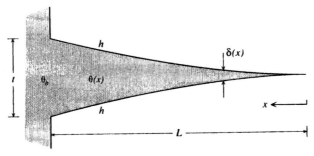

Figure P5.5

5.6 Figure P5.6 shows a simplified model of an electronic circuit board cooled by a laminar fully developed flow in a parallel-wall channel of fixed length L. The walls of the channel are insulated. The board substrate has a sufficiently high thermal conductivity so that the board temperature T_w may be assumed uniform in the longitudinal direction. The pressure difference Δp is fixed and the fluid inlet temperature is T_0. Determine the channel spacing D for which the thermal conductance $\dot{Q}/(T_w - T_0)$ is maximum, where $\dot{Q}$ is the total heat transfer rate removed by the stream from the board. Also, obtain an expression for the corresponding maximum thermal conductance.

5.7 The electronic circuit board shown in Figure P5.7 is thin and long enough to be modeled as a surface with uniform heat flux q''. The heat generated by the circuitry is removed by the fully developed laminar flow in the channel formed by the board and a parallel wall above it. That wall and the underside of the board are insulated. The length L is specified, and the inlet temperature of the coolant is T_0. The circuit board reaches its highest temperature (T_h) at the trailing edge. The maximum temperature is fixed by electrical design. The designer would like to build as much circuitry and as many components into the board as possible. This objective is equivalent to seeking a board and channel

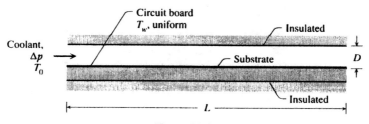

Figure P5.6

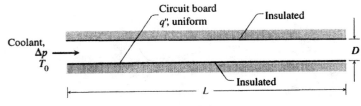

Figure P5.7

design that ensures the removal of the largest rate of heat generated by the board ($q''L$). The lone degree of freedom is the spacing D.

(a) Determine the spacing D that maximizes the rate at which heat is removed by the stream, and an expression for the corresponding maximum heat transfer rate.

(b) Compare these results with the results of Problem 5.6 (with $T_w = T_h$) in which the board was made isothermal by bonding it to a high-conductivity substrate. Why is the maximum heat transfer rate higher when the board is isothermal? Is the increase in heat transfer significant enough to justify the use of a high-conductivity substrate? Discuss.

5.8 Figure P5.8 shows the cross section through water channels intended to serve as a heat sink in the substrate of an integrated circuit [14–16]. The water removes the heat generated by the circuit. The channels are etched into the high-conductivity substrate (silicon) and are capped with a cover plate that is a relatively poor thermal conductor. Water is

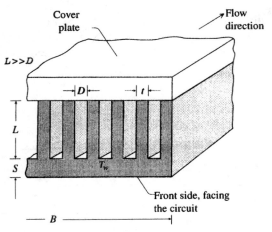

Figure P5.8

pumped through the channels, and the flow is laminar and fully developed.

(a) Determine the fin thickness t that maximizes the heat transfer rate from the substrate (T_w) to the water flow (local bulk temperature T_f).

(b) Derive a formula for the total heat transfer rate q', per unit length in the direction of flow, removed by all the streams.

6

APPLICATIONS WITH
THERMODYNAMICS AND
HEAT AND FLUID FLOW

In this chapter we consider a class of design problems that require the use of
thermodynamics (the first law *and* the second law) in addition to heat transfer
and fluid flow principles. Our objectives are to show the interplay of principles
drawn from these branches of engineering within a design analysis context
and to illustrate the trade-off nature of design. Only thermal aspects are considered in this chapter; economic issues are left to Chapters 7–9.

The presentation begins with a review of heat exchanger fundamentals in
Section 6.1. Then, in Section 6.2 we consider the trade-off between thermal
and fluid flow irreversibilities. In Section 6.3 the preliminary design of the
case study cogeneration system preheater is considered. The chapter concludes in Section 6.4 with additional applications featuring the use of thermodynamics and heat transfer in elementary design analysis settings.

6.1 HEAT EXCHANGERS

The heat exchanger is a multifaceted engineering system whose design involves not only the calculation of the heat transfer rate across the heat exchanger surface but also the pumping power needed to circulate the fluids,
the flow arrangement (e.g., counterflow vs. crossflow), the construction of the
actual hardware, and the ability to disassemble the apparatus for periodic
cleaning. It is beyond the scope of this book to present a comprehensive
treatment of heat exchanger design. This is covered by the heat exchanger
monographs [1–3]. Here we discuss only a few aspects of heat exchanger
analysis and design especially relevant to the objectives of this book. In Sec-

tion 9.3 the design of heat exchanger networks using *pinch analysis* is considered.

Overall Heat Transfer Coefficient. Let the subscripts h and c represent the two sides, or streams (hot and cold), of the heat exchanger surface shown in Figure 6.1 If $R_{t,w}$ is the conduction thermal resistance of the wall separating the two fluids, the overall thermal resistance of the heat exchanger surface is

$$\frac{1}{U_c A_c} = \frac{1}{\gamma_h h_{e,h} A_h} + R_{t,w} + \frac{1}{\gamma_c h_{e,c} A_c} \tag{6.1}$$

The overall heat transfer coefficient U_c is based on the cold-side area A_c. Alternatively, the left side of Equation 6.1 can be labeled $1/U_h A_h$, in which U_h is the overall heat transfer coefficient based on the hot-side area A_h.

Each of the *effective* heat transfer coefficients $h_{e,h}$ and $h_{e,c}$ accounts for two effects: the convective film resistance $(1/h)$ and the thermal resistance across the scale deposited on the surface in contact with fluid (r_s). The effective coefficient is defined as $1/h_e = r_s + 1/h$. Recommended values for the scale (or fouling) factor r_s can be found in heat transfer textbooks [4].

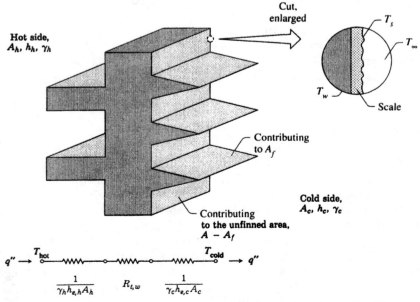

Figure 6.1   Heat exchanger surface with fins and scale on both sides [4].

The *overall surface efficiency factor* for each side of the wall (γ_h, γ_c) is

$$\gamma = 1 - \frac{A_f}{A}(1 - \eta) \tag{6.2}$$

where A_f is the finned area, that is, the portion contributed to A by the exposed surfaces of the fins. The fin efficiency η is introduced by Equation 4.10.

Effectiveness–NTU Relations. Let C_h and C_c represent the capacity flow rates of the two streams flowing along the sides of the heat exchanger surface, where C denotes $\dot{m}c_p$. The *number of heat exchanger* (or *heat transfer*) *units* is defined by

$$\text{NTU} = \frac{UA}{C_{\min}} \tag{6.3}$$

where $C_{\min}$ is the smaller of the two capacity rates. The heat exchanger *effectiveness* is the ratio between the actual heat transfer rate between the two streams, $\dot{Q}$, and the thermodynamic maximum value of that heat transfer rate:

$$\varepsilon = \frac{\text{actual heat transfer rate}}{\text{maximum heat transfer rate}} = \frac{\dot{Q}}{\dot{Q}_{\max}} \tag{6.4}$$

In a *counterflow* heat exchanger the effectiveness definition of Equation 6.4 reduces to

$$\varepsilon = \frac{C_h(T_{h,\text{in}} - T_{h,\text{out}})}{C_{\min}(T_{h,\text{in}} - T_{c,\text{in}})} = \frac{C_c(T_{c,\text{out}} - T_{c,\text{in}})}{C_{\min}(T_{h,\text{in}} - T_{c,\text{in}})} \tag{6.5}$$

where the subscripts *in* and *out* indicate the inlet and outlet temperatures of the hot (h) and cold (c) streams. The effectiveness–NTU relation is given by

$$\varepsilon = \frac{1 - \exp[-\text{NTU}(1 - C_{\min}/C_{\max})]}{1 - (C_{\min}/C_{\max})\exp[-\text{NTU}(1 - C_{\min}/C_{\max})]} \tag{6.6}$$

in which $C_{\max}$ is the larger of the two capacity rates. Two extreme cases of Equation 6.6 are used in Sections 6.2.4 and 6.4.1, respectively: the *balanced counterflow heat exchanger*, where $C_{\min} = C_{\max}$ and $\varepsilon = \text{NTU}/(1 + \text{NTU})$, and the heat exchanger in which the $C_{\max}$ stream experiences a change of phase at nearly constant pressure, where $C_{min}/C_{max} = 0$ and $\varepsilon = 1 - \exp(-NTU)$. The effectiveness–NTU relations for other flow arrangements can be found in Reference 4 and other heat transfer textbooks. In general, ε is a function of NTU, the capacity flow rate ratio C_{min}/C_{max}, and the flow arrangement (e.g., parallel flow, crossflow, mixed or unmixed streams, and single-pass or multipass).

Pressure Drop. The pressure difference required to push a stream through its heat exchanger channel is generally a complicated function of flow parameters and channel geometry. As shown in Figure 6.2, the pressure drop over a sufficiently straight portion of the channel is $\Delta p_s = p_b - p_c$. The pressure drop can be evaluated using Equation 4.40, provided that the fluid density does not vary appreciably along the passage. When the density varies, the contribution of Equation 4.40 is complemented by an effect due to the acceleration or deceleration of the stream, and Δp_s is evaluated from

$$\Delta p_s = f \frac{4L}{D_h} \frac{\rho V^2}{2} + G^2 \left(\frac{1}{\rho_{out}} - \frac{1}{\rho_{in}} \right) \tag{6.7}$$

In this expression G is the *mass velocity* of the stream, $G = \dot{m}/A = \rho_{in} V_{in} = \rho_{out} V_{out}$, and A is the channel cross-sectional area. In the first term on the

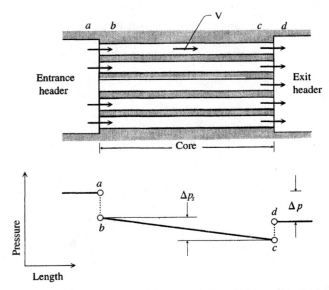

Figure 6.2 The pressure variation of a stream flowing through the core of a heat exchanger [4].

right side, ρ is the density averaged between the inlet and outlet, namely, $\rho = \frac{1}{2}(\rho_{in} + \rho_{out})$, and V is the corresponding average velocity $V = G/\rho$.

Referring again to Figure 6.2, the total pressure drop Δp consists of Δp_s plus contributions made by other geometric features (sudden contractions, enlargements, bends, protuberances). As considered in the discussion of Equation 2.31, the pressure drop contribution made by each feature of this kind is included in the total Δp calculation as the product of the dynamic pressure $\frac{1}{2}\rho V^2$ and a particular *loss coefficient,* which can be found in the heat exchanger literature [3]. In the case of Figure 6.2 there are only two such contributions, one due to the contraction at plane a–b:

$$p_a - p_b = (1 - \sigma^2_{a-b})\tfrac{1}{2}\rho_b V^2_b - K_c \tfrac{1}{2}\rho_b V^2_b \qquad (6.8)$$

and the other due to the enlargement of the flow cross section at plane c–d:

$$p_d - p_c = (1 - \sigma^2_{c-d})\tfrac{1}{2}\rho_c V^2_c - K_e \tfrac{1}{2}\rho_c V^2_c \qquad (6.9)$$

where the σ's are contraction ratios (defined as σ = flow cross-sectional area/frontal area): $\sigma_{a-b} = A_b/A_a$ and $\sigma_{c-d} = A_c/A_d$. The contraction loss coefficient K_c and the enlargement loss coefficient K_e for bundles of parallel tubes are presented in Figure 6.3. The Reynolds number Re in this figure is based on the tube diameter and on the mean velocity through the narrower channel: the velocity *downstream* of a contraction and the velocity *upstream* of an enlargement.

The total pressure drop experienced by the stream in Figure 6.2 is $\Delta p = (p_a - p_b) + \Delta p_s - (p_d - p_c)$. In special cases where the density is essentially constant from a to d (i.e., when $V_b = V_c = V$) and where the contraction ratio equals the enlargement ratio ($\sigma_{a-b} = \sigma_{c-d}$), the total pressure drop formula reduces to

$$\Delta p = K_c \tfrac{1}{2}\rho V^2 + \Delta p_s + K_e \tfrac{1}{2}\rho V^2 \qquad (6.10)$$

The three terms on the right side represent, in order, the contraction loss, the straight-section pressure drop, and the enlargement loss.

Compact Heat Exchanger Surfaces. The pressure drop and heat transfer in many passages with complicated geometries have been measured and presented in dimensionless form. The sample shown in Figure 6.4 refers to the passage external to a bundle of finned tubes. The figure gives the variations with Reynolds number of the friction factor f and the *Colburn j_H factor* (Equation 6.21). The heat transfer surface area is increased by the fins, which occupy much of the space that would have been left open between the tubes. Many geometries of this kind have been designed in order to increase the heat transfer area per unit volume (heat transfer area density):

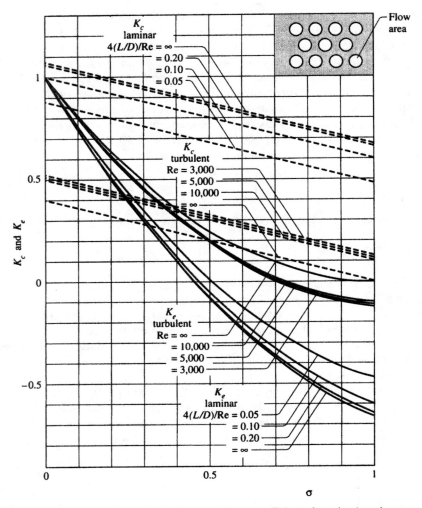

Figure 6.3 Abrupt contraction and enlargement loss coefficients for a heat exchanger core with multiple circular-tube passages [5].

$$\alpha = \frac{A}{V} \tag{6.11}$$

As the heat transfer coefficient is lower for a gas flow than for a liquid flow (Figure 4.3), it is most beneficial to place the fins on the gas side of a gas–liquid heat exchanger. This would mean that in Figure 6.4 liquid flows inside the tubes, and gas flows over the outer surface. As indicated on the figure, the gas flows perpendicular to the tubes.

The other parameters that describe the geometry of a flow passage such as shown in Figure 6.4 are

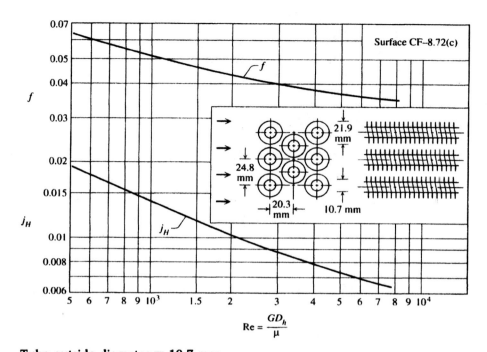

Tube outside diameter = 10.7 mm
Fin pitch = 343 per meter
Flow passage hydraulic diameter D_h = 4.43 mm
Fin thickness, average (fins are tapered slightly) = 0.48 mm, copper
Free-flow area/frontal area, σ = 0.494
Heat transfer area/total volume, α = 446 m²/m³
Fin area/total area, A_f/A = 0.876

Figure 6.4 Pressure drop and heat transfer data for the flow passage through a bundle of staggered finned tubes [6].

$$\sigma = \frac{A_c}{A_{fr}} = \frac{\text{minimum free-flow area}}{\text{frontal area}} \qquad (6.12)$$

$$\frac{A_f}{A} = \frac{\text{fin area}}{\text{total heat transfer area}} \qquad (6.13)$$

and the hydraulic diameter

$$D_h = 4\frac{A_c L}{A} \qquad (6.14)$$

The dimensionless parameter σ accounts for the contraction and enlargement experienced by the stream, cf. Equations 6.8 and 6.9. The area ratio A_f/A is the same as in Equation 6.2 for the overall surface efficiency γ. In Equation 6.14, L is the length of the heat exchanger in the direction of flow, and A_c, the minimum free-flow area, refers to the smallest cross section encountered by the fluid. In Figure 6.4, this area occurs between two adjacent tubes aligned on the vertical: A_c is coplanar with the two centerlines.

The mass flow rate of the stream that flows through this complicated passage is constant in the direction of flow:

$$\dot{m} = \rho A_{fr} V \tag{6.15}$$

where V is the average velocity based on the frontal area. On the other hand, the mass velocity G is based on the maximum velocity V_{max}, which is the velocity through the minimum free-flow area:

$$G = \rho V_{max} = \frac{\dot{m}}{A_c} \tag{6.16}$$

where $A_c = \sigma A_{fr}$. The Reynolds number used on the abscissa of Figure 6.4 is also based on the maximum velocity:

$$Re = \frac{V_{max} D_h}{\nu} = \frac{G D_h}{\mu} \tag{6.17}$$

The total pressure drop across a heat exchanger core constructed as shown in Figure 6.2, namely the difference between the pressure on the left (the inlet) and the pressure on the right (the outlet), is given by [3] as

$$\Delta p = \frac{G^2}{2\rho_{in}} \left[f \frac{A}{A_c} \frac{\rho_{in}}{\rho} + (1 + \sigma^2) \left(\frac{\rho_{in}}{\rho_{out}} - 1 \right) \right] \tag{6.18}$$

In this equation ρ is the average density evaluated at the average temperature, $\frac{1}{2}(T_{in} + T_{out})$. As an alternative to the formula for density given under Equation 6.7, the average density can also be estimated by averaging the fluid specific volume $(1/\rho)$ between the inlet and outlet values,

$$\frac{1}{\rho} = \frac{1}{2} \left(\frac{1}{\rho_{in}} + \frac{1}{\rho_{out}} \right) \tag{6.19}$$

The friction factor f that multiplies the first term in the square brackets of Equation 6.18 is obtained from Figure 6.4. The factor f accounts for fluid friction against solid walls *and* for the entrance and exit losses. The second term in the square brackets of Equation 6.18 accounts for the acceleration or deceleration of the stream. This contribution is negligible when the density is essentially constant along the passage.

Using Equation 6.14, the area ratio multiplying f in Equation 6.18 can be expressed as $A/A_c = 4L/D_h$, and thus is proportional to the length of the flow passage. When the volume V is specified, the ratio A/A_c needed in Equation 6.18 can be estimated by writing

$$\frac{A}{A_c} = \frac{\alpha V}{\sigma A_{\mathrm{fr}}} \tag{6.20}$$

Charts such as Figure 6.4 also contain the information necessary for calculating the average heat transfer coefficient h for the particular flow passage. This is expressed in nondimensional terms as the Colburn j_H factor:

$$j_H = \mathrm{St}\ \mathrm{Pr}^{2/3} \tag{6.21}$$

in which the Stanton number (St) is based on the mass velocity G given by Equation 6.16:

$$\mathrm{St} = \frac{h}{Gc_p} = \frac{h}{\rho c_p V_{\mathrm{max}}} \tag{6.22}$$

In Figure 6.4 and many like it for other compact surfaces, the factors j_H and f exhibit almost the same dependence on Reynolds number [4]. In Figure 6.4, the ratio $j_H/(f/2)$ decreases only by one third (from 0.6 to 0.4) as Re increases from 500 to 8000. Accordingly, even in a complicated passage the ratio $j_H/(f/2)$ is relatively constant and of order 1.

The calculation of the heat transfer between the two fluids that interact in crossflow in Figure 6.4 also requires an estimate of the average heat transfer coefficient for the internal surface of the tubes. This estimate can be obtained by following the method outlined in Section 4.3.2.

6.2 THE TRADE-OFF BETWEEN THERMAL AND FLUID FLOW IRREVERSIBILITIES

In this section we turn our attention to the second law aspects of the operation of heat exchangers. In these systems the destruction of exergy is due to two mechanisms: the heat transfer across a finite temperature difference and the flow with friction through ducts or around fins. We shall refer to these mechanisms as the heat transfer irreversibility and the fluid flow irreversibility, respectively.

We have previously considered the *separate* effects of heat transfer and friction on the exergy destruction in a heat exchanger (Section 3.5.2). The present section aims at considering their *simultaneous* effect on exergy destruction–or, equivalently, on entropy generation. In particular, we show that the heat transfer and fluid friction irreversibilities tend to *compete* with one another, and that the total rate of entropy generation (rate of exergy destruction) can be minimized by adjusting the size of one irreversibility against the other. These adjustments can be made by properly selecting the physical dimensions of the solid parts (fins, ducts, heat exchanger surface). It must be understood, however, that the result is at best a *thermodynamic optimum*. As discussed in Section 1.2, constraints such as cost, size, and reliability enter into the determination of truly optimal designs.

6.2.1   Local Rate of Entropy Generation

The starting point for an indepth thermodynamic study of the effect of heat transfer and friction in a flowing fluid is an expression for the local rate of entropy generation. Such an expression can be obtained by applying an entropy balance to a control volume consisting of a volume element in the flow field. After reduction, the following expression for the local *volumetric* rate of entropy generation is obtained [7]:

$$\dot{S}'''_{\text{gen}} = \frac{k}{T^2}(\nabla T)^2 + \frac{\mu}{T}\Phi \tag{6.23}$$

The first term on the right represents the rate of entropy generation associated with heat transfer, and the second term is the rate of entropy generation associated with fluid friction. The product of the function Φ and the viscosity μ denotes the *viscous dissipation,* or rate of irreversible conversion of mechanical energy to internal energy via viscous effects. Expressions for $\dot{S}'''_{\text{gen}}$ in rectangular, cylindrical, and spherical coordinates are provided in Reference 7.

Using expressions for the variations of temperature and velocity in the flow, the distribution of the local rate of entropy generation can be obtained in principle from Equation 6.23. Illustrations are provided in Reference 7. Owing to various complexities, however, the temperature and velocity variations are seldom known for cases of practical interest, as, for example, when the flow is turbulent and/or the flow geometry is complicated. Thus, the evaluation of the local volumetric rate of entropy generation from Equation 6.23 is precluded. But in many of these instances the variation in the entropy generation can be evaluated using average heat transfer and friction data measured along the solid boundaries confining the flow. We consider such applications in Sections 6.2.2 and 6.2.3. In other cases, an overall entropy generation rate for the overall heat exchanger suffices. Examples are provided in Sections 6.2.4 and 6.2.5.

6.2.2 Internal Flows

In this section we consider entropy generation in a duct of arbitrary geometry with heat transfer and friction at the wall. The rate of entropy generation *per unit length*, $\dot{S}'_{gen}$, is evaluated using correlations for average heat transfer and fluid friction.

Consider the general heat exchanger passage shown in Figure 6.5, in which the cross-sectional area A and wetted perimeter p_w are arbitrary. The heat transfer rate per unit length is q' ($=q''p_w$) and the mass flow rate is $\dot{m}$. Both q' and $\dot{m}$ are known. At steady state, the heat transfer q' crosses the temperature gap ΔT formed between the wall temperature ($T + \Delta T$) and the bulk temperature of the stream (T). The stream flows with friction in the x direction, hence, the pressure gradient $(-dp/dx) > 0$.

Taking a passage of length dx as the control volume, energy and entropy balances reduce to give, respectively,

$$\dot{m} \frac{dh}{dx} = q' \qquad (6.24)$$

$$\dot{S}'_{gen} = \dot{m} \frac{ds}{dx} - \frac{q'}{T + \Delta T} \qquad (6.25)$$

where $\dot{S}'_{gen}$ is the entropy generation rate per unit duct length. Combining these statements with Equation 2.34b, we have [8]

$$\dot{S}'_{gen} = \frac{q' \, \Delta T}{T^2(1 + \Delta T/T)} + \frac{\dot{m}}{\rho T}\left(-\frac{dp}{dx}\right) \cong \frac{q' \, \Delta T}{T^2} + \frac{\dot{m}}{\rho T}\left(-\frac{dp}{dx}\right) \quad (6.26)$$

Note that the denominator of the first term on the right side has been simplified by assuming that ΔT is considerably smaller than the absolute temperature T.

By study of the $\dot{S}'_{gen}$ expression, Equation 6.26, and of many like it for

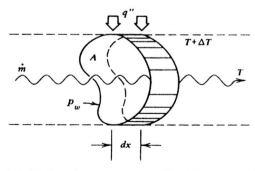

Figure 6.5 Heat exchanger passage with arbitrary geometry [7].

other simple devices, it can be concluded that a proposed design change (e.g., making the passage narrower) induces changes of opposite signs in the two terms of the expression: The heat transfer irreversibility (first term) decreases while the fluid flow irreversibility increases. Accordingly, a trade-off exists between the two irreversibility contributions, and there is a design for which the overall measure of exergy destruction ($\dot{S}'_{gen}$) is a minimum.

The trade-off becomes clearer if Equation 6.26 is rewritten, using the friction factor f, Stanton number (St $= \bar{h}/\rho c_p U$, Equation 4.46), mass velocity ($G = \dot{m}/A$), Reynolds number (Re $= GD_h/\mu$), and hydraulic diameter ($D_h = 4A/p_w$), in the form

$$\dot{S}'_{gen} = \underbrace{\frac{(q')^2 D_h}{4T^2 \dot{m} c_p \text{St}}}_{\dot{S}'_{gen,\Delta T}} + \underbrace{\frac{2\dot{m}^3 f}{\rho^2 T D_h A^2}}_{\dot{S}'_{gen,\Delta p}} \tag{6.27}$$

For turbulent conditions St and f usually increase simultaneously [4] as the designer seeks to improve the thermal contact between wall and fluid. Accordingly, the first term on the right of Equation 6.27 would tend to decrease as the second term on the right increases, and conversely. Finally, as both q' and $\dot{m}$ are regarded as fixed, we note that the heat-exchanger passage has two degrees of freedom: the wetted perimeter p_w and the cross-sectional area A, or any other pair of independent parameters such as (Re, D_h) or (G, D_h).

Example 6.1 For a straight tube of diameter D, determine the variation in the rate of entropy generation per unit of length using average heat transfer and friction data for turbulent, fully developed flow, and discuss the result.

Solution

MODEL

1. The duct cross section is circular, therefore $p_w = \pi D$, $A = \frac{1}{4}\pi D^2$ and $D_h = D$.
2. The heat transfer rate q' and the mass flow rate $\dot{m}$ are specified.
3. The flow is turbulent and fully developed.

ANALYSIS. For a tube of diameter D, Equation 6.27 can be rewritten with the relations

$$\mathrm{Nu}_D = \frac{\bar{h}D}{k} = \mathrm{St}\ \mathrm{Re}_D\mathrm{Pr}, \qquad \mathrm{Re}_D = \frac{4\dot{m}}{\pi\mu D} \tag{1}$$

to give

$$\dot{S}'_{\mathrm{gen}} = \frac{(q')^2}{\pi T^2 k\ \mathrm{Nu}_D} + \frac{32\dot{m}^3 f}{\pi^2 \rho^2 T D^5} \tag{2}$$

Then, with the following correlations for Nu_D and f for fully developed turbulent pipe flow [4]

$$\mathrm{Nu}_D \cong 0.023\ \mathrm{Re}_D^{0.8}\mathrm{Pr}^{0.4} \qquad (0.7 < \mathrm{Pr} < 160,\ \mathrm{Re}_D > 10^4) \tag{3}$$

$$f \cong 0.046\ \mathrm{Re}_D^{-0.2} \qquad (2 \times 10^4 < \mathrm{Re}_D < 10^6) \tag{4}$$

Equation 2 can be written in terms of Re_D as

$$\dot{S}'_{\mathrm{gen}} = \frac{13.84(q')^2}{kT^2\ \mathrm{Pr}^{0.4}}\ \mathrm{Re}_D^{-0.8} + \frac{0.0446\mu^5}{\rho^2 T\dot{m}^2}\ \mathrm{Re}_D^{4.8} \tag{5a}$$

Alternatively, this result can be expressed in the form

$$\dot{S}'_{\mathrm{gen}} = a\ \mathrm{Re}_D^{-0.8} + b\ \mathrm{Re}_D^{4.8} \tag{5b}$$

where a and b are the coefficients of the corresponding terms in Equation 5a.

By inspection of Equations 5a, b, the trade-off nature of the heat transfer and fluid friction irreversibilities is very clearly indicated. Thus, the first term on the right referring to the heat transfer irreversibility varies *inversely* with the Reynolds number; the second term referring to the effect of friction varies *directly* with the Reynolds number. This points to a value of the Reynolds number for which $\dot{S}'_{\mathrm{gen}}$ is a minimum.

As noted in Section 2.1.4, the value of the entropy generation rate usually does not have much significance by itself. The significance is usually determined through comparison. In the present application, the *minimum* entropy generation rate provides a suitable measure. Thus, solving $d\dot{S}'_{\mathrm{gen}}/d\mathrm{Re} = 0$, we find that the entropy generation rate is minimum when the Reynolds number (or tube diameter) is

$$\mathrm{Re}_{\mathrm{opt}} = 2.023\ \mathrm{Pr}^{-0.071}B_0^{0.358} \tag{6}$$

where

$$B_0 = \dot{m}q' \; \frac{\rho}{\mu^{5/2}(kT)^{1/2}} \tag{7}$$

and the subscript D has been dropped for simplicity here and in Equation 8. Thus, the effect of Re on $\dot{S}'_{gen}$ can be expressed as

$$\frac{\dot{S}'_{gen}}{\dot{S}'_{gen,min}} = \underbrace{0.856 \left(\frac{Re}{Re_{opt}} \right)^{-0.8}}_{\dot{S}'_{gen,\Delta T}} + \underbrace{0.144 \left(\frac{Re}{Re_{opt}} \right)^{4.8}}_{\dot{S}'_{gen,\Delta p}} \tag{8}$$

where $\dot{S}'_{gen,min} = \dot{S}'_{gen}(Re_{opt})$ (see also N_S, Equation 6.31a). The nondimensional entropy generation rate given by Equation 8 is plotted in Figure E6.1, where ϕ is the ratio

$$\phi = \frac{\dot{S}'_{gen,\Delta p}}{\dot{S}'_{gen,\Delta T}} \tag{9}$$

The figure shows that the heat transfer term dominates when $Re_D < Re_{D,opt}$, whereas the pressure drop term dominates when $Re_D > Re_{D,opt}$. Note that since $\dot{m}$ is fixed, the abscissa $Re_D/Re_{D,opt}$ is the same as D_{opt}/D.

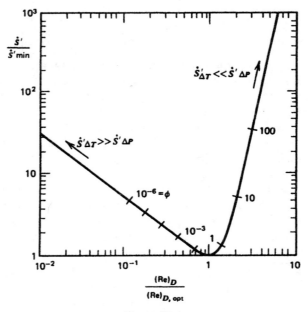

Figure E6.1

6.2.3 External Flows

Another significant class of convective heat transfer arrangements is comprised of solid bodies located within flowing streams. This includes as a particular case fins of various geometries that extend into the heat transfer fluid circulating through the finned-surface heat exchanger. In any member of this class we expect the irreversibility to be due to heat transfer between the solid and fluid, and also to frictional drag as the fluid flows over the surface.

Consider the external flow of Figure 6.6. A body of arbitrary shape and surface area A is suspended in a flowing gas (or liquid) for which the free-stream velocity and absolute temperature are U_∞ and T_∞, respectively. The average wall temperature is $\overline{T}_w$. The heat transfer rate between the body and stream is $\dot{Q}\,(=\overline{h}A(\overline{T}_w - T_\infty))$. F_D denotes the drag force: the sum of all forces distributed over A and oriented opposite the direction of flow. Applying mass, energy and entropy balances to an appropriately selected control volume enclosing the body, we obtain after reduction [7],

$$\dot{S}_{\text{gen}} = \frac{\dot{Q}(\overline{T}_w - T_\infty)}{T_\infty^2} + \frac{F_D U_\infty}{T_\infty} \tag{6.28}$$

In writing Equation 6.28 we have assumed that the surface-averaged temperature difference between the body and the free stream $(\overline{T}_w - T)$ is much smaller than either absolute temperature, $\overline{T}_w$ or T_∞. The two terms on the right side of Equation 6.28 represent, in order, the contribution due to heat transfer

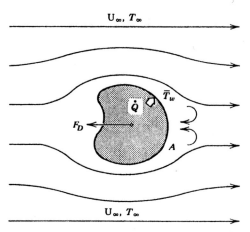

Figure 6.6 Convective heat transfer in external flow [7].

across a finite temperature difference, and the contribution due to fluid friction.

To illustrate the application of Equation 6.28, consider the pin fin represented by the cylinder in crossflow (diameter D, length W) shown in Figure 6.7. We assume the fin efficiency is greater than 0.9, and so the variation of temperature along the cylinder can be neglected. The total heat transfer rate (the heat transfer duty) is fixed, $\dot{Q} = h\pi DW(\bar{T}_w - T_\infty)$ where $\bar{T}_w$ is the perimeter averaged temperature. The heat transfer coefficient h can be estimated based on Equation 4.30. The total drag force is $F_D = \frac{1}{2}C_D DW\rho U_\infty^2$, for which the relationship between the drag coefficient C_D and Re_D is available in Figure 4.4. After making these substitutions in Equation 6.28, we obtain

$$\frac{\dot{S}_{gen}/W}{\mu U_\infty^2/T_\infty} = \frac{B^2 \mathrm{Pr}^{1/3}}{\pi \mathrm{Nu}_D(\mathrm{Re}_D, \mathrm{Pr})} + \tfrac{1}{2}\mathrm{Re}_D C_D(\mathrm{Re}_D) \qquad (6.29)$$

where $\mathrm{Re}_D = U_\infty D/\nu$, and B is a nondimensional heat transfer duty parameter of the cylinder,

$$B = \frac{\dot{Q}/W}{U_\infty(k\mu T_\infty \,\mathrm{Pr}^{1/3})^{1/2}} \qquad (6.30)$$

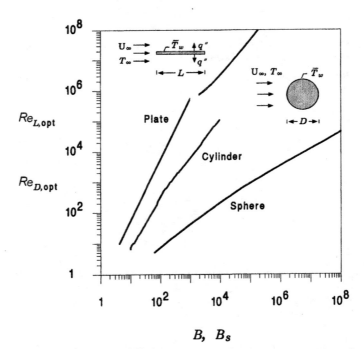

Figure 6.7 Reynolds number for minimum entropy generation for external forced convection over a cylinder, sphere, and flat plate [9].

In the Re_D range $10^2–10^5$ the drag coefficient is practically constant, while h is approximately proportional to $Re_D^{-1/2}$ (Equation 4.30). This means that the entropy generation rate behaves as $\dot{S}_{gen} \cong a\,Re_D^{-1/2} + b\,Re_D$, where a and b are constants (groups of physical quantities) that can be deduced from Equation 6.29. Thus, as for the case of Example 6.1, the trade-off nature of the heat transfer and fluid friction irreversibilities is clearly drawn: The first term on the right of Equation 6.29 refers to the heat transfer irreversibility and varies *inversely* with the Reynolds number; the second term on the right refers to friction and varies *directly* with the Reynolds number. As in Example 6.1, these variations suggest a minimum rate of entropy generation that provides a convenient normalizing parameter. Thus, for $Pr \geq 0.7$, the value of the Reynolds number corresponding to the minimum rate of entropy generation varies with the duty parameter B as shown in Figure 6.7. The corresponding nondimensional entropy generation rate, called the *entropy generation number N_S* in Reference 7, takes the form

$$N_S = \frac{\dot{S}_{gen}}{\dot{S}_{gen,min}} \cong \frac{2}{3}\left(\frac{Re_D}{Re_{D,opt}}\right)^{-1/2} + \frac{1}{3}\left(\frac{Re_D}{Re_{D,opt}}\right) \qquad (6.31a)$$

or,

$$N_S = \frac{\dot{S}_{gen}}{\dot{S}_{gen,min}} \cong \frac{2}{3}\left(\frac{D}{D_{opt}}\right)^{-1/2} + \frac{1}{3}\left(\frac{D}{D_{opt}}\right) \qquad (6.31b)$$

Equation 6.31b is plotted in Figure 6.8.

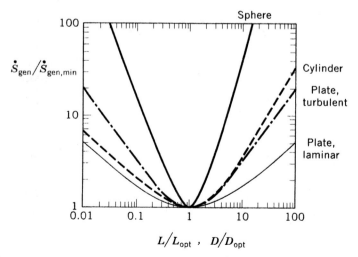

Figure 6.8 Nondimensional entropy generation versus D/D_{opt} and L/L_{opt} for various body configurations [9].

Table 6.1   Parameters in Equation 6.32 for various geometries with external forced convection

$$\frac{\dot{S}_{gen}}{\dot{S}_{gen,min}} \cong a\left(\frac{X}{X_{opt}}\right)^{-\alpha} + b\left(\frac{X}{X_{opt}}\right)^{\beta}$$

Geometry	X	a	b	α	β	Range
Cylinder[a]	D	$\frac{2}{3}$	$\frac{1}{3}$	$\frac{1}{2}$	1	$10^2 < Re_D < 10^5$, $Pr \geq 0.7$
Sphere[b]	D	$\frac{4}{7}$	$\frac{3}{7}$	$\frac{3}{2}$	2	$10^3 < Re_D < 10^5$, $Pr \geq 0.7$
Flat plate[c]	L	$\frac{1}{2}$	$\frac{1}{2}$	$\frac{1}{2}$	$\frac{1}{2}$	$Re_L < 5 \times 10^5$, $Pr \geq 0.7$
Flat plate[c]	L	$\frac{1}{2}$	$\frac{1}{2}$	$\frac{4}{5}$	$\frac{4}{5}$	$5 \times 10^5 < Re_L < 10^8$, $Pr \geq 0.7$

[a] B defined by Equation 6.30.
[b] $B_S = \dot{Q}/\nu(k\mu T_\infty Pr^{1/3})^{1/2}$.
[c] B defined by Equation 6.30, where $\dot{Q}/W = 2Lq''$, Figure 6.7 (inset).

Similar results are developed for other geometries in Reference 9. In each case the nondimensional entropy generation rate exhibits the same form as in Equation 6.31b:

$$N_S = \frac{\dot{S}_{gen}}{\dot{S}_{gen,min}} \cong a\left(\frac{X}{X_{opt}}\right)^{-\alpha} + b\left(\frac{X}{X_{opt}}\right)^{\beta} \qquad (6.32)$$

where X denotes the characteristic length of the body. Table 6.1 provides the parameters of this expression corresponding to the geometries of Figures 6.7 and 6.8.

Although minimum entropy generation by itself rarely would be a design objective, it is often instructive to know the effect of various design parameters on the entropy generated, at least approximately. Figures 6.7 and 6.8 show this for simple geometries in external forced convection. For each geometry, we see that the heat transfer irreversibility dominates when $X/X_{opt} < 1$, whereas fluid friction dominates when $X/X_{opt} > 1$.

6.2.4   Nearly Ideal Balanced Counterflow Heat Exchangers

The trade-off between heat transfer and fluid flow irreversibilities is apparent once more if we consider *balanced* counterflow heat exchangers. That is, heat exchangers for which the capacity flow rates are the same on the two sides of the heat transfer surface:

$$(\dot{m}c_p)_1 = (\dot{m}c_p)_2 = \dot{m}c_p \qquad (6.33)$$

The two sides are indicated by the subscripts 1 and 2. For simplicity, the same fluid is assumed on each side, and the fluid is modeled as an ideal gas with constant c_p. With reference to Figure 6.9, we write T_1 and T_2 for the fixed (given) inlet temperatures of the two streams, and p_1 and p_2 for the respective inlet pressures. Applying an entropy balance, the entropy generation rate of the overall heat exchanger is

$$\dot{S}_{gen} = (\dot{m}c_p)_1 \ln \frac{T_{1,\text{out}}}{T_1} + (\dot{m}c_p)_2 \ln \frac{T_{2,\text{out}}}{T_2} - (\dot{m}R)_1 \ln \frac{p_{1,\text{out}}}{p_1} - (\dot{m}R)_2 \ln \frac{p_{2,\text{out}}}{p_2}$$

$$(6.34)$$

The outlet temperatures $T_{1,\text{out}}$ and $T_{2,\text{out}}$ can be eliminated by using the heat exchanger effectiveness, Equation 6.5, namely $\varepsilon = (T_1 - T_{1,\text{out}})/(T_1 - T_2)$. If we restrict consideration to the limit of small ΔT and Δp:

$$1 - \varepsilon \ll 1, \qquad \frac{p_1 - p_{1,\text{out}}}{p_1} \ll 1, \qquad \frac{p_2 - p_{2,\text{out}}}{p_2} \ll 1 \qquad (6.35)$$

Equation 6.34 can be written in nondimensional form as an entropy generation number [10]:

$$N_S = \frac{\dot{S}_{gen}}{\dot{m}c_p} = (1 - \varepsilon)\frac{(T_2 - T_1)^2}{T_1 T_2} + \frac{R}{c_p}\left[\left(\frac{\Delta p}{p}\right)_1 + \left(\frac{\Delta p}{p}\right)_2\right] \qquad (6.36)$$

(balanced, nearly ideal)

In this form, it is clear that the overall entropy generation rate is composed of contributions from three sources of irreversibility: the stream-to-stream heat transfer, the pressure drop along the first stream, Δp_1, and the pressure drop along the second stream, Δp_2. As shown by subsequent developments, the

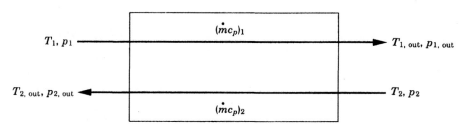

Figure 6.9 Counterflow heat exchanger with specified capacity rates and inlet conditions.

heat transfer irreversibility term can be split into two terms, each describing the contribution made by one side of the heat transfer surface. This allows the roles of *design constraints* and *degrees of design freedom* to be illustrated simply within an idealized but still complex context as follows:

Assuming the thermal resistance of the wall is negligible, the overall thermal resistance can be written as

$$\frac{1}{\overline{h}A_1} = \frac{1}{\overline{h}_1 A_1} + \frac{1}{\overline{h}_2 A_2} \tag{6.37}$$

where A_1 and A_2 are the heat transfer surface areas swept by each stream, and $\overline{h}_1$ and $\overline{h}_2$ are heat transfer coefficients based on these respective areas. On the left side of Equation 6.37, $\overline{h}$ is the overall heat transfer coefficient based on A_1. The summation of thermal resistances (Equation 6.37) is equivalent to writing

$$\frac{1}{\text{NTU}} = \frac{1}{\text{NTU}_1} + \frac{1}{\text{NTU}_2} \tag{6.38}$$

where

$$\text{NTU} = \frac{\overline{h}A_1}{\dot{m}c_p}, \qquad \text{NTU}_1 = \frac{\overline{h}_1 A_1}{\dot{m}c_p}, \qquad \text{NTU}_2 = \frac{\overline{h}_2 A_2}{\dot{m}c_p} \tag{6.39}$$

In a balanced counterflow heat exchanger, the effectiveness–NTU relation is particularly simple: $\varepsilon = \text{NTU}/(1 + \text{NTU})$, and even simpler when $1 - \varepsilon \ll 1$, namely $\varepsilon = 1 - 1/\text{NTU}$. Combining the latter expression for ε with Equations 6.36 and 6.38, we obtain

$$N_S = \underbrace{\frac{\tau^2}{\text{NTU}_1} + \frac{R}{c_p}\left(\frac{\Delta p}{p}\right)_1}_{N_{S,1}} + \underbrace{\frac{\tau^2}{\text{NTU}_2} + \frac{R}{c_p}\left(\frac{\Delta p}{p}\right)_2}_{N_{S,2}} \tag{6.40}$$

where $N_{S,1}$ and $N_{S,2}$ may be called *one-sided entropy generation numbers* and

$$\tau = \frac{|T_2 - T_1|}{(T_1 T_2)^{1/2}} \tag{6.41}$$

The one-side entropy generation numbers $N_{S,1}$ and $N_{S,2}$ have the same analytical form; therefore, we concentrate on one of them ($N_{S,1}$) and keep in mind that the analysis can be repeated identically for the other ($N_{S,2}$).

The heat transfer and fluid friction contributions to $N_{S,1}$ are coupled through the geometric parameters of the heat exchanger duct (passage) on side 1 of the heat exchanger surface. This coupling is brought to light by rewriting $N_{S,1}$ in terms of the nondimensional ratio $(4L/D_h)_1$, using the identity

$$\text{NTU}_1 = \left(\frac{4L}{D_h}\right)_1 \text{St}_1 \tag{6.42}$$

Thus

$$N_{S,1} = \frac{\tau^2}{\text{St}_1}\left(\frac{D_h}{4L}\right)_1 + \frac{R}{c_p}g_1^2 f_1\left(\frac{4L}{D_h}\right)_1 \tag{6.43}$$

where $L_1 = A_1/p_{w,1}$ is the length of the passage, f_1 and St_1 are defined according to

$$f_1(\text{Re}_1) = \frac{\rho D_{h,1}}{2G_1^2}\frac{\Delta p_1}{L_1}, \qquad \text{St}_1(\text{Re}_1, \text{Pr}) = \frac{\bar{h}_1}{c_p G_1} \tag{6.44}$$

and the dimensionless mass velocity is

$$g_1 = \frac{G_1}{(2\rho p_1)^{1/2}} \tag{6.45}$$

The first term on the right side of Equation 6.43 varies *inversely* with the ratio $4L/D_h$, whereas the second term varies *directly* with the same ratio. Accordingly, when the mass velocity and Reynolds number are fixed, there is a value of the ratio $4L/D_h$ for which $N_{S,1}$ is a minimum [10]:

$$\left(\frac{4L}{D_h}\right)_{1,\text{opt}} = \frac{\tau}{g_1[(R/c_p)f_1\text{St}_1]^{1/2}}, \qquad N_{S,1,\text{min}} = 2\tau\left(\frac{R}{c_p}\right)^{1/2}g_1\left(\frac{f_1}{\text{St}_1}\right)^{1/2} \tag{6.46}$$

The main features of this minimum are shown qualitatively in Figure 6.10. The minimum entropy generation rate is proportional to g_1. Furthermore, in the case of the most common heat exchanger surfaces, the group $(f_1/\text{St}_1)^{1/2}$ is only a weak function of Re_1. This means that the mass velocity and minimum rate of entropy generation on one side of a heat exchanger surface are proportional.

Area Constraint. The one-side irreversibility depends on parameters fixed by the fluid type and the inlet conditions: τ, R/c_p, Pr, and parameters that depend on the size and geometry of the heat exchanger passage: $(4L/D_h)_1$,

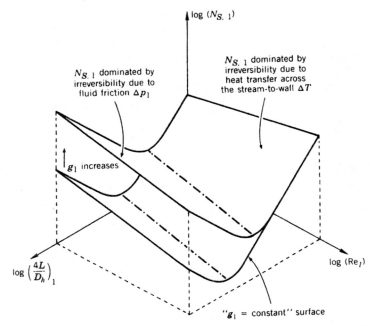

Figure 6.10 The entropy generation rate of a *nearly ideal* balanced counterflow heat exchanger [10].

Re_1, and g_1. How many of these three parameters are degrees of freedom depends on the number of design constraints. One important constraint concerns the heat transfer area A_1 [7]. The minimization of irreversibility subject to constant area might be important in cases where the cost of the heat exchanger surface is a major consideration.

In dimensionless form, the constant-area constraint can be expressed as

$$a_1 = \frac{A_1}{\dot{m}} (2\rho p_1)^{1/2} \quad \text{(const)} \tag{6.47}$$

where a_1 is the dimensionless area of side 1 of the surface. It is easy to show that $a_1 g_1 = (4L/D_h)_1$, and that Equation 6.43 can be written as

$$N_{S,1}(g_1, Re_1) = \frac{\tau^2}{a_1 g_1 St_1} + \frac{R}{c_p} a_1 f_1 g_1^3 \tag{6.48}$$

The only degrees of freedom, therefore, are the Reynolds number and the mass velocity. Minimizing the entropy generation number subject to a fixed (known) Reynolds number yields the optimal mass velocity:

$$g_{1,\text{opt}} = \left[\frac{\tau^2}{(3R/c_p)a_1^2 f_1 \text{St}_1} \right]^{1/4} , \quad N_{S,1,\min} = \left[\frac{256\tau^6(R/c_p)f_1}{27a_1^2\text{St}_1^3} \right]^{1/4} \quad (6.49)$$

Volume Constraint. This constraint might be important in the design of heat exchangers for applications where space is limited [7]. The dimensionless volume constraint can be written as

$$v_1 = V_1 \frac{8p_1}{v\dot{m}} \quad (\text{const}) \quad (6.50)$$

where V_1 is the volume of the passage (duct) on side 1. Noting that V_1 equals L_1 times the cross-sectional flow area of the passage, we have also $v_1 g_1^2 = (4L/D_h)_1\text{Re}_1$. This allows $N_{S,1}$ to be expressed in terms of only g_1 and Re_1 as degrees of freedom:

$$N_{S,1} = \frac{\tau^2 \text{Re}_1}{v_1 g_1^2 \text{St}_1} + \frac{R}{c_p} \frac{v_1 f_1 g_1^4}{\text{Re}_1} \quad (6.51)$$

If we regard Re_1 as fixed, the optimal mass flow rate and corresponding minimum irreversibility are

$$g_{1,\text{opt}} = \left[\frac{\tau^2\text{Re}_1^2}{2(R/c_p)v_1^2 f_1 \text{St}_1} \right]^{1/6} , \quad N_{S,1,\min} = \left[\frac{27\tau^4(R/c_p)\text{Re}_1 f_1}{4v_1 \text{St}_1^2} \right]^{1/3} \quad (6.52)$$

which are the counterparts of Equations 6.49.

Combined Area and Volume Constraint. When the area A_1 and the volume V_1 of the heat exchanger passage are constrained simultaneously, there is only one degree of freedom left for optimizing the thermodynamic performance of the passage [11]. Combining Equations 6.43, 6.47, and 6.50 yields

$$N_{S,1} = \frac{\tau^2 v_1}{a_1^2 \text{St}_1 \text{Re}_1} + \frac{R}{c_p} \frac{a_1^4 f_1 \text{Re}_1^3}{v_1^3} \quad (6.53)$$

where Re_1 is a variable. In commercial pipes at large Reynolds numbers f_1 and St_1 are relatively insensitive to changes in Re_1. In such applications $N_{S,1}$ of Equation 6.53 is a minimum when

$$\text{Re}_{1,\text{opt}} = \frac{v_1}{a_1^{3/2}} \left[\frac{\tau^2}{3(R/c_p)f_1 \text{St}_1} \right]^{1/4} \tag{6.54}$$

6.2.5 Unbalanced Heat Exchangers

Let us now consider entropy generation in limiting cases involving *unbalanced* heat exchangers: heat exchangers for which the capacity rates differ. The unbalance is described by the ratio of the two capacity rates:

$$\omega = \frac{(\dot{m}c_p)_1}{(\dot{m}c_p)_2} > 1 \tag{6.55}$$

Both counterflow and parallel-flow heat exchangers are considered.

As noted before, there are three sources of irreversibility commonly found in heat exchangers. These are associated with the stream-to-stream temperature difference and the pressure drops of the two streams. This is brought out by Equation 6.34, and for the case of balanced counterflow heat exchangers by the special forms given by Equations 6.36 and 6.40. Referring to Equation 6.40, note that the entropy generation rate of balanced counterflow heat exchangers vanishes as NTU $\rightarrow \infty$, $\Delta p_1 \rightarrow 0$, $\Delta p_2 \rightarrow 0$. This behavior is not exhibited by unbalanced heat exchangers, however, as will now be shown.

For an unbalanced *counterflow heat exchanger*, the exit temperature of the cold stream approaches the inlet temperature of the hot stream as NTU $\rightarrow \infty$: $T_{2,\text{out}} \rightarrow T_1$. Then, with an energy balance for the heat exchanger we also have

$$T_{1,\text{out}} = T_1 - \frac{1}{\omega}(T_1 - T_2) \tag{6.56}$$

With these relations, Equation 6.54 gives on reduction [10]

$$N_{S,\text{imbalance}} = \frac{\dot{S}_{\text{gen}}}{(\dot{m}c_p)_2} = \ln\left\{ \left[1 - \frac{1}{\omega}\left(1 - \frac{T_2}{T_1} \right) \right]^\omega \frac{T_1}{T_2} \right\} \tag{6.57}$$

The analysis of a *parallel-flow heat exchanger* with two streams proceeds by combining Equation 6.34 with the appropriate effectiveness–NTU relation [4]. The resulting expression is

$$N_{S,\text{imbalance}} = \frac{\dot{S}_{\text{gen}}}{(\dot{m}c_p)_2} = \ln\left\{ \left(\frac{T_2}{T_1} \right)^\omega \left[1 + \left(\frac{T_1}{T_2} - 1 \right) \frac{\omega}{1+\omega} \right]^{1+\omega} \right\} \tag{6.58}$$

In writing Equations 6.57 and 6.58 we have assumed not only that NTU $\rightarrow \infty$ but also that $\Delta p_1 = \Delta p_2 = 0$. These equations show clearly that for unbalanced heat exchangers the entropy generation rate does not vanish even under such idealized circumstances. In the limit of extreme unbalance, $\omega \to \infty$, Equations 6.57 and 6.58 become

$$N_{S,\text{imbalance}} = \frac{T_2}{T_1} - 1 - \ln \frac{T_2}{T_1} \qquad (\omega \to \infty) \qquad (6.59)$$

In this limit, stream 1 is so large that its temperature remains equal to T_1 from inlet to outlet; seen from the outside, it behaves like a stream that condenses or evaporates isobarically.

We also note that when the two streams and their inlet conditions are given for an unbalanced heat exchanger, the irreversibility of the parallel-flow arrangement is consistently greater than the irreversibility of the counterflow scheme. Figure 6.11 shows the behavior of the respective entropy generation numbers, and how they both approach the value indicated by Equation 6.59 as the flow unbalance ratio ω increases. Taking the limit of Equation 6.58 as $\omega \to 1$, it is easy to see that the irreversibility of the parallel-flow arrangement is finite even in the balanced case (see also $\omega = 1$ in Figure 6.11), indicating that the counterflow arrangement is thermodynamically superior to the parallel-flow arrangement.

6.3 AIR PREHEATER PRELIMINARY DESIGN

In this section we consider the preliminary design of the air preheater included in the case study cogeneration system shown in Figure 1.7. Two configura-

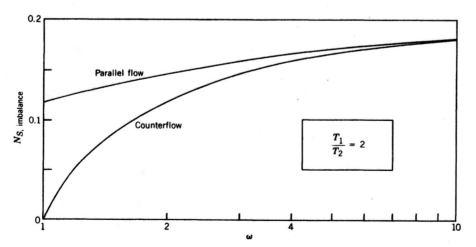

Figure 6.11 The entropy generation number for an unbalanced heat exchanger in parallel flow and counterflow [11].

tions are considered: a shell-and-tube counterflow heat exchanger and a plate-fin crossflow heat exchanger. The presentation is aimed at illustrating several of the heat exchanger principles introduced thus far, and at highlighting some of the design issues attending the specification of a heat exchanger for a particular duty. The present discussion is related only to the concept development stage of the life-cycle design flow chart shown in Figure 1.2. We do not consider here, for example, the special materials that might be required owing to the high preheater service temperatures. This is an issue that would be addressed at the detailed design stage.

6.3.1   Shell-and-Tube Counterflow Heat Exchanger

A shell-and-tube counterflow heat exchanger is shown schematically in Figure 6.12a. The figure is labeled with data from Table 1.2. Compressed air flows through n parallel tubes with inner diameter D, outer diameter D_o, and length L. The tubes are mounted in a shell of cross-sectional (frontal) area A_{fr}, as illustrated in Figure 6.12b. Combustion products flow through the tube-to-tube spaces, parallel to the tubes.

The present discussion is aimed at determining the number of tubes required and the heat exchanger dimensions. In particular, five values are to be determined: n, D, D_o, L, and A_{fr}. Although tubes are commercially available at only certain nominal diameters and wall thicknesses, the present considerations are simplified by taking $D_o/D = 1.2$. This leaves four values to be determined: n, D, L, and A_{fr}. As the temperatures and pressures are specified at the preheater inlets and exits, we find in the discussion to follow that the design must satisfy three conditions, one for heat transfer and two for fluid flow. Accordingly, with four unknowns and three equations, there is just one degree of freedom.

As elements of the model used in the present discussion, we list the following:

- No significant heat transfer occurs from the outer surface of the shell.
- The tubes are straight with smooth inner and outer surfaces.
- The flow of air and the flow of combustion products are each fully developed.
- The air and combustion products behave as ideal gases with constant (average) specific heats.
- The pressure drops due to contraction and enlargement are negligible.
- The thermal resistance of the tube wall is negligible.

Heat Transfer Relations.   We begin with the heat transfer aspects of the heat exchanger. The average air temperature is (603.7 K + 850 K)/2 $\cong$ 727 K. The specific heat of air at this temperature is $c_{p,\text{air}} = 1.135$ kJ/kg·K. The capacity rate on the high-pressure side (the tube side) of the heat exchanger is then

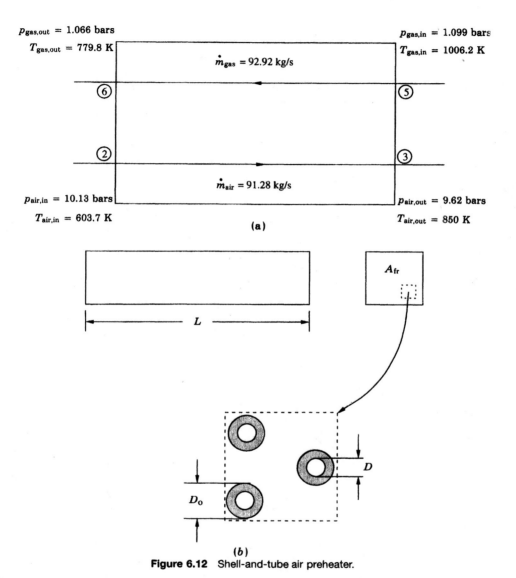

$p_{gas,out} = 1.066$ bars

$T_{gas,out} = 779.8$ K

$\dot{m}_{gas} = 92.92$ kg/s

$p_{gas,in} = 1.099$ bars

$T_{gas,in} = 1006.2$ K

⑥ ⑤

② ③

$\dot{m}_{air} = 91.28$ kg/s

$p_{air,in} = 10.13$ bars

$T_{air,in} = 603.7$ K

$p_{air,out} = 9.62$ bars

$T_{air,out} = 850$ K

(a)

L

A_{fr}

D

D_o

(b)

Figure 6.12 Shell-and-tube air preheater.

$$C_{air} = \dot{m}_{air}c_{p,air} = (91.28 \text{ kg/s})(1.135 \text{ kJ/kg·K}) = 103.6 \text{ kW/K}$$

The capacity rate on the low-pressure side can be deduced from an energy balance on the overall preheater:

$$C_{air}(T_{out} - T_{in})_{air} = C_{gas}(T_{in} - T_{out})_{gas}$$

where the subscript "gas" denotes the combustion products. The result is

$$C_{gas} = (103.6 \text{ kW/K}) \frac{850 - 603.7}{1006.2 - 779.8} = 112.7 \text{ kW/K}$$

which means that the counterflow heat exchanger is only slightly unbalanced. In the discussion to follow, we take

$$C_{gas} = C_{max}, \qquad C_{air} = C_{min}, \qquad \frac{C_{min}}{C_{max}} = 0.919$$

Next, using Equation 6.5, the effectiveness is given by

$$\varepsilon = \frac{T_{air,out} - T_{air,in}}{T_{gas,in} - T_{air,in}} = \frac{850 - 603.7}{1006.2 - 603.7} = 0.612$$

Solving Equation 6.6, we calculate the required NTU:

$$\text{NTU} = \frac{\ln\left[\dfrac{1 - \varepsilon C_{min}/C_{max}}{1 - \varepsilon}\right]}{1 - C_{min}/C_{max}} = 1.484$$

With Equation 6.3 the overall thermal conductance is

$$UA = C_{min} \text{ NTU} = (103.6 \text{ kW/K}) \, 1.484 = 153.7 \text{ kW/K}$$

This value is related to the thermal conductances on the two sides of the heat transfer surface by Equation 6.1:

$$\frac{1}{UA} = \frac{1}{hn \, \pi DL} + \frac{1}{h_o n \, \pi D_o L} \tag{6.60}$$

where h and h_o are the air-side and gas-side heat transfer coefficients. We will return to Equation 6.60 after we analyze the pressure drops.

Pressure Drop Relations. The pressure drop on the air side can be estimated from Equation 6.7 as

$$\Delta p_{air} = f \, \frac{2L}{D} \frac{G^2}{\rho_{air}} + G^2 \left(\frac{1}{\rho_{out}} - \frac{1}{\rho_{in}}\right)_{air} \tag{6.61}$$

Figure 6.12a provides $\Delta p_{air} = (10.13 - 9.62)$ bars $= 0.51$ bar. Using the ideal-gas equation of state, we have

$$\frac{\rho_{out}}{\rho_{in}} = \frac{p_{out}T_{in}}{p_{in}T_{out}} = 0.675$$

and

$$\rho_{in} = \frac{p_{in}}{R_{air}\,T_{in}} = \frac{10.13 \times 10^5 \text{ N/m}^2}{(0.287 \text{ kJ/kg·K})(603.7 \text{ K})} = 5.85 \text{ kg/m}^3$$

Thus

$$\rho_{out} = 0.675\rho_{in} = 3.95 \text{ kg/m}^3$$

and

$$\rho_{air} = \tfrac{1}{2}(\rho_{in} + \rho_{out})_{air} = 4.90 \text{ kg/m}^3$$

Applying Equation 6.16

$$G = \frac{\dot{m}_{air}}{n(\pi D^2/4)} = (116.2 \text{ kg/s})\frac{1}{nD^2}$$

Since the pipes are assumed to be smooth, Figure 4.7 gives $f = f(\text{Re}_D)$, where

$$\text{Re}_D = \frac{GD}{\mu_{air}}.$$

In this expression, $\mu_{air} = 4.1 \times 10^{-5}$ kg/s·m (from [4]) at the average temperature 727 K. Substituting these quantities into Equation 6.61 gives after reduction

$$18.38n^2 \left(\frac{D}{1 \text{ m}}\right)^4 = f\,\frac{2L}{D} + 0.4 \tag{6.62}$$

where $f = f(\text{Re}_D)$, and

$$\text{Re}_D = \frac{2.84 \times 10^6 \text{ m}}{nD}$$

The pressure drop on the gas side can be analyzed in the same way beginning with an application of Equation 6.7:

$$\Delta p_{gas} = f_o \frac{2L}{D_h} \frac{G_o^2}{\rho_{gas}} + G_o^2 \left(\frac{1}{\rho_{out}} - \frac{1}{\rho_{in}} \right)_{gas} \tag{6.63}$$

From Figure 6.12a, $\Delta p_{gas} = (1.099 - 1.066)$ bars $= 0.033$ bar. Using the ideal gas equation of state

$$\frac{\rho_{out}}{\rho_{in}} = \frac{p_{out}T_{in}}{p_{in}T_{out}} = 1.252$$

and

$$\rho_{out} = \frac{p_{out}}{R_{gas}T_{out}} = 0.465 \text{ kg/m}^3$$

we obtain

$$\rho_{in} = 0.37 \text{ kg/m}^3$$

$$\rho_{gas} = \tfrac{1}{2}(\rho_{in} + \rho_{out})_{gas} = 0.42 \text{ kg/m}^3$$

The area available for the flow of the gas equals the total frontal area less the area occupied by the n tubes:

$$A_{flow} = A_{fr} - n \frac{\pi D_o^2}{4} \tag{6.64}$$

We also have

$$G_o = \frac{\dot{m}_{gas}}{A_{flow}}$$

and

$$D_h = \frac{4A_{flow}}{n \pi D_o}$$

As the tubes are smooth, Figure 4.7 gives $f_o = f_o(\mathrm{Re}_{D_h})$, where

$$\mathrm{Re}_{D_h} = \frac{GD_h}{\mu_{gas}}$$

In this expression, we approximate the viscosity of the gas as the viscosity of air at the average gas temperature, 900 K. That is, $\mu_{gas} = 4.57 \times 10^{-5}$ kg/s·m (from [4]).

Substituting these quantities into Equation 6.63 gives, after reduction,

$$0.16 \left(\frac{A_{\text{flow}}}{1 \text{ m}^2}\right)^2 = f_o \frac{2L}{D_h} - 0.228 \qquad (6.65)$$

where $f_o = f_o(\text{Re}_{D_h})$ and

$$\text{Re}_{D_h} = (2.03 \times 10^6 \text{ m}) \frac{D_h}{A_{\text{flow}}}$$

Collecting Results. The development thus far has provided one heat-transfer-related equation: Equation 6.60, and two fluid-flow-related equations: Equations 6.62 and 6.65. These equations can be reduced further to obtain a final set of three equations as follows:

The heat transfer coefficients needed in Equation 6.60 can be estimated based on the Colburn analogy (Equation 4.46):

Air Side	Gas Side
$h = c_{p,\text{air}} G \text{ St}$	$h_o = c_{p,\text{gas}} G_o \text{St}_o$
$\text{St} = \frac{1}{2} f \text{ Pr}_{\text{air}}^{-2/3}$	$\text{St}_o = \frac{1}{2} f_o \text{Pr}_{\text{gas}}^{-2/3}$
$\text{Pr}_{\text{air}} = 0.71$	$\text{Pr}_{\text{gas}} = 0.72$
$c_{p,\text{air}} = 1.135 \text{ kJ/kg·K}$	$c_{p,\text{gas}} = \dfrac{C_{\text{gas}}}{\dot{m}_{\text{gas}}} = 1.213 \text{ kJ/kg·K}$

Since $UA = 153.7$ kW/K, Equation 6.60 then becomes

$$1 = 0.59 \frac{D}{fL} + 0.58 \frac{A_{\text{flow}}}{n f_o L D} \qquad (6.66)$$

Another simplification follows by noting that Re_D and Re_{D_h} are proportional:

$$\frac{\text{Re}_{D_h}}{\text{Re}_D} = \frac{(2.03 \times 10^6 \text{ m})(D_h/A_{\text{flow}})}{(2.84 \times 10^6 \text{ m})(1/nD)} = 0.715 \frac{nD}{A_{\text{flow}}} D_h$$

$$= 0.715 \frac{nD}{A_{\text{flow}}} \frac{4A_{\text{flow}}}{n \pi (1.2) D} = 0.76$$

In the Re_D range 10^4–10^6, the friction factor varies as $\text{Re}_D^{-0.2}$; therefore f and f_o are proportional:

$$\frac{f_o}{f} = \left(\frac{\text{Re}_{D_h}}{\text{Re}_D}\right)^{-0.2} = (0.76)^{-0.2} = 1.056$$

In conclusion, Equation 6.66 can be written as

$$1 = 0.59 \frac{D}{fL} + 0.55 \frac{A_{\text{flow}}}{nDfL} \tag{6.67}$$

Turning to Equation 6.65, it can be expressed as

$$0.16\tilde{A}^3 n^2 \left(\frac{D}{1\ \text{m}}\right)^4 = f\frac{2L}{D} - 0.3\tilde{A} \tag{6.68}$$

where $\tilde{A}$ is the dimensionless cross-sectional area of the gas stream,

$$\tilde{A} = \frac{A_{\text{flow}}}{nD^2} \tag{6.69}$$

Eliminating the group $2fL/D$ between Equations 6.62 and 6.68 gives

$$n\left(\frac{D}{1\ \text{m}}\right)^2 = \left(\frac{0.4 + 0.3\tilde{A}}{18.4 - 0.16\tilde{A}^3}\right)^{1/2} \tag{6.70}$$

Expressing Equation 6.67 in terms of $\tilde{A}$ yields

$$\frac{fL}{D} = 0.59 + 0.55\tilde{A} \tag{6.71}$$

Finally, combining Equations 6.62 and 6.71,

$$n\left(\frac{D}{1\ \text{m}}\right)^2 = (0.086 + 0.06\tilde{A})^{1/2} \tag{6.72}$$

Design Solutions. With Equations 6.67–6.72 the required numerical work is simplified greatly. Thus, Equations 6.70 and 6.72 give

$$\tilde{A} = 4.38$$
$$nD^2 = 0.59\ \text{m}^2 \tag{6.73}$$

Substituting these values into Equation 6.69,

$$A_{\text{flow}} = 2.59 \text{ m}^2$$

Hence, from Equation 6.64

$$A_{\text{fr}} = A_{\text{flow}} + n \frac{\pi(1.2D)^2}{4} = 3.26 \text{ m}^2$$

In conclusion, the frontal area and the gas flow cross section are fixed: They do not vary with the lone degree of freedom recognized at the start of this solution.

The remaining design parameters (n, D, L) can vary, but only one can be chosen independently. They must satisfy Equation 6.71 (which now reads $fL/D = 3.0$) and Equation 6.73. When Re_D falls in the range 10^4–10^6, $f = 0.046 \, Re_D^{-0.2}$. The following numerical results are obtained by treating D as the independent parameter:

D (m)	n	L (m)	Re_D	L/D	A (m²)	U (W/m²·K)
0.01	5910	5.6	4.8×10^4	563	1254	123
0.02	1477	12.9	9.6×10^4	647	1442	107

These values show that the total heat transfer area $A = n \pi D_o L$ and overall heat transfer coefficient U are relatively insensitive to D or the number of tubes n. However, the flow length L increases almost proportionally with the tube diameter D. The volume of the heat exchanger $(A_{\text{fr}} L)$ increases too, because the frontal area is fixed. The weight of the heat exchanger is controlled by the weight of the tubes: The tube material volume $n \pi (D_o^2 - D^2)L$ increases as the group nD^2L, which increases abruptly as D increases. Note again that there is one degree of freedom (e.g., D) and that a relatively small D would be selected if a small volume or small weight is desired for the heat exchanger. Still, the final selection in this case would be governed by costs.

6.3.2 Plate-Fin Crossflow Heat Exchanger

In this section we illustrate an alternative preliminary design for the air preheater by choosing a plate-fin crossflow configuration: Figure 6.13. The flow rates, temperatures, and pressures at the inlets and outlets of the air and gas streams are the same as indicated in Figure 6.12. The calculation of related quantities such as the capacity rates and average properties (ρ, c_p, μ, Pr) is given in the preceding section and is not repeated here. Once again, the heat exchanger model is based on a set of assumptions:

- The heat transfer from the overall heat exchanger to the ambient is negligible.

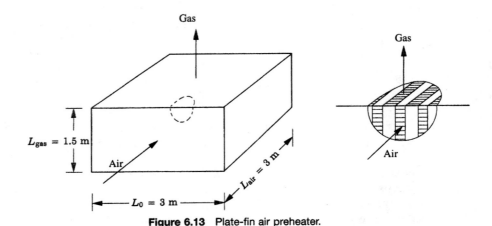

Figure 6.13 Plate-fin air preheater.

- The air and combustion gases behave as ideal gases with constant (average) specific heats.
- The thermal resistance of the plate between the air and gas sides is negligible.

In the present case the pressure drops due to contraction and enlargement are *not* assumed to be negligible, however.

The calculations in this section differ markedly from those performed for the shell-and-tube design. There we started with the temperature and pressure drop constraints, and calculated the heat exchanger size that meets the specifications. In the plate-fin crossflow arrangement, the calculations are necessarily iterative: We start with a certain set of surface designs and overall heat exchanger size and calculate the temperature and pressure drop characteristics of the assumed apparatus. If the calculated characteristics do not correspond to the specifications, we change the initial assumption (e.g., one or more dimensions of the overall heat exchanger), and repeat the calculations. The initial assumption is the degree of freedom in this design problem. We return to this aspect in the last paragraph of this section.

Geometric Characteristics. We illustrate this procedure by assuming as shown in Figure 6.13 that $L_0 = L_{air} = 3$ m, $L_{gas} = 1.5$ m, giving an overall size of $3 \times 3 \times 1.5$ m^3, with 3×1.5 m^2 as the frontal area of the air stream, and 3×3 m^2 as the frontal area of the gas stream:

$$V = 13.5 \text{ m}^3, \qquad A_{fr,air} = 4.5 \text{ m}^2, \qquad A_{fr,gas} = 9 \text{ m}^2$$

Note that we allowed a larger frontal area for the gas stream than for the air stream because the gas stream is less dense. Note also that we chose these overall dimensions and not dimensions several times smaller or several times larger because we had the benefit of knowing the numerical results (orders of magnitude) tabulated at the end of the shell-and-tube preliminary design. As surface designs we select, for purposes of illustration, the louvered plate-fin surface $\frac{3}{8}$–6.06 for the air side, and the plain plate-fin surface 11.1 for the gas side. The geometric characteristics of these surfaces are found in Tables 9-3a, b of Reference 3:

	Air Side	Gas Side
Plate spacing, b (mm)	6.4	6.4
Hydraulic radius, r_h (mm), $= \frac{1}{4}D_h$	1.11	0.77
Fin thickness, δ (mm)	0.15	0.15
Transfer area per volume between plates, β (m²/m³)	840	1204
Fin area per total area	0.64	0.756

The plate between the air side and gas side has the thickness $a = 3$ mm and is made of high-temperature alloy steel with a thermal conductivity $k = 20.8$ W/m·K.

The numerical work follows the steps detailed in Reference 3, and thus our presentation can be succinct. The ratios of the total heat transfer of one side to total heat exchanger volume are [3, p. 36]:

$$\alpha_{air} = \frac{b_{air}\beta_{air}}{b_{air} + b_{gas} + 2a} = 400.3 \text{ m}^{-1}$$

$$\alpha_{gas} = \frac{b_{gas}\beta_{gas}}{b_{air} + b_{gas} + 2a} = 574.1 \text{ m}^{-1}$$

(6.74)

The heat transfer areas of the air and gas sides are

$$A_{air} = \alpha_{air}V = 5404 \text{ m}^2, \qquad A_{gas} = \alpha_{gas}V = 7750 \text{ m}^2 \qquad (6.75)$$

The ratios of the flow cross-sectional area to the frontal area are

$$\sigma_{air} = \frac{A_{flow,air}}{A_{fr,air}} = \alpha_{air}r_{h,air} = 0.445$$

$$\sigma_{gas} = \frac{A_{flow,gas}}{A_{fr,gas}} = \alpha_{gas}r_{h,gas} = 0.443$$

(6.76)

The corresponding flow cross-sectional areas are then

$$A_{flow,air} = \sigma_{air}A_{fr,air} = 2.00 \text{ m}^2, \qquad A_{flow,gas} = \sigma_{gas}A_{fr,gas} = 3.99 \text{ m}^2 \qquad (6.77)$$

Friction Factors and Heat Transfer Coefficients. The flow areas determined above are needed for calculating the mass velocities

$$G_{air} = \frac{\dot{m}_{air}}{A_{flow,air}} = 45.6 \text{ kg/m}^2 \cdot \text{s}$$

$$G_{gas} = \frac{\dot{m}_{gas}}{A_{flow,gas}} = 23.3 \text{ kg/m}^2 \cdot \text{s} \qquad (6.78)$$

and the respective Reynolds numbers (note $D_h = 4r_h$):

$$Re_{air} = \frac{D_{h,air}G_{air}}{\mu_{air}} = 4935$$

$$Re_{gas} = \frac{D_{h,gas}G_{gas}}{\mu_{gas}} = 1570 \qquad (6.79)$$

With these values for the Reynolds numbers we obtain from Figures 10-38 and 10-26 of Reference 3

$$St_{air}Pr_{air}^{2/3} = 0.0069, \qquad St_{air} = 0.009, \qquad f_{air} = 0.037$$

and

$$St_{gas}Pr_{gas}^{2/3} = 0.0044, \qquad St_{gas} = 0.057, \qquad f_{gas} = 0.0145$$

The convective heat transfer coefficients are thus

$$h_{air} = St_{air}G_{air}c_{p,air} = 466 \text{ W/m}^2 \cdot \text{K}$$

$$h_{gas} = St_{gas}G_{gas}c_{p,gas} = 162 \text{ W/m}^2 \cdot \text{K} \qquad (6.80)$$

The numerical calculations for the fin efficiency η_f and the overall surface efficiency factor γ are given in Reference 3:

$$\eta_{f,air} = 0.665, \qquad \eta_{f,gas} = 0.85$$

and

$$\gamma_{air} = 0.786, \qquad \gamma_{gas} = 0.887$$

The overall heat transfer coefficient is given by Equation 6.1 where the plate resistance is neglected:

$$\frac{1}{U_{air}} = \frac{1}{\gamma_{air}h_{air}} + \frac{1}{\gamma_{gas}h_{gas}A_{gas}/A_{air}} = \frac{1}{132 \text{ W/m}^2 \cdot \text{K}} \qquad (6.81)$$

Collecting Results. The number of heat transfer units is given by

$$\text{NTU} = \frac{A_{\text{air}} U_{\text{air}}}{C_{\text{min}}} = 6.9 \tag{6.82}$$

With this NTU value and $C_{\text{min}}/C_{\text{max}} = 0.919$, the effectiveness–NTU chart for a crossflow heat exchanger with both fluids unmixed (e.g., Figure 9.24 in [4]) gives, approximately, $\varepsilon \cong 0.8$. This effectiveness is greater than the value fixed by the inlet and outlet temperatures specified in Figure 6.12: $\varepsilon = 0.612$. In the present case, the outlet temperatures of the two streams are calculated with Equation 6.5:

$$T_{\text{air,out}} \cong 926 \text{ K}, \qquad T_{\text{gas,out}} = 710 \text{ K}$$

Because of the higher effectiveness, the air outlet temperature is higher than the 850 K temperature specified in Figure 6.12. Similarly, the outlet temperature of the products of combustion is lower than the specified 779.8 K.

The pressure drop along each of the streams is calculated with a formula obtained by adding Equations 6.7 through 6.9 (see also Equation 2-26 in [3]):

$$\frac{\Delta p}{p_{\text{in}}} = \frac{G^2}{2\rho_{\text{in}}\, p_{\text{in}}} \left[(K_c + 1 - \sigma^2) + 2\left(\frac{\rho_{\text{in}}}{\rho_{\text{out}}} - 1\right) \right. $$
$$\left. + f\, \frac{A}{A_{\text{flow}}}\, \frac{\rho_{\text{in}}}{\rho} - (1 - \sigma^2 - K_e)\, \frac{\rho_{\text{in}}}{\rho_{\text{out}}} \right] \tag{6.83}$$

The terms inside the square brackets represent, in order, the effects of entrance (contraction), flow acceleration, core friction, and exit (enlargement). In the core friction term, ρ is the average density of the stream. The ratio of flow area to frontal area is listed in Equations 6.76. The ratio of heat transfer area to flow area is obtained from Equations 6.75 and 6.77:

$$\frac{A_{\text{air}}}{A_{\text{flow,air}}} = 2700, \qquad \frac{A_{\text{gas}}}{A_{\text{flow,gas}}} = 1942$$

The contraction and enlargement loss coefficients can be estimated approximately using Figure 5-4 of Reference 3, in which we use the σ and Re values of each stream:

$$K_{c,\text{air}} = 0.45, \qquad K_{e,\text{air}} = 0.25$$
$$K_{c,\text{gas}} = 0.5, \qquad K_{e,\text{gas}} = 0.22$$

Substituting these values into Equation 6.83 written for each of the two streams, gives

$$\left(\frac{\Delta p}{p_{in}}\right)_{air} = 0.02, \qquad \left(\frac{\Delta p}{p_{in}}\right)_{gas} = 0.17$$

While calculating these relative pressure drops we learn that the core friction term accounts for roughly 98% of the total value. The contraction and enlargement contributions are on the order of 1% for the air stream and 5% for the gas stream; however, these two effects enter with opposite signs in Equation 6.83 and effectively cancel each other.

The pressure drops calculated in this first iteration do not match the values shown in Figure 6.12:

$$\left(\frac{\Delta p}{p_{in}}\right)_{air} = 0.05 \quad , \quad \left(\frac{\Delta p}{p_{in}}\right)_{gas} = 0.05$$

This mismatch, together with that noted between the calculated effectiveness (0.8) and the specified effectiveness (0.612) means that the core dimensions assumed at the start of these calculations (Figure 6.13) are not the appropriate ones. Subsequent iterations show that they are of the correct order of magnitude, however. In the next iteration, the designer might try a somewhat different set of core overall dimensions:

- A longer dimension in the direction of the air stream (e.g., $L_{air} = 5$ m), so that the air pressure drop would increase from 2% closer to the specified 5%.
- A shorter dimension aligned with the gas stream (e.g., $L_{gas} = 0.75$ m), to reduce the gas pressure drop from 17% toward the specified 5%.
- A third dimension that keeps the overall volume relatively unchanged, but a little smaller (e.g., $L_0 = 3$ m). Accordingly, the new heat transfer area, NTU, and ε will be somewhat smaller, increasing the chance for a better match between the calculated outlet temperature and the values specified in Figure 6.12.

This and subsequent iterations can be executed on a computer by properly accounting for the data that must be used from Reference 3 (e.g., curve fitting these data).

6.3.3 Closure

The preliminary preheater results considered in this section, together with possible alternative configurations, would be subject to screening in the concept development stage (Section 1.5.3) until the preferred configuration for the detailed design stage is identified. At that stage, additional considerations such as materials of construction, fabrication, maintenance, and costs would enter. Such considerations always play a part, even when the component in question is to be obtained from a vendor, for it is by such means that the specifications to be met by the vendor are determined.

6.4 ADDITIONAL APPLICATIONS

In this section, a number of additional applications are presented featuring the use of thermodynamics together with concepts of heat transfer. The objective is to reinforce understanding of the interplay of principles drawn from these branches of engineering within a design analysis context and to illustrate further the trade-off nature of design. To allow these objectives to be met simply, the presentation features idealized modeling and thermodynamic optima only. The results generally have only qualitative significance. Further applications of this modeling approach can be found in the literature [7, 11, 12], where the method is known as entropy generation minimization.

6.4.1 Refrigeration

As a first case, let us consider the vapor compression refrigeration system shown in Figure 6.14. Denoting the refrigerant mass flow rate by $\dot{m}$, the compressor power is

$$\dot{W} = \dot{m}(h_2 - h_1) = \frac{\dot{W}_s}{\eta_c} \tag{6.84}$$

where η_c is the *isentropic compressor efficiency* and $\dot{W}_s = \dot{m}(h_{2s} - h_1)$. The expansion across the valve is a *throttling* process:

$$h_3 = h_4 \tag{6.85}$$

The refrigerant is cooled in the condenser from the superheated vapor state 2 to the saturated liquid state 3 by contact with a stream of ambient air at temperature T_H and capacity rate $(\dot{m}c_p)_H$, labeled C_H. Although the refrigerant temperature varies from T_2 to T_3, most of the heat transfer between refrigerant and ambient air occurs during the condensation of the refrigerant, that is, when the refrigerant temperature is at T_3. The condenser heat transfer rate $\dot{Q}_H$ can be approximated in terms of a heat exchanger effectiveness based on the condensation temperature:

$$\dot{Q}_H = \dot{m}(h_2 - h_3) = \varepsilon_H C_H (T_3 - T_H) \tag{6.86}$$

where

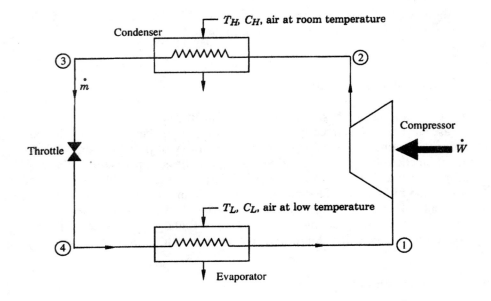

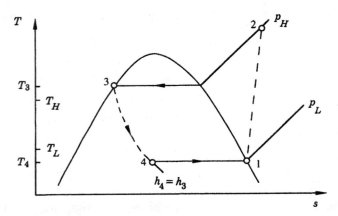

Figure 6.14   Refrigeration plant model with vapor compression cycle and irreversibilities due to heat transfer, compression, and throttling.

$$\varepsilon_H = 1 - \exp\left(-\frac{U_H A_H}{C_H}\right) \tag{6.87}$$

Similarly, the heat transfer to the refrigerant in the evaporator is

$$\dot{Q}_L = \dot{m}(h_1 - h_4) = \varepsilon_L C_L (T_L - T_4) \tag{6.88}$$

where

$$\varepsilon_L = 1 - \exp\left(-\frac{U_L A_L}{C_L}\right) \tag{6.89}$$

Condenser and evaporator pressure drops are ignored in the present discussion. Equations 6.84 and 6.88 can be used to evaluate the *coefficient of performance* (COP):

$$\text{COP} = \frac{\dot{Q}_L}{\dot{W}} \tag{6.90}$$

Owing to size, cost, and other considerations, the heat transfer areas A_H and A_L cannot be specified arbitrarily but are typically constrained. A constraint that allows the trade-off nature of design to be demonstrated simply is the assumption of constant total heat transfer area:

$$A_H + A_L = A \quad \text{(const)}$$

or $\tag{6.91}$

$$y = A_H/A, \quad 1 - y = A_L/A$$

This constraint is invoked elsewhere [13–15], where the question of how to divide the constant area A between A_H and A_L to maximize the coefficient of performance is considered.

To focus on the roles of the heat transfer areas A_H and A_L in a simplified context, consider the idealized schematic of Figure 6.15, where we have neglected the temperature variation of the refrigerant through the condenser and have denoted the constant temperature of the condensing stream by T_{max}. This simplification is consistent with the one made in the ε_H model of Equation 6.87. The evaporation temperature (T_{min} in Figure 6.15) represents the constant refrigerant temperature between states 4 and 1. We also take the isentropic compressor efficiency to be 100% and ignore the irreversibility of the expansion across the valve. That is, the cycle executed by the refrigerant is assumed to be *internally reversible*. Accordingly, we may write

$$\frac{\dot{Q}_H}{T_{max}} = \frac{\dot{Q}_L}{T_{min}} \tag{6.92}$$

The rest of the refrigerator model remains unchanged. The heat exchangers are described by

$$\dot{Q}_H = \varepsilon_H C_H (T_{max} - T_H) \tag{6.93}$$

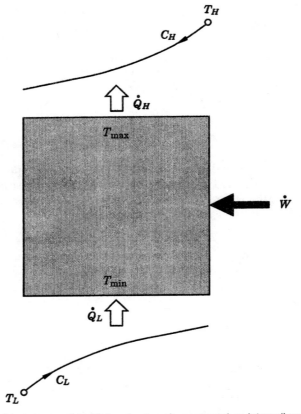

Figure 6.15 Refrigerator model with two heat exchangers and an internally reversible cycle.

$$\dot{Q}_L = \varepsilon_L C_L (T_L - T_{min}) \tag{6.94}$$

with ε_H and ε_L given by Equations 6.87 and 6.89. By inspecting Figure 6.15, we note that the C_H stream is colder than T_{max}, and the C_L stream is warmer than T_{min}.

Introducing the energy balance $\dot{W} + \dot{Q}_L = \dot{Q}_H$, Equations 6.92 through 6.94 give

$$\frac{T_H}{\dot{Q}_H} = \frac{T_L}{\dot{Q}_L} - \left(\frac{1}{\varepsilon_H C_H} + \frac{1}{\varepsilon_L C_L}\right) \tag{6.95}$$

Regarding $\dot{Q}_L$ as fixed, maximization of the COP is equivalent to minimizing the expression in brackets on the right side of Equation 6.95 [14]. Using Equations 6.87 and 6.89, this expression becomes

$$\frac{1}{\varepsilon_H C_H} + \frac{1}{\varepsilon_L C_L} = \frac{(C_H)^{-1}}{1 - \exp(-yN)}$$
$$+ \frac{(C_L)^{-1}}{1 - \exp[-(1 - y)N (U_L/U_H)/(C_L/C_H)]} \quad (6.96)$$

where y is the area ratio $y = A_H/A$ and N is the number of heat transfer units based on the total area and the room temperature parameters,

$$N = \frac{U_H A}{C_H} \quad (6.97)$$

Specifying the ratios C_L/C_H and U_L/U_H, each of which is less than unity, and the value of N, the right side of Equation 6.96 varies only with the area ratio y. Minimizing this expression with respect to y gives the optimal area ratio $A_{H,opt}/A$, which is plotted versus C_L/C_H in Figure 6.16 for sample values of N and U_L/U_H. The figure shows that the optimal A_H is in general not equal to $\frac{1}{2}A$–that is, not equal to A_L. The optimal way of dividing A between the two heat exchangers depends on the total amount of heat exchanger area available and the relative external characteristics of the two heat exchangers, that is, the ratios C_L/C_H and U_L/U_H.

6.4.2 Power Generation

In this section we use a simplified model of an extraterrestrial power plant [4, 11] to consider certain trade-offs existing in the design of any such plant. An example of a power plant for use in space is presented in Figure 6.17. The working fluid executes a closed Brayton cycle. It is heated by a heat source (isotope, solar, or nuclear) and cooled by a radiator. Optimization studies for this power plant configuration are reported in Reference 16. A principal conclusion is that the solar heat source is more suitable than the isotope and nuclear heat sources when the power plant size ($\dot{W}$) is in the range 2–100 kW. Among the degrees of freedom in the design of the solar power plant of Figure 6.17 are: (a) turbine inlet temperature, (b) compressor inlet temperature, (c) extent of recuperation used in the cycle, (d) pressure drops along ducts between components of the plant, (e) compressor specific speed, (f) compressor pressure ratio, (g) alternator rotor speed, and (h) working fluid. The design of such a power plant is complicated further by the need to incorporate energy storage in the cycle to ensure power generation during eclipse periods.

To gain insights about trade-offs in this application simply, consider the model shown in Figure 6.18. The solar collector of temperature T_H and area A_H is heated by the sun at T_s and cooled by the cold background at T_∞. The radiator of temperature T_L and area A_L is also cooled by the background.

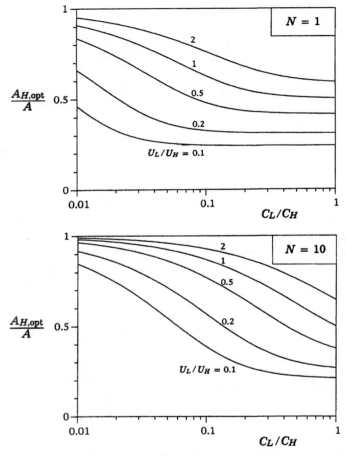

Figure 6.16 Optimal area ratio A_H/A for the heat exchangers of an idealized refrigeration system.

The working fluid of the power plant is heated in the collector and cooled in the radiator. As in Section 6.4.1, we assume (1) that the cycle executed by the working fluid is internally reversible, which means that the rate of entropy generation within the cycle is zero:

$$\frac{\dot{Q}_L}{T_L} - \frac{\dot{Q}_H}{T_H} = 0 \tag{6.98}$$

and (2) that the total area of the collector and radiator is limited:

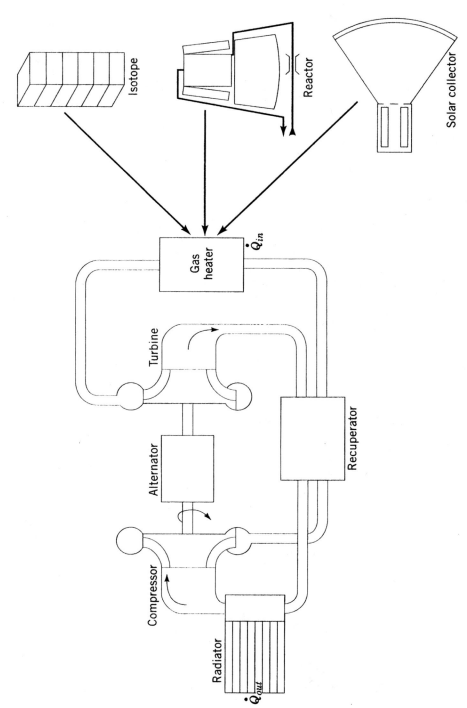

Figure 6.17 Brayton cycle power plant for use in space [16] (Reprinted with permission from the American Society of Mechanical Engineers.)

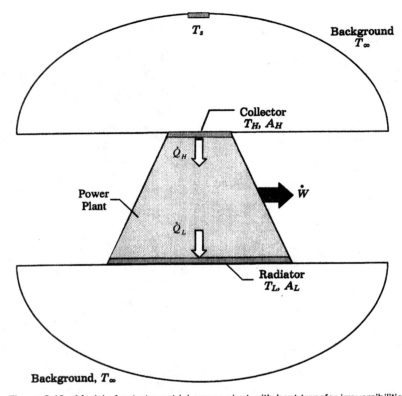

Figure 6.18 Model of extraterrestrial power plant with heat transfer irreversibilities.

$$A_H + A_L = A \quad \text{(const)} \tag{6.99}$$

The objective is to maximize the power output of the plant, $\dot{W}$. The power output is obtained by combining Equation 6.98 with the energy balance $\dot{Q}_H = \dot{W} + \dot{Q}_L$:

$$\dot{W} = \dot{Q}_H \left(1 - \frac{T_L}{T_H}\right) \tag{6.100}$$

To calculate $\dot{Q}_H$, we model the collector, sun, and background as three black surfaces forming the enclosure shown above A_H. Then, with relations from Section 4.4,

$$\dot{Q}_H = A_H F_{Hs}\sigma(T_s^4 - T_H^4) - A_H F_{H\infty}\sigma(T_H^4 - T_\infty^4) \tag{6.101}$$

where F_{Hs} and $F_{H\infty}$ are the collector–sun and collector–background view factors. The relation between them is

$$F_{H\infty} = 1 - F_{Hs} \tag{6.102}$$

Note further that T_∞^4 is negligible relative to T_H^4, because $T_\infty \cong 4$ K, which is much less than the expected order of magnitude of T_H. With these observations, Equation 6.101 becomes

$$\dot{Q}_H = \sigma A_H (F_{Hs}T_s^4 - T_H^4) \tag{6.103}$$

For the enclosure with two black surfaces formed under the radiator A_L in Figure 6.18, we have

$$\dot{Q}_L = \sigma A_L (T_L^4 - T_\infty^4) \tag{6.104}$$

in which T_∞^4 can be neglected relative to T_L^4, giving

$$\dot{Q}_L = \sigma A_L T_L^4 \tag{6.105}$$

The next step is to eliminate $\dot{Q}_H$ and $\dot{Q}_L$ between Equations 6.98, 6.103, and 6.105. The result is

$$\zeta - 1 = \frac{A_L}{A_H}\left(\frac{T_L}{T_H}\right)^3 \tag{6.106}$$

where

$$\zeta = F_{Hs}\left(\frac{T_s}{T_H}\right)^4 \tag{6.107}$$

If the area constraint (Equation 6.99) is expressed in terms of the area ratio y,

$$A_H = yA, \qquad A_L = (1 - y)A \tag{6.108}$$

we find that Equations 6.106, 6.103, and 6.100 become, in order,

$$\frac{T_L}{T_H} = \left[\frac{y}{1 - y}(\zeta - 1)\right]^{1/3} \tag{6.109}$$

$$\dot{Q}_H = \sigma A F_{Hs} T_s^4 y \left(1 - \frac{1}{\zeta}\right) \tag{6.110}$$

$$\tilde{W} = \frac{\dot{W}}{\sigma A F_{Hs} T_s^4} = y \left(1 - \frac{1}{\zeta}\right) \left\{ 1 - \left[\frac{y}{1-y}(\zeta - 1) \right]^{1/3} \right\} \tag{6.111}$$

The dimensionless power output $\tilde{W}$ emerges as a function of two parameters, ζ and y. The variable $\tilde{W}$ can be maximized numerically with respect to both ζ and y, and the results are [11]

$$\zeta_{\mathrm{opt}} = 1.538, \quad y_{\mathrm{opt}} = 0.35, \quad \tilde{W}_{\mathrm{max}} = 0.0414 \tag{6.112}$$

Accordingly, the collector accounts for roughly one third of the total surface available, which means that A_H is about half the size of A_L. Furthermore, by substituting $T_s = 5762$ K in ζ_{opt} and Equations 6.107 and 6.106, we find that

$$T_{H,\mathrm{opt}} = 5174 F_{Hs}^{1/4} \text{ K} \tag{6.113}$$

$$T_{L,\mathrm{opt}} = 3423 F_{Hs}^{1/4} \text{ K}$$

For a flat-plate collector the view factor F_{Hs} is approximately 10^{-4}, and Equations 6.113 give $T_{H,\mathrm{opt}} \doteq 520$ K and $T_{L,\mathrm{opt}} \doteq 340$ K. If the solar radiation is concentrated into the collector, F_{Hs} and the resulting collector and radiator temperatures will be higher. Though idealized, the present model does yield temperatures roughly in the range of those reported in [16] on the basis of a more detailed model of a system *with* a solar radiation concentrator.

6.4.3 Exergy Storage by Sensible Heating

The process of exergy storage by sensible heating [17], shown schematically in Figure 6.19, has a number of applications. The system consists of a large liquid bath of mass M and specific heat c placed in an insulated vessel. The system is charged with exergy when the hot gas entering the system is cooled while flowing through a heat exchanger immersed in the bath. The gas is eventually discharged to the atmosphere. The bath temperature T as well as the gas outlet temperature T_{out} gradually rise, approaching the hot gas inlet temperature T_∞. When the system is discharged, the stored exergy can be retrieved, at least in part.

Let us study this system using the following model: The bath is filled with liquid such as water or oil. The liquid is modeled as incompressible. The bath is well mixed thermally, so that at any given time its temperature is uniform, $T(t)$. The initial bath temperature equals the environment temperature T_0. The gas obeys the ideal-gas model with a constant specific heat c_p.

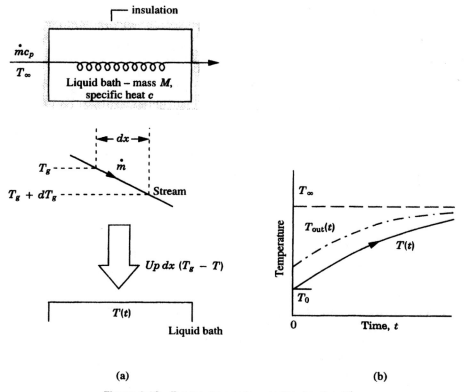

Figure 6.19 Exergy storage by sensible heating [7].

The time dependence of the bath temperature $T(t)$ and gas outlet temperature $T_{out}(t)$ can be derived with reference to Figure 6.19a as follows: Writing an energy balance for an incremental length dx, we have

$$Up\, dx(T_g - T) + \dot{m}c_p\, dT_g = 0 \qquad (6.114)$$

where p is the local heat transfer area per unit length (the wetted perimeter) and U is the overall heat transfer coefficient based on p. Integrating over the gas flow path, from $x = 0$ $(T_g = T_\infty)$ to $x = L$ $(T_g = T_{out})$,

$$\frac{T_{out}(t) - T(t)}{T_\infty - T(t)} = \exp(-NTU) \qquad (6.115)$$

where $NTU = UpL/(\dot{m}c_p)$.

Applying an energy balance to the liquid bath as a system, we have

$$Mc \frac{dT}{dt} = \dot{m}c_p \, (T_\infty - T_{out}) \tag{6.116}$$

where both T and T_{out} are functions of time. Combining Equations 6.115 and 6.116 and integrating from $t = 0$ ($T = T_0$) to any time t yields

$$\frac{T(t) - T_0}{T_\infty - T_0} = 1 - \exp(-y\theta) \tag{6.117}$$

$$\frac{T_{out}(t) - T_0}{T_\infty - T_0} = 1 - y \exp(-y\theta) \tag{6.118}$$

where

$$y = 1 - \exp(-\text{NTU}), \qquad \theta = \frac{\dot{m}c_p}{Mc} t \tag{6.119}$$

As expected, both T and T_{out} approach T_∞ asymptotically. This means that the ability to increase the temperature of the storage material increases with increasing the charging time θ and the number of heat transfer units NTU. It is shown next that there is a charging time for which the liquid bath stores the most *exergy* per unit of exergy drawn from the high-temperature gas source at T_∞.

Figure 6.20 illustrates the two sources of irreversibility in this application. First, the heat transfer between the hot stream and the cold bath takes place across a finite temperature difference. Second, the gas stream exhausted into the atmosphere is eventually cooled down to T_0, again by heat transfer across

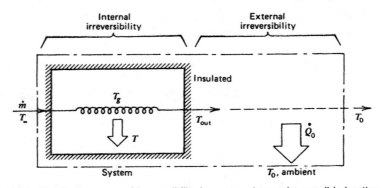

Figure 6.20   The two sources of irreversibility in exergy storage by sensible heating [7, 17].

a finite temperature difference. A third irreversibility source, neglected in the present analysis, but treated in Reference 17, is the frictional pressure drop on the gas side of the heat exchanger. The combined effect of these irreversibilities is a basic characteristic of all exergy storage systems: Only a fraction of the exergy brought in by the hot stream is stored in the liquid bath. Similarly, on discharge, only a fraction of the stored exergy can be retrieved.

Applying an entropy balance to the system defined by the dashed boundary of Figure 6.20, the rate of entropy generation is

$$\dot{S}_{gen} = \dot{m}c_p \ln \frac{T_0}{T_\infty} + \frac{\dot{Q}_0}{T_0} + \frac{d}{dt}(Mc \ln T) \tag{6.120}$$

where $\dot{Q}_0 = \dot{m}c_p(T_{out} - T_0)$ and $p_\infty = p_0$. We are interested in the entropy generated during the time interval 0 to t; this quantity can be calculated by integrating Equation 6.120 and using $T_{out}(t)$ from Equation 6.118:

$$\frac{1}{Mc} \int_0^t \dot{S}_{gen} \, dt = \theta\left(\ln \frac{T_0}{T_\infty} + \frac{T_\infty - T_0}{T_0}\right)$$

$$+ \ln\left\{1 + \frac{T_\infty - T_0}{T_0}[1 - \exp(-y\theta)]\right\}$$

$$- \frac{T_\infty - T_0}{T_0}[1 - \exp(-y\theta)] \tag{6.121}$$

This result is more instructive if it is expressed as the ratio of destroyed exergy, $T_0 \int_0^t \dot{S}_{gen} \, dt$, divided by the exergy of the gas drawn from the hot gas supply. Using Equation 3.13 together with ideal-gas relations, the exergy of the gas drawn from the supply is

$$\dot{m}e_\infty = \dot{m}c_p\left(T_\infty - T_0 - T_0 \ln \frac{T_\infty}{T_0}\right) \tag{6.122}$$

In writing this, we recognize that only the *thermal* component of the exergy associated with the ideal gas stream is relevant here. Thus, we define the *entropy generation number* as the ratio

$$N_S = \frac{T_0}{\dot{m}te_\infty} \int_0^t \dot{S}_{\text{gen}} \, dt =$$

$$e_\infty \left\{ 1 - \frac{\tau[1 - \exp(-y\theta)] - \ln\{1 + \tau[1 - \exp(-y\theta)]\}}{\theta[\tau - \ln(1 + \tau)]} \right\} \quad (6.123)$$

where τ is the dimensionless overall temperature difference,

$$\tau = \frac{T_\infty - T_0}{T_0} \quad (6.124)$$

Figure 6.21 shows how N_S depends on the charging time θ, and the heat transfer area (NTU). It is clear that for any given τ and NTU there exists a time θ_{opt} when N_S reaches its minimum. Away from this minimum, N_S approaches unity. In the limit $\theta \to 0$, the entire exergy content of the hot stream is destroyed by heat transfer to the liquid bath, which is initially at the same temperature as the atmosphere. In the limit $\theta \to \infty$, the gas stream leaves the bath as hot as it enters (at $T_{\text{out}} \cong T_\infty$), and its exergy content is destroyed

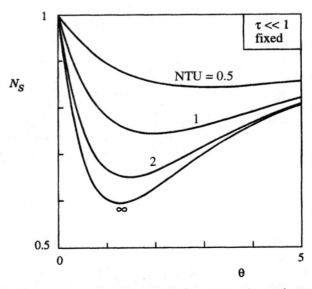

Figure 6.21 The fraction of exergy destroyed during exergy storage by sensible heating [7, 17].

entirely by direct heat transfer to the atmosphere. In Figure 6.21, the smallest N_S occurs when θ is of order 1. This means that the optimal charging time t_{opt} is of the same order of magnitude as $Mc/(\dot{m}c_p)$. In other words, we would terminate the heating process when the thermal inertia of the hot gas used ($\dot{m}c_p t$) is comparable with the thermal inertia of the liquid bath (Mc).

The time θ_{opt} depends on NTU and τ, as shown by Figure 6.22, obtained by maximizing Equation 6.123 numerically. Finally, the constant NTU curves of Figure 6.21 show that the fraction of destroyed exergy is greater than 60% for cases of practical interest. The complete process in which the heating (storage) phase is followed by a cooling (removal) phase is considered elsewhere [18].

6.4.4 Concluding Comment

The applications treated in this section—refrigeration, power generation, and exergy storage—are physically diverse. Still, the presentations are linked by a common philosophy: Fundamental trade-offs and insights can be identified using simple models of the systems and processes of interest. Though generally more qualitative than quantitative in character, information developed by such means can be used for preliminary decision making or as a point of departure for more indepth study.

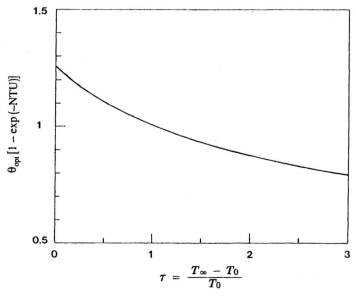

Figure 6.22 The time for maximum exergy storage by sensible heating [7, 17].

6.5 CLOSURE

In Chapters 2–6 various design analysis and modeling features related to thermodynamics, heat transfer, and fluid flow have been considered. These considerations are generally necessary but not sufficient for thermal system life-cycle design. Figure 1.2 indicates that several other considerations play a role. In particular, the *thermal modeling* principles introduced thus far must be complemented by *economic modeling* principles, for the design that minimizes cost is usually the one of interest. The presentation of this book continues, therefore, with a discussion of economic modeling in Chapters 7 and 8, and optimization in Chapter 9.

REFERENCES

1. S. Kakac, A. E. Bergles, and F. Mayinger, eds., *Heat Exchangers: Thermal-Hydraulic Fundamentals and Design,* Hemisphere, Washington, D.C., 1981.

2. R. K. Shah and A. C. Mueller, Heat exchangers, in W. M. Rohsenow, J. P. Hartnett, and E. N. Ganic, eds., *Handbook of Heat Transfer Applications,* 2nd ed., McGraw-Hill, New York, 1985, Chapter 4.

3. W. M. Kays and A. L. London, *Compact Heat Exchangers,* 2nd ed., McGraw-Hill, New York, 1964.

4. A. Bejan, *Heat Transfer,* Wiley, New York, 1993.

5. W. M. Kays and A. L. London, Convective heat-transfer and flow-friction behavior of small cylindrical tubes—circular and rectangular cross sections, *Trans. ASME,* Vol. 74, 1952, pp. 1179–1189.

6. A. L. London, W. M. Kays, and D. W. Johnson, Heat transfer and flow-friction characteristics of some compact heat-exchanger surfaces, *Trans. ASME,* Vol. 74, 1952, pp. 1167–1178.

7. A. Bejan, *Entropy Generation through Heat and Fluid Flow,* Wiley, New York, 1982.

8. A. Bejan, General criterion for rating heat exchanger performance, *Int. J. Heat Mass Transfer,* Vol. 21, 1978, pp. 655–658.

9. A. J. Fowler and A. Bejan, Correlation of optimal sizes of bodies with external forced convection heat transfer, *Int. Comm. Heat Mass Transfer,* Vol. 21, 1994, pp. 17–27.

10. A. Bejan, The concept of irreversibility in heat exchanger design: counterflow heat exchangers for gas-to-gas applications, *J. Heat Transfer,* Vol. 99, 1977, pp. 374–380.

11. A. Bejan, *Advanced Engineering Thermodynamics,* Wiley, New York, 1988.

12. A. Bejan, *Entropy Generation Minimization,* CRC Press, Boca Raton, FL, 1995.

13. A. Bejan, Theory of heat transfer-irreversible refrigeration plants, *Int. J. Heat Mass Transfer,* Vol. 32, 1989, pp. 1631–1639.

14. S. A. Klein, Design considerations for refrigeration cycles, *Int. J. Refrig.,* Vol. 15, 1992, pp. 181–185.

15. V. Radcenco, J. V. C. Vargas, A. Bejan, and J. S. Lim, Two design aspects of defrosting refrigerators, *Int. J. Refrig.*, Vol. 18, 1995, pp. 76–86.
16. W. G. Baggenstoss and T. L. Ashe, Mission design drivers for closed Brayton cycle space power conversion configurations, *J. Eng. Gas Turbines Power,* Vol. 114, 1992, pp. 721–726.
17. A. Bejan, Two thermodynamic optima in the design of sensible heat units for energy storage, *J. Heat Transfer,* Vol. 100, 1978, pp. 708–712.
18. R. J. Krane, A second law analysis of the optimum design and operation of thermal energy storage systems, *Int. J. Heat Mass Transfer,* Vol. 30, 1987, pp. 43–57.

PROBLEMS

6.1 A counterflow oil cooler lowers the temperature of a 2-kg/s stream of oil from a 110°C inlet to an outlet temperature of 50°C. The oil specific heat is 2.25 kJ/kg·K. The coolant is a 1-kg/s stream of cold water with an inlet temperature of 20°C. The overall heat transfer coefficient has the value $U = 400$ W/m²·K. Calculate the necessary heat exchanger area A, the total stream-to-stream heat transfer rate, and the water outlet temperature.

6.2 The heat transfer surface of a counterflow heat exchanger is characterized by $U = 600$ W/m²·K and $A = 10$ m². On the hot side, a 1-kg/s stream of water flows with an inlet temperature of 90°C. On the cold side, the surface is cooled by a 4-kg/s stream of water with an inlet temperature of 20°C. Calculate the stream-to-stream heat transfer rate and the two outlet temperatures.

6.3 A heat exchanger lowers the temperature of a 1-kg/s stream of hot water from 80 to 40°C. The available coolant is a 1-kg/s stream of cold water, with an inlet temperature of 20°C. The overall stream-to-stream heat transfer coefficient is $U = 800$ W/m²·K. Calculate the required heat transfer area A by assuming that the two streams are in parallel flow. Explain why the parallel-flow arrangement cannot perform the assigned function of the heat exchanger. How would you change the design so that the heat exchanger can perform its function?

6.4 A shell-and-tube heat exchanger contains 50 tubes, each with an inside diameter of 2 cm and length of 5 m. The header is shaped as a disk with a diameter of 30 cm. Air at 1 atm flows through the tubes with a total flow rate of 500 kg/h. The air inlet and outlet temperatures are 100 and 300°C, respectively. Calculate the total pressure drop across the tube bundle by evaluating, in order, the pressure drops due to friction in the tube, acceleration, abrupt contraction, and abrupt enlargement. Compare the frictional pressure drop estimate with the total pressure drop.

6.5 As shown in Figure P6.5, a heat exchanger transfers heat per unit length q' to an ideal-gas stream flowing through a stack of n parallel-plate channels. The total thickness of the stack, D, is fixed. The plate-to-plate spacing of each channel is D/n, the mass flow rate through each channel is $\dot{m}/n$, and the heat transfer rate transmitted to each channel stream is q'/n. Assuming that the flow regime is laminar, determine the optimal number of channels (n_{opt}) so that the overall entropy generation rate for the stack ($\dot{S}'_{gen}$) is minimized.

6.6 We wish to determine whether less exergy is destroyed when we drive a car with all windows closed and the air conditioner on than when we drive with the windows open and the air conditioner off. The car can be modeled as a blunt body with frontal area $A = 5 \ m^2$ traveling through air with the velocity U_∞. It is known that in the flow regime of interest, the drag coefficients are $C_D \cong 0.47$ when all windows are closed and $C_D \cong 0.51$ with the windows open. The air conditioner consumes electrical power at a constant rate, $\dot{W}_{AC} = 746 \ W$ (i.e., 1 hp).

6.7 A counterflow heat exchanger consists of two concentric pipes, such that a stream of hot oil flows through the inner pipe and a stream of water flows through the annular space. The following data are given:

	Oil Stream	Water Stream
Mass flow rate, $\dot{m}$	2500 kg/h	2000 kg/h
Specific heat, c	1.67 J/g·K	4.15 J/g·K
Inlet temperature, T_{in}	77°C	32°C
Outlet temperature, T_{out}	55°C	—

The pressure drops of both streams can be ignored. Determine in order: the water outlet temperature, $T_{out,water}$; the effectiveness, ε; the number of heat transfer units, NTU; the entropy generation rate, $\dot{S}_{gen}$.

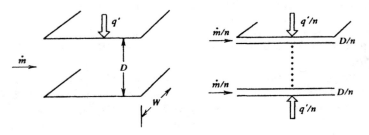

Figure P6.5

Consider next a proposal to increase the heat transfer area by 20%, by simply lengthening the concentric-pipe arrangement by 20%, while maintaining the pipe diameters, the flow rates, the fluid properties, and the two inlet temperatures the same. Determine the new NTU, ε, $T_{out,oil}$, $T_{out,water}$, and $\dot{S}_{gen}$ values.

6.8 In the simple model shown in Figure P6.8, the irreversibilities of the actual refrigeration plant are assumed to be located in the two heat exchangers, $(UA)_H$ above room temperature (T_H, fixed) and $(UA)_L$ below the refrigeration load temperature (T_L, fixed). An internally reversible cycle completes the model. The heat transfer rates $\dot{Q}_H$ and $\dot{Q}_L$ are given by $\dot{Q}_H = (UA)_H(T_{HC} - T_H)$ and $\dot{Q}_L = (UA)_L(T_L - T_{LC})$. Derive an expression for $(UA)_{min}$ when the power input $\dot{W}$ is fixed. Show that this case is characterized by $(UA)_H = (UA)_L$. Comment on the way in which the other degrees of freedom influence $(UA)_{min}$ and the exergetic (second-law) efficiency.

6.9 With reference to the refrigeration plant shown in Figure P6.8b, minimize $UA = (UA)_H + (UA)_L$ while regarding the refrigeration load as

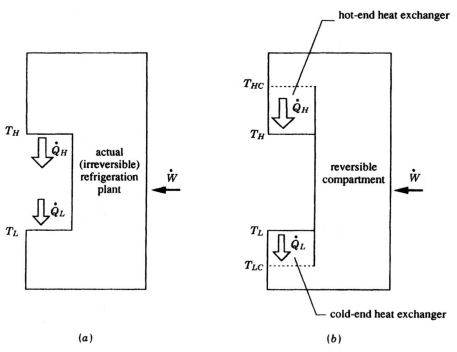

(a) (b)

Figure P6.8

specified. Show that in this case $(UA)_{min}$ is divided equally between the two heat exchangers. Derive an expression for the exergetic (second-law) efficiency and comment on the effect of $(UA)_{min}$ on the efficiency.

6.10 Using the power plant model shown in Figure P6.10, minimize the total heat exchanger size UA when the power output $\dot{W}$ is specified, the heat source and environment temperatures T_H and T_L are fixed, and the heat transfer rates are $\dot{Q}_H = (UA)_H (T_H - T_{HC})$ and $\dot{Q}_L = (UA)_L (T_{LC} - T_L)$. Show that in this case there are two degrees of freedom, and the case is characterized by $(UA)_H = (UA)_L$. Show further that the thermal efficiency of the optimized power plant is $\eta = 1 - (T_L/T_H)^{1/2}$.

6.11 As shown in Figure P6.11, a power plant consists of two components that communicate with thermal reservoirs at T_H and T_L: an internally reversible power cycle and a thermal conductance that accounts for the leakage of heat to the ambient (from T_H to T_L) *around* the power producing components of the engine. The bypass heat leak $\dot{Q}_c$ is proportional to the temperature difference across it, $\dot{Q}_c = C(T_H - T_L)$.

(a) Is the power plant operating reversibly? Explain.

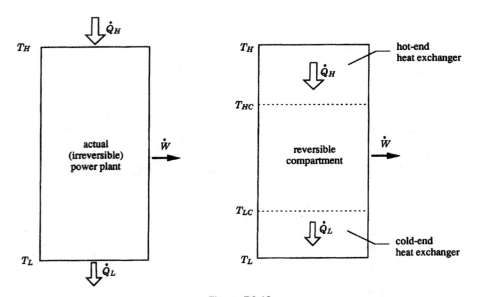

Figure P6.10

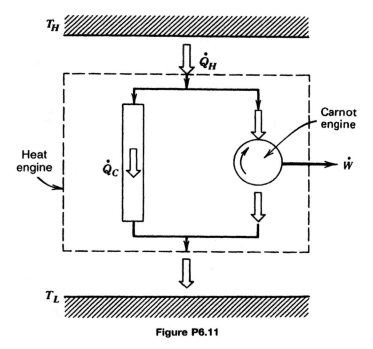

Figure P6.11

(b) For a fixed input $\dot{Q}_H$, determine an expression for the entropy generated in the heat engine, $\dot{S}_{gen}$.

(c) If the hot-end temperature T_H is a design variable, determine the optimal temperature $T_{H,opt}$ such that the power output $\dot{W}$ is maximum.

(d) Show that the optimum determined in part (c) corresponds to a design in which the overall entropy generation rate is minimum.

6.12 Figure P6.12 shows an installation for storing exergy by melting a phase-change material at temperature T_m. The exergy source is a gas (or liquid) stream with a mass flow rate $\dot{m}$ and initial temperature T_∞. The exergy stored by melting can be calculated by imagining that the melted material is returned to its original state (i.e., solidified) by contact with an internally reversible power plant that operates between T_m and the ambient T_0. Determine the melting point T_m for which the power output $\dot{W}$ is maximum.

6.13 A shell-and-tube heat exchanger consists of n parallel tubes of diameter D, as shown in Figure P6.13. Fluid 1 flows through the tubes, while fluid 2 flows parallel to the tubes, through the tube-to-tube spaces. The two fluids, their properties, and mass flow rates ($\dot{m}_1$, $\dot{m}_2$) are given. In

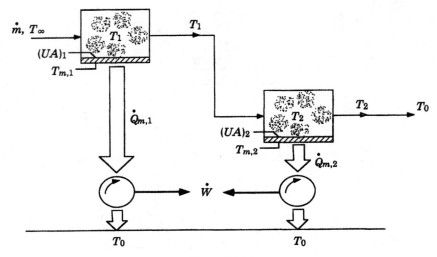

Figure P6.12

the cross section shown in the figure, T_1 and T_2 are the bulk temperatures of the two streams, and T_w is the temperature of the tube wall. The number of tubes n and the frontal area A_{fr} are also given. The thickness and thermal resistance of the tube wall can be neglected.

Obtain an expression for the thermal resistance $(T_1 - T_2)/q'$, where q' is the heat transfer rate per unit of heat exchanger length (from fluid 1, through the n tubes, to fluid 2). Determine the optimal tube diameter that minimizes the thermal resistance. To simplify the analysis, use the Colburn analogy and assume that the friction factors (f_1 inside tubes, f_2 outside tubes) are insensitive to changes in D.

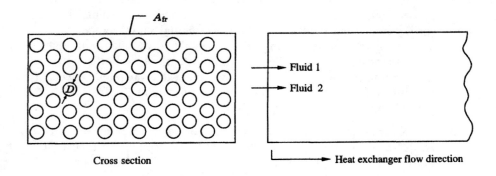

Figure P6.13

7

ECONOMIC ANALYSIS

The successful completion of a thermal design project requires estimation of the major costs involved in the project [e.g., total capital investment, fuel costs, operating and maintenance (O&M) expenses, and cost of the final products] considering various assumptions and predictions referring to the economic, technological, and legal environments and using techniques from engineering economics. This chapter outlines the economic considerations for (a) estimating capital investment costs, (b) calculating the main product costs under realistic assumptions for inflation, cost escalation, cost levelization, depreciation, taxes, and plant financing, and (c) evaluating the profitability of alternative investments.

One of the most important factors affecting the selection of a design option for a thermal system is the cost of the final product(s). The *cost* of an item is the amount of money paid to acquire or produce it. The *market price* of an item is, in general, affected not only by the production cost of the item and the desired profit but also by other factors such as demand, supply, competition, regulation, and subsidies. In designing a thermal system we are primarily interested in production costs, and we use market prices only to value the system by-products.

The total cost of an item consists of fixed costs and variable costs. The term *fixed costs* identifies those costs that do not depend strongly on the production rate. Costs for depreciation, taxes on facilities, insurance, maintenance, and rent belong to this category. *Variable costs* are those costs that vary more or less directly with the volume of output. These include the costs for materials, labor, fuel, and electric power.

Good cost estimation is a key factor in successfully completing a design project. Cost estimates should be made during all stages of design to provide a basis for decision making at each stage. The discussion in this chapter, however, will focus on the cost estimation to be conducted after a base-case flow sheet (flow diagram) has been developed, the flow rates and thermodynamic states have been determined, and the quantities and conditions of materials as well as the necessary equipment and unit operations have been specified. The discussion considers both the general case of a new-system design and the expansion or retrofitting of an existing system.

Each company has its own preferred approach for conducting an economic analysis and calculating the cost of the main product. In this chapter, the *revenue requirement method* [1] is detailed. With this approach, the cost of the main product can be calculated through the following four steps:

Step 1. Estimate the total capital investment.

Step 2. Determine the economic, financial, operating, and market input parameters for the detailed cost calculation.

Step 3. Calculate the total revenue requirement.

Step 4. Calculate the levelized product cost.

These steps are discussed in Sections 7.1–7.4, respectively.

7.1 ESTIMATION OF TOTAL CAPITAL INVESTMENT

This section considers the first of the four steps: estimation of the total capital investment. In contrast to fuel costs and O&M costs, which are continuous or repetitive in nature, an investment cost is a one-time cost. In an economic analysis, therefore, investment costs are treated differently than fuel and O&M expenses, which are considered in Section 7.2.7.

The capital needed to purchase the land, build all the necessary facilities, and purchase and install the required machinery and equipment for a system is called the *fixed-capital investment* (FCI). The fixed-capital investment represents the total system cost assuming a so-called overnight construction, that is, a zero-time design and construction period. The *total capital investment* (TCI) is the sum of the fixed-capital investment and other outlays. For brevity, it is convenient to employ acronyms for economic parameters, such as FCI and TCI used here, and we frequently do so in this chapter. Table 7.1 contains a list of these acronyms.

Table 7.2 shows a general list of items to be considered in the estimation of the total capital investment for a new system. Cost estimates for fixed-capital investment consist of two major cost elements: direct and indirect costs. *Direct costs* are the costs of all permanent equipment, materials, labor, and other resources involved in the fabrication, erection, and installation of the permanent facilities. *Indirect costs* do not become a permanent part of the facilities but are required for the orderly completion of the project. Other

Table 7.1   List of acronyms and abbreviations used in Chapter 7

ACRS	Accelerated cost-recovery system
ADJ	Adjustments to book depreciation in calculating the balance at the beginning of a year
AFUDC	Allowance for funds used during construction
ANCF	Average annual net cash inflow
AP	Average annual profit
ARR	Average annual rate of return
ATCF	Average annual total cash inflow
BBY	Balance at the beginning of a year
BCR	Benefit–cost ratio
BD	Book depreciation
BL	Book life
BPV	By-product value
CC	Carrying charges
CCF	Capitalized-cost factor
CEAF	Common-equity allowance for funds used during construction
CELF	Constant-escalation levelization factor
CI	Cost index
CPVCE	Capitalized present value of cash expenses
CRF	Capital-recovery factor
DC	Direct costs
DITX	Deferred income taxes
EBCR	Eckstein's benefit–cost ratio
EYTXBV	End-year tax book value
FC	Fuel cost
FCI	Fixed-capital investment
IC	Indirect costs
ITX	Income taxes
LRD	Cost of licensing, research, and development
LHV	Lower heating value
MACRS	Modified accelerated cost-recovery system
MAR	Minimum acceptable return
MPUC	Main-product unit cost
MPQ	Main-product quantity
NBCR	Net benefit–cost ratio
NP	Net present value
OFSC	Offsite costs
O&M	Operating and maintenance
OMC	Operating and maintenance costs
ONSC	Onsite costs
OTXI	Other taxes and insurance
PEC	Purchased-equipment cost
PFI	Plant-facilities investment
PVAE	Present value of annual cash expenses
RCEAF	Recovery of common-equity allowance for funds used during construction
ROI	Return on investment
SPCAF	Single-payment compound-amount factor

(continued)

Table 7.1 *Continued*

SPDF	Single-payment discount factor
SUC	Startup costs
TCC	Total capitalized cost
TCI	Total capital investment
TCR	Total capital recovery
TDI	Total depreciable investment
TL	Tax life
TNI	Total net investment
TRR	Total revenue requirement
TXD	Tax depreciation
TXI	Taxable income
USCAF	Uniform-series compound-amount factor
USPWF	Uniform-series present-worth factor
USSFF	Uniform-series sinking-fund factor
WC	Working capital

Table 7.2 Breakdown of total capital investment (TCI)[a]

I. Fixed-capital investment (FCI)
  A. Direct costs (DC)
    1. Onsite costs (ONSC)
      · Purchased-equipment cost (PEC; 15–40% of FCI)
      · Purchased-equipment installation (20–90% of PEC; 6–14% of FCI)
      · Piping (10–70% of PEC; 3–20% of FCI)
      · Instrumentation and controls (6–40% of PEC; 2–8% of FCI)
      · Electrical equipment and materials (10–15% of PEC; 2–10% of FCI)
    2. Offsite costs (OFSC)
      · Land (0–10% of PEC; 0–2% of FCI)
      · Civil, structural, and architectural work (15–90% of PEC; 5–23% of FCI)
      · Service facilities (30–100% of PEC; 8–20% of FCI)
  B. Indirect costs (IC)
    1. Engineering and supervision (25–75% of PEC; 6–15% of DC; 4–21% of FCI)
    2. Construction costs including contractor's profit (15% of DC; 6–22% of FCI)
    3. Contingencies (8–25% of the sum of the above costs; 5–20% of FCI)
II. Other outlays
  A. Startup costs (5–12% of FCI)
  B. Working capital (10–20% of TCI)
  C. Costs of licensing, research, and development
  D. Allowance for funds used during construction (AFUDC)

[a] Total capital investment = fixed-capital investment + other outlays
                            = direct costs + indirect costs + other outlays

outlays consist of the startup costs, working capital, costs of licensing, research and development, and allowance for funds used during construction (interest incurred during construction).

The following subsections outline briefly the estimation of total capital investment. More information is provided in [2–13].

7.1.1 Cost Estimates of Purchased Equipment

The *onsite costs* of Table 7.2 correspond to the installed equipment cost (including supporting equipment and connections) for all items shown on the process flow diagram. These items are built within a specific geographic area called battery limits. Therefore, the terms *inside-battery-limits costs* and *outside-battery-limits costs* are used as synonyms for onsite costs and offsite costs, respectively. Over the battery limits, raw materials, utilities (electricity, water, steam, refrigeration, etc.), chemicals, and catalysts are imported and manufactured products, by-products, and utilities (to be treated for recycling) are exported. The *offsite costs* of Table 7.2 include costs associated with the production and distribution of utilities, roads, general offices, wastewater-treating facilities, and storage facilities for raw materials and finished products.

Estimating the cost of purchased equipment (including spare parts and components) is the first step in any detailed cost estimation. The type of equipment and its size, the range of operation, and the construction materials have been determined during the flow diagram development. It is apparent that the accuracy of cost estimates depends on the amount and quality of the available information and the budget and time available for making estimates.

The best cost estimates for purchased equipment can be obtained directly through vendors' quotations. Shipping and installation charges need to be estimated separately. For large projects, vendors' quotations should be obtained at least for the most expensive equipment items. The next best source of cost estimates are cost values from past purchase orders, quotations from experienced professional cost estimators, or calculations using the extensive cost databases often maintained by engineering companies or company engineering departments. In addition, some commercially available software packages can assist with cost estimation; however, the quality of cost estimates obtained through software packages may not necessarily be better than the quality of data received from various estimating charts as discussed next.

Estimating Charts. When vendor quotations are lacking, or the cost or time requirements to prepare cost estimates are unacceptably high, the purchase prices of various equipment items can be obtained from the literature [2–8 and 11–16] where they are usually given in the form of estimating charts. These charts have been obtained through the correlation of a large number of cost and design data. Figure 7.1 presents a typical chart for steam boilers [3]. This figure gives the *purchased-equipment base cost* (C_B) for various boiler

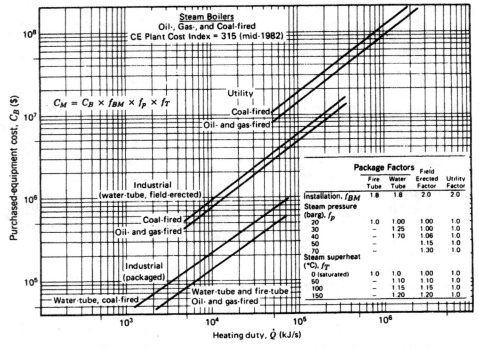

Figure 7.1 Purchased-equipment costs and factors for module, pressure, and temperature for steam boilers. (From [3], with permission from John Wiley & Sons, Inc.)

types in a log-log plot as a function of the heating duty. The base cost for an equipment item is the purchase cost of this item fabricated from the most common or base material (usually carbon steel) and designed to operate at a given (usually low) temperature and pressure range. The base cost is usually given for a specific *design type* of an equipment item if different types are available. As an exception to this rule, Figure 7.1 gives the base cost for various boiler types.

Often the estimating charts or separate plots contain information on the effects of equipment features, material, temperature, and pressure on costs. These effects are considered through factors such as the *design-type factor* (f_d), *material factor* (f_m), *temperature factor* (f_T), and *pressure factor* (f_p), which correct the base cost of purchased equipment for the corresponding variable. In addition to these factors, the so-called *bare module factors* (f_{BM}) can be applied [7, 11].

A module includes the major piece of equipment in question plus all of the supporting equipment and connections. The bare module factor times the total base cost of purchased equipment within the module gives the sum of the direct and indirect costs for that module. That is, the module cost (C_M) may be obtained from expressions such as

$$C_M = \underline{C_B f_d f_m f_T f_p} f_{BM} \tag{7.1}$$

or

$$C_M = \underline{C_B[(f_d + f_T + f_p)f_m} + f_{BM} - 1] \tag{7.2}$$

depending on the estimating chart used. Here, C_B is the base cost of purchased equipment. The C_B value is obtained from the appropriate cost chart. The underlined quantities on the right side of Equations 7.1 and 7.2 represent the purchased cost (C_{PE}) for the equipment item being considered. The values of the factors f_m, f_T, f_p, and f_{BM} for several equipment items can be found in the literature [3, 4, 7, 12]. When more than one chart is used, different relations might apply to different charts. The relation to be used in conjunction with these factors for calculating C_M is typically given together with the estimating chart.

Figure 7.2 illustrates graphically the procedure for obtaining a bare module factor for use with Equation 7.2. The numbers given in parenthesis refer to a shell-and-tube heat exchanger [7]. The calculation of the modular factor starts by assigning 100% to the purchased-equipment cost and then moving downward and to the right. After estimating the direct material costs and the ratio of the direct labor costs to the direct material costs, the total direct costs are computed. Estimation of the indirect cost factor leads to the total module cost and the bare module factor. For a shell-and-tube heat exchanger, the calculated value of the modular factor f_{BM} in Equation 7.2 is 3.18.

The accuracy of estimating charts is often relatively poor. The deviation between values obtained from charts and actual vendor quotations may vary by a factor of 2 [4]. Charts should, therefore, show a broad band rather than a line, to indicate the anticipated cost range. However, a single value representing an average or typical cost is convenient for the estimator, and most references show just a single line for each equipment category in the estimating chart. Though often lacking in accuracy, estimating charts are extremely useful tools in *preliminary* cost estimates. An alternative to the estimating charts is a table that gives the component unit cost together with the appropriate scaling exponent (Equation 7.3) and the bare module factor (Appendix B in [12]).

The equipment cost reported in the literature occasionally includes the associated installation cost. If the sum of the equipment and installation cost is known, and no additional information is provided, the purchased-equipment cost can be obtained by dividing the sum by an average factor of 1.45.

Effect of Size on Equipment Cost. In a typical cost-estimating chart, when all available cost data are plotted versus the equipment size on a log-log plot, the data correlation results in a straight line (in a given capacity range). The slope of this line, α, represents an important cost-estimating parameter (scaling exponent) as shown by the relation

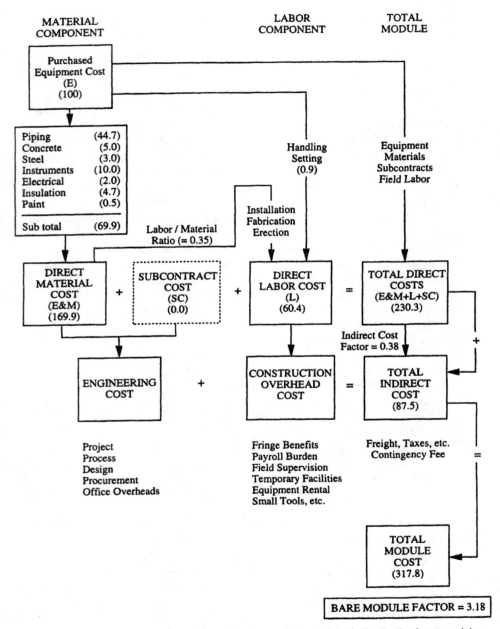

Figure 7.2 Schematic representation of the procedure used to estimate the bare module factor [7, 11, 12]. The values in parenthesis are for shell-and-tube heat exchangers (from [7]).

$$C_{\mathrm{PE},Y} = C_{\mathrm{PE},W} \left(\frac{X_Y}{X_W}\right)^{\alpha} \tag{7.3}$$

This equation allows the purchase cost of an equipment item ($C_{\mathrm{PE},Y}$) at a given capacity or size (as expressed by the variable X_Y) to be calculated when the purchase cost of the same equipment item ($C_{\mathrm{PE},W}$) at a different capacity or size (expressed by X_W) is known. The variable X in Equation 7.3 is the primary design variable or combination of variables that characterizes the size of the equipment item in question. For thermal process equipment, the scaling exponent α is usually less than unity, expressing the fact that the percentage increase (or decrease) in equipment cost is smaller than the percentage increase (or decrease) in equipment size. In the absence of other cost information, an exponent value of 0.6 may be used. This approach is known as the *six-tenths rule*.

The scaling exponent α for the same equipment item might change with the reference year and with varying size. In the latter case, the exponent is usually assumed to remain constant within a given size range of the equipment. Table 7.3 shows the primary variable X and typical values of the exponent α for equipment items commonly used in thermal design projects. When Equation 7.3 is used, the size limits for which this correlation is valid should not be exceeded.

Cost Indices. All cost data used in an economic analysis must be brought to the same *reference year:* the year used as a basis for the cost calculations. For cost data based on conditions at a different time, this is done with the aid of an appropriate *cost index* (CI):

Cost at the reference year = original cost

$$\times \left(\frac{\text{cost index for the reference year}}{\begin{array}{c}\text{cost index for the year when}\\ \text{the original cost was obtained}\end{array}}\right) \tag{7.4}$$

The cost index is an inflation indicator used to correct the cost of equipment items, material, labor, and supplies to the date of the estimate. Existing cost indicators include the following: Chemical Engineering (CE) Plant Cost Index (based on construction costs for chemical plants, reported monthly in the magazine *Chemical Engineering*); Marshall and Swift (M&S) Equipment Cost Index (based on construction costs for various chemical process industries, listed in *Chemical Engineering* and in the *Oil and Gas Journal*); Nelson-Farrar Refinery Cost Index (based on construction costs in the petroleum industry); and Engineering-News-Record (ENR) Construction Cost Index (based on general industrial construction, published in the magazine *Engineering News Record*). In addition, the U.S. Bureau of Labor Statistics pub-

Table 7.3   Typical values for the scaling exponent α in Equation 7.3 for various equipment items [2–4, 7, 12]

Equipment	Variable X	Size Range	Exponent α
Air cooler	Surface area	28–650 m²	0.75
Blower (axial, centrifugal and reciprocating; w/o drive)	Power	0.05–8 MW	0.95
Blower (rotary; w/o drive)	Power	0.05–1.2 MW	0.60
Boiler (industrial, packaged; coal, gas or oil)	Heat duty	1.5–80 MW	0.73
Boiler (industrial, field erected; coal, gas, or oil)	Heat duty	5–350 MW	0.78
Boiler (utility; coal, gas or oil)	Heat duty	50–2000 MW	0.85
Compressor (axial, centrifugal and reciprocating; w/o drive)	Power	0.05–8 MW	0.95
Compressor (rotary; w/o drive)	Power	0.05–1.2 MW	0.60
Cooling tower	Cooling water rate	0.05–9 m³/s	0.93
Driver for pump, fan, compressor or blower	Power		
Gas-turbine drive		0.07–7.5 MW	0.70
Gas-engine drive		0.07–7.5 MW	0.70
Motor drive		0.04–7.5 MW	0.93
Turbine drive		0.15–7.5 MW	0.43
Dryer (drum; vacuum)	Surface area	1.0–10 m²	0.76
Dryer (drum; atmospheric)	Surface area	1.0–10 m²	0.40
Duct	Cross-sectional area	0.05–10 m²	0.55
Electric motor	Power	3.5–15 kW	0.80
		15–150 kW	1.00
		0.15–6 MW	0.40
Evaporator	Surface area	10–1000 m²	0.54
Fan (centrifugal)	Volumetric rate	0.5–5.0 std m³/s	0.44
		9.5–33 std m³/s	1.17
Fired heater	Heat duty	0.5–10 MW	0.78
Gas-turbine system	Net power	0.01–15 MW	0.65
		70–200 MW	0.89

Table 7.3 *Continued*

Equipment	Variable X	Size Range	Exponent α
Heat exchanger	Surface area		
Double pipe		0.2–6 m²	0.16
Flat plate		15–1500 m²	0.40
Shell and tube		15–400 m²	0.66
Spiral plate		2–200 m²	0.43
Spiral tube		2–20 m²	0.60
Hopper (industrial)	Capacity	25–2000 m³	0.68
Internal combustion engines	Power	0.007–10 MW	0.81
Piping	Pipe diameter	5–75 cm	0.95
Process vessel	Tray stack height	1–120 m	0.97
Pump (reciprocating; including motor)	Power	0.02–0.3 kW	0.25
		0.3—20 kW	0.45
		20–200 kW	0.84
Pump (centrifugal; including motor)	Power	0.02–0.3 kW	0.23
		0.3–20 kW	0.37
		20–200 kW	0.48
Pump (vertical; including motor)	Circulating capacity	0.06–20 m³/s	0.76
Pump (turbine)	Power	0.5–300 kW	0.45
Reactor	Volume	0.2–4 m³	0.55
Separator (centrifugal)	Capacity	1.4–7 m³	0.49
Stack	Height	10–150 m	1.20
Steam turbine	Power		
Noncondensing		0.1–15 MW	0.50
Condensing		50–600 MW	0.90
Storage tank	Volume	0.07–150 m³	0.30
Storage vessel	Volume	150–19 × 10³ m³	0.65
Tank (flat head)	Volume	0.4–40 m³	0.57
Tank (glass lined)	Volume	0.4–4 m³	0.49
Transformer	Capacity	0.2–50 MVA	0.39

lishes monthly cost indices for materials and labor in various groups of industries in the *Monthly Labor Review.*

For thermal design projects, use of the Marshall and Swift Equipment Cost Index is recommended for single equipment items. The Chemical Engineering (CE) Cost Index is recommended for total plants, or groups of components. For example, Figure 7.1 shows a CE plant cost index of 315 that corresponds to a mid-1982 cost level. If costs estimated from this figure are to be converted to mid-1992 values (CE equipment cost index = 392), the costs would be multiplied by the ratio 392/315.

Cost indices applied to single equipment items should be used with caution when considering time intervals of 10 years or more; they should not be applied to equipment items for which the prices have been significantly affected by technology improvements or competition. For an entire industry branch and large groups of equipment items, however, cost indices provide a reasonable estimate of the average cost escalation with time.

7.1.2   Estimation of the Remaining FCI Direct Costs

Thus far, we have considered the purchased-equipment cost component of the fixed-capital investment. The remaining direct cost components listed in Table 7.2 are now discussed. These costs may be estimated either through calculations based on detailed flow diagrams and drawings or through a *factor method*. As illustrated in the paragraphs to follow, the factor method calculates the cost components of the fixed-capital investment in terms of a percentage of the purchased-equipment costs (% of PEC). Alternatively, a percentage of fixed-capital investment (% of FCI) can be used, as shown in Table 7.2; this approach is not discussed in detail here, however. The percentages given in Table 7.2 are average values based on experience from various plants in the chemical process industry. More details are given in the literature [2–5].

Purchased-Equipment Installation. The installation cost covers the freight and insurance for the transportation from the factory, the costs for labor, unloading, handling, foundations, supports, and all other construction expenses related directly to the erection and necessary connections of the purchased equipment. This cost is needed only when the economic analysis is conducted separately for single equipment items or small groups of items. If the module factors are available, they should be used directly. As mentioned previously, these factors, which vary with the equipment type, already include the installation costs. In general, the installation costs for equipment vary from 20 to 90% of the purchased-equipment cost [2–4]. In the absence of other information, an average value of about 45% may be used.

For estimates of the complete system costs, installation is often not considered separately but is included indirectly in other cost components such as piping, instrumentation, and electrical equipment [4]. Finally, the expenses for equipment insulation, required when high or low temperatures are involved, are usually included in either the installation or the piping costs. Insulation costs are between 2 and 8% of the purchased-equipment cost [2].

Piping. The cost for piping includes the material and labor costs of all items required to complete the erection of all the piping used directly in the system. This cost represents 10–70% of the purchased-equipment cost [2, 4, 6]. The low end of this range (10–20%) applies to plants with only solids handling, and the high end (50–70%) to plants with fluids handling and with considerable recycling and heat exchange. For a process involving solids only, solids and fluids, or fluids only, an average value of 16, 31, and 66% of purchased-equipment cost, respectively, can be used to estimate the piping costs [2].

Instrumentation and Controls. The factor used to calculate these costs tends to increase as the degree of automation increases, and to decrease with increasing total cost. Factors between 2 and 30% of the purchased-equipment cost have customarily been used. However, the increasing use of computers and more complex control systems often dictates a higher value for this factor: A more typical range of values of the cost factor for instrumentation and controls might be 6–40% of the purchased-equipment cost. For conventional steam power plants, the low end values (6–10%) apply. In the absence of other information, an average value of about 20% may be assumed for this factor.

Electrical Equipment and Materials. This cost, which includes materials and installation labor for substations, distribution lines, switch gears, control centers, emergency power supplies, area lighting, and so forth is usually 10–15% of the purchased-equipment cost, the average value being about 11%. In specific systems (e.g., electrolytic installations), however, this factor could become as high as 40%.

Land. The cost of land strongly depends on the location and, unlike other costs considered in this section, usually does not decrease with time. If land is to be purchased, the cost may be up to 10% of the purchased-equipment cost.

Civil, Structural, and Architectural Work. This category includes the total cost for all buildings, including services, as well as the costs for roads, sidewalks, fencing, landscaping, yard improvements, and so forth. As Table 7.4 indicates, the cost factor for this category varies widely depending on whether the work refers to a new system at a new site or to an expansion of an existing site.

Service Facilities. The cost of service facilities (or auxiliary facilities) includes all costs for supplying the general utilities required to operate the system such as fuel(s), water, steam, and electricity (assuming that these util-

Table 7.4 Cost for civil, structural, and architectural work as a percentage of purchased-equipment cost (adapted from [2])

Type of Process Plant	New Plant at New Site (%)	New Unit at Existing Site (%)	Expansion of an Existing Site (%)
Solid processing (e.g., coal briquetting) plant	83	40	30
Solid–fluid processing (e.g., shale oil) plant	62	44	22
Fluid processing (e.g., distillation) plant	60	20–33[a]	21

[a]The lower number applies to petroleum refining and related industries.

ities are not generated in the main process), refrigeration, inert gas, and sewage. This category also includes the costs of waste disposal, environmental control (e.g., air pollution monitoring and wastewater treatment), fire protection, and the equipment required for shops, first aid, and cafeteria. The total cost of service facilities may range from 30 to 100% of the purchased-equipment cost. An average value of 65% may be assumed for this factor in the absence of other information.

7.1.3   Indirect Costs of FCI

Let us turn now to the indirect cost categories listed in Table 7.2: engineering and supervision, construction, and contingencies. As for the direct costs considered in Section 7.1.2, the indirect costs can be estimated as percentages of the purchased-equipment cost or the fixed-capital investment. Frequently, a percentage of the total direct costs is also used for indirect costs.

Engineering and Supervision. The capital investment for engineering and supervision includes the cost for developing the detailed plant design and drawings, and the costs associated with cost engineering, scale models, purchasing, engineering supervision and inspection, administration, travel, and consultant fees. The costs for engineering and supervision may be 25–75% of the purchased-equipment cost. An average value of about 30% of the purchased-equipment cost or 8% of the total direct costs for the plant can often be used [2].

Construction (Including Contractor's Profit). The capital investment for construction includes all expenses for temporary facilities and operations, tools and equipment, home office personnel located at the construction site, insurance, and so forth. These costs are in addition to the construction charges discussed previously. Also the contractor's fee (profit), which is usually negotiable, is included in this category.

In some estimates, the costs for construction might be included in a single category under engineering, supervision, and construction or in the category of equipment installation. The cost for construction, including contractor's profit, averages about 15% of the total direct cost.

Contingencies. Cost estimates are based on assumptions for cost and productivity, which may vary significantly from the actual values. In addition, unpredictable events due to weather, work stoppages, sudden price changes, and transportation difficulties might affect the actual costs. Finally, design changes might become necessary after completion of the design process. All these uncertainties and risks are considered through a contingency factor, which normally ranges between 5 and 20% of the fixed-capital investment.

The contingency factor depends on the complexity, size, and uniqueness of the plant. When using a new process, a separate contingency factor for this process should be used in addition to the overall system contingency factor. The contingency factors to be used in electric power generation processes are discussed in Reference 1.

7.1.4 Other Outlays

The discussion to this point has concerned the direct and indirect costs listed in Table 7.2 that make up the fixed-capital investment category. Let us now consider the Table 7.2 category labeled *other outlays.*

Startup Costs. These costs are mainly associated with design changes that have to be made after completion of construction but before the system can operate at design conditions. The startup costs include labor, materials, equipment, and overhead expenses to be used only during startup time, plus the loss of income while the system is not operating or operating at only partial capacity during the same period.

Depending on the tax situation of the company, the startup costs may be presented either as part of the total capital investment or as a one-time-only expenditure in the first year of system operation. In the first of these approaches, the startup costs are said to be *capitalized.* Companies often consider the fraction of the startup costs that is allocated to equipment modifications as part of the capital investment, whereas the funds used for additional workforce and materials needed to start up the system are considered first-year operating expenses [10].

For simplicity of the analysis presented here, we consider all startup costs as part of the indirect costs. These costs might run from 5 to 12% of the fixed-capital investment. For electric power plants, the startup costs are the sum of the following unescalated[1] costs [1]: (a) one month of fixed operating and maintenance costs, (b) one month of variable operating costs calculated at full load, (c) one week of fuel at full load, and (d) 2% of the *plant-facilities investment.* The plant-facilities investment (PFI) is equal to the fixed-capital investment minus land costs. In the absence of other information, an average value of about 10% of the fixed-capital investment may be used for the startup costs.

Working Capital. The working capital represents the funds required to sustain plant operation, that is, to pay for the operating expenses before payment is received through the sale of the plant product(s). The working capital consists of the total amount of money invested in (a) raw materials, fuels, and supplies carried in stock, (b) finished products in stock and semifinished products in the process of being manufactured, (c) accounts receivable, (d) cash kept on hand for operating expenses (raw-material purchases, wages, fringe

---

[1]See Section 7.2.2 for cost escalation.

benefits, etc.), taxes, and other current obligations, and (e) accounts payable [2]. The working capital depends on the average length of time required for the product to be manufactured and reach the customer, as well as on the time it takes to receive payment for the products sold. The costs associated with the initial charge of catalysts, solvents, adsorbents, and so forth can be considered either as a separate outlay, or, as done here, as part of the working capital.

According to Reference 1, the working capital for power plants is calculated from the sum of the unescalated expenses representing 2 months of fuel and variable operating costs at full load, and 3 months of labor costs plus a contingency of 25% of the total of the above three items. The initial investment analysis is simplified if we assume that the working capital is related to the total capital investment: The working capital usually represents 10–20% of the total capital investment. Therefore, in the absence of other information, a factor of 15% may be used.

Licensing, Research, and Development. If a lump-sum payment with or without royalties is required, costs associated with licensing and costs incurred in the past for research and development directly related to the system (or a process of the system) being considered must be added to the total capital investment—that is, capitalized.

Allowance for Funds Used During Construction (AFUDC). The time period between the beginning of design and system startup may be 1–5 years. During this period, parts of the investment must be released to finance design studies, civil engineering work, purchase and installation of equipment and so forth. Thus, various amounts of money are disbursed without obtaining any revenue. As discussed later, this money may come from the company resources, direct loans, or a combination of loans and company resources.

The allowance for funds used during construction (sometimes called interest during construction) represents the time value of money (Section 7.2.1) during construction, and is based on an interest rate (cost-of-money rate) equal to the *weighted cost of capital* [1]. This interest is compounded on an annual basis (end of year) during the construction period for all funds spent during the year or previous years. When the total design and construction time period is known, the allocation of fixed-capital investment to the individual years of this period is estimated, and the interest rate is given, the allowance for funds used during construction can be calculated using the formulas presented in Section 7.2.

7.1.5 Simplified Relationships

The object of this section is to provide simplified expressions for estimating the total capital investment required for a new system or an expansion. The expressions are obtained using typical values for the various cost categories of Table 7.2 discussed in Sections 7.1.1 through 7.1.4.

According to Table 7.2, the total capital investment (TCI) is the sum of the fixed-capital investment (FCI), startup costs (SUC), working capital (WC), costs of licensing, research, and development (LRD), and allowance for funds used during construction (AFUDC):

$$TCI = FCI + SUC + WC + LRD + AFUDC \qquad (7.5)$$

whereas the fixed-capital investment is the sum of the direct cost (DC) and the indirect costs (IC):

$$FCI = DC + IC \qquad (7.6)$$

The direct costs include the onsite costs (ONSC) and the offsite costs (OFSC):

$$DC = ONSC + OFSC \qquad (7.7)$$

The onsite costs can be estimated directly from correlations [7]. The offsite costs may be 100–200% (average value 120%) of the onsite costs for the construction of a new facility or between 40 and 50% (average value 45%) of the onsite costs for an expansion of an existing facility. Accordingly

$$OFSC = \begin{cases} 1.20 \ ONSC & \text{(new system)} \qquad (7.8a) \\ 0.45 \ ONSC & \text{(expansion)} \qquad (7.8b) \end{cases}$$

From the discussion of Section 7.1.3 we have

$$WC = 0.15 \ TCI \qquad (7.9)$$
$$SUC = 0.10 \ FCI \qquad (7.10)$$

If we further assume that

$$LRD + AFUDC = 0.15 \ FCI \qquad (7.11)$$

and

$$IC = 0.25 \ DC \qquad (7.12)$$

we obtain from Equations 7.5, 7.9, 7.10, and 7.11

$$TCI = 1.47 \ FCI \qquad (7.13)$$

Then with Equations 7.6, 7.7, 7.12 and 7.13 we have

$$TCI = 1.84 \; DC$$
$$= 1.84(ONSC + OFSC) \qquad (7.14)$$

Introducing Equations 7.8, Equation 7.14 gives

$$TCI = \begin{cases} 4.05 \; ONSC & \text{(new system)} & (7.15a) \\ 2.67 \; ONSC & \text{(expansion)} & (7.15b) \end{cases}$$

The FCI for a new system is usually in the range of 280–550% of the PEC with the average value being about 430%. For a system expansion the FCI averages about 283% of the PEC. That is,

$$FCI = \begin{cases} 4.30 \; PEC & \text{(new system)} & (7.16a) \\ 2.83 \; PEC & \text{(expansion)} & (7.16b) \end{cases}$$

With these, Equation 7.13 gives

$$TCI = \begin{cases} 6.32 \; PEC & \text{(new system)} & (7.17a) \\ 4.16 \; PEC & \text{(expansion)} & (7.17b) \end{cases}$$

These calculations show that to obtain a rough estimate of the total capital investment for a new system we may use either Equation 7.17a after the purchased-equipment costs have been estimated or Equation 7.15a after the module costs have been calculated according to Reference 7. For a system expansion, Equations 7.17b or 7.15b may be used, respectively. When any of these equations is applied, it is important to remember the underlying assumptions. In general, with the above simplified relationships we can only obtain order-of-magnitude cost estimates. Since the total capital investment is several times larger than the sum of equipment base costs, the cost estimation of purchased equipment becomes very important for the accuracy of total capital investment estimation.

Complete Plant Cost Estimation. Some charts in the literature [4, 12–14] give the average cost of complete plants as a function of plant size. These charts are, however, based on relatively old data and have a high potential for error because of changes in technology, design requirements (environmental controls, energy costs, etc.) and plant size range. For several types of complete plants, Table 7.5 gives the capacity exponent α for use in Equation 7.3.

7.1.6 Cogeneration System Case Study

To illustrate the concepts discussed in this section, Table 7.6 provides an estimate of the total capital investment of the cogeneration system presented in Figure 1.7. Gas-turbine systems are supplied as packaged systems, and the cost breakdown among their components is usually not quoted. However, to illustrate the relative cost contributions of the components to the total, we assume that vendors have provided in mid-1994 dollars the purchased-equipment costs shown at the top of Table 7.6. The remaining components of direct costs and the indirect costs are estimated using the factors shown in parentheses.

Table 7.5 Typical values for the scaling exponent α in Equation 7.3 for entire plants [4, 7, 12–14]

Plant	Capacity Variable	Capacity Range	Exponent α
Air (liquid)	Product flow rate	70–4000 t/d[a]	0.66
Air plant (packaged)	Compressed dried air rate	0.1–100 Nm³/s	0.70
Ammonia	Product flow rate	450–1800 t/d	0.58
Argon	Product flow rate	25–500 Nm³/h	0.89
Carbon black	Product flow rate	45–900 t/d	0.67
Carbon dioxide (gas or liquid)	Product flow rate	85–920 t/d	0.72
Cogeneration plant	Net power	5–150 MW	0.75
Electric power plant	Net power	1.0–1000 MW	0.80
Ethanol	Product flow rate	(4–40) × 10³ m³/yr	1.00
		(40–400) × 10³ m³/yr	0.90
Liquified natural gas	Product flow rate	(1–20) × 10³ t/d	0.68
Methanol	Product flow rate	(1.2–18) × 10⁶ t/yr[a]	0.78
Natural gas purification	Product flow rate	18–270 t/d	0.75
Nitrogen (liquid)	Product flow rate	70–4000 t/d	0.66
Oxygen (gaseous)	Product flow rate	35–900 t/d	0.59
Oxygen (liquid)	Product flow rate	500–2700 t/d	0.37
Refinery (complete)	Feed flow rate	(9–120) × 10³ bbl/d[a]	0.86
Refrigeration unit	Cooling load	0.05–10 MW	0.70
Refuse-to-electricity	Net power	10–150 MW	0.75
Sour gas treating	Feed flow rate	0.5–16 Nm³/d	0.84
Sulfuric acid	Product flow rate	85–1000 t/d	0.56
SNG from coal	Product flow rate	(0.5–5.5) × 10⁶ Nm³/d	0.75
Wastewater treatment plant	Water flow rate	0.005–5 m³/s	0.67
Water desalination	Water flow rate	0.05–3 m³/s	0.89

[a]The symbols t/d and t/yr refer to metric tons per day and metric tons per year, respectively. The symbol bbl/d means barrels per day.

Table 7.6   Estimate of total capital investment for the case study cogeneration system (all costs are expressed in thousands of mid-1994 dollars)

I.	**Fixed capital investment**	
	A. Direct costs	
	1. Onsite costs	
	Purchased-equipment costs	
	• Air compressor	3,735
	• Air preheater	936
	• Combustion chamber	338
	• Gas turbine	3,739
	• Heat-recovery steam generator	1,310
	• Other plant equipment	942
	Total purchased-equipment cost (PEC)	**11,000**
	Purchased-equipment installation (33% of PEC)	3,630
	Piping (35% of PEC)	3,850
	Instrumentation and controls (12% of PEC)	1,320
	Electrical equipment and materials (13% of PEC)	1,430
	Total onsite costs	**21,230**
	2. Offsite costs	
	Land	500
	Civil, structural, and architectural work (21% of PEC)	2,310
	Service facilities (35% of PEC)	3,850
	Total offsite costs	**6,660**
	Total direct costs	**27,890**
	B. Indirect costs	
	Engineering and supervision (8% of DC)	2,231
	Construction costs and contractor's profit (15% of DC)	4,183
	Sum	34,304
	Contingency (15% of the above sum)	5,146
	Total indirect costs	**11,560**
	Fixed-capital investment	**39,450**
II.	**Other outlays**	
	Startup costs (Section 7.2.9)	1,247
	Working capital (Section 7.2.9)	1,924
	Costs of licensing, research and development	0
	Allowance for funds used during construction (Section 7.2.9)	3,353
	Total other outlays	**6,524**
Total capital investment		**45,974**

Referring to Table 7.6, the fixed-capital investment for this project is estimated at 39.45×10^6 mid-1994 dollars. The plant-facilities investment, which is the difference between fixed-capital investment and land costs, is 38.95×10^6 mid-1994 dollars for this project. The calculation of other outlays is discussed in Section 7.2.8. The total capital investment of the cogeneration system is 45.974×10^6 mid-1994 dollars.

For the cogeneration system we assume that all values shown in Table 7.6 are either given or calculated as indicated in the table. If, however, the simplified relationships of Section 7.1.5 would be applied to results shown in Table 7.6, we would obtain the following values for total capital investment depending on whether Equation 7.15b or Equation 7.17b is used:

$$\text{TCI} = 2.67 \ \text{ONSC} = (2.67)(21{,}230 \times 10^3) = \$56.684 \times 10^6$$

or

$$\text{TCI} = 4.16 \ \text{PEC} = (4.16)(11{,}000 \times 10^3) = \$45.760 \times 10^6$$

Thus, for this particular cogeneration system, the simplified Equation 7.17b gives a good estimate of the total capital investment, whereas the value predicted by Equation 7.15b is 23% higher than the value calculated in Table 7.6.

7.2 PRINCIPLES OF ECONOMIC EVALUATION

In the previous section methods have been discussed for estimating the capital investment costs of equipment and complete systems. This section presents a brief review of engineering-economics principles. The use of these principles is illustrated in Section 7.2.8 by application to the cogeneration system case study. The principles are also used in Sections 7.3 and 7.4 in estimating the cost of a product and in Section 7.5 in evaluating the profitability of various investments, projects, and expenditures. If further elaboration is required on any point, readers should consult the literature [e.g., 2, 5, 13, 17, 18].

7.2.1 Time Value of Money

Decisions about capital expenditures generally require consideration of the *earning power of money:* A dollar in hand today is worth more than a dollar received one year from now because the dollar in hand now can be invested for the year. Thus, as the cost evaluation of a project requires comparisons of money transactions at various points in time, we need methods that will enable us to account for the value of money over time. In this section such methods are considered; the principal results elicited are listed in Table 7.7.

Future Value. If P dollars (*present value*) are deposited in an account earning i percent interest per time period and the interest is compounded at the end of each of n time periods, the account will grow to a *future value* (F)

$$F = P(1 + i)^n \tag{7.18}$$

Interest is the compensation paid for the use of borrowed money. The interest rate is usually stated as a percentage; in equations, however, it is expressed as a decimal (e.g., 0.07 instead of 7%). Instead of the term interest rate, we will use the terms *rate of return* for an investment made and *annual cost of money* for borrowed capital.

Compounding Frequency. In engineering economy, the unit of time is usually taken as the year. If compounding occurs p times per year ($p \geq 1$) for a total number of n years ($n \geq 1$), and i is the annual rate of return, Equation 7.18 becomes

$$F = P\left(1 + \frac{i}{p}\right)^{np} \tag{7.19}$$

Here, np is the number of periods and i/p is the rate of return per period. In this case, the annual rate of return i is known as the nominal rate of return. The *effective rate of return* is the annual rate of return that would yield the same results if compounding were done once a year instead of p times per year. The effective rate of return, which is higher than the nominal rate of return, is obtained by eliminating F/P from Equations 7.18 and 7.19 as

$$i_{\text{eff}} = \left(1 + \frac{i}{p}\right)^{p} - 1 \tag{7.20}$$

If *continuous compounding* of money ($p \to \infty$) is used, the future value is calculated from

$$F = Pe^{in} \tag{7.21}$$

It is apparent that in the case of continuous compounding the effective rate of return becomes

$$i_{\text{eff}} = e^{i} - 1 \tag{7.22}$$

In Equations 7.21 and 7.22, i is the nominal annual rate of return and n is the total number of years.

If the time is less than one year, the *simple interest* formula can be used to calculate the future value:

$$F = P(1 + n i_{\text{eff}}) \tag{7.23}$$

where n is now a fraction of a year and i_{eff} is the annual effective rate of return.

Equations 7.19 and 7.21 can be expressed in the same form as Equation 7.18:

$$F = P(1 + i_{eff})^n \qquad (7.24)$$

where i_{eff} is calculated according to either Equation 7.20, if compounding occurs p times per year, or Equation 7.22 in the case of continuous compounding. The term $(1 + i_{eff})^n$, referred to as the *single-payment compound-amount factor* (SPCAF), is listed in Table 7.7.

Unless otherwise indicated, the terms *interest, rate of return,* and *annual cost of money* refer to their effective values. Also, to simplify calculations, when the cost of money is calculated for one or more years plus a fractional part of a year, Equation 7.24 is applied with a noninteger exponent. A more precise approach would be to use the simple cost-of-money formula (Equation 7.23) for the fractional part and the compound cost-of-money formula (Equation 7.24) for the integer part of the entire time period being considered.

Table 7.7 Summary of basic formulas and factors used in economic analysis

Formula[a]	Factor[b]	Equation
$\dfrac{F}{P} = (1 + i_{eff})^n$	Single-payment compound-amount factor (SPCAF)	(7.24)
$\dfrac{P}{F} = \dfrac{1}{(1 + i_{eff})^n}$	Single-payment present-worth factor or single-payment discount factor (SPDF)	(7.25)
$\dfrac{F}{A} = \dfrac{(1 + i_{eff})^n - 1}{i_{eff}}$	Uniform-series compound-amount factor (USCAF)	
$\dfrac{A}{F} = \dfrac{i_{eff}}{(1 + i_{eff})^n - 1}$	Uniform-series sinking fund factor (USSFF)	(7.26)
$\dfrac{P}{A} = \dfrac{(1 + i_{eff})^n - 1}{i_{eff} (1 + i_{eff})^n}$	Uniform-series present-worth factor (USPWF)	(7.27)
$\dfrac{A}{P} = \dfrac{i_{eff} (1 + i_{eff})^n}{(1 + i_{eff})^n - 1}$	Capital-recovery factor (CRF)	(7.28)
$\dfrac{C_K}{P_0} = \dfrac{(1 + i_{eff})^n}{(1 + i_{eff})^n - 1}$	Capitalized-cost factor (CCF)	(7.32)
$\dfrac{A}{P_0} = \dfrac{k(1 - k^n)}{1 - k} \text{CRF}$	Constant-escalation levelization factor (CELF)	(7.34)

[a] In these expressions, cost-of-money compounding and ordinary annuities are assumed. The exponent n denotes the number of years and i_{eff} is the effective rate of return. The factor k is defined by Equation 7.35.
[b] Tabulated values for these factors are provided in the literature [2, 17, 18].

Present Value. When evaluating projects, we often need to know the present value of funds that we will spend or receive at some definite periods in the future. The *present value* (or *present worth*) of a future amount is the amount that if deposited at a given rate of return and compounded would yield the actual amount received at a future date. From Equation 7.24 we see that a given future amount F has a present value P:

$$P = F \frac{1}{(1 + i_{\text{eff}})^n} \qquad (7.25)$$

The term $1/(1 + i_{\text{eff}})^n$, called the *single-payment present-worth factor* or the *single-payment discount factor* (SPDF), is listed in Table 7.7. Since the difference between the future value and the present value is often called *discount,* in this case the term i_{eff} is called effective *discount rate.*

Annuities. An *annuity* is a series of equal-amount money transactions occurring at equal time intervals (periods). Usually, the time period corresponds to one year. Money transactions of this type can be used, for instance, to pay off a debt or accumulate a desired amount of capital. Annuities are used in this chapter to calculate the levelized costs of the final product, fuel, and so forth (see Sections 7.2.8 and 7.4). An *annuity term* is the time from the beginning of the first time interval to the end of the last time interval.

The most common type of annuity is the *ordinary annuity,* which involves money transactions occurring at the end of each time interval. If A dollars are deposited at the end of each period in an account earning i_{eff} percent per period (effective rate of return per period), the future sum F (*amount of the annuity* or *future value of the annuity*) accrued at the end of the nth period is

$$F = A \frac{(1 + i_{\text{eff}})^n - 1}{i_{\text{eff}}} \qquad (7.26)$$

The term $[(1 + i_{\text{eff}})^n - 1]/i_{\text{eff}}$ is called the *uniform-series compound-amount factor* (USCAF), and the reciprocal term $i_{\text{eff}}/[(1 + i_{\text{eff}})^n - 1]$ is called the *uniform-series sinking fund factor* (USSFF). These terms are also listed in Table 7.7. Multiplication of the USSFF with a known future sum F yields the amount of the equivalent annuity A.

The *present value* or *present worth of an annuity* is defined as the amount of money that would have to be invested at the beginning of the annuity term (present time) at an effective compound rate of return per period i_{eff}, to yield a total amount at the end of the annuity term equal to the amount of the annuity.

The next two entries of Table 7.7, Equations 7.27 and 7.28, are obtained by combining Equation 7.26 with Equation 7.25. That is

$$\frac{P}{A} = \frac{(1 + i_{eff})^n - 1}{i_{eff}(1 + i_{eff})^n} \tag{7.27}$$

The expression on the right side of this equation is called the *uniform-series present-worth factor* (USPWF). The reciprocal of this factor is the *capital-recovery factor* (CRF):

$$\text{CRF} = \frac{A}{P} = \frac{i_{eff}(1 + i_{eff})^n}{(1 + i_{eff})^n - 1} \tag{7.28}$$

The CRF is used to determine the equal amounts A of a series of n money transactions, the present value of which is P.

Until now we have considered ordinary annuities where the money transaction occurs at the end of each time interval. If, however, the transaction occurs at the beginning of each time interval of the annuity term, Equations 7.26, 7.27, and 7.28 are replaced by Equations 7.29, 7.30, and 7.31, respectively:

$$F = A \frac{(1 + i_{eff})^{n+1} - (1 + i_{eff})}{i_{eff}} \tag{7.29}$$

$$P = A \frac{(1 + i_{eff})^n - 1}{i_{eff}(1 + i_{eff})^{n-1}} \tag{7.30}$$

$$\text{CRF} = \frac{i_{eff}(1 + i_{eff})^{n-1}}{(1 + i_{eff})^n - 1} \tag{7.31}$$

For brevity, these formulas and the corresponding factors are not shown in Table 7.7.

Capitalized Cost. An asset (e.g., a piece of equipment) of fixed-capital cost C_{FC} will have a finite economic life of n years. The *economic life* (or *book life*) of an asset is the best estimate of the length of time that the asset can be used. The *salvage value* of an asset is the estimated economic worth of the asset at the end of its economic life.

Engineers often want to determine the total cost of an asset under conditions permitting perpetual replacement of the asset without considering inflation. The so-called *capitalized cost* (C_K) is defined in engineering economics as the first cost of the asset plus the present value of the indefinite annuity that corresponds to the perpetual replacement of the asset every n years. Assuming that the renewal cost of the asset remains constant (no inflation) at $C_{FC} - S$, and that both the useful life of the asset and the rate of return remain constant, the present value of the indefinite annuity is calculated from Equation 7.25 as

$$(C_K - C_{FC}) = (C_K - S)/(1 + i_{eff})^n$$

That is, the capitalized cost C_K is in excess of the fixed-capital cost C_{FC} by an amount which, when compounded at an effective rate of return i_{eff} for n years, will have a future value of C_K minus the salvage value S of the asset. Solving the last equation for C_K, we obtain for the capitalized cost

$$C_K = \left[C_{FC} - \frac{S}{(1 + i_{eff})^n} \right]\left[\frac{(1 + i_{eff})^n}{(1 + i_{eff})^n - 1} \right] \qquad (7.32)$$

The second factor in square brackets on the right side of Equation 7.32, called the *capitalized-cost factor* (CCF), is listed in Table 7.7 where P_0 denotes the present value at the beginning of the time period. P_0 is the first factor in square brackets on the right side of Equation 7.32. The capitalized-cost factor is equal to the capital-recovery factor of an ordinary annuity (Equation 7.28) divided by the effective rate of return. Multiplication of the present value of an asset (or expenditure) by the capitalized-cost factor yields the capitalized cost C_K of the asset (or expenditure). The capitalized cost is often used in decision-making processes to compare the total cost of competing equipment options with different economic lives (Section 7.5.5).

In accounting, the term capitalized cost describes an expenditure that results in benefits extending over more than one year. For instance, the expenditures associated with other outlays (Section 7.1.4 and Table 7.2) are part of the total capital investment; capitalization in this sense does not mean that the expenditures can automatically be written off for tax purposes (Section 7.2.5).

The use of the term capitalized cost is more meaningful in accounting than in engineering economics where the term merely characterizes a special case of present-value calculation referring to an infinite project life. However, because the term capitalized cost is encountered very often in the literature of both engineering economics and accounting, it is important to be familiar with the different meanings that may be attached to it.

7.2.2 Inflation, Escalation, and Levelization

Inflation. *General price inflation* is the rise in price levels associated with an increase in available currency and credit without a proportional increase in available goods and services of equal quality [1]. The *consumer price index,* which is tabulated by the federal government, is a composite price index that measures general inflation.

When inflation occurs, costs change every year. Cost changes in past years are considered using appropriate cost indices (e.g., Equation 7.4). For future years a varying annual inflation rate can be used, but such a rate always represents a prediction. For simplicity we assume a constant average annual inflation rate (r_i) for future years.

Escalation. The *real escalation rate* of an expenditure is the annual rate of expenditure change caused by factors such as resource depletion, increased demand, and technological advances [1]. The first two factors lead to a positive real escalation rate whereas the third factor results in a negative rate. The real escalation rate (r_r) is independent and exclusive of inflation.

The *nominal* (or *apparent*) *escalation rate* (r_n) is the total annual rate of change in cost and includes the effects of both the real escalation rate and inflation:

$$(1 + r_n) = (1 + r_r)(1 + r_i) \tag{7.33}$$

To simplify calculations, we assume that all costs except fuel costs and the values of by-products change annually with the constant average inflation rate r_i; that is, we take $r_r = 0$. Since fuel costs are expected over a long period of future years to increase on the average faster than the predicted inflation rate, a positive real escalation rate for fuel costs may be appropriate for the economic analysis of many thermal systems.

Levelization. Cost escalation applied to an expenditure (e.g., fuel costs or O&M costs) over an n-year period results in a nonuniform cost schedule in which the expenditure at any year is equal to the previous year expenditure multiplied by $(1 + r_n)$, where r_n is the constant rate of change, the nominal escalation rate. This results in a geometrically increasing series of expenditures [17, 18]. The *constant-escalation levelization factor* (CELF) is used to express the relationship between the value of the expenditure at the beginning of the first year (P_0) and an equivalent annuity (A), which is now called a *levelized value.* The levelization factor depends on both the effective annual cost-of-money rate, or discount rate, i_{eff}, and the nominal escalation rate r_n:

$$\frac{A}{P_0} = \text{CELF} = \frac{k(1 - k^n)}{1 - k} \, \text{CRF} \tag{7.34}$$

where

$$k = \frac{1 + r_n}{1 + i_{\text{eff}}} \tag{7.35}$$

and the variables CRF and r_n are determined from Equations 7.28 and 7.33, respectively. Equation 7.34 assumes that all transactions are made at the end of their respective years and P_0 is the cost at the beginning of the first year. Equation 7.34 represents the final entry of Table 7.7.

The concept of levelization is general and is defined as the use of time-value-of-money arithmetic to convert a series of varying quantitites to a financially equivalent constant quantity (annuity) over a specified time interval. We apply the concept of levelization in Sections 7.3 and 7.4 to calculate the levelized fuel and O&M costs, the levelized total revenue requirements and the levelized total cost of the main product of a thermal system. This concept is also illustrated in the following example.

Example 7.1   Determine the levelized value (A) of a 5-year series of payments ($n = 5$) that increase at an annual nominal escalation rate (r_n) of 4%. The average annual nominal discount rate (i_{eff}) is 10% and the estimated cost of the payment at the beginning of the first year (P_0) is $1000.

Solution   To gain familiarity with concepts introduced thus far, we will first solve this problem using Equations 7.18, 7.25, and 7.28, and then present a more direct solution using the constant-escalation levelization factor, Equation 7.34, together with Equation 7.35.

The payment C_m in year m is (Equation 7.18)

$$C_m = P_0(1 + r_n)^m, \qquad m = 1, 2, ..., 5$$

The present value of that payment, P_m, is (Equation 7.25)

$$P_m = C_m \frac{1}{(1 + i_{\text{eff}})^m}$$

The following table summarizes the calculations

Year	$(1 + r_n)^m$	C_m	$1/(1 + i_{eff})^m$	P_m
1	1.040	1040.00	0.909	945.45
2	1.082	1081.60	0.826	893.88
3	1.125	1124.86	0.751	845.13
4	1.170	1169.86	0.683	799.03
5	1.217	1216.65	0.621	755.45

$$\sum_{m=1}^{n} P_m = 4238.94$$

The levelized value A is then obtained from the sum of present values using the capital recovery factor, Equation 7.28

$$A = \text{CRF} \sum_{m=1}^{n} P_m = \frac{i_{eff}(1 + i_{eff})^n}{(1 + i_{eff})^n - 1} \sum_{m=1}^{n} P_m$$

The CRF for $n = 5$ and $i_{eff} = 10\%$ is CRF $= 0.2638$. Thus $A = (0.2638)(4238.94) = \$1118.2$.

Alternatively, the levelized value A can be obtained using Equations 7.34 and 7.35. From Equation 7.35

$$k = \frac{1 + r_n}{1 + i_{eff}} = \frac{1 + 0.04}{1 + 0.10} = 0.9455$$

Substituting in Equation 7.34

$$\text{CELF} = \frac{k(1 - k^n)}{1 - k} \text{CRF} = \frac{0.9455(1 - 0.9455^5)}{1 - 0.9455} (0.2638)$$

$$= (4.239)(0.2638) = 1.1182$$

and $A = P_0 \text{CELF} = (1000)(1.1182) = \$1118.2$.

7.2.3 Current versus Constant Dollars

The *real interest rate* (*real cost-of-money rate, real return rate,* or *real discount rate*) is the money paid for the use of capital that does not include an adjustment for the anticipated general price-inflation rate in the economy. This adjustment is included, however, in the *nominal interest rate* (*nominal cost-of-money rate, nominal return rate,* or *nominal discount rate*). The real interest rate is based on the potential *real* (i.e., excluding inflation) earning power of money. The relation between the real interest rate (i_r) and the nominal interest rate (i_n) is

$$(1 + i_n) = (1 + i_r)(1 + r_i) \tag{7.36a}$$

where r_i is the general inflation rate. Using this equation we can calculate, for instance, the real cost of money when the nominal cost of money is 12% and the inflation rate is 5%:

$$i_r = \frac{i_n - r_i}{1 + r_i} = \frac{0.12 - 0.05}{1.05} = 0.067 \text{ or } 6.7\%$$

An economic analysis can be conducted either in *current dollars* by including the effect of inflation in projections of capital expenditures, fuel costs, and O&M costs, or in *constant dollars* by excluding inflation and considering only real escalation rates in cost projections and the real cost of money. Each approach has its strengths and weaknesses [1]: A current-dollar analysis uses estimated costs that will more closely approximate the actual costs when they occur but appears to overemphasize fuel and O&M costs. In addition, such an analysis may obscure some real cost trends and result in levelized values considerably higher than today's values. A constant-dollar analysis gives a clear picture of real cost trends, enables engineers to get a better understanding of the costs involved, and makes levelized costs appear close to today's values. Such a method, however, appears to understate all cost values and leads to cash flows that may be significantly less than the actual values.

In general, a current-dollar analysis gives the impression that the project being analyzed is more costly than we would expect based on today's cost values, whereas a constant-dollar analysis presents the project as less costly than it ultimately will be. The choice between current or constant dollars depends on the purpose of the analysis. The results of studies involving no more than 10 years are best presented in current dollars. Longer term studies may be best presented in constant dollars so that the effect of many years of inflation does not distort the costs to the point that they bear no resemblance to today's cost values.

Let us consider now the relation between constant and current dollars at the jth year after the reference year used in the cost calculations. For simplicity we assume that the nominal and real escalation rates and the general inflation rate remain constant. The cost of an asset in the jth year expressed in constant (subscript cs) and current (subscript cu) dollars, respectively, is given by

$$C_{j,\text{cs}} = C_0(1 + r_r)^j$$

$$C_{j,\text{cu}} = C_0(1 + r_n)^j$$

where C_0 is the cost of the same asset in the reference year. From these equations, we obtain

$$C_{j,cs} = C_{j,cu} \frac{(1 + r_r)^j}{(1 + r_n)^j}$$

and with Equation 7.33

$$C_{j,cs} = \frac{C_{j,cu}}{(1 + r_i)^j} \qquad\qquad (7.36b)$$

This equation is used in Section 7.3.5.

7.2.4 Time Assumptions

In an economic analysis, all available cost numbers (e.g., land costs, total plant facilities investment, other outlays, O&M costs, fuel costs, and by-product values) must be escalated to the date they are expended. Unless otherwise indicated, the following assumptions are made here:

- Land costs incur at the beginning of the first year of the design and construction period.
- The total capital investment is allocated to the individual years of the design and construction period. The expenditures for each year are incurred in the middle of the year.
- The startup costs are expended in the middle of the last year of design and construction.
- The working capital and the costs of licensing, research, and development are escalated to the end of the last year of design and construction.
- The allowance for funds used during construction (AFUDC) is paid annually during the design and construction period; the sum of AFUDC is calculated at the end of the last year of this period.
- The costs for fuel, operation, and maintenance are incurred in the middle of each year of the system economic life.
- The revenues from the sale of products are received in the middle of each year of the system economic life.

For the economic analysis of a thermal system we must register the date of reference for each cost number and specify (a) the beginning and length of the design and construction period, (b) the anticipated economic life, and (c) the life for tax purposes. The beginning of commercial operation (beginning of economic-life period) is assumed to coincide with the end of the design and construction period.

7.2.5 Depreciation

Depreciation reflects the fact that the value of an asset tends to decrease with age (or use) due to physical deterioration, technological advances, and other factors that ultimately will lead to the retirement of the asset. In addition, depreciation is a mechanism for repaying the original amount obtained from debt holders if the debt is to be retired. Finally, depreciation is an important accounting concept serving to reduce taxes during plant operation. In that respect, depreciation is not strictly related to the physical or economic lifetime of an asset. The asset life used for tax purposes (as determined by statute) could be shorter than the asset's anticipated economic life.

There are many methods for depreciating the value of an asset. Some of these methods—*straight-line, sum-of-the-years-digits,* and *declining-balance methods*—give no consideration to interest costs, whereas others (*sinking-fund* and *present-worth methods*) take into account the interest on investment. The present-worth method is based on the reduction with time of future profits obtainable with the property. Table 7.8 summarizes the mathematical relationships that can be used to calculate the depreciation allocation at the end of a year of the property life, and the cumulative depreciation allocation at the end of a year. More details are given in [2, 17, 18].

Table 7.8 Summary of selected tax depreciation methods (adapted from [17])[a]

Method	Depreciation Allocation at the End of Year z (DP_z)	Cumulative Depreciation Allocation at the End of Year z $\left(CDP_z = \sum_{i=1}^{z} DP_i\right)$
Straight line	$\dfrac{C_0 - S}{n}$	$(C_0 - S)\dfrac{z}{n}$
Sum-of-the-years digits	$(C_0 - S)\left[\dfrac{2(n + 1 - z)}{n(n + 1)}\right]$	$(C_0 - S)\left[\dfrac{z(2n + 1 - z)}{n(n + 1)}\right]$
Double declining balance ($n \geq 3$)	$C_0\left(\dfrac{2}{n}\right)\left(\dfrac{n - 2}{n}\right)^{z-1}$	$C_0\left[1 - \left(\dfrac{n - 2}{n}\right)^{z}\right]$
125% declining balance	$C_0\left(\dfrac{1.25}{n}\right)\left(\dfrac{n - 1.25}{n}\right)^{z-1}$	$C_0\left[1 - \left(\dfrac{n - 1.25}{n}\right)^{z}\right]$
Sinking fund	$(C_0 - S)\left[\dfrac{i(1 + i)^{z-1}}{(1 + i)^n - 1}\right]$	$(C_0 - S)\left[\dfrac{(1 + i)^z - 1}{(1 + i)^n - 1}\right]$

[a]C_0 = total depreciable investment (TDI) at the beginning of the economic life period (dollars)

S = salvage value of the property at the end of the (tax or economic) life considered in the depreciation (dollars)

n = tax life or economic life considered in the depreciation calculations (years)

i = interest rate (decimal ratio)

z = attained age of the property (years) ($1 \leq z \leq n$)

Note: Prior to the last year of ownership, the cumulative depreciation must not exceed $(C_0 - S)$.

The difference between the original cost of a property and the cumulative depreciation at the end of a year is defined as the *book value* at the end of that year. It represents the worth of the property as shown in the accounting records. Since the tax depreciation is, in general, different than the depreciation used for company internal purposes, a tax book value and a company book value can be calculated separately. It is advisable for a company to keep two separate sets of books, one for meeting income tax requirements and one for meeting company requirements.

The *accelerated cost recovery system* (ACRS) gives a set of statutory annual percentage factors for determining depreciation allowances based on a statutory class life period for a property. The percentage factors for specific life periods were established in the United States by federal income tax regulations that went into effect in 1981 (ACRS) and 1987 (*modified accelerated cost recovery system*, or MACRS). It should be noted that according to the MACRS, there is one more year of tax depreciation than the tax life. Thus, if the tax life of the property is 15 years, depreciation will be considered for 16 years. Assuming application of MACRS, the annual tax depreciation (TXD) and the end-year tax book value (EYTXBV) are obtained from the following equations:

$$TXD_j = TDI\, f_{MACRS,j}, \qquad j = 1, ..., TL + 1 \qquad (7.37a)$$

$$TXD_j = 0, \qquad j = TL + 2, ..., BL \qquad (7.37b)$$

$$EYTXBV_0 = TDI \qquad (7.38a)$$

$$EYTXBV_j = EYTXBV_{j-1} - TXD_j, \qquad j = 1, ..., TL + 1 \qquad (7.38b)$$

$$EYTXBV_j = 0, \qquad j = TL + 2, ..., BL \qquad (7.38c)$$

Here the subscript j refers to the jth year of system life; TL and BL are the tax life and book life, respectively, of the system; and TDI and $f_{MACRS,j}$ are the total depreciable investment at the beginning of the economic life and the MACRS factor in the jth year, respectively. More details about ACRS, MACRS, and depreciation in general are given in the literature [2, 17, 18].

The declining-balance method, the sum-of-the-years-digits method, and the ACRS or MACRS give greater depreciation costs in the early life years of the property than in the later years; this is normally advantageous for a company. The annual depreciation amount is constant when the straight-line method is used, whereas it increases with time when the sinking-fund and present-worth methods are applied. The final choice of the depreciation method depends on the property involved and on company practices.

Most expenditures that result in benefits extending over more than one year must be capitalized, and the cost must be allocated over the property life. Expenses associated with fuel costs and O&M costs are not depreciated but allocated to the year of acquisition. In addition, certain costs (e.g., land and working capital) represent *nondepreciable expenditures* for income tax purposes because ideally they should be completely recovered when the system shuts down.

In this chapter, we use only a simplified version of depreciation and accounting. Accordingly, the depreciation values from the analysis we present in Section 7.3 cannot be used for comparison with actual accounting systems.

7.2.6 Financing and Required Returns on Capital

The money to cover the total capital requirement of an investment can come through the following sources:

- Borrowing capital, for instance, by selling bonds (*debt financing*)
- The sale of common and preferred stock (*equity financing*)
- Existing funds of the company (*self-financing*)
- A combination of these

In the present discussion we assume, for simplicity, that the money to cover the total capital requirement comes through debt and equity financing. The return on investment for the investors is the cost of money for the company. Thus, the returns on debt and equity represent the amount that investors (stockholders and debtors, respectively) are paid for the use of their money. This cost applies only to funds that the company currently uses. It is not an obligation on investment capital that was raised in the past and has since been removed from the asset accounts through depreciation. Depreciation funds, if not used to pay back investors, can be reinvested in other projects, which are then responsible for earning a return on the invested money. Therefore, the cost-of-money calculation is based on the undepreciated investment: the amount remaining in the asset accounts after depreciation has been deducted.

The average cost of money in a project depends on the fractions of the total capital requirement financed through debt, preferred stock, and common stock and on the required return on each type of financing. For example, if the total capital requirement of the cogeneration system case study is covered using debt (50%), preferred stock (15%), and common stock (35%), and the minimum acceptable returns on investment are 10, 11.7, and 15%, respectively, the average annual rate of the cost of money is calculated as

$$i = (0.10)(0.50) + (0.117)(0.15) + (0.15)(0.35) = 0.12 \ (12.0\%)$$

This calculation is used in Section 7.2.9 (see entry 4 of Table 7.9).

The average rate of the cost of money (discount rate) calculated in this way is the *before-tax rate*. The *after-tax discount rate* (i_{at}) reflects the effect of the deductibility of debt return on the federal income tax calculation for the company, and is calculated using

$$i_{at} = i - f_d i_d t \tag{7.39}$$

where i is the before-tax discount rate, f_d and i_d represent the fraction of the total capital requirement financed through debt, and the corresponding rate of return, respectively, and t is the total income tax rate. Using data from the previous calculation, $f_d = 0.5$, $i_d = 0.10$, and a total income tax rate of 38% (calculated in Section 7.2.8 and listed in Table 7.9 as entry 5a), the after-tax discount rate for the cogeneration system becomes

$$i_{at} = 0.12 - (0.5)(0.10)(0.38) = 0.101 \ (10.1\%)$$

The decision to use a before-tax or an after-tax discount rate affects the results of an economic evaluation. In this book, before-tax discount rates are used.

7.2.7 Fuel, Operating, and Maintenance Costs

Fuel costs are usually part of the operating and maintenance costs. However, because of the importance of fuel costs in thermal systems, in this presentation fuel costs are considered separately from the O&M costs. The operating and maintenance costs can be divided into fixed and variable costs. The fixed O&M costs are composed of costs for operating labor, maintenance labor, maintenance materials, overhead, administration and support, distribution and marketing, research and development, and so forth. The variable operating costs depend on the *average annual system capacity factor,* which determines the equivalent average number of hours of system operation per year at full load. The variable operating costs consist of the costs for operating supplies other than fuel costs (e.g., raw water and limestone), catalysts, chemicals, and disposing of waste material.

The fuel costs and the variable operating costs can be easily calculated from the flow diagrams. Once we know the flow of a raw-material stream or of a utility, we simply multiply the flow by its unit cost and by the average total time of operation per year to obtain the contribution of the flow being considered to the total annual costs. Information on the estimation of fixed O&M costs is given in the literature [1–5, 9].

7.2.8 Taxes and Insurance

Income taxes are calculated by multiplying the income tax rate by the taxable income, which is the difference between total revenue and all tax-deductible expenditures. Income tax rates have varied significantly in recent years [2]. In the examples considered here a 34% federal income tax rate and a 6% state income tax rate are assumed. Since state income taxes are deductible for federal income tax purposes, the total (combined) income tax rate is

$$t = (1 - 0.06)(0.34) + 0.06 \simeq 0.38 \ (38\%)$$

This is listed in Table 7.9 as entry 5a.

Table 7.9 **Parameters and assumptions used in the calculation of total revenue requirement for the cogeneration system (all monetary values are expressed in mid-1994 dollars)**

Parameter (units)	Value
1a. Average general inflation rate (1994–2017) (%)	5.0
b. Average nominal escalation rate of all (except fuel) costs (1994–2017) (%)	5.0
c. Average nominal escalation rate of natural gas costs (1994–2017) (%)	6.0
2a. Beginning of the design and construction period	Jan. 1, 1996
b. Date of commercial operation	Jan. 1, 1998
3a. Plant economic life (years)	20
b. Plant life for tax purposes (years)	15
4. Plant financing fractions and required returns on capital:	

	Common Equity	Preferred Stock	Debt	
Type of financing				
Financing fraction (%)	35.0	15.0	50.0	
Required annual return (%)	15.0	11.7	10.0	
Resulting average cost of money (%)			12.0	

Parameter (units)	Value
5a. Average combined income tax rate (1994–2017) (%)	38.0
b. Average property tax rate (1994–2017) [% of PFI (in end-1997 dollars)]	1.50
c. Average insurance rate (1994–2017) [% of PFI (in end-1997 dollars)]	0.50
6. Average capacity factor (%)	85.0
7. Labor positions for operating and maintenance	30
8. Average labor rate ($/h)	28.00
9. Annual fixed operating and maintenance costs ($10^6\$$)	3.8
10. Annual variable operating and maintenance costs at full capacity ($10^3\$$)	350
11. Unit cost of fuel ($/GJ–LHV)	3.0
12. Allocation of plant-facilities investment to the individual years of design and construction (%)	
Jan. 1–Dec. 31, 1996	40.0
Jan. 1–Dec. 31, 1997	60.0

Assumptions

MACRS depreciation used for tax purposes (see Section 7.2.5).
Straight-line depreciation used for company purposes.
Net salvage value equal to zero.
No investment tax credit and no grants in aid of construction.
Revenue levelizing occurs with before-tax average cost of money.

Tax-deductible expenditures include fuel costs, O&M charges, return on debt, and investment cost recovery (depreciation calculated for tax purposes). The tax code allows for tax accounting procedures that result in greater cost-recovery tax deductions in the early life years of property than with straight-line depreciation. As a result, parts of income tax payments are deferred to later years. In any year of the economic life of a system, the difference between the income taxes actually paid and the income taxes that would have been paid if a straight-line depreciation had been used is called the *deferred income tax.*

The deferred income tax could either *flow through* and decrease the revenue requirements (and consequently the product costs) in the early years or be *normalized* [1]. In the normalization method, deferred income taxes are accumulated in a reserve account, and used—along with retained equity earnings and book depreciation—as a source of internally generated cash to pay for new investment items. In this case, the revenue requirements are calculated as if income taxes were paid based on a straight-line tax depreciation. The deferred income tax funds stay with the company until they are used to pay the tax obligation later in the life of the investment. The application of the normalization method is illustrated in Section 7.3 (Table 7.13) for the cogeneration system case study.

Depending on the location, the annual property taxes are usually between 1 and 4% of the plant-facilities investment [2]. The annual insurance costs are typically between 0.5 and 1.5% of the plant facilities investment. Design engineers can contribute to a reduction in insurance costs by understanding the different types of insurance available, the legal responsibilities of a company with regard to accidents and emergencies, and other factors that must be considered in obtaining adequate insurance.

7.2.9 Cogeneration System Case Study

To illustrate concepts introduced thus far, let us suppose that in the summer of 1994 a company considering construction of the case study cogeneration system initiates a detailed economic analysis of the proposal. Table 7.6 summarizes the total capital investment. Table 7.9 summarizes other aspects of the economic analysis.

The cost values given in Tables 7.6 and 7.9 are expressed in mid-1994 dollars. All cost items except fuel are assumed to increase with the general inflation rate, which is taken as 5.0% per year (entries 1a and 1b of Table 7.9). It is anticipated that the average annual increase in the cost of natural gas will be 6.0% (entry 1c of Table 7.9), which corresponds to an annual real escalation rate for fuel of about 0.95%.

The company managers estimate that design and construction could start in January 1996 and would last for 2 years (entry 2). The economic life is estimated to be 20 years (entry 3a), that is, from January 1, 1998, to December 31, 2017. The system life for tax purposes is 15 years (entry 3b). It is also assumed that when the system is retired, the cost of removal will equal the salvage value, resulting in zero salvage. Entries 4 and 5 of Table 7.9 are discussed in Sections 7.2.6 and 7.2.8, respectively. Let us briefly review the remaining entries.

The average capacity factor for the cogeneration system is estimated as 85% (entry 6 of Table 7.9), which means that the system will operate at full load 7446 h out of the total available 8760 h per year. The company managers assume that 30 labor positions will be required for operation and maintenance (entry 7) at an average labor rate of $28.00 per hour (entry 8). The average number of working hours per labor position is 2080 h per year. Thus, the annual direct labor costs are $1.747 $\times$ 10^6. Based on these numbers, the annual fixed O&M costs and the annual variable O&M costs at full capacity are estimated by management to be $3.8 $\times$ 10^6 and $0.35 $\times$ 10^6, respectively (entries 9 and 10 of Table 7.9). All these cost values are given in mid-1994 dollars. The annual variable O&M costs at the given capacity factor of 85% are 297.5 $\times$ 10^3 mid-1994 dollars. To calculate the corresponding costs for the first year of operation, these costs are escalated at a nominal escalation rate of 5% per year to the middle of 1998. The escalated annual values for the fixed and variable O&M costs are $4.619 $\times$ 10^6 and 0.362 $\times$ 10^6, respectively.

The cost of natural gas (approximated as methane) is $3.0 per GJ of lower heating value (entry 11 of Table 7.9). The lower heating value of the fuel is LHV = 50.01 MJ/kg. The fuel mass flow rate is 1.6419 kg/s (Table 1.2). The annual fuel cost (FC) is then

$$FC = (\$0.003/MJ)(50.01 \text{ MJ/kg})(1.6419 \text{ kg/s})(7446 \text{ h/year})(3600 \text{ s/h})$$

$$= 6.603 \times 10^6 \text{ mid-1994 dollars/year}$$

To calculate the annual fuel costs for the first year of commercial operation (mid-1998 dollars), we escalate this number at an annual escalation rate of 6% to the middle of 1998,

$$FC = (6.603 \times 10^6)(1.06)^4 = \$8.336 \times 10^6$$

Having calculated the fuel and O&M costs, we can now estimate the start-up costs and the working capital. As discussed in Section 7.1.4, the startup costs for a power plant are the sum of the following unescalated costs: (a) one month of fixed O&M costs, (b) one month of variable operating costs calculated at full load, (c) one week of full-load fuel, and (d) 2% of the plant facilities investment. Thus,

$$SUC = (3.8 \times 10^6)/12 + (0.35 \times 10^6)(0.85)/12 + (6.603 \times 10^6)/52$$

$$+ (0.02)(38.95 \times 10^6)$$

$$= \$1.247 \times 10^6 \text{ (mid-1994 dollars)}$$

or after escalation to the middle of 1997

$$SUC = \$1.444 \times 10^6 \text{ (mid-1997 dollars)}$$

Similarly, the working capital is the sum of the unescalated expenses representing 2 months of fuel and variable operating costs at full load, and 3 months of labor costs plus a contingency of 25% of the total of the above three items. Neglecting the variable operating costs for the cogeneration system, we obtain

$$WC = [(6.603 \times 10^6)/6 + (1.747 \times 10^6)/4](1.25)$$

$$= 1.924 \times 10^6 \quad \text{(mid-1994 dollars)}$$

or after escalation to the end of 1997

$$WC = 2.282 \times 10^6 \text{ (end-1997 dollars)}$$

The mid-1994 dollar values of startup costs and working capital calculated here are shown in Table 7.6 under the heading other outlays.

The plant-facilities investment (PFI) for this project is estimated in Section 7.1.6 as 38.95×10^6 mid-1994 dollars. According to entry 12 of Table 7.9, 40% of this amount must be escalated at an annual rate of 5% to the middle of 1996, when it is expended, whereas the remaining 60% of the PFI must be escalated to the middle of 1997. The system-financing fractions given as entry 4 of Table 7.9 are assumed to apply to any investment expenditure at any year. As summarized in Table 7.10, the allowance for funds used during construction is calculated separately for each type of financing using the corresponding returns on investment. The reason for distinguishing among the different types of financing in the AFUDC calculation is that the common equity AFUDC is not part of the net depreciable investment but instead is assumed to be recovered as common equity at the end of plant economic life. The calculated amount for the total AFUDC is 4.985×10^6 end-1997 dollars. The AFUDC value given in Table 7.6 (namely, $\$3.353 \times 10^6$) is obtained by discounting the end-1997 AFUDC to the middle of 1994 using the average discount rate of 12.0%. It is apparent that the AFUDC contribution to the total investment increases with increasing rates of return on investment and with increasing length of the design and construction period. For instance, the AFUDC is a very significant cost component in the construction of nuclear power plants.

Finally, we calculate the total nondepreciable and depreciable capital investments, and the total net capital investment for the cogeneration system (all values in thousands of dollars). We will first calculate the total net outlay (total cash expended) for the plant, which is equal to the total capital investment (Equation 7.5) minus AFUDC and any grants in aid of construction that may be projected. The date at which each expenditure is incurred is given in parentheses.

Table 7.10 Calculation of the allowance for funds used during construction (end-1997 values) of the cogeneration system (all values are rounded and given in thousands of dollars)

Design and Construction Year	Calendar Year	Plant-Facilities Investment			Common Equity		Preferred Equity		Debt	
		In Mid-1994 Dollars	Amount of Escalation	Escalated Investment	Escalated Investment	AFUDC	Escalated Investment	AFUDC	Escalated Investment	AFUDC
1	1996	15,580	1,597	17,177	6,012	1,402	2,577	465	8,588	1,320
2	1997	23,370	3,684	27,054	9,469	685	4,058	231	13,527	660
	Subtotals[a]	38,950	5,281	44,231	15,481	2,087	6,635	696	22,115	1,980
		(A)	(B)	(C)	(D)	(E)	(F)	(G)	(H)	(L)
AFUDC for the cost of land ($538 expended on Jan. 1, 1996)						61		20		56
AFUDC for startup costs ($1,444 expended on July 1, 1997)						37		12		35
Totals						2,185		728		2,072

Total AFUDC = 2185 + 728 + 2072 = $4985 (end-1997 dollars)
or total AFUDC = $3,353 (mid-1994 dollars)

[a](C) = (A) + (B) = (D) + (F) + (H); (D) = 0.35 (C); (F) = 0.15 (C); (H) = 0.50 (C).

Cost of land (1/1/96)	$ 538
Escalated PFI from Table 7.10 (6/30/96 and 6/30/97)	44,231
Startup costs from this section, (6/30/97)	1,444
Working capital from this section (12/31/97)	2,282
Costs of licensing, research and development from Table 7.6	0
Grants in aid of construction	(−)0
Total net outlay	**$48,495**

The total net outlay is expressed in mixed-year dollars. The contributions to the allowance for funds used during construction calculated in Table 7.10, are summarized as follows:

Common equity AFUDC	$ 2,185
Preferred equity AFUDC	728
Debt AFUDC	2,072
Total AFUDC	**$4,985**

Now we have the necessary data to calculate the total capital investment:

Total net outlay	$48,495
Total AFUDC	4,985
Total capital investment	**$53,480**

The total nondepreciable and depreciable capital investments are calculated next:

Cost of land	$ 538
Working capital	2,282
Common equity AFUDC	2,185
Total nondepreciable capital investment	**$5,005**

Total capital investment	$53,480
Total nondepreciable capital investment	(−)5,005
Total depreciable capital investment	**$48,475**

If an *investment tax credit* may be taken, this must be subtracted from the total capital investment to provide the net capital investment:

Total capital investment	$53,480
Investment tax credit	(−)0
Total net capital investment	**$53,480**

After calculating the total depreciable investment, the annual tax depreciation amount and the end-year tax book value can be calculated using Equations 7.37 and 7.38 and the MACRS factors for a life period of 15 years (tax life) and for property put in service after 1986 [2]. The results are summarized in Table 7.11.

7.3 CALCULATION OF REVENUE REQUIREMENTS

The annual *total revenue requirement* (total product cost) for a system is the revenue that must be collected in a given year through the sale of all products to compensate the system operating company for all expenditures incurred in the same year and to ensure sound economic plant operation. Figure 7.3 shows the major cost categories considered in the calculation of the total revenue requirement. Although the terms *expenses* and *carrying charges* are sometimes used differently by engineers and accountants, we will adopt the definition of expenses as shown in Figure 7.3: the sum of fuel costs and operating and maintenance costs. Expenses include goods and services that are used in a short period of time, usually less than one year. Expenses are paid for directly from revenue and, therefore, are not capitalized.

The carrying charges shown in Figure 7.3 represent obligations associated with an investment. Unlike expenses, the costs associated with an investment are not paid directly from revenue when incurred because this would require

Table 7.11 Statutory percentages for use in the MACRS for a life period of 15 years, annual tax depreciation amount, and tax book value at the end of each year for the cogeneration system

Year of Commercial Operation	Calendar Year	MACRS Depreciation Factor (%)	Annual Tax Depreciation ($)	End-Year Tax Book Value ($)
0	1997	—	—	48,475,000
1	1998	5.00	2,423,750	46,051,250
2	1999	9.50	4,605,125	41,446,125
3	2000	8.55	4,144,612	37,301,513
4	2001	7.70	3,732,575	33,568,938
5	2002	6.93	3,359,318	30,209,620
6	2003	6.23	3,019,992	27,189,628
7	2004	5.90	2,860,025	24,329,603
8	2005	5.90	2,860,025	21,469,578
9	2006	5.91	2,864,873	18,604,705
10	2007	5.90	2,860,025	15,744,680
11	2008	5.91	2,864,872	12,879,808
12	2009	5.90	2,860,025	10,019,783
13	2010	5.91	2,864,873	7,154,910
14	2011	5.90	2,860,025	4,294,885
15	2012	5.91	2,864,872	1,430,013
16	2013	2.95	1,430,013	0
Totals		**100.00**	48,475,000	

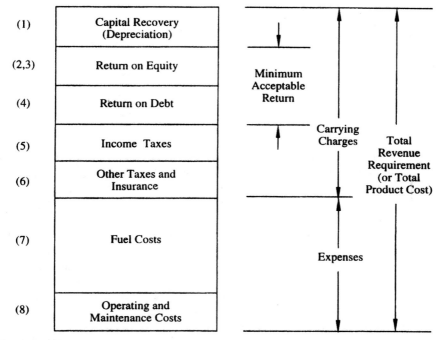

Figure 7.3 Revenue categories (cost categories) for the revenue requirement method of economic analysis of thermal systems [1].

present customers to pay for items that will be used to supply customers many years into the future. The carrying charges obligation remains until the system is retired at the end of its estimated economic life.

The minimum acceptable return (MAR) shown in Figure 7.3 is the minimum return on investment that the company expects from the project being considered. This usually means that other equally sound investments yielding the MAR are available and could compete with this project.

Costs associated directly with existing investments are sometimes referred to as *sunk costs* in decision-making procedures because they cannot be affected by a current course of action except for the costs that may be recovered through salvage. For example, capital recovery, return on equity, return on debt, and property taxes must be considered in the total product cost calculation regardless of the actual production rate. Although irrelevant for future-oriented economic studies, sunk costs are still important for the calculation of carrying charges. With the exception of income taxes and perhaps insurance, all other components of carrying charges are independent of the actual use of the system.

The annual total revenue requirement (TRR) is equal to the sum of the following eight annual amounts: total capital recovery (TCR); minimum return on investment (ROI) for common equity (subscript ce), preferred stock (subscript ps) and debt (subscript d); income taxes (ITX); other taxes and insurance (OTXI); fuel costs (FC); and operating and maintenance costs (OMC):

$$TRR_j = TCR_j + ROI_{j,ce} + ROI_{j,ps} + ROI_{j,d}$$
$$\text{(9)} \qquad \text{(1)} \qquad \text{(2)} \qquad \text{(3)} \qquad \text{(4)}$$

$$+ ITX_j + OTXI_j + FC_j + OMC_j \quad (7.40)$$
$$\text{(5)} \qquad \text{(6)} \qquad \text{(7)} \qquad \text{(8)}$$

The subscript j refers to the jth year of system operation. The numbers given below each term of Equation 7.40 key the variables to the categories of Figure 7.3. If the actual annual revenues exceed the annual TRR value calculated from Equation 7.40, the actual rate of return on equity becomes higher than the minimum value assumed in the calculation.

The *year-by-year analysis,* which illustrates the projected costs and revenue requirements for every year of the system economic life, is used to calculate the levelized revenue requirement and the levelized cost of the main product. In addition, this analysis is very useful for comparing technical alternatives (see Section 7.5). The cogeneration system considered previously will be used to demonstrate the year-by-year analysis. The results are summarized in Table 7.12 in which each column numbered 1 through 9 corresponds to a term of Equation 7.40. The following sections provide details concerning the calculation of the values provided by these columns of data. Additional examples are given in References 1, 19, 20.

7.3.1 Total Capital Recovery

The net investment must be recovered during the system economic life. Table 7.13 shows the year-by-year calculation of the total capital recovery for the cogeneration system. Column D of this table corresponds to column 1 of Table 7.12. The annual book depreciation (BD), shown in column A of Table 7.13, is calculated from the book life (BL) and the total depreciable investment (TDI) using the straight-line method:

$$BD_j = \frac{TDI}{BL}, \qquad j = 1, ..., BL \quad (7.41)$$

The index j refers to the jth year of book life. The deferred income taxes (DITX) for the jth year of tax life (TL), shown in column B of Table 7.13, are based on the difference between the annual tax depreciation (TXD), found via Equation 7.37a and Table 7.11 (using the MACRS method), and the annual book depreciation (BD), determined using Equation 7.41:

$$DITX_j = (TXD_j - BD_j)t, \qquad j = 1, ..., TL + 1 \quad (7.42)$$

Table 7.12 Year-by-year revenue requirement analysis for the cogeneration system (all values are round numbers given in thousand dollars)

Year	Calendar Year	(1) Capital Recovery	(2) Return on Common Equity	(3) Preferred Stock Dividends	(4) Interest on Debt	(5) Income Taxes	(6) Other Taxes and Insurance	(7) Fuel Cost	(8) O&M Costs	(9) Total Revenue Requirement (current dollars)	Total Revenue Requirement (constant dollars)
1	1998	2,533	2,808	939	2,674	2,363	885	8,336	4,981	25,517	20,993
2	1999	3,362	2,688	892	2,540	1,432	885	8,836	5,230	25,865	20,266
3	2000	3,187	2,526	830	2,365	1,470	885	9,366	5,491	26,120	19,491
4	2001	3,030	2,372	772	2,199	1,497	885	9,928	5,766	26,448	18,796
5	2002	2,888	2,227	716	2,040	1,515	885	10,524	6,054	26,849	18,172
6	2003	2,760	2,089	663	1,889	1,527	885	11,155	6,357	27,323	17,613
7	2004	2,699	1,958	612	1,744	1,476	885	11,825	6,674	27,872	17,111
8	2005	2,699	1,830	562	1,602	1,367	885	12,534	7,008	28,486	16,655
9	2006	2,701	1,702	512	1,460	1,256	885	13,286	7,359	29,160	16,237
10	2007	2,699	1,574	462	1,318	1,149	885	14,083	7,726	29,896	15,854
11	2008	2,701	1,446	413	1,176	1,038	885	14,928	8,113	30,699	15,505
12	2009	2,699	1,318	363	1,034	931	885	15,824	8,518	31,571	15,186
13	2010	2,701	1,190	313	892	820	885	16,773	8,944	32,518	14,897
14	2011	2,699	1,062	263	749	713	885	17,780	9,392	33,542	14,634
15	2012	2,701	934	213	607	602	885	18,847	9,861	34,650	14,398
16	2013	2,155	806	163	465	1,039	885	19,977	10,354	35,845	14,185
17	2014	1,612	707	123	351	1,496	885	21,176	10,872	37,221	14,028
18	2015	1,612	636	92	263	1,434	885	22,447	11,415	38,784	13,921
19	2016	1,612	565	62	175	1,372	885	23,793	11,986	40,450	13,828
20	2017	1,612	494	31	88	1,310	885	25,221	12,586	42,225	13,747

Table 7.13 Year-by-year capital-recovery schedule for the cogeneration system (all values are round numbers and are given in thousand dollars)

Year of Commercial Operation	Calendar Year	(A) Annual Book Depreciation	(B) Deferred Income Taxes	(C) Recovery of Common Equity AFUDC	(D) Total Capital Recovery
1	1998	2,424	0	109	2,533
2	1999	2,424	829	109	3,362
3	2000	2,424	654	109	3,187
4	2001	2,424	497	109	3,030
5	2002	2,424	356	109	2,888
6	2003	2,424	227	109	2,760
7	2004	2,424	166	109	2,699
8	2005	2,424	166	109	2,699
9	2006	2,424	168	109	2,701
10	2007	2,424	166	109	2,699
11	2008	2,424	168	109	2,701
12	2009	2,424	166	109	2,699
13	2010	2,424	168	109	2,701
14	2011	2,424	166	109	2,699
15	2012	2,424	168	109	2,701
16	2013	2,424	−378	109	2,155
17	2014	2,424	−921	109	1,612
18	2015	2,424	−921	109	1,612
19	2016	2,424	−921	109	1,612
20	2017	2,424	−921	109	1,612
Subtotal		48,475	0	2,185	50,660
Cost of land and working capital					2,820
Total investment					53,480

Note: Cost of land and working capital are recovered as common equity at the end of economic life.

where t is the total income tax rate. The deferred income taxes for the years (TL + 2) through BL are obtained from

$$\text{DITX}_j = -\frac{\sum_{k=1}^{\text{TL}+1} \text{DITX}_k}{\text{BL} - (\text{TL} + 1)}, \qquad j = \text{TL} + 2, ..., \text{BL} \qquad (7.43)$$

The common-equity allowance for funds used during construction (CEAF), which is not considered in the net depreciable investment, is recovered using a constant annual amount (RCEAF), shown in column C of Table 7.13, and obtained through

$$\text{RCEAF}_j = \frac{\text{CEAF}}{\text{BL}}, \qquad j = 1, ..., \text{BL} \qquad (7.44)$$

where CEAF is the common-equity AFUDC at the end of the design and construction period, Table 7.10.

The total capital recovery (TCR) for the jth year of book life is the sum of book depreciation (Equation 7.41), deferred income taxes (Equations 7.42 and 7.43), and recovery of the common-equity AFUDC (Equation 7.44):

$$TCR_j = BD_j + DITX_j + RCEAF_j, \qquad j = 1, ..., BL \qquad (7.45)$$

The TCR values are shown in column D of Table 7.13 and column 1 of Table 7.12. As Table 7.13 shows, the sum of total annual capital recovery values plus working capital and cost of land is equal to the total plant investment. Working capital and cost of land are not recovered during the system economic life and appear as common equity at the end of the book life, as shown in the last common equity balance of Table 7.14.

7.3.2 Returns on Equity and Debt

Table 7.14 shows the year-by-year distribution of capital recovery among debt, preferred stock, and common equity for the cogeneration system. With the aid of this table we calculate the balance at the beginning of each year for each type of financing. This balance is the basis for obtaining the returns on equity and the debt interest in Table 7.12, columns 2–4.

The year-by-year distribution of capital recovery (Table 7.14) is obtained as follows: The total net investment (TNI) is distributed at the beginning of the first year of book life among debt, preferred stock, and common equity using the corresponding financing fractions (Table 7.9) f_d ($=0.5$), f_{ps} ($=0.15$), and f_{ce} ($=0.35$), where the subscripts d, ps, and ce refer to debt, preferred stock, and common equity, respectively. This assumption determines the balances at the beginning of the first year ($BBY_{1,x}$) for the xth type of financing:

$$BBY_{1,x} = TNI \, f_x, \qquad x = d, ps, ce \qquad (7.46)$$

We assume further that deferred taxes are distributed among each type of financing according to the corresponding fraction. The columns labeled adjustment in Table 7.14 for debt and preferred stock consist of the appropriate portions of deferred income taxes. The adjustment column for common equity also includes the recovery of common-equity allowance for funds used during construction, Table 7.13:

$$ADJ_{j,d} = DITX_j f_d, \qquad j = 1, ..., BL \qquad (7.47a)$$

$$ADJ_{j,ps} = DITX_j f_{ps}, \qquad j = 1, ..., BL \qquad (7.47b)$$

$$ADJ_{j,ce} = DITX_j f_{ce} + RCEAF_j, \qquad j = 1, ..., BL \qquad (7.47c)$$

Table 7.14 Year-by-year distribution of capital recovery for the cogeneration system (all values are round numbers and are given in thousand dollars)

Year	Calendar Year	Debt			Preferred Stock			Common Equity			Total Capital Recovery
		Balance Beginning of Year	Book Depreciation	Adjustment	Balance Beginning of Year	Book Depreciation	Adjustment	Balance Beginning of Year	Book Depreciation	Adjustment	
1	1998	26,740	1,337	0	8,022	401	0	18,718	686	109	2,533
2	1999	25,403	1,337	414	7,621	401	124	17,923	686	399	3,362
3	2000	23,651	1,337	327	7,095	401	98	16,838	686	338	3,187
4	2001	21,987	1,337	249	6,596	401	75	15,814	686	283	3,030
5	2002	20,402	1,337	178	6,121	401	53	14,845	686	234	2,888
6	2003	18,887	1,337	113	5,666	401	34	13,926	686	189	2,760
7	2004	17,437	1,337	83	5,231	401	25	13,052	686	167	2,699
8	2005	16,017	1,337	83	4,805	401	25	12,199	686	167	2,699
9	2006	14,597	1,337	84	4,379	401	25	11,346	686	168	2,701
10	2007	13,176	1,337	83	3,953	401	25	10,492	686	167	2,699
11	2008	11,756	1,337	84	3,527	401	25	9,640	686	168	2,701
12	2009	10,335	1,337	83	3,101	401	25	8,786	686	167	2,699
13	2010	8,916	1,337	84	2,675	401	25	7,933	686	168	2,701
14	2011	7,495	1,337	83	2,248	401	25	7,080	686	167	2,699
15	2012	6,075	1,337	84	1,822	401	25	6,227	686	168	2,701
16	2013	4,654	1,337	−189	1,396	401	−57	5,373	686	−23	2,155
17	2014	3,506	1,337	−461	1,052	401	−138	4,710	686	−213	1,612
18	2015	2,629	1,337	−461	789	401	−138	4,238	686	−213	1,612
19	2016	1,753	1,337	−461	526	401	−138	3,765	686	−213	1,612
20	2017	876	1,337	−461	263	401	−138	3,293	686	−213	1,612
21	2018	0	0	0	0	0	0	2,820	0	0	0
Totals			**26,740**	**0**		**8,022**	**0**		**13,713**	**2,185**	**50,660**

The columns labeled book depreciation in Table 7.14 are calculated for each type of financing (debt, preferred stock, or common equity) using straight-line depreciation from

$$BD_{j,x} = \frac{BBY_{1,x} - \sum_{k=1}^{BL} ADJ_{k,x}}{BL}, \qquad j = 1, ..., BL, \quad x = d, ps, ce \quad (7.48)$$

with $BBY_{1,x}$ according to Equation 7.46. The balance at the beginning of the jth year for each type of financing is

$$BBY_{j,x} = BBY_{j-1,x} - (BD_{j,x} + ADJ_{j,x}),$$

$$j = 2, ..., BL, \quad x = d, ps, ce \qquad (7.49)$$

The balance at the beginning of the year immediately following the last year of plant operation (calendar year 2018 for the cogeneration system) is zero for debt and preferred stock but equal to the value of land and working capital ($\$2.82 \times 10^6$) for common equity. The sum of the totals for book depreciation and adjustments for all types of financing is equal to the total capital recovery ($\$50.66 \times 10^6$). The annual capital recovery values shown in the last columns of Tables 7.13 and 7.14 are repeated in the first column of Table 7.12.

According to the total revenue requirement method of economic analysis, the depreciation funds are assumed to be used to pay back investors for the principal. This assumption is used here because it helps to compare alternative investments. In an expanding company, however, these funds would likely be reinvested.

For each investment type, the return-on-investment (ROI) calculation for any year of system operation is based on the outstanding investment: the balance at the beginning of the year (BBY) (Equation 7.49 and Table 7.14) for the corresponding investment. That is, the return on investment for year j is

$$ROI_{j,x} = BBY_{j,x}i_x, \qquad j = 1, ..., BL, \quad x = d, ps, ce \qquad (7.50)$$

Here i_x is the annual (nominal or real) rate of return for the xth investment type. As discussed in Section 7.2.9, for the cogeneration system, i_x is 10.0, 11.7, and 15.0% for debt, preferred stock, and common equity, respectively. It is apparent that the annual amount of return on each type of investment declines with the increasing number of years of operation. For the cogeneration system the annual ROI values are given in columns 2–4 of Table 7.12.

7.3.3 Taxes and Insurance

Next, let us consider columns 5 and 6 of Table 7.12 concerning taxes and insurance. The annual taxable income (TXI) is determined by subtracting all annual tax-deductible expenditures from the total annual revenue requirement (TRR). The tax-deductible expenditures for the jth year of operation consist of interest on debt ($ROI_{j,d}$), other taxes and insurance ($OTXI_j$), fuel costs (FC_j), operating and maintenance costs (OMC_j), and tax depreciation (TXD_j). Accordingly

$$TXI_j = TRR_j - ROI_{j,d} - OTXI_j - FC_j - OMC_j - TXD_j$$

or with Equation 7.40 the taxable income in the jth year is obtained as

$$TXI_j = TCR_j + ROI_{j,ce} + ROI_{j,ps} + ITX_j - TXD_j \qquad (7.51)$$

where TXD_j is found using Equations 7.37a and 7.37b. Note that the income taxes to be paid during the jth year (ITX_j) are part of the same year's taxable income. The income taxes are

$$ITX_j = t\,TXI_j \qquad (7.52)$$

where t is the total income tax rate. Equations 7.51 and 7.52 lead to

$$ITX_j = \frac{t}{1-t}\,(TCR_j + ROI_{j,ce} + ROI_{j,ps} - TXD_j) \qquad (7.53)$$

Combining Equations 7.42, 7.45, and 7.53 we obtain the following relation, which can be used alternatively for calculating the income taxes of the jth year:

$$ITX_j = \frac{t}{1-t}\,(ROI_{j,ce} + ROI_{j,ps} + RCEAF_j) - DITX_j \qquad (7.54)$$

The variables ROI, RCEAF, and DITX are calculated from Equations 7.50, 7.44, and 7.42 (or 7.43), respectively. Column 5 of Table 7.12 shows the annual amounts of income taxes calculated for the cogeneration system according to the preceding equation.

The annual sum of other taxes (property taxes) and insurance costs (OTXI) may be calculated as a constant percentage of the escalated plant-facilities investment (PFI):

$$OTXI_j = PFI\,f_{OTXI}, \qquad j = 1, \ldots, BL \qquad (7.55)$$

The underlying assumption is that, since property taxes increase with time and insurance costs decrease with time, the sum of the two remains constant. For the cogeneration system, the value of the factor f_{OTXI} is assumed to be 2% (entries 5b and 5c in Table 7.9). The OTXI values are shown in column 6 of Table 7.12.

7.3.4 Fuel, Operating, and Maintenance Costs

The fuel costs and the total O&M costs for the first year of operation of the cogeneration system are determined in Section 7.2.9 as 8.336×10^6 and 4.981×10^6, respectively. In the year-by-year analysis these costs are escalated at an annual escalation rate of 6 and 5%, respectively. The values shown in columns 7 and 8 of Table 7.12 demonstrate the significance of fuel and O&M costs, particularly in the last years of plant operation.

7.3.5 Total Revenue Requirement (TRR)

The last two columns in Table 7.12 present the annual total revenue requirement in current and constant dollars. According to Equation 7.40, the current-dollar annual values are calculated by adding all the values shown in the previous columns (1–8) for the same year (round-off accounts for slight differences). The total revenue requirement in constant dollars for the jth year of operation of the cogeneration system $(TRR_{j,cs})$ can be calculated directly from the total revenue requirement in current dollars for the same year $(TRR_{j,cu})$ using Equation 7.36b, which becomes

$$TRR_{j,cs} = \frac{TRR_{j,cu}}{(1 + r_i)^{(3.0+j)}} \tag{7.56}$$

Here, r_i is the annual general inflation rate. The exponent $(3.0 + j)$ in Equation 7.56 converts the dollars in the middle of the jth year of operation into mid-1994 dollars. This equation is used to obtain the total revenue requirement in the absence of inflation; the values are shown in the last column of Table 7.12. These values indicate that the total revenue requirement (expressed in constant dollars) decreases with the increasing number of years of plant operation. In other words, if users are to be charged according to the annual total revenue requirement, they will pay (in constant dollars) more in the first years than in the last years of operation.

7.4 LEVELIZED COSTS AND COST OF THE MAIN PRODUCT

When evaluating the cost effectiveness and considering design modifications of a thermal system, it is necessary to compare the annual values of carrying charges, fuel costs, and O&M expenses (see Sections 8.3.1, 9.2 and 9.6). As

Table 7.12 illustrates, however, these cost components may vary significantly within the economic life. In general, carrying charges decrease while fuel and O&M costs increase with increasing years of operation. Therefore, levelized annual values for all cost components should be used when considering design modifications.

The calculation of levelized costs is discussed in Section 7.2.2. The levelized value of the annual total revenue requirement is obtained by first calculating the sum of present values of the 20 annual values shown in Table 7.12 and then applying Equation 7.27 to convert this value into an equivalent annuity. The average cost of money of 12% (Section 7.2.6) is used as the discount rate in these operations. Figure 7.4 illustrates the time scale used for calculating the annual levelized costs. The levelized annual total revenue requirement (TRR_L) for the 20 years of operation of the cogeneration system is calculated to be $TRR_{L,cu} = \$28.798 \times 10^6$ on a current-dollar basis, and $TRR_{L,cs} = \$17.088 \times 10^6$ on a constant-dollar basis. To simplify the following presentation, only the calculation of current-dollar values will be discussed. The constant-dollar values are calculated in a similar way using real escalation rates and real rates of return.

The levelized annual fuel costs in current dollars ($FC_{L,cu}$) are calculated using Equations 7.28, 7.34, and 7.35:

$$k_F = \frac{1 + 0.06}{1 + 0.12} = 0.94643$$

$$CRF_{cu} = \frac{0.12 \, (1.12)^{20}}{1.12^{20} - 1} = 0.13388$$

$$FC_{L,cu} = \frac{8.336 \times 10^6}{1.06} \; \frac{0.94643 \, (1 - 0.94643^{20})(0.13388)}{1 - 0.94643} = \$12.416 \times 10^6$$

The underlined factor in the last equation represents the fuel costs at the beginning of the first time period used for levelization (middle of 1997), Figure 7.4. Similarly, the levelized annual operating and maintenance costs ($OMC_{L,cu}$) are

$$k_{OM} = \frac{1 + 0.05}{1 + 0.12} = 0.9375$$

$$OMC_{L,cu} = \frac{4.981 \times 10^6}{1.05} \; \frac{0.9375 \, (1 - 0.9375^{20}) \, (0.13388)}{1 - 0.9375} = \$6.906 \times 10^6$$

The levelized annual carrying charges ($CC_{L,cu}$) are then calculated as

$$CC_{L,cu} = TRR_{L,cu} - FC_{L,cu} - OMC_{L,cu} = (28.798 - 12.416 - 6.906) \times 10^6$$

or

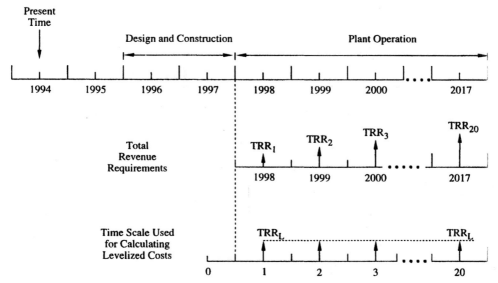

Figure 7.4   Illustration of the time scales used for the year-by-year analysis and the calculation of levelized annual costs for the case study cogeneration system.

$$CC_{L,cu} = \$9.476 \times 10^6$$

The results of the levelized cost calculation in current and constant dollars for a 20-year period are summarized in Table 7.15.

When a capital investment is based on long-term economic criteria, such as the criteria used for the 20 years of cogeneration system operation, some caution is necessary, particularly if a decision is sensitive to the late-years results. Some of the projected late-years benefits might never materialize because of current unforeseen changes in the technological, economic, and legal environment (e.g., system modifications caused by technical obsolescence and new governmental regulations). In this case, a shorter time period, ten years, for example, should be considered in calculating the levelized costs that will be used for optimization or decision-making purposes. The year-by-year analysis is still based on the entire assumed book life and tax life, but the levelized costs are computed using only the first 10 years of operation.

For this reason, Table 7.15 also shows the levelized costs in current and constant dollars for the first 10 years of cogeneration system operation. A comparison of the values obtained for the 10- and 20-year levelization periods demonstrates that the longer the levelization period, the smaller the relative contribution of carrying charges to the total revenue requirement. The relative contribution of fuel costs and O&M expenses increases with increasing length of the levelization period. The thermoeconomic evaluation presented in Chapter 8 uses the levelized current-dollar values obtained for the 10-year levelization period.

Table 7.15 Levelized annual costs and total revenue requirement in current and constant dollars for levelization time periods of 20 and 10 years for the cogeneration system (all values are round numbers and are given in thousands of dollars)

	Current Dollars; 20-Year Period	Relative Contribution (%)	Constant Dollars; 20-Year Period	Current Dollars; 10-Year Period	Relative Contribution (%)	Constant Dollars; 10-Year Period
Carrying charges	9,476	32.9	5,623	10,527	39.1	7,202
Fuel costs	12,416	43.1	7,367	10,411	38.7	7,124
O&M costs	6,906	24.0	4,098	5,989	22.2	4,098
Total revenue requirement	28,798	100.0	17,088	26,927	100.0	18,424

If a system produces only one main product, its unit cost (MPUC) can be calculated directly from the annual total revenue requirement (TRR), the annual total value of the by-products (BPV) produced in the same plant and the main-product quantity (MPQ):

$$MPUC = \frac{TRR - BPV}{MPQ} \qquad (7.57)$$

This equation can be applied equally to any specific year of operation or to the levelized values of TRR and BPV. For example, if we consider the steam of the cogeneration system as a by-product and assume that the company projects an annually increasing selling price of the steam, the 20-year levelized value of which is \$0.035/kg, the levelized total annual value of steam is obtained as

$$BPV_L = (0.035 \ \$/kg)(14 \ kg/s)(3600 \ s/h)(7446 \ h/yr)$$

$$= \$13.135 \times 10^6/yr$$

The electrical energy developed by the cogeneration system per year is

$$MPQ = (30,000 \ kW)(7446 \ h/yr) = 223.38 \times 10^6 \ kWh/yr$$

The levelized unit cost of electricity for the 20-year period can be calculated now from Equation 7.57:

$$MPUC_L = \frac{28.798 \times 10^6 - 13.135 \times 10^6}{223.38 \times 10^6} = \$0.0701/kWh = 7.01\cent/kWh$$

This value represents the *levelized cost* for a 20-year period assuming average annual nominal escalation rates for the fuel costs and the O&M expenses of 6 and 5%, respectively, for the entire economic life of the cogeneration system. Some of these assumptions might not materialize in the future, however. We should also keep in mind that levelized costs are not *directly* comparable to actual costs at any given year of plant operation.

In a conventional economic analysis of a thermal system that generates more than one product we need to know the selling *prices* of all but one product in order to calculate the *cost* associated with this product. In other words, a conventional economic analysis does not provide criteria for apportioning the carrying charges, fuel costs, and O&M expenses to the various products generated in the same system. This point is discussed further in Chapter 8 where it is demonstrated that mass or energy should not be used as a basis for such thermal system costing purposes. The appropriate variable is exergy.

7.5   PROFITABILITY EVALUATION AND COMPARISON OF ALTERNATIVE INVESTMENTS

Before capital is invested in a project, it is necessary to estimate the expected profit from the investment. Most capital expenditure decisions involve choosing the "best" of a number of alternative projects, solutions, or courses of action that, usually, are mutually exclusive. Thus, calculating the profitability of an investment and choosing the best alternative are important objectives of an economic analysis.

In profitability calculations we consider profits and costs that will occur in the future. The associated risks and uncertainties may be significant. The analysis of investments and decisions under risk and uncertainty (see, e.g., [21, 22]) is outside the scope of this book. Here, we present only a *deterministic investment analysis* based on the following assumptions of certainty [21]: (a) There is a *perfect capital market:* the supply of funds is unrestricted. (b) There is complete certainty about investment outcomes. (c) Investment projects are indivisible. (d) The profitability of one project does not in any way affect the profitability of any other project.

In this section we discuss briefly the criteria most commonly used in business practice to evaluate the profitability of an investment and compare alternative investments. The following simplified case study will be used to illustrate the application of each method.

A company considers two mutually exclusive alternative projects that yield the same service. Project I has a total initial capital investment of $350,000 and a salvage value at the end of its 7-year economic life of $28,000. The initial capital investment for project II is $400,000 and the salvage value at the end of its 8-year economic life is $40,000. The working capital investment is $20,000 and $30,000 for projects I and II, respectively. The minimum attractive rate of return is 15%. The annual cash expenses (cash outflows) are constant for each year at $52,000 and $44,000 for projects I and II, respectively. For project I the annual net cash flow after all expenses, taxes, and insurance have been paid is $108,000 for the first 3 years and $115,000 for the remaining 4 years of its economic life. Project II has a constant annual net cash inflow of $120,000. All costs and profits occur at the end of each year. Table 7.16 summarizes the two projects and shows schematically the cash flows associated with the two investments. The downward arrow at the beginning of the economic life denotes the sum of the initial capital and working capital committed to the project. At the end of the economic life, as denoted by the shorter upward arrow, the salvage value and the working capital are recovered. The company, which uses straight-line depreciation, wants to know which investment, if any, should be made. Each method dis-

Table 7.16   Case study alternatives of Section 7.5

	Project I	Project II
Economic life (years)	7	8
Total initial capital investment ($)	350,000	400,000
Salvage value ($)	28,000	40,000
Working capital investment ($)	20,000	30,000
Annual cash expenses ($)	52,000	44,000
Annual net cash inflow ($)	108,000 (first 3 years) 115,000 (last 4 years)	120,000

Cash flow diagrams:
Project I:

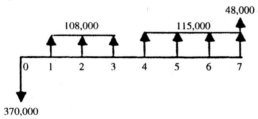

Project II:

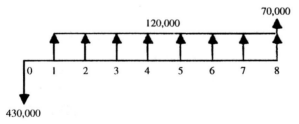

Assumptions:
Minimum attractive rate of return (after taxes) = 15%
Straight-line depreciation is employed.
Costs of land and startup are neglected.

cussed in this section will be used to answer this question. Initially, we will ignore the 1-year difference in the economic life of the two projects. This effect is discussed in Section 7.5.3.

7.5.1   Average Rate of Return and Payback Period

Let us begin by calculating some average decision variables for each project: the average rate of return and the payback period.

Average Rate of Return. The arithmetic average annual net cash inflows (ANCF) for the two projects without considering the time value of money are

$$ANCF_I = \frac{(3)(108,000) + (4)(115,000)}{7} = \$112,000$$

$$ANCF_{II} = \$120,000$$

The total depreciable investment (TDI) is the difference between the initial capital investment and the salvage value:

$$TDI_I = 350,000 - 28,000 = \$322,000$$

$$TDI_{II} = 400,000 - 40,000 = \$360,000$$

The annual book depreciation (BD) calculated according to the straight-line formula (Table 7.10 or Equation 7.41) is obtained by dividing the TDI by the project economic life:

$$BD_I = \$322,000/7 = \$46,000$$

$$BD_{II} = \$360,000/8 = \$45,000$$

The average annual profit (AP) is the difference between annual net cash flow (ANCF) and annual depreciation (BD):

$$AP_I = 112,000 - 46,000 = \$66,000$$

$$AP_{II} = 120,000 - 45,000 = \$75,000$$

The *average rate of return* (ARR) on the initial investment is the ratio between the average annual profit and the total initial investment including working capital:

$$ARR_I = \frac{66,000}{350,000 + 20,000} = 17.8\%$$

$$ARR_{II} = \frac{75,000}{400,000 + 30,000} = 17.4\%$$

Although the average rate of return method is sometimes used to evaluate and compare projects, it could give misleading results because the time value of money is not included in the calculation. If a decision would be made based on the calculated ARR values, however, investment I should be recommended.

Note that both projects have the same average annual total (gross) cash inflow (ATCF), which is the sum of the annual average net cash inflows and the annual cash outflows:

$$ATCF_I = 112,000 + 52,000 = \$164,000$$

$$ATCF_{II} = 120,000 + 44,000 = \$164,000$$

Payback Period. The *payback* (or *payout*) *period* is defined as the length of time required for the cash inflows received from a project to recover the original cash outlays required by the initial investment. Mathematically, the payback period τ_{PB} is defined by the relation [21]

$$\sum_{z=0}^{\tau_{PB}} Y_z = 0 \qquad (7.58)$$

where Y_z is the net cash flow to or from the project in the zth year. The payback period is often used as a limit rather than as a direct criterion. Thus, a company might establish some maximum payback period for a given class of projects and then reject all projects having payback periods greater than the maximum. In addition, the payback period may be used to rank various project alternatives: The projects with the shortest payback periods are given the highest preference.

The payback period for each project of the case study can be calculated by dividing the total depreciable investment (TDI) by the average annual net cash inflow (ANCF):

$$\tau_{PB} = \frac{TDI}{ANCF}$$

that is,

$$\tau_{PB,I} = \frac{322,000}{112,000} = 2.9 \text{ years}$$

$$\tau_{PB,II} = \frac{360,000}{120,000} = 3.0 \text{ years}$$

Based on this criterion, investment I, which has the least payback period, should be recommended.

Calculation of the payback period is relatively simple. However, use of this popular criterion should be avoided because it does not consider the time value of money and discriminates against long-lived projects. Let us now calculate the payback period by applying Equation 7.58, using *actual* (instead of average) cash flows (including working capital) and ignoring the tax deductibility of investments. The following relations are obtained:

Project I:

$$-350{,}000 - 20{,}000 + 108{,}000 + 108{,}000 + 108{,}000 + x_I\,115{,}000 = 0$$

where x_I represents the part of the fourth year of project I required to recover completely the original cash outlays associated with project I. Thus,

$$x_I = 0.4 \quad \text{and} \quad \tau_{PB,I} = 3.4 \text{ years}$$

Project II:

$$-400{,}000 - 30{,}000 + (\tau_{PB,II})(120{,}000) = 0$$

$$\tau_{PB,II} = 3.6 \text{ years}$$

Once again, Project I would be favored. However, the calculation of the payback period using only actual cash flows has an *additional* drawback: It completely neglects cash flows that occur after the payout year.

7.5.2 Methods Using Discounted Cash Flows

We now consider three evaluation methods that are based on *rational criteria* because they consider the time value of money and all cash flow streams during the life of a project. Each method uses the net present value (NP_0) of all net cash flows associated with each project alternative. The formula for calculating the net present value is

$$NP_0 = \sum_{z=0}^{BL} Y_z(1 + i)^{-z} \tag{7.59}$$

where Y_z is the net cash flow at the end of zth time period. BL and i are the book life and the effective discount rate, respectively.

Net Present Value Method. When the *net present value method* is used for project selection, the following rules apply: accept any project for which the present value is positive; reject any project with negative present value; the projects with the highest present values are given the highest preference among various alternatives; if two projects are mutually exclusive, accept the one having the greater present value.

For the projects of Table 7.16, the net present values are

$$NP_{0,I} = -350,000 - 20,000 + 108,000 \left(\frac{1}{1+i} + \frac{1}{(1+i)^2} + \frac{1}{(1+i)^3} \right)$$

$$+ 115,000 \left(\frac{1}{(1+i)^4} + \frac{1}{(1+i)^5} + \frac{1}{(1+i)^6} + \frac{1}{(1+i)^7} \right)$$

$$+ (20,000 + 28,000) \frac{1}{(1+i)^7}$$

For $i = 15\%$ we obtain $NP_{0,I} = \$110,511$.

$$NP_{0,II} = -400,000 - 30,000 + 120,000 \left[\frac{1}{1+i} + \frac{1}{(1+i)^2} + \cdots \right.$$

$$\left. + \frac{1}{(1+i)^8} \right] + (30,000 + 40,000) \frac{1}{(1+i)^8}$$

For $i = 15\%$ the net present value of project II is $NP_{0,II} = \$131,362$. When the minimum attractive rate of return of 15% is used, both investments are profitable, but project II, which has the highest net present value, should be recommended.

The net present value method implies that each net cash flow at the end of each time period is *reinvested* at an annual rate of return equal to the discount rate used in Equation 7.59. This implicit assumption is not necessary in the revenue requirement method (Section 7.3), which is also based on the general net present value concept. In the revenue requirement method we assume that the depreciation funds are used to *pay back* investors, and thus, there is no need to consider the reinvestment of the depreciation funds. Since the revenue requirement method does not consider any profit above the minimum return on investment, which is also paid back to the investors, no additional assumptions about the reinvestment of net cash flows are necessary. When the revenue requirement method is used to compare alternative projects, the project with the lowest total levelized revenue requirement is selected.

Benefit–Cost Ratio Method. The *benefit–cost ratios* may also be used to evaluate the profitability of an investment. Three such ratios are [21]

$$\text{Benefit–cost ratio} \quad BCR = \frac{NP_0 + C_0}{C_0} \tag{7.60}$$

$$\text{Net benefit–cost ratio} \quad NBCR = \frac{NP_0}{C_0} \tag{7.61}$$

$$\text{Eckstein's benefit–cost ratio} \quad EBCR = \frac{b_1 + b_2}{C_0 + b_2} \tag{7.62}$$

where NP_0 is the net present value of all cash flows of a project at time $\tau = 0$ (Equation 7.59), C_0 is the total investment cost at $\tau = 0$, b_1 is the difference in present values of cash inflows and outflows from year $z = 1$ to BL, and b_2 represents the present value of cash outflows from year $z = 1$ to BL. The sum $b_1 + b_2$ represents the present value of cash inflows.

Let us evaluate the benefit–cost ratios for the two projects of Table 7.16. With Equation 7.60, we have

$$BCR_I = \frac{110,511 + 370,000}{370,000} = 1.30$$

$$BCR_{II} = \frac{131,362 + 430,000}{430,000} = 1.31$$

Equation 7.61 yields

$$NBCR_I = \frac{110,511}{370,000} = 0.30$$

$$NBCR_{II} = \frac{131,362}{430,000} = 0.31$$

Finally, to apply Equation 7.62, we require

$$b_1 = NP_0 + C_0$$

and

$$b_2 = A_{exp} \frac{(1 + i)^n - 1}{i(1 + i)^n}$$

where A_{exp} represents the constant annual cash expenses. Thus, we obtain

$$b_{1,I} = \$480,511$$

$$b_{1,II} = \$561,362$$

$$b_{2,I} = 52,000 \frac{1.15^7 - 1}{(0.15)(1.15^7)} = \$216,342$$

$$b_{2,II} = 44,000 \frac{1.15^8 - 1}{(0.15)(1.15^8)} = \$197,442$$

and

$$\mathrm{EBCR_I} = \frac{480{,}511 + 216{,}342}{370{,}000 + 216{,}342} = 1.19$$

$$\mathrm{EBCR_{II}} = \frac{561{,}362 + 197{,}442}{430{,}000 + 197{,}442} = 1.21$$

All three benefit–cost ratios applied to the case study suggest that project II should be recommended. The benefit–cost ratio method provides a criterion based on the net present value (or on the present value of cash inflows) *per dollar outlay,* which is a criterion expressing the cost efficiency of a project. In decision making, however, there is no particular advantage in using the benefit–cost ratio method over the net-present-value method. The net-present-value calculation is straightforward and somewhat simpler.

Internal Rate of Return Method. Both the net present value and benefit–cost methods use an interest rate that is usually based on the company's cost of money but is external to the specific project being considered. The internal rate of return method seeks to avoid the arbitrary choice of an interest rate; instead, it calculates an interest rate, initially unknown, that is internal to the project. The procedure is to determine the interest rate $i*$, called the *internal rate of return,* that makes the net present value of an investment zero. That is

$$\mathrm{NP_0} = \sum_{z=0}^{\mathrm{BL}} Y_z(1 + i*)^{-z} = 0 \tag{7.63}$$

The interest rate $i*$ is found iteratively. A project is selected if the calculated internal rate of return is greater than the minimum acceptable rate of return. The projects with the highest internal rates of return are given the highest preference among various alternatives.

The internal rate of return does not express the rate of return on the *initial* investment. It represents instead the rate of interest earned on the *time-varying, unrecovered* balances of an investment such that the final investment balance is zero at the end of the project life [21]. The internal rate of return method assumes reinvestment of cash flows at the calculated project internal rate of return.

For the two projects of the case study, the following internal rates of return are calculated: $i_I* = 24.0\%$ and $i_{II}* = 23.4\%$. Thus, investment I is recommended by the internal rate of return method. Previously, we found that project II is recommended by the net present value method. The inconsistency in ranking projects I and II when the net present value and the internal rate of return methods are used is caused by the different assumptions associated with these methods.

7.5.3   Different Economic Lives

In the previous illustrations of the profitability–evaluation methods (Sections 7.5.1 and 7.5.2) using the present case study, we ignored the difference in economic life between the two projects. However, the comparison of feasible alternatives should be conducted on an equivalent basis that includes a *realistic* and *identical* analysis period. Therefore, the previously obtained results for the case study need to be modified before a final conclusion is reached with respect to their ranking.

Unequal lives among feasible alternatives tend to complicate their direct comparison. For example, both the net present value and the value of the internal rate of return depend on the project life. In such a case we need a procedure that will put the feasible alternatives on a comparable basis by using a meaningful and identical *planning horizon* (or *planning period*) for the analysis. The following approaches can be used.

Repeatability Approach.   According to this approach, the planning horizon is equal to a common multiple of the lives of all alternatives considered simultaneously. For each feasible alternative, all economic assumptions applying to the initial life span will also apply to all succeeding life spans. This approach can become very involved mathematically and is not always realistic. For the case study projects, which have economic lives of 7 and 8 years, respectively, a time period of 56 years should be used in which project I will be repeated 8 times, and project II 7 times.

Cotermination Approach.   In this case the planning horizon may be any meaningful length of time: less than or equal to the life of the shortest lived alternative, greater than or equal to the longest alternative life, or between the shortest and longest lives. In applying this method we may need to (a) assume that some or all of the alternatives are repeated, and (b) calculate the salvage value at the end of the planning period for each alternative.

For example, if we select the planning horizon to be equal to the life of the shortest lived alternative (7 years) for the two project alternatives, we calculate first the salvage value of project II at the end of the seventh year using straight-line depreciation and $BD_{II} = \$45,000$ found previously:

$$S_{II,7} = 400,000 - (7)(45,000) = \$85,000$$

The net present value of project II for the 7-year period is

$$NP_{0,II,7} = -400,000 - 30,000 + 120,000 \left[\frac{1}{1+i} + \frac{1}{(1+i)^2} + \cdots \right.$$

$$\left. + \frac{1}{(1+i)^7} \right] + (30,000 + 85,000) \frac{1}{(1+i)^7}$$

and with $i = 15\%$ we obtain $NP_{0,II,7} = \$112,483$. When this value is compared to the previously calculated value of $NP_{0,I,7} = \$110,511$, we can conclude that investment II should still be recommended; the advantage is now smaller, however.

The internal rate of return of project II for the 7-year period is calculated by equating $NP_{0,II,7}$ to zero and determining the internal rate of return i^* iteratively. We obtain a value of $i^*_{II,7} = 22.7\%$. When compared with the 8-year value, $i_{II,8} = 23.4\%$ and with the internal rate of return of the first project $i^*_{I,7} = 24.0\%$, note that $i^*_{I,7} > i^*_{II,8} > i^*_{II,7}$; thus, when the internal rate of return is used as the criterion, the 7-year planning horizon makes project I look more attractive than when unequal lives were used.

Next, to solve the same problem for a planning horizon of 8 years, which is equal to the life of the longest alternative, we assume repetition of investment I using the same economic assumptions as in the initial life span. Using $BD_I = \$46,000$ found previously, the salvage value of project I at the end of the eighth year (first year of the repeated investment) is

$$S_{I,8} = 350,000 - 46,000 = \$304,000$$

The net present value of project I for the 8-year period is

$$NP_{0,I,8} = -350,000 - 20,000 + 108,000 \left(\frac{1}{(1 + i)} + \frac{1}{(1 + i)^2} + \frac{1}{(1 + i)^3} \right.$$

$$\left. + \frac{1}{(1 + i)^8} \right) + 115,000 \left(\frac{1}{(1 + i)^4} + \frac{1}{(1 + i)^5} \right.$$

$$\left. + \frac{1}{(1 + i)^6} + \frac{1}{(1 + i)^7} \right) - \frac{350,000 - 28,000}{(1 + i)^7}$$

$$+ (20,000 + 304,000) \frac{1}{(1 + i)^8}$$

With $i = 15.0\%$ we obtain $NP_{0,I,8} = \$112,635$. Previously we calculated $NP_{0,II,8} = \$131,362$. Accordingly, the net present value calculated for 8 years suggests that project II should be recommended.

The internal rate of return of project I for the 8-year period is calculated from the previous equation by equating $NP_{0,I,8}$ to zero. The new value, $i^*_{I,8} = 23.5\%$, is slightly higher than $i^*_{II,8}$ ($=23.4\%$) but smaller than $i^*_{I,7}$ ($=24.0\%$). We conclude that an 8-year planning horizon increases the attractiveness of project II. The above calculations illustrate clearly the effect of planning horizon on the comparison of alternative investments.

Capitalized-Cost Approach. In this approach we assume perpetual replacement of each investment and calculate the total capitalized cost (TCC). The capitalized cost is discussed at the end of Section 7.2.1. The total capitalized cost includes the capitalized cost for the original investment, C_K (Equation 7.32), plus the capitalized present value of the cash expenses CPVCE, plus the working capital WC:

$$\text{TCC} = C_K + \text{CPVCE} + \text{WC} \tag{7.64}$$

The capitalized present value of cash expenses is calculated by first determining the present value of all annual cash expenses, PVAE.

$$\text{PVAE} = \sum_{z=1}^{n} C_z \frac{1}{(1 + i)^z} \tag{7.65a}$$

If the annual cash expenses C_z remain constant in time, as occurs in both projects of Table 7.16, Equation 7.65a becomes

$$\text{PVAE} = C_z \frac{(1 + i)^n - 1}{i(1 + i)^n} \tag{7.65b}$$

Then the PVAE is converted into a capitalized present value (CPVCE) by multiplying PVAE with the appropriate capitalized cost factor (CCF), Table 7.7:

$$\text{CPVCE} = \text{PVAE} \times \text{CCF} = \text{PVAE} \frac{(1 + i)^n}{(1 + i)^n - 1} \tag{7.66}$$

Combining Equations 7.65b and 7.66, we obtain for the capitalized present value of cash expenses for projects I and II:

$$\text{CPVCE} = \frac{C_z}{i} \tag{7.67}$$

Finally, Equation 7.64 combined with Equations 7.32 and 7.67 becomes

$$\text{TCC} = \left[C_0 - \frac{S}{(1 + i)^n} \right] \left[\frac{(1 + i)^n}{(1 + i)^n - 1} \right] + \frac{C_z}{i} + \text{WC} \tag{7.68}$$

Thus the total capitalized cost for projects I and II is

$$\text{TCC}_I = \left[350,000 - \frac{28,000}{(1 + 0.15)^7} \right] \left[\frac{(1 + 0.15)^7}{(1 + 0.15)^7 - 1} \right] + \frac{52,000}{0.15} + 20,000$$

$$= \$910,640$$

$$\text{TCC}_{II} = \left[400,000 - \frac{40,000}{(1 + 0.15)^8} \right] \left[\frac{(1 + 0.15)^8}{(1 + 0.15)^8 - 1} \right] + \frac{44,000}{0.15} + 30,000$$

$$= \$898,173$$

The total capitalized cost based on a rate of return of 15% is lower for investment II than for investment I; therefore investment II should be recommended.

7.5.4 Discussion of Profitability–Evaluation Methods

Under the assumptions discussed at the beginning of Section 7.5, we have presented five criteria for evaluating the economic acceptability of a project and for comparing alternative investments. These criteria are (a) average rate of return, (b) payback period, (c) net present value, (d) benefit–cost ratio, and (e) internal rate of return. All of these criteria are used in practice depending on company preferences and on the projects to be evaluated. The first two criteria are popular and easy to implement. However, they do not consider the time value of money, and they may neglect some cash flow streams late in the life of a project. Therefore, these criteria should be avoided, particularly in projects with long economic lives.

The methods using discounted cash flows lead to rational criteria for evaluating and comparing alternative investments. Among them, the net present value is the preferred criterion for evaluating the profitability of a project and ranking alternative investments, particularly if the interest (discount) rate can be specified or secured. The benefit–cost ratio is a special case of net present value and involves only additional calculations. The internal rate of return might introduce some ambiguity [21], and thus should be used cautiously. Before using any of the above criteria, we must understand the associated assumptions and examine whether these assumptions are acceptable for the project being evaluated.

For the comparison of projects with different economic lives, a realistic and identical time period (planning horizon) should be used. The conclusions elicited may differ with different specifications of the identical time period, however. When the repeatibility approach becomes very involved, the capitalized-cost approach should be used. However, in specific projects with relatively short lives, the cotermination approach may be preferred.

7.6 CLOSURE

This chapter begins with a brief review of methods for estimating the total capital investment of a thermal system. These methods are not always accurate and are often based partially on approximate percentages or factors applicable to a particular component, plant, or process. Such factors should be used only when more accurate data are not available.

After a review of principles of engineering economics, a detailed application of the revenue requirement method to the cogeneration system of Figure 1.7 is illustrated. This method, which is a special case of the general net present value method, is used here as an example of an economic analysis for a thermal design project. The levelized costs associated with carrying charges, fuel, and operation and maintenance are calculated in current and constant dollars and for different time periods. Levelized costs are used as input data for the thermoeconomic evaluation presented in Chapter 8 and the thermoeconomic optimization discussed in Chapter 9. Chapter 7 concludes with a discussion of criteria for evaluating the economic feasibility of a project and comparing alternative investments.

An economic analysis generally involves more uncertainties than a thermodynamic analysis. This fact should not deter engineers from conducting detailed economic analyses using the best available data, however. Sensitivity studies are recommended to investigate the effect of major assumptions (e.g., cost of money, inflation rate, and real escalation rate of fuels) on the results of an economic analysis.

REFERENCES

1. *Technical Assessment Guide (TAG™)*, Electric Power Research Institute, TR-100281, Vol. 3, Revision 6, 1991.

2. M. S. Peters and K. D. Timmerhaus, *Plant Design and Economics for Chemical Engineers*, 4th ed., McGraw-Hill, New York, 1991.

3. G. D. Ulrich, *A Guide to Chemical Engineering Process Design and Economics*, Wiley, New York, 1984.

4. D. E. Garrett, *Chemical Engineering Economics*, Van Nostrand Reinhold, New York, 1989.

5. F. J. Valle-Riestra, *Project Evaluation in the Chemical Process Industries*, McGraw-Hill, New York, 1983.

6. A. Chauvel, et al., *Manual of Economic Analysis of Chemical Processes*, McGraw-Hill, New York, 1976.

7. K. M. Guthrie, *Process Plant Estimating, Evaluation and Control*, Craftsman, Solana Beach, CA, 1974.

8. J. H. Perry and C. H. Chilton, *Chemical Engineers' Handbook*, 5th ed., McGraw-Hill, New York, 1973.

9. K. K. Humphreys, *Jelen's Cost and Optimization Engineering*, 3rd ed., McGraw-Hill, New York, 1991.

10. J. M. Douglas, *Conceptual Design of Chemical Processes*, McGraw-Hill, New York, 1988.

11. K. M. Guthrie, Data and techniques for preliminary capital cost estimating, *Chem. Eng.*, March 24, 1969, pp. 114–142.

12. W. D. Baasel, *Preliminary Chemical Engineering Plant Design,* 2nd ed., Van Nostrand Reinhold, New York, 1990.

13. *Modern Cost Engineering: Methods and Data,* compiled and edited by *Chemical Engineering,* McGraw-Hill, New York, 1979.

14. K. M. Guthrie, Capital and operating costs for 54 chemical processes, *Chem. Eng.,* June 15, 1970, pp. 140–156.

15. *Sources and Production Economics of Chemical Products,* compiled and edited by *Chemical Engineering,* McGraw-Hill, New York, 1974, pp. 121–180.

16. O. P. Kharbanda, *Process Plant and Equipment Cost Estimation,* Craftsman, Solana Beach, CA, 1979.

17. G. W. Smith, *Engineering Economy,* 4th ed., Iowa State University Press, Ames, 1987.

18. E. P. DeGarmo, W. G. Sullivan, and J. A. Bontadelli, *Engineering Economy,* 9th ed., Macmillan, New York, 1992.

19. G. Tsatsaronis and M. Winhold, *Thermoeconomic Analysis of Power Plants,* EPRI AP-3651, RP2029-8, Final Report, Electric Power Research Institute, Palo Alto, CA, August 1984.

20. G. Tsatsaronis, M. Winhold, and C. G. Stojanoff, *Thermoeconomic Analysis of a Gasification-Combined-Cycle Power Plant,* EPRI AP4734, RP2029-8, Final Report, Electric Power Research Institute, Palo Alto, CA, August 1986.

21. L. E. Bussey, *The Economic Analysis of Industrial Projects,* Prentice-Hall, Englewood Cliffs, NJ, 1978.

22. R. Schlaifer, *Analysis of Decisions under Uncertainty,* McGraw-Hill, New York, 1969.

PROBLEMS

7.1 The purchased cost of a shell-and-tube heat exchanger with 100 m^2 of heating surface area was \$32,720 in 1990. What is the expected purchased cost of a similar heat exchanger with 200 m^2 in 1990 and 1993? The Marshall Swift Equipment Index for 1990 and 1993 is 915.1 and 964.2, respectively.

7.2 The heat exchanger of Problem 7.1 is a U-tube heat exchanger having a carbon steel shell and stainless steel tubes and is rated at a design pressure of 35 bars. The purchased cost for the heat exchanger is calculated from

$$C_{\mathrm{PE}} = C_B(f_d + f_p)f_m$$

The values of the material factor (f_m) in the surface area range of 100–500 m^2 are 2.81 for carbon steel/stainless steel and 8.95 for carbon steel/titanium used as shell–tube materials. The pressure factor (f_p) is 0.52 for design pressures between 25 and 50 bars and 0.55 for pressures between 50 and 65 bars. The base cost (C_B) is calculated for a floating-head shell-and-tube heat exchanger using carbon steel for both shell and tubes and rated at a design pressure below 10 bars. The base cost of a heat exchanger with 100 m² of heating surface was $8500 in 1990.

(a) What will be the purchased cost of a carbon steel/titanium U-tube heat exchanger rated at 55 bars and having a heating surface area of 150 m² in 1993?

(b) If Equation 7.2 with $f_T = 0$ applies and the bare module factor is 3.18, calculate the module cost for the heat exchanger described in (a).

7.3 Determine the module cost for a 500,000-gal stainless steel vertical storage tank in 1993. The base cost of a carbon-steel tank in 1989 dollars is

$$C_B = (V)^{0.63}$$

where V is the tank capacity in gallons. Assume that Equation 7.2 applies with $f_T = f_p = 0$, $f_d = 1$, $f_m = 3.2$ (for stainless steel) and $f_{BM} = 2.52$.

7.4 Using a nominal annual interest rate of 15%, find the effective annual interest rate if compounding is (1) annual, (2) quarterly, (3) monthly, and (4) continuous. For uniform annual payments of $10,000 for 10 years, compare the compound amount accumulated at the end of 10 years under (1) through (4).

7.5 For the cogeneration system case study, calculate the levelized annual costs associated with carrying charges, operation and maintenance, and fuel using the following simplified model:

(1) The purchased-equipment costs should be calculated from the functions and constants given in Appendix B. These values are expressed in mid-1994 dollars.

(2) The average annual inflation rate is 4.0%; the average nominal escalation rate of all (except fuel) costs is 4.0%, and the average nominal escalation rate of the natural gas (CH_4) cost is 4.2%. The economic life is 20 years, the average annual cost of money is 12.0%, and the average capacity factor is estimated to be 90%. The unit cost of fuel is 3.0 mid-1994 dollars per GJ lower heating value.

(3) The effects of taxes, insurance, and financing of capital expenditures are neglected.

(4) Overnight construction is assumed (i.e., the allowance for funds used during construction is zero).

(5) Startup costs, working capital, and costs of licensing, research, and development are assumed to be zero.

(6) The total capital investment (TCI) is four times higher than the sum of purchased-equipment costs expressed in mid-1997 dollars.

$$\text{TCI} = 4.0 \sum_{k=1}^{5} (\text{PEC})_k$$

(7) Commercial operation of the plant starts on January 1, 1998. The purchased-equipment costs will be escalated with the inflation rate.

(8) The annual operating and maintenance expenses (expressed in mid-1997 dollars) are estimated to be 20% of the total purchased-equipment costs expressed in mid-1997 dollars:

$$Z^{\text{OM}} = 0.20 \sum_{k=1}^{5} (\text{PEC})_k$$

(9) The annual levelized carrying charges are obtained by multiplying the total capital investment by the capital-recovery factor.

(10) The annual levelized operating and maintenance costs are obtained by multiplying the O&M costs in mid-1997 dollars by the constant escalation levelization factor.

Calculate in current dollars for a 20-year period the annualized levelized (a) carrying charges, (b) operating and maintenance costs, and (c) fuel costs.

7.6 Derive Equations 7.29 and 7.30.

7.7 Using the values from Table 7.12 for the total revenue requirement of the cogeneration system, calculate the levelized total revenue requirement in constant dollars for a 10-year levelization period. Then calculate the levelized values in constant dollars for fuel costs, operating and maintenance costs, and carrying charges for the same time period. Compare the calculated values with the corresponding values given in Table 7.15.

7.8 For the cogeneration system of Figure 1.7 draw a figure showing the end-year tax book values calculated according to the following depreciation methods:
- MACRS (see Table 7.11)
- Straight line
- Double declining balance with a switch to the straight-line method at the end of the ninth year
- Sum-of-the-years digits
- Sinking fund

Assume $C_0 = \$48.475 \times 10^6$, $S = 0$, $n = 15$, and $i = 12\%$. Compare the results obtained from the various methods.

7.9 A community is studying the feasibility of installing a solid-waste energy conversion plant. Calculate the amount the community must charge in dollars per 1000 Btu of energy output in order to break even over the 240-month economic life of the plant. At the beginning of plant operation, the total net capital investment is 40×10^6. The salvage value at the end of the plant economic life is zero. The monthly fuel, operating and maintenance costs are 800×10^3. These costs are expected to increase by 0.1% per month. A periodic overhaul costing 1.0×10^6 each should be conducted every 10 months. The first overhaul is done at the end of the tenth month and the last one at the end of month 240. The overhaul costs are expected to increase each time by 4%. The energy output of the plant at the design capacity is 18×10^{10} Btu/month and the cost of money is 1% per month.

7.10   A company is considering the construction of a small building, requiring a $50,000 downpayment and a $450,000, 30-year loan at 10% with equal annual payments. The cost savings through the construction are estimated at $60,000 for the first year and are expected to increase 5% per year thereafter. The operating and maintenance costs are estimated at 40% of the cost savings and should increase at the same 5% annual rate. The resale value of the building in 30 years is estimated at $1,000,000. Find the internal rate of return on this investment over the 30-year life.

7.11   In the design of a thermal system, two mutually exclusive alternatives (A and B) are under consideration. These alternatives are as follows:

	Alternative A	Alternative B
Investment cost ($)	50,000	120,000
Salvage value ($)	10,000	20,000
Annual costs ($)	9,000	5,000
Useful life (years)	4	10

If a perpetual service life is assumed, which of these alternatives should be recommended if the annual cost of money is 10%?

7.12   Another boiler must be added to the steam-generating system of a small plant. Two boiler manufacturers have submitted bids from which the following information is obtained:

	Boiler A	Boiler B
First cost ($)	50,000	120,000
Salvage value ($)	10,000	20,000
Useful life (years)	20	40
Annual operating costs ($)	9,000	6,000

If the minimum acceptable rate of return is 10%, which boiler should be purchased?

8

THERMOECONOMIC
ANALYSIS
AND EVALUATION

Thermoeconomics is the branch of engineering that combines exergy analysis and economic principles to provide the system designer or operator with information not available through conventional energy analysis and economic evaluations but crucial to the design and operation of a cost-effective system. We can consider thermoeconomics as exergy-aided cost minimization. Since the thermodynamic considerations of thermoeconomics are based on the exergy concept, the term *exergoeconomics* can also be used to describe the combination of exergy analysis and economics [1, 2].

Chapters 2 and 3 have outlined the principles for conducting detailed thermodynamic evaluations of thermal systems. In particular, techniques have been developed for evaluating the thermodynamic inefficiences of these systems: exergy destructions and exergy losses. However, we often need to know how much such inefficiencies cost. Knowledge of these costs is very useful for improving the cost effectiveness of the system, that is, for reducing the costs of the final products produced by the system.

In addition, if a system has more than one product, as for example the cogeneration system shown in Figure 1.7, we would want to know the production costs for each product. This is a common problem in chemical plants where electrical power, chilled water, compressed air, and steam at various pressure levels are generated in one department and sold to another. The plant operator wants to know the true cost at which each of the utilities is generated; these costs are then charged to the appropriate final products according to the type and amount of each utility used to generate a final product. In the design of a thermal system, such cost allocation assists in pinpointing cost-ineffective processes and operations and in identifying technical options that might improve the cost effectiveness of the system.

Accordingly, the objective of a thermoeconomic analysis might be (a) to calculate separately the costs of each product generated by a system having more than one product, (b) to understand the cost formation process and the flow of costs in the system, (c) to optimize specific variables in a single component, or (d) to optimize the overall system.

In this chapter, we discuss the basic elements of thermoeconomics, which include cost balances, means for costing exergy transfers, and several thermoeconomic variables used in the evaluation and optimization of the design of thermal systems. We also present some advanced costing techniques of thermoeconomics.

Because of the variation of costs from year to year, when we evaluate the design of a thermal system from the cost viewpoint, we use the cost levelization approach (e.g., Table 7.15 for the cogeneration system case study). Therefore, the cost values used throughout this chapter and Chapter 9 are levelized costs. For conciseness, the term levelized is often omitted in the following discussions, however.

8.1 FUNDAMENTALS OF THERMOECONOMICS

Cost accounting in a company is concerned primarily with (a) determining the actual cost of products or services, (b) providing a rational basis for pricing goods or services, (c) providing a means for allocating and controlling expenditures, and (d) providing information on which operating decisions may be based and evaluated [3]. This frequently calls for the use of cost balances. In a conventional economic analysis, a cost balance is usually formulated for the overall system (subscript tot) operating at steady state:

$$\dot{C}_{P,\text{tot}} = \dot{C}_{F,\text{tot}} + \dot{Z}_{\text{tot}}^{\text{CI}} + \dot{Z}_{\text{tot}}^{\text{OM}} \tag{8.1}$$

The cost balance expresses that the *cost rate* associated with the product of the system ($\dot{C}_P$) equals the total rate of expenditures made to generate the product, namely the fuel cost rate ($\dot{C}_F$) and the cost rates associated with capital investment ($\dot{Z}^{\text{CI}}$) and operating and maintenance ($\dot{Z}^{\text{OM}}$). Here and throughout this chapter, the terms fuel and product are used in the sense introduced in Section 3.5.3. When referring to a single stream associated with a fuel or product, the expression *fuel stream* or *product stream* is used. The rates $\dot{Z}^{\text{CI}}$ and $\dot{Z}^{\text{OM}}$ are calculated by dividing the annual contribution of capital investment and the annual operating and maintenance (O&M) costs, respectively, by the number of time units (usually hours or seconds) of system operation per year. The sum of these two variables is denoted by $\dot{Z}$:

$$\dot{Z} = \dot{Z}^{\text{CI}} + \dot{Z}^{\text{OM}} \tag{8.2}$$

Finally, note that in Chapters 8 and 9, the variable $\dot{C}$ denotes a cost rate associated with an exergy stream: stream of matter, power, or heat transfer while the variable $\dot{Z}$ represents all remaining costs.

8.1.1 Exergy Costing

For a system operating at steady state there may be a number of entering and exiting material streams as well as both heat and work interactions with the surroundings. Associated with these transfers of matter and energy are exergy transfers into and out of the system and exergy destructions caused by the irreversibilities within the system. Since exergy measures the true thermodynamic value of such effects, and costs should only be assigned to commodities of value, it is meaningful to use exergy as a basis for assigning costs in thermal systems. Indeed, thermoeconomics rests on the notion that exergy is the only rational basis for assigning costs to the interactions that a thermal system experiences with its surroundings and to the sources of inefficiencies within it. We refer to this approach as *exergy costing.*

In exergy costing a cost is associated with each exergy stream. Thus, for entering and exiting streams of matter with associated rates of exergy transfer $\dot{E}_i$ and $\dot{E}_e$, power $\dot{W}$, and the exergy transfer rate associated with heat transfer $\dot{E}_q$ we write, respectively

$$\dot{C}_i = c_i \dot{E}_i = c_i(\dot{m}_i e_i) \tag{8.3a}$$

$$\dot{C}_e = c_e \dot{E}_e = c_e(\dot{m}_e e_e) \tag{8.3b}$$

$$\dot{C}_w = c_w \dot{W} \tag{8.3c}$$

$$\dot{C}_q = c_q \dot{E}_q \tag{8.3d}$$

Here c_i, c_e, c_w, and c_q denote *average costs per unit of exergy* in dollars per gigajoule (\$/GJ) or cents per thousand Btu (¢/10^3 Btu).

Exergy costing involves cost balances usually formulated for each component separately. A cost balance applied to the kth system component shows that the sum of cost rates associated with all exiting exergy streams equals the sum of cost rates of all entering exergy streams plus the appropriate charges due to capital investment ($\dot{Z}_k^{CI}$) and operating and maintenance expenses ($\dot{Z}_k^{OM}$). The sum of the last two terms is denoted by $\dot{Z}_k$. Accordingly, for a component receiving a heat transfer and generating power, we would write

$$\sum_e \dot{C}_{e,k} + \dot{C}_{w,k} = \dot{C}_{q,k} + \sum_i \dot{C}_{i,k} + \dot{Z}_k \tag{8.4a}$$

This equation simply states that the total cost of the exiting exergy streams equals the total expenditure to obtain them: the cost of the entering exergy streams plus the capital and other costs. Note that when a component receives power (as in a compressor or a pump) the term $\dot{C}_{w,k}$ would move with its positive sign to the right side of this expression. The term $\dot{C}_{q,k}$ would appear with its positive sign on the left side if there is a heat transfer from the component. Cost balances are generally written so that all terms are positive.

Introducing the cost rate expressions of Equation 8.3, Equation 8.4a becomes

$$\sum_e (c_e \dot{E}_e)_k + c_{w,k} \dot{W}_k = c_{q,k} \dot{E}_{q,k} + \sum_i (c_i \dot{E}_i)_k + \dot{Z}_k \qquad (8.4b)$$

The exergy rates ($\dot{E}_e$, $\dot{W}$, $\dot{E}_q$, and $\dot{E}_i$) exiting and entering the kth component are calculated in an exergy analysis conducted at a previous stage. The term $\dot{Z}_k$ is obtained by first calculating the capital investment and O&M costs associated with the kth component and then computing the levelized values (see Section 7.4) of these costs per unit of time (year, hour, or second) of system operation.

The variables in Equation 8.4b are the levelized costs per unit of exergy for the exergy streams associated with the kth component ($c_{e,k}$, $c_{w,k}$, $c_{q,k}$, and $c_{i,k}$). In analyzing a component, we may assume that the costs per exergy unit are known for all entering streams. These costs are known from the components they exit or, if a stream enters the overall system consisting of all components under consideration, from the purchase cost of this stream. Consequently, the unknown variables to be calculated from a cost balance for the kth component are the costs per exergy unit of the exiting material streams ($c_{e,k}$) and, if power or useful heat are generated in that component, the cost per unit of exergy associated with the transfer of power ($c_{w,k}$) or heat ($c_{q,k}$).

To become familiar with exergy costing, we will now consider three elementary cases. At the conclusion of the present section, exergy costing is also illustrated in Example 8.1 by application to the gas turbine cogeneration system case study.

Steam or Gas Turbine. Consider the simple adiabatic turbine shown in Figure 8.1a. For this case, Equation 8.4a reads

$$\dot{C}_e + \dot{C}_w = \dot{C}_i + \dot{Z} \qquad (8.5a)$$

where $\dot{Z}$ represents the sum of charges associated with the turbine's capital investment and operating and maintenance costs. With Equations 8.3a, 8.3b, and 8.3c, Equation 8.5a becomes

$$c_e \dot{E}_e + c_w \dot{W} = c_i \dot{E}_i + \dot{Z} \qquad (8.5b)$$

where $\dot{W}$, $\dot{E}_i$, and $\dot{E}_e$ would be known from a prior exergy analysis (Chapter 3), $\dot{Z}$ would be known from a previous economic analysis (Chapter 7), and the

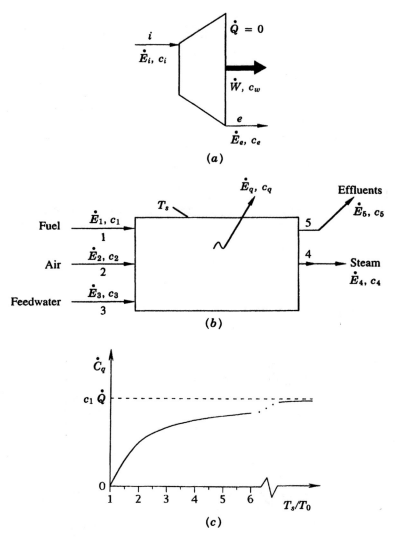

Figure 8.1 Elementary exergy costing examples. (a) Simple steam turbine at steady state. (b) Boiler at steady state. (c) Costing of the boiler exergy loss associated with heat transfer.

cost per exergy unit of the stream entering the turbine (c_i) would be known from the application of exergy costing to the components upstream of the turbine. This leaves the two exergy unit costs c_e and c_w as the unknowns. To determine the *two* unknowns, *one auxiliary relation* is required to supplement Equation 8.5b.

Since the purpose of a turbine is to generate power, all costs associated with the purchase and operation of the turbine should be charged to the power.

Both the exergy rate spent to generate the power and the exergy rate exiting the turbine were supplied to the working fluid in the components upstream of the turbine at the *same* average cost per exergy unit, c_i. In agreement with accounting practice, this cost would change *only if* exergy were *added* to the working fluid during the turbine expansion. Therefore, the cost per unit of exergy of the working fluid remains constant:

$$c_e = c_i \tag{8.5c}$$

With this auxiliary relation all costs associated with the purchase and operation of the turbine (capital and other costs plus cost of the exergy spent in the turbine) are charged to the generated power. Inserting Equation 8.5c into Equation 8.5b, we obtain the cost per exergy unit of the generated power:

$$c_w = \frac{c_i(\dot{E}_i - \dot{E}_e) + \dot{Z}}{\dot{W}} \tag{8.5d}$$

Boiler. Consider next the boiler shown in Figure 8.1*b*. For this case the cost balance, Equation 8.4a, reads

$$\dot{C}_4 + \dot{C}_5 + \dot{C}_q = \dot{C}_1 + \dot{C}_2 + \dot{C}_3 + \dot{Z} \tag{8.6a}$$

In accordance with the arrow on the figure, heat is transferred from the boiler, and the associated cost rate appears on the left side of the cost balance. Similarly, the cost rate associated with the effluents appears on the left side of the cost balance. Here, the exergy transfers associated with the heat transfer and the effluents are regarded as losses.

With Equations 8.3a, 8.3b, and 8.3d, Equation 8.6a becomes

$$c_4\dot{E}_4 + c_5\dot{E}_5 + c_q\dot{E}_q = c_1\dot{E}_1 + c_2\dot{E}_2 + c_3\dot{E}_3 + \dot{Z} \tag{8.6b}$$

where $\dot{E}_1$, $\dot{E}_2$, $\dot{E}_3$, $\dot{E}_4$, $\dot{E}_5$, $\dot{E}_q$, and $\dot{Z}$ would be known from prior exergetic and economic analyses. The costs per exergy unit of the entering streams (c_1, c_2, and c_3) may be regarded as known. This leaves the unit costs c_4, c_5, and c_q as the unknowns. To determine the values of these *three* unknowns, *two* auxiliary relations are required to supplement Equation 8.6b.

As the purpose of the boiler is to generate steam at specified values of temperature and pressure, all costs associated with owning and operating the boiler should be charged to the steam. It is reasonable then to focus on the remaining exiting streams (the losses) when considering appropriate auxiliary relations. Means for costing exergy losses are considered in Section 8.1.4. Here, we write simply

$$c_5 = c_1 \tag{8.6c}$$

$$c_q = c_1 \tag{8.6d}$$

That is, we assume that the exergy losses are covered by the supply of additional fuel to the boiler and that the average unit cost of supplying the fuel remains constant with varying exergy losses. Inserting Equations 8.6c and 8.6d in Equation 8.6b and solving, the cost per exergy unit of the steam exiting the boiler is

$$c_4 = \frac{c_1(\dot{E}_1 - \dot{E}_5 - \dot{E}_q) + c_2\dot{E}_2 + c_3\dot{E}_3 + \dot{Z}}{\dot{E}_4} \tag{8.6e}$$

It is instructive to consider the exergy loss associated with heat transfer in greater detail. Thus, with Equation 3.10b, the monetary loss associated with heat transfer to the surroundings is

$$\dot{C}_q = c_1\dot{E}_q$$

$$= c_1\dot{Q}\left(1 - \frac{T_0}{T_s}\right) \tag{8.7}$$

where T_s is the temperature at which the heat transfer from the boiler occurs. Equation 8.7 is shown graphically in Figure 8.1c. The plot indicates that the cost of the exergy loss associated with heat transfer is greater for heat transfer at higher temperatures than at low temperatures. The plot also shows that the cost rapidly approaches zero as the heat transfer usefulness (expressed by the associated exergy) goes to zero. This behavior is what we would expect from a rational costing scheme: one that accounts *properly* for value.

Referring again to Figure 8.1c, the limiting value approached asymptotically by the costing curve at large (and physically unreasonable) temperature ratios is $c_1\dot{Q}$. With conventional practices, this is closely the value that would be assigned to the heat loss if energy, rather than exergy, were used as the basis for costing. It is particularly important to note that such energy costing would charge the same for a given value of the heat loss ($\dot{Q}$) regardless of the temperature at which it occurs. However, it is irrational to charge as much for a heat loss occurring, for instance, close to the ambient temperature, where the usefulness of the heat transfer is negligible, as for the same heat transfer occurring at a higher temperature, where the potential usefulness may be considerable. This example illustrates clearly that energy-based costing can lead to misevaluations. Reference 2 presents additional examples that show that exergy, rather than mass or energy, should serve as a basis for costing applications.

Finally, note that in the foregoing discussion we have implied that the cost per exergy unit of fuel (c_1) is approximately the same as the cost per energy unit for the same fuel to the boiler. This follows from the fact that the chemical exergy of most fuels is about equal to the fuel higher heating value (see Equation 3.17a).

Simple Cogeneration System. Consider next the simple cogeneration system shown in Figure 8.2 consisting of a boiler and a backpressure steam turbine. This system has two useful product streams: power and low-pressure steam. The object is to calculate the cost at which each of the two product streams is generated for each specified turbine exit pressure.

Several idealizations are made to simplify the calculations: The feedwater and combustion air are assumed to enter the boiler with negligible exergy and cost, the effluents exit the boiler with negligible cost, heat transfer can be ignored, and the turbine *polytropic efficiency* is constant at 80% regardless of the turbine exit pressure.[1] As shown in the figure, exergy enters the boiler at a rate of 100 MW with a unit cost $c_1 = \$4.0/GJ$. Additional state and cost data are given in the figure. Exergy is destroyed within the boiler at a rate of 60 MW. The capital and other expenses for the boiler $(\dot{Z}_b)$ are 0.3 dollars per second of system operation at the design capacity. The rate of exergy destruction, the generated power, and the capital and other costs for the turbine $(\dot{Z}_t)$ vary with the backpressure setting (p_4). Figure 8.2 shows the relation used to calculate $\dot{Z}_t$ as a function of the power $\dot{W}$ generated in the turbine. Table 8.1 summarizes the thermodynamic and cost data for the steam turbine of Figure 8.2 obtained at various turbine exit pressures.

The cost rate balances for the boiler and turbine are, respectively,

$$c_2 \dot{E}_2 + c_3 \dot{E}_3 = c_1 \dot{E}_1 + \dot{Z}_b \tag{8.8a}$$

$$c_4 \dot{E}_4 + c_w \dot{W} = c_2 \dot{E}_2 + \dot{Z}_t \tag{8.9a}$$

Solving Equation 8.8a for c_2, and using $c_3 = 0$ as noted previously, we obtain

$$c_2 = \frac{c_1 \dot{E}_1 + \dot{Z}_b}{\dot{E}_2} \tag{8.8b}$$

With data from Figure 8.2

---

[1] For the effluents exiting the boiler we assume, $c_3 = 0$. Note that this costing approach differs from that of Equation 8.6c. For discussion, see Section 8.1.4 (Equations 8.20 and 8.21a). For a discussion of the *polytropic efficiency*, see e.g., G. Rogers and Y. Mayhew, *Engineering Thermodynamics*, 4th ed., Longman, Harlow, UK, 1992, pp. 474, 478.

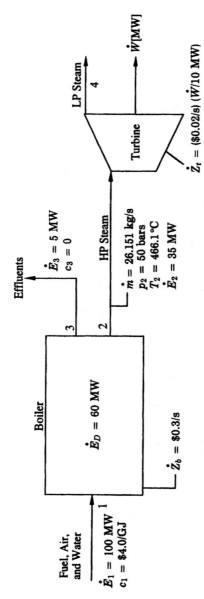

Figure 8.2 Simple cogeneration system consisting of a boiler and a backpressure turbine.

Table 8.1 Thermodynamic[a] and cost[b] data for the turbine of Figure 8.2[c]

p_4 (bars)	T_4 (°C)	$(\dot{m}h_4)$ (MW)	$(\dot{m}s_4)$ (kW/K)	$\dot{W}$ (MW)	$\dot{E}_D$ (MW)	$\dot{E}_4$ (MW)	$\dot{Z}_t$ ($/s)	c_w ($/GJ)	c_4^* (¢/kg)
50	466.1	−329.909	271.682	0	0	35.000	0	0	2.677
40	435.8	−331.389	272.206	1.480	0.156	33.364	0.0030	24.135	2.552
30	398.4	−333.211	272.888	3.302	0.360	31.338	0.0066	24.179	2.397
20	349.0	−335.632	273.845	5.723	0.645	28.632	0.0114	24.246	2.190
9	261.9	−339.912	275.756	10.003	1.215	23.782	0.0200	24.435	1.819
5	205.2	−342.694	277.160	12.785	1.633	20.582	0.0256	24.555	1.574
2	128.3	−346.434	279.433	16.525	2.311	16.164	0.0330	24.797	1.236
1	99.6	−348.906	281.134	18.997	2.818	13.185	0.0380	24.997	1.008

[a]Thermodynamic data obtained using $T_0 = 298.15$ K and

$$\dot{W} = \dot{m}(h_2 - h_4) \quad \text{(energy balance)}$$

$$\dot{E}_D = T_0 \dot{S}_{gen} = T_0 \dot{m}(s_4 - s_2) \quad \text{(Equations 2.24 and 3.8c)}$$

$$\dot{E}_4 = \dot{E}_2 - \dot{W} - \dot{E}_D \quad \text{(exergy balance)}$$

$(\dot{m}h_2) \equiv (\dot{m}h_4)$ at $p_4 = 50$ bars.

$(\dot{m}s_2) \equiv (\dot{m}s_4)$ at $p_4 = 50$ bars.

[b]Cost data obtained using

$$\dot{Z}_t = \$0.02/\text{s} \, [\dot{W}/10] \, (\dot{W} \text{ in MW}) \quad \text{(see Figure 8.2)}$$

$$c_w = \frac{c_2(\dot{E}_2 - \dot{E}_4) + \dot{Z}_t}{\dot{W}} \quad \text{(Equation 8.9d)}$$

$$c_4^* = c_4 \dot{E}_4/\dot{m} \quad \text{(Equation 8.9c)}$$

[c]The data shown in Figure 8.2 and in this table are for illustration purposes only and do not necessarily reflect typical values.

$$c_2 = \frac{(\$4.0/GJ)(100 \times 10^{-3}\ GJ/s) + (\$0.3/s)}{(35 \times 10^{-3}\ GJ/s)}$$

$$= \$20.0/GJ$$

This cost value can be expressed on a per unit of mass basis using

$$c_2^* = c_2 e_2 = c_2 \left(\frac{\dot{E}_2}{\dot{m}}\right) \tag{8.8c}$$

Substituting values

$$c_2^* = (\$20.0/GJ) \left(\frac{35 \times 10^{-3}\ GJ/s}{26.151\ kg/s}\right)$$

$$= \$26.77 \times 10^{-3}/kg = 2.677\ \text{¢}/kg$$

As Equation 8.9a involves *two* unknowns, c_4 and c_w, *one* auxiliary relation is required to determine each quantity. With the same reasoning as for Equation 8.5c, we set

$$c_4 = c_2 \tag{8.9b}$$

That is, on an exergy basis the unit cost of the steam exiting the turbine equals the unit cost of the steam entering the turbine. Accordingly, $c_4 = \$20.0/GJ$. This can be expressed on a per unit of mass basis using the specific exergy of the steam at state 4:

$$c_4^* = c_4 e_4 = c_4 \left(\frac{\dot{E}_4}{\dot{m}}\right) \tag{8.9c}$$

As a sample calculation, consider the backpressure $p_4 = 9$ bars. With $\dot{E}_4$ from Table 8.1

$$c_4^* = c_4 \left(\frac{\dot{E}_4}{\dot{m}}\right)$$

$$= (\$20.0/GJ) \left(\frac{23.782 \times 10^{-3}\ GJ/s}{26.151\ kg/s}\right)$$

$$= \$18.19 \times 10^{-3}/kg = 1.819\ \text{¢}/kg$$

which is the value listed in Table 8.1 corresponding to this backpressure setting.

Next, introducing Equation 8.9b into Equation 8.9a and solving for the unit cost of power, we obtain

$$c_w = \frac{c_2(\dot{E}_2 - \dot{E}_4) + \dot{Z}_t}{\dot{W}} \qquad (8.9d)$$

Using the data from Table 8.1 for $p_4 = 9$ bars,

$$c_w = \frac{(\$20.0/GJ)[(35.0 - 23.782) \times 10^{-3} \text{ GJ/s}] + (\$0.02/s)}{10 \times 10^{-3} \text{ GJ/s}}$$

$$= \$24.435/GJ$$

which is the value listed in Table 8.1 corresponding to this backpressure setting.

By repeating this procedure for backpressures varying from 50 to 1 bar, all the cost data reported in Table 8.1 can be calculated. Then, using the values for c_4^* reported in the table, Figure 8.3 giving the cost of the low-pressure steam per unit of mass as a function of the turbine exhaust pressure can be developed.

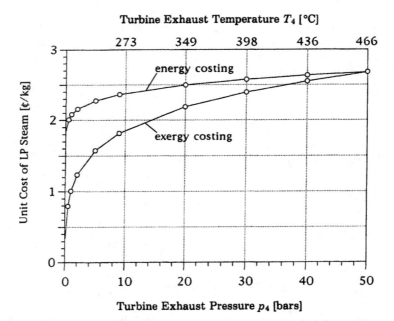

Figure 8.3 Cost of low-pressure steam per unit of mass, as a function of the turbine exhaust conditions for the system of Figure 8.2.

The exergy costing curve of Figure 8.3 shows that high-pressure (and high-temperature) steam is valued more highly per unit of mass than low-pressure (low-temperature) steam. The figure also shows that the cost per unit of mass rapidly approaches zero as the thermodynamic usefulness of steam (expressed by its exergy) goes to zero. This is in fact the behavior that would be expected from a rational costing scheme.

A second cost curve is shown in Figure 8.3 for comparison purposes. This curve is obtained by applying a similar procedure but based on energy[2] rather than exergy. Unlike exergy costing, *energy costing* does not make distinctions about the usefulness of energy transfers. The irrationality of energy costing is evident from the behavior of the energy costing curve at low turbine exhaust pressures. At a pressure of 1 bar, for example, the unit cost on an energy basis for steam is closely 2 ¢/kg even though such steam has very limited usefulness. This example illustrates once again the significant misevaluations that can occur when energy is used as the basis for costing. For additional discussion see Reference 4.

8.1.2 Aggregation Level for Applying Exergy Costing

The level at which the cost balances are formulated affects the results of a thermoeconomic analysis. Accordingly, in thermal design it is recommended that the lowest possible aggregation level be used. This level is usually represented by the individual components (heat exchangers, turbines, reactors, etc.). Even in cases where the available information is insufficient for applying exergy costing at the component level, it is generally preferable to make appropriate assumptions that enable exergy costing to be applied at the component level than to consider only groups of components. Depending on the component, it may even be appropriate to distinguish among the various processes (chemical reactions, heat transfer, pressure drop, etc.) taking place within the component. For this, the costing approaches discussed in Section 8.4 may be useful.

To illustrate the effect of the aggregation level, refer again to the simple cogeneration system of Figure 8.2, recalling that in the previous discussion of this case cost balances were formulated individually for the boiler and steam turbine (Equations 8.8a and 8.9a). Now let us consider an overall system, as shown in Figure 8.4, for the case where the back pressure is $p_4 = 9$ bars.

The cost rate balance for this sytem is

---

[2]When energy costing is applied, one might write for entering and exiting material streams, power, and heat transfer, respectively. $\dot{C}_i = k_i \dot{m}_i(h_i - h_{i,0})$; $\dot{C}_e = k_e \dot{m}_e(h_e - h_{e,0})$; $\dot{C}_w = k_w \dot{W}$; $\dot{C}_q = k_q \dot{Q}$. Here, k_i, k_e, k_w, and k_q denote average costs per unit of energy; $h_{i,0}$ and $h_{e,0}$ are the specific enthalpy at ambient conditions (T_0 and p_0) of the entering and exiting streams, respectively.

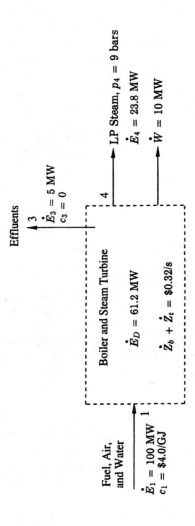

Figure 8.4 System employed to discuss aggregation level for exergy costing. (Compare with Figure 8.2.)

$$\dot{C}_1 + (\dot{Z}_b + \dot{Z}_t) = \dot{C}_4 + \dot{C}_w + \cancel{\dot{C}_3} \qquad (8.10a)$$

Using exergy costing, we have

$$c_1 \dot{E}_1 + (\dot{Z}_b + \dot{Z}_t) = c_4 \dot{E}_4 + c_w \dot{W} \qquad (8.10b)$$

With $\dot{E}_1$, $\dot{E}_4$, $\dot{W}$, $\dot{Z}_b$, $\dot{Z}_t$, and c_1 known from Figure 8.2 or Table 8.1, this equation involves *two* unknowns: the costs per exergy unit associated with power (c_w) and low-pressure steam (c_4). Therefore, *one* auxiliary relation is required before these unknowns can be determined. However, without explicitly considering the nature of the actual components making up the overall system, as we did in the previous discussion, the rationale for introducing an auxiliary relation is much less clear-cut. With only the information in Figure 8.4 as a guide, it might be decided to share the fuel, capital, and other costs between the generated power and the steam by assuming that each unit of exergy for the two product streams has the same cost, whether it is power or steam. That is,

$$c_4 = c_w \qquad (8.10c)$$

With this auxiliary relation, Equation 8.10b gives

$$c_4 = c_w = \frac{c_1 \dot{E}_1 + (\dot{Z}_b + \dot{Z}_t)}{\dot{E}_4 + \dot{W}} \qquad (8.10d)$$

Substituting values

$$c_4 = c_w = \frac{(\$4.0/\text{GJ})(100 \times 10^{-3} \text{ GJ/s}) + (\$0.3/\text{s} + \$0.02/\text{s})}{33.8 \times 10^{-3} \text{ GJ/s}}$$

$$= \$21.3/\text{GJ}$$

In the previous discussion of this case, we calculated $c_4 = \$20.0/\text{GJ}$ and $c_w = \$24.4/\text{GJ}$ when cost balances were formulated separately for the boiler and the steam turbine. With the present approach, in which only the overall system is considered, the unit cost of the steam is higher because it now bears part of the burden for spent fuel, capital, and other costs associated with the steam turbine. On the other hand, the unit cost of the generated power is lower for the aggregated system because power is no longer required to bear the full burden of these costs.

By considering only the aggregated system, we do not take into account important information related to the *actual production processes* and, thus, to the actual cost formation process within the system. Equations 8.10 could represent any system that generates steam and electricity. For example, they could apply to a system where the low-pressure steam is obtained by throttling high-pressure boiler steam and the steam turbine is a condensing turbine. From this discussion it is apparent that the quality of the results and recommendations from a thermoeconomic analysis depends on the aggregation level at which exergy costing is applied.

8.1.3 Cost Rates, Auxiliary Relations, and Average Costs Associated with Fuel and Product

As a part of the discussion of exergetic efficiency in Section 3.5.3, we introduced the concepts of *product* and *fuel*. Recall that the product is defined according to the purpose of owning and operating the component under consideration. The fuel represents the resources expended in generating the product. Both the product and the fuel are expressed in terms of exergy. Table 3.3 provides the definitions of fuel ($\dot{E}_F$) and product ($\dot{E}_P$) for selected components.

The *cost rates associated with the fuel* ($\dot{C}_F$) and *product* ($\dot{C}_P$) of a component are obtained simply by replacing the exergy rates ($\dot{E}$) given in Table 3.3 by cost rates ($\dot{C}$). Table 8.2 summarizes the definitions of cost rates associated with the fuel and product of the components contained in Table 3.3. A comparison of Tables 3.3 and 8.2 demonstrates that the cost rate associated with the fuel (or product) of a component contains the cost rates of the same streams used in the same order and with the same sign as in the definition of the exergy of fuel (or product).

Let us consider Table 8.2 in detail, beginning with the components involving only one exit stream: compressor, pump, fan, mixing unit, gasifier, and combustion chamber. As noted in the table, for these components the cost rate balance may be solved for the cost per exergy unit of the exiting stream without the need for auxiliary relations. As a rule, $n - 1$ auxiliary relations are required for components with n *exiting* exergy streams. This includes the above cases having one exit stream for which no auxiliary relation is required.

Continuing the discussion of Table 8.2, consider the case of a turbine with one extraction. The product and fuel for this case are identified in Table 3.3 as $\dot{W}$ and ($\dot{E}_1 - \dot{E}_2 - \dot{E}_3$), respectively. The associated cost rates shown in Table 8.2 are then $\dot{C}_P = \dot{C}_w$ and $\dot{C}_F = (\dot{C}_1 - \dot{C}_2 - \dot{C}_3)$, respectively. A cost rate balance reads

$$\dot{C}_w + \dot{C}_2 + \dot{C}_3 = \dot{C}_1 + \dot{Z} \tag{8.11a}$$

or in terms of the fuel and product cost rates

$$c_w \dot{W} = (c_1 \dot{E}_1 - c_2 \dot{E}_2 - c_3 \dot{E}_3) + \dot{Z} \tag{8.11b}$$

With three exiting exergy streams, two auxiliary relations are required to determine the unit cost of power c_w (the variables c_1 and $\dot{Z}$ are known, but

Table 8.2 Cost rates associated with fuel and product as well as auxiliary thermoeconomic relations for selected components at steady-state operation

Component	Compressor, Pump, or Fan	Turbine or Expander	Heat Exchanger[a]	Mixing Unit	Gasifier or Combustion Chamber	Boiler
Schematic	(compressor diagram: $\dot{w}$ in at 3, 1 in, 2 out)	(turbine diagram: 1 in, $\dot{w}$ out at 4, 2 and 3 out)	(heat exchanger: hot stream 3 in, 4 out; cold stream 1 in, 2 out)	(mixing unit: cold 1, hot 2 in, 3 out)	(combustion chamber: oxidant 2, fuel 1 in, reaction products 3 out)	(boiler: flue gas 4, feedwater 5, main steam 6, cold reheat 7, hot reheat 8, coal 1, air 2, ash 3)
Cost rate of product ($\dot{C}_P$)	$\dot{C}_2 - \dot{C}_1$	$\dot{C}_w$	$\dot{C}_2 - \dot{C}_1$	$\dot{C}_3$	$\dot{C}_3$	$(\dot{C}_6 - \dot{C}_5) + (\dot{C}_8 - \dot{C}_7)$
Cost rate of fuel ($\dot{C}_F$)	$\dot{C}_w$	$\dot{C}_1 - \dot{C}_2 - \dot{C}_3$	$\dot{C}_3 - \dot{C}_4$	$\dot{C}_1 + \dot{C}_2$	$\dot{C}_1 + \dot{C}_2$	$(\dot{C}_1 + \dot{C}_2) - (\dot{C}_3 + \dot{C}_4)$
Auxiliary thermoeconomic relations	None	$c_2 = c_3 = c_1$	$c_4 = c_3$	None	None	$\dfrac{\dot{C}_6 - \dot{C}_5}{\dot{E}_6 - \dot{E}_5} = \dfrac{\dot{C}_8 - \dot{C}_7}{\dot{E}_8 - \dot{E}_7}$ For c_3 and c_4 see Section 8.1.4 and Equations 8.14
Variable calculated from cost balance	c_2	c_w	c_2	c_3	c_3	c_6 or c_8

[a] These definitions assume that the purpose of the heat exchanger is to heat the cold stream ($T_1 \geq T_0$). If the purpose of the heat exchanger is to provide cooling ($T_3 \leq T_0$), then exergy is removed from the cold stream, and the following relations should be used: $\dot{C}_P = \dot{C}_4 - \dot{C}_3$; $\dot{C}_F = \dot{C}_1 - \dot{C}_2$; and $c_2 = c_1$. The variable c_4 is calculated from the cost balance.

the variables c_2 and c_3 are unknown). Using the rationale presented in Section 8.1.1, we formulate the following two auxiliary relations:

$$c_2 = c_1 \quad \text{and} \quad c_3 = c_1$$

That is, the generated power is expected to bear the full burden of the costs associated with owning and operating the turbine. With the two auxiliary relations the cost rate balance gives

$$c_w = \frac{c_1\dot{E}_1 - c_2\dot{E}_2 - c_3\dot{E}_3 + \dot{Z}}{\dot{W}}$$

$$= \frac{c_1(\dot{E}_1 - \dot{E}_2 - \dot{E}_3) + \dot{Z}}{\dot{W}}$$

Using an exergy balance, the fuel term in the above expression can be rewritten to give

$$c_w = \frac{c_1(\dot{W} + \dot{E}_D) + \dot{Z}}{\dot{W}}$$

or

$$c_w = c_1 + \frac{(c_1\dot{E}_D + \dot{Z})}{\dot{W}} \tag{8.11c}$$

where $\dot{E}_D$ is the rate of exergy destruction within the turbine. Accordingly, we see that the increase in unit cost between the fuel (c_1) and product (c_w) is caused, in addition to capital investment and O&M costs, by exergy destruction (and exergy loss, when present).

Turning next to the heat exchanger case of Table 8.2, the product and fuel identified in Table 3.3 are $\dot{E}_2 - \dot{E}_1$ and $\dot{E}_3 - \dot{E}_4$, respectively. The associated cost rates listed in Table 8.2 are then $\dot{C}_P = (\dot{C}_2 - \dot{C}_1)$ and $\dot{C}_F = (\dot{C}_3 - \dot{C}_4)$, respectively. A cost rate balance then reads

$$\dot{C}_2 + \dot{C}_4 = \dot{C}_1 + \dot{C}_3 + \dot{Z} \tag{8.12a}$$

or in terms of the fuel and product cost rates

$$(c_2\dot{E}_2 - c_1\dot{E}_1) = (c_3\dot{E}_3 - c_4\dot{E}_4) + \dot{Z} \tag{8.12b}$$

Here, c_1, c_3, $\dot{E}_1$, $\dot{E}_2$, $\dot{E}_3$, $\dot{E}_4$, and $\dot{Z}$ are known, but c_2 and c_4 are unknown. With two exiting streams and one cost-rate balance available, one auxiliary relation is required to determine the unit costs of these streams. Since the purpose of such a heat exchanger is to heat the cold stream and exergy is *removed* from the hot stream, the average cost per exergy unit for the hot stream remains constant and is equal to the average cost at which each exergy unit of stream 3 was supplied in upstream components. Thus, we write

$$c_4 = c_3$$

With this auxiliary relation, stream 2 bears the full burden of the costs associated with owning and operating the heat exchanger. The cost rate balance, Equation 8.12b, becomes

$$c_2 = \frac{c_1 \dot{E}_1 + c_3(\dot{E}_3 - \dot{E}_4) + \dot{Z}}{\dot{E}_2} \tag{8.12c}$$

On the other hand, if the purpose of the heat exchanger of Table 8.2 is to provide cooling, then $c_2 = c_1$, and Equation 8.12b is solved for c_4. Stream 4 is then burdened with all costs associated with the heat exchanger.

Finally, we consider the boiler of Table 8.2. In Table 3.3 the product and fuel for this case are identified as $(\dot{E}_6 - \dot{E}_5) + (\dot{E}_8 - \dot{E}_7)$ and $(\dot{E}_1 + \dot{E}_2) - (\dot{E}_3 + \dot{E}_4)$, respectively. The associated cost rates given in Table 8.2 are then $\dot{C}_P = (\dot{C}_6 - \dot{C}_5) + (\dot{C}_8 - \dot{C}_7)$ and $\dot{C}_F = (\dot{C}_1 + \dot{C}_2) - (\dot{C}_3 + \dot{C}_4)$, respectively. A cost rate balance reads

$$\dot{C}_3 + \dot{C}_4 + \dot{C}_6 + \dot{C}_8 = \dot{C}_1 + \dot{C}_2 + \dot{C}_5 + \dot{C}_7 + \dot{Z} \tag{8.13a}$$

or in terms of the fuel and product cost rates

$$(c_6 \dot{E}_6 - c_5 \dot{E}_5) + (c_8 \dot{E}_8 - c_7 \dot{E}_7) = (c_1 \dot{E}_1 + c_2 \dot{E}_2) - (c_3 \dot{E}_3 + c_4 \dot{E}_4) + \dot{Z} \tag{8.13b}$$

Here, $\dot{E}_1$ through $\dot{E}_8$, c_1, c_2, c_5, c_7, and $\dot{Z}$ are known, but c_3, c_4, c_6, and c_8 are unknown. Thus, *three* auxiliary relations are required to calculate the *four* unknowns. As discussed next, the relations are provided by Equation 8.14a (or Equation 8.14c), Equation 8.14b (or Equation 8.14d), and Equation 8.15.

Streams 3 (ash) and 4 (flue gas) are discharged to the natural environment either directly (without additional expenses) or after using ash handling equipment (with an associated cost rate of $\dot{Z}_{ah}$) and flue-gas desulfurization equipment (with an associated cost rate of $\dot{Z}_{ds}$). In the former case, we may set

$$c_3 = 0 \tag{8.14a}$$

and

$$c_4 = 0 \tag{8.14b}$$

whereas in the latter case the additional equipment items must be included in the thermoeconomic analysis, and it is appropriate to write

$$c_3 = -\dot{Z}_{ah}/\dot{E}_3 \tag{8.14c}$$

and

$$c_4 = -\dot{Z}_{ds}/\dot{E}_4 \tag{8.14d}$$

Equations 8.14c and 8.14d result in *negative cost rates* for streams 3 and 4 ($\dot{C}_3 < 0$ and $\dot{C}_4 < 0$). Through these negative cost rates, all costs associated with the final disposal of streams 3 and 4 are charged to the useful streams exiting the boiler—that is, to the main steam and the reheated steam.

Since exergy is supplied *simultaneously* to two different product streams of the boiler, the following third auxiliary relation is appropriate:

$$\frac{\dot{C}_6 - \dot{C}_5}{\dot{E}_6 - \dot{E}_5} = \frac{\dot{C}_8 - \dot{C}_7}{\dot{E}_8 - \dot{E}_7} \tag{8.15}$$

The numerators in this equation indicate, respectively, the cost increases from boiler feedwater (inlet) to superheated main steam (outlet) and from cold reheat to hot reheat. The denominators represent the corresponding exergy increases. Equation 8.15 simply states that a unit of exergy is supplied to each product stream in the boiler at the same average cost.

The discussion of the selected components presented in Table 8.2 illustrates the following general principles applied in the formulation of auxiliary relations:

1. When the *product* definition for a component involves a *single* exergy stream, as for example, in a turbine, the unit cost of this exiting stream is calculated from the cost balance. The auxiliary relations are formulated for the remaining *exiting* exergy streams that are used in the definition of fuel or (rarely) in the definition of exergy loss associated with the component being considered.

2. When the *product* definition for a component involves m exiting exergy streams, $m - 1$ auxiliary relations referring to these product streams must be formulated. In a boiler with one reheating (Table 8.2), for example, we have $m = 2$ and, thus, one auxiliary relation involving streams 6 and/or 8 is required (Equation 8.15). In the absence of information about the production process of each of the m streams, it may be assumed that each unit of exergy is supplied to each product stream at the same average cost.

3. When the *fuel* definition for a component involves the difference between the entering and exiting states of the same stream of matter, as in the cases of a turbine and a heat exchanger in Table 8.2, the average

cost per exergy unit remains constant for this stream. This cost changes only when exergy is *supplied* to the stream, which then becomes part of the product definition.

After introducing the cost rates associated with the fuel ($\dot{C}_F$) and product ($\dot{C}_P$) as discussed above and illustrated in Table 8.2, we can define the *average costs per exergy unit of fuel and product* for a component. The average unit cost of the fuel ($c_{F,k}$) for the kth component is defined by

$$c_{F,k} = \frac{\dot{C}_{F,k}}{\dot{E}_{F,k}} \tag{8.16}$$

and expresses the average cost at which each exergy unit of fuel is supplied to the kth component [5–7]. Similarly, the average unit cost of the product ($c_{P,k}$) for the kth component is the average cost at which each exergy unit of the product of the kth component is generated:

$$c_{P,k} = \frac{\dot{C}_{P,k}}{\dot{E}_{P,k}} \tag{8.17}$$

As discussed in Sections 8.1.4, 8.2, and 8.3, the average costs per exergy unit of fuel and product are used in the thermoeconomic evaluation of a component.

8.1.4 Costing of Exergy Loss Streams

Finally, we introduce the *cost rate associated with exergy loss* ($\dot{C}_L$) that represents the monetary loss associated with the rejection of exergy (exergy loss) from a system to its surroundings. The exergy loss ($\dot{E}_{L,k}$) of the kth component of the system might consist of exergy loss associated with heat transfer to the surroundings, streams of matter rejected to the surroundings and not further used within the overall system being analyzed or in another system. Using the cost rates associated with fuel, product, and exergy loss for the kth component, the cost rate balance becomes

$$\dot{C}_{P,k} = \dot{C}_{F,k} - \dot{C}_{L,k} + \dot{Z}_k \tag{8.18}$$

With Equations 8.16 and 8.17, Equation 8.18 can be written as follows:

$$c_{P,k}\dot{E}_{P,k} = c_{F,k}\dot{E}_{F,k} - \dot{C}_{L,k} + \dot{Z}_k \tag{8.19}$$

From this equation, it is apparent that the cost rate of the exergy loss stream $(\dot{C}_{L,k})$ affects the cost rate associated with the product $(\dot{C}_{P,k})$ of the component.

The simplest approach to costing of an exergy loss associated with the kth component is to set

$$\dot{C}_{L,k} = 0 \qquad (8.20)$$

When $\dot{C}_{L,k}$ is zero in Equation 8.19, the product as expressed by $c_{P,k}$ bears the full burden of the costs associated with owning and operating the kth component. In many cases, Equation 8.20 may be applied to cost balances formulated for the overall system when the purpose of the thermoeconomic analysis is to calculate the costs of the final products or to evaluate or optimize the overall system.

Equation 8.20 should be applied only to the streams *finally discharged* to the natural environment. If additional expenses are necessary for cleanup before a stream of matter representing an exergy loss is discharged to the natural environment, these expenses and the associated equipment must be included in the system being considered. This case is discussed in Section 8.1.3 using the boiler of Table 8.2 as an example (see Equations 8.14).

When the purpose of the thermoeconomic analysis is to understand the cost formation process and the cost flow in the system, to evaluate the performance of a single component, or to optimize specific design variables in a single component, *all* exergy loss streams should be costed as if they were to be further used by the system. In so doing, we calculate the *monetary loss* $(\dot{C}_{L,k})$ associated with the exergy loss $(\dot{E}_{L,k})$ in the kth component

$$\dot{C}_{L,k} = c_{F,k}\dot{E}_{L,k} \qquad (8.21a)$$

This equation assumes that the exergy loss is covered through the supply of additional fuel $(\dot{E}_{F,k})$ to the kth component and that the average cost $(c_{F,k})$ of supplying the fuel exergy unit remains constant with varying exergy loss in the kth component. This approach is used in Section 8.1.1 to evaluate the cost penalty associated with a heat loss (see Equation 8.7).

Alternatively, we might assume that the exergy loss results in a reduction of the product $\dot{E}_{P,k}$ and that the average cost of generating the product remains practically constant with varying exergy loss in the kth component. We may then calculate the monetary loss associated with the exergy loss from

$$\dot{C}_{L,k} = c_{P,k}\dot{E}_{L,k} \qquad (8.21b)$$

As this approach overestimates the cost penalty associated with exergy loss, it is not recommended.

When Equation 8.21a or 8.21b is used, all cost penalties associated with exergy losses must be charged to the final products of the system and recovered through the sale of the products. To illustrate this, refer to the cogeneration system of Figure 1.7. If a positive monetary rate would be calculated for stream 7 from Equation 8.21a, this rate must be distributed between net power and steam generated by the total system, when calculating the final total cost associated with each product (see Example 8.1).

In general, very few components have exergy losses that, for costing purposes, need to be distinguished from the exergy destruction (see Section 3.5.2). The concept of exergy loss is normally applicable to the overall system rather than to a single component having an exiting stream that is not further used in the overall system. This component should not be penalized for such loss, particularly if the exiting stream has been used in more than one component, or if it leaves the overall system with the lowest allowable temperature, pressure, and chemical exergy values. In such a case, Equation 8.21a could be used in cost optimization studies involving the entire system. For example, the heat-recovery steam generator (HRSG) of the cogeneration system of Figure 1.7 should not be penalized for stream 7 leaving the overall system, particularly since the exergy of the combustion gases is also used in the gas turbine and the air preheater. (The HRSG would not become more efficient if the designer would use part of the exergy of stream 7 elsewhere.) The exergy rate $\dot{E}_7$ represents a loss for the overall system but not for the HRSG.

Example 8.1   Using data from Tables 3.1, 7.6, and 7.15 for the cogeneration system case study, calculate the levelized cost rates and the levelized costs per unit of exergy associated with all streams in the system.

Solution

MODEL

1. All data are for operation at steady state. The exergy flow rates from Table 3.1 apply.
2. Calculations are based on the levelized annual costs in current dollars for a levelization time period of 10 years (Table 7.15).
3. The annual carrying charges and operating and maintenance costs are apportioned among the system components according to the contribution of each component to the sum of purchased-equipment costs (Table 7.6).
4. A zero unit cost is assumed for air entering the air compressor ($c_1 = 0$) and water entering the HRSG ($c_8 = 0$).

ANALYSIS   The following cost balances and auxiliary relations are formulated for each component of the cogeneration system:

Air Compressor (ac)

$$\dot{C}_1 + \dot{C}_{11} + \dot{Z}_{ac} = \dot{C}_2 \qquad (1)$$

$$\dot{C}_1 = 0 \text{ (assumption 4)} \qquad (2)$$

where subscript 11 denotes the power input to the compressor. The associated exergy flow rate is $\dot{E}_{11} = \dot{W}_{ac} = 29.662$ MW. The value of $\dot{C}_{11}$ is calculated from the gas turbine analysis (see Equation (9)).

Air Preheater (ph)

$$\dot{C}_2 + \dot{C}_5 + \dot{Z}_{ph} = \dot{C}_3 + \dot{C}_6 \qquad (3)$$

The auxiliary relation for a heat exchanger, the purpose of which is to heat the cold stream, is that the cost per exergy unit on the hot side (fuel side) remains constant ($c_6 = c_5$). Thus,

$$\frac{\dot{C}_6}{\dot{E}_6} = \frac{\dot{C}_5}{\dot{E}_5} \qquad (4)$$

Combustion Chamber (cc)

$$\dot{C}_3 + \dot{C}_{10} + \dot{Z}_{cc} = \dot{C}_4 \qquad (5)$$

The levelized cost rate of stream 10, which is supplied to the total system from outside, is obtained by dividing the annual levelized fuel costs from Table 7.15 (10.411×10^6) by the average total number of hours of system operation per year (7446 h per year). Thus,

$$\dot{C}_{10} = \$1398/h \qquad (6)$$

Gas Turbine (gt)

$$\dot{C}_4 + \dot{Z}_{gt} = \dot{C}_5 + \dot{C}_{11} + \dot{C}_{12} \qquad (7)$$

where subscript 12 denotes the net power generated by the turbine: $\dot{W}_{net} = \dot{W}_{12} = 30$ MW. The auxiliary relation for a gas turbine is that on the fuel side ($\dot{E}_4 - \dot{E}_5$) the cost per exergy unit remains constant ($c_5 = c_4$). Thus,

$$\frac{\dot{C}_5}{\dot{E}_5} = \frac{\dot{C}_4}{\dot{E}_4} \qquad (8)$$

Ignoring the losses during the transmission of power from the gas turbine to the air compressor, the cost per exergy unit of this power ($\dot{W}_{ac}$) is equal to the cost per exergy unit of the net power exported from the system ($c_{11} = c_{12}$). Thus,

$$\frac{\dot{C}_{11}}{\dot{W}_{11}} = \frac{\dot{C}_{12}}{\dot{W}_{12}} \tag{9}$$

Heat-Recovery Steam Generator (hrsg)

$$\dot{C}_6 + \dot{C}_8 + \dot{Z}_{\text{hrsg}} = \dot{C}_7 + \dot{C}_9 \tag{10}$$

Here, the cost per exergy unit on the fuel side of the HRSG remains constant ($c_7 = c_6$). Thus,

$$\frac{\dot{C}_7}{\dot{E}_7} = \frac{\dot{C}_6}{\dot{E}_6} \tag{11}$$

The above relation allows us to calculate the monetary loss associated with the exergy loss of stream 7. This cost is apportioned between the two product streams 9 and 12 later in the discussion. For the water stream entering the HRSG we have from assumption 4 of the model

$$\dot{C}_8 = 0 \tag{12}$$

Before the system of linear Equations (1)–(12) is solved for the unknown variables $\dot{C}_1$ through $\dot{C}_{12}$, it is necessary to calculate the values of the $\dot{Z}$ terms appearing in the cost balances. According to assumption 3 of the model, the term $\dot{Z}_k$ for the kth component is calculated with the aid of the annual levelized carrying charges ($\$10.527 \times 10^6$) and the annual levelized operating and maintenance costs ($\$5.989 \times 10^6$) for the total system from

$$\dot{Z}_k = \frac{(10.527 + 5.989) \times 10^6}{(11.0 \times 10^6)(7446)} \, \text{PEC}_k \tag{13}$$

PEC_k is the purchased-equipment cost of the kth component expressed in mid-1994 dollars (Table 7.6). The purchased-equipment cost for the overall system is 11.0×10^6 mid-1994 dollars and is used in the denominator of Equation (13). The number 7446, also in the denominator of Equation (13), represents the total annual number of hours of system operation at full load. Using Equation (13), we obtain the values of 753, 188, 68, 753, 264, and 190 (all in levelized dollars per hour) for the air compressor, air preheater, combustion chamber, gas turbine, heat-recovery steam generator, and other plant equipment, respectively. The $\dot{Z}$ value for other system equipment is charged directly to product streams 9 and 12 later in the thermoeconomic analysis.

Solving the linear system consisting of Equations (1)–(12), we obtain the values of the cost flow rates and the unit costs associated with each stream given in the following table. Note that the highest exergy unit cost is achieved at stream 2 exiting the air compressor where all exergy available at the exit is supplied by mechanical power, which is the most expensive fuel in the system. Also note that the cost per exergy unit is considerably higher for steam (stream 9) than for the net power (stream 12).

Stream No.	$\dot{E}$ (MW)	$\dot{C}$ ($/h)	c ($/GJ)
1	0	0	0.00
2	27.538	2756	27.80
3	41.938	3835	25.40
4	101.454	5301	14.51
5	38.782	2026	14.51
6	21.752	1137	14.51
7	2.773	145	14.51
8	0.062	0	0.00
9	12.810	1256	27.23
10	84.994	1398	4.57
11	29.662	2003	18.76
12	30.000	2026	18.76

$\dot{E}$, exergy rates; $\dot{C}$, levelized cost rates; and c, levelized costs per exergy unit.

The cost associated with other plant equipment must be apportioned to the two product streams on the basis of an estimate of the equipment contribution to the generation of each product stream. Here, for simplicity, we divide the cost rate associated with other plant equipment ($190/h) equally between steam and net power and obtain the *adjusted* cost rates

$$\dot{C}'_9 = \dot{C}_9 + \dot{Z}_{other}/2 = 1256 + 95 = \$1351/h$$

and

$$\dot{C}'_{12} = \dot{C}_{12} + \dot{Z}_{other}/2 = 2026 + 95 = \$2121/h$$

These values satisfy the cost rate balance for the total system: the sum of $\dot{C}'_9$, $\dot{C}'_{12}$, and $\dot{C}_7$ must be equal to the sum of carrying charges, fuel costs, and O&M expenses calculated on a per hour basis.

The monetary loss associated with the exergy loss of stream 7 is $145/h. This cost rate also must be charged to steam and net power. In the absence of other criteria, the exergy values $\dot{E}_9 - \dot{E}_8$ and $\dot{W}_{12}$ may be used as weighting factors for apportioning the cost rate $\dot{C}_7$ between steam and electricity, respectively. Application of these weighting factors results in additional costs of $43/h and $102/h for streams 9 and 12, respectively. After this final adjustment we obtain the *final* cost rates

$$\dot{C}_9'' = \dot{C}_9' + \$43/h = \$1394/h$$

and

$$\dot{C}_{12}'' = \dot{C}_{12}' + \$102/h = \$2223/h$$

COMMENTS

1. Other assumptions might be used to apportion the cost of other plant equipment and the cost of stream 7 between the product streams (steam and net power). Also, assumption 3 of the model can be revised if additional information is available.
2. If Equation (11) were replaced by $\dot{C}_7 = 0$ in solving the system of linear equations, all costs associated with the rejected stream 7 ($145/h) would be charged to the steam (stream 9), but as discussed at the end of Section 8.1.4 this approach is not recommended.
3. As the uncertainty associated with cost data is larger than with thermodynamic data, economic calculations generally do not have the same accuracy as thermodynamic calculations. This may result in small discrepancies in cost balances owing to round-off.

8.1.5 Non-Exergy-Related Costs for Streams of Matter

The costing approach exemplified by Equations 8.3a and 8.3b assumes that costs associated with material streams are related only to the exergy rate of each respective stream. However, non-exergy-related costs can affect the total cost rate associated with material streams. Examples include the cost rates associated with a treated water stream at the outlet of a water treatment unit, an oxygen or nitrogen stream at the outlet of an air separation unit, a limestone stream supplied to a gasifier or fluidized-bed reactor, iron feedstock supplied to a metallurgical process, and an inorganic chemical fed into a chemical reactor.

Accordingly, when significant non-exergy-related costs occur, the total cost rate associated with material stream j, denoted by $\dot{C}_j^{TOT}$, is given by

$$\dot{C}_j^{TOT} = \dot{C}_j + \dot{C}_j^{NE} \tag{8.22}$$

where $\dot{C}_j$ is the cost rate directly related to the exergy of stream j (e.g., Equations 8.3a and 8.3b) whereas $\dot{C}_j^{NE}$ is the cost rate due to nonexergetic effects.

The term $\dot{C}_j^{NE}$ represents a convenient way for charging non-exergy-related costs from one component to another component that should bear such costs. In such a case the value of $\dot{C}_j^{NE}$ remains unchanged as the jth material stream crosses intermediate components. For example, consider an air separation unit supplying an oxygen stream that is preheated before it is used in a gasifier. The cost of generating the oxygen stream in the air separation unit is charged directly through a $\dot{C}^{NE}$ cost rate to the gasifier and does not affect the cost per exergy unit and the exergy-related cost rate of the oxygen stream as this is preheated in a heat exchanger. More details about $\dot{C}^{NE}$ are given in Reference 7.

8.1.6 Closure

When conducting a thermoeconomic analysis of a given system, a cost balance is formulated for each component. This balance, combined with appropriate auxiliary thermoeconomic relations, results in a system of linear algebraic equations, which is solved for the unknown values of cost rates or of cost per exergy unit. As illustrated by Example 8.1, the solution of the system of linear algebraic equations provides the cost per exergy unit and the cost rates associated with each exergy stream. This information is sufficient for (a) calculating separately the cost at which each product stream (e.g., steam and electricity in a cogeneration system) is generated by the system and (b) understanding the cost formation process and the flow of costs in the system. Accordingly, each exergy stream is characterized by a cost per exergy unit and a cost rate, in addition to the exergy rate of the stream. In a thermoeconomic evaluation and optimization, the information on stream costs is used as information to calculate *thermoeconomic variables associated with each system component.* The next two sections discuss such thermoeconomic variables.

8.2 THERMOECONOMIC VARIABLES FOR COMPONENT EVALUATION

The following quantities, known as thermoeconomic variables, play a central role in the thermoeconomic evaluation and optimization of thermal systems: (a) the average unit cost of fuel, $c_{F,k}$ (Equation 8.16); (b) the average unit cost of product, $c_{P,k}$ (Equation 8.17); (c) the cost rate of exergy destruction, $\dot{C}_{D,k}$; (d) the relative cost difference, r_k; and (e) the exergoeconomic factor, f_k. In this section, three of these variables are discussed: $\dot{C}_{D,k}$, r_k, and f_k. In Section 8.3, all five of the thermoeconomic variables listed are applied to the design evaluation of a new system and the performance evaluation of an existing system.

8.2.1 Cost of Exergy Destruction

In the cost balance formulated for a component (e.g., Equations 8.4 and 8.19), there is no cost term directly associated with exergy destruction. Accordingly, the cost associated with the exergy destruction in a component or process is a *hidden cost*, but a very important one, that can be revealed only through a thermoeconomic analysis.

The effect of exergy destruction can be demonstrated by combining Equations 3.28 and 8.19:

$$\dot{E}_{F,k} = \dot{E}_{P,k} + \dot{E}_{L,k} + \dot{E}_{D,k} \tag{3.28}$$

$$c_{P,k}\dot{E}_{P,k} = c_{F,k}\dot{E}_{F,k} - \dot{C}_{l,k} + \dot{Z}_k \tag{8.19}$$

to eliminate $\dot{E}_{F,k}$ and obtain

$$c_{P,k}\dot{E}_{P,k} = c_{F,k}\dot{E}_{P,k} + (c_{F,k}\dot{E}_{L,k} - \dot{C}_{L,k}) + \dot{Z}_k + c_{F,k}\dot{E}_{D,k} \tag{8.23a}$$

or to eliminate $\dot{E}_{P,k}$ and obtain

$$c_{P,k}\dot{E}_{F,k} = c_{F,k}\dot{E}_{F,k} + (c_{P,k}\dot{E}_{L,k} - \dot{C}_{L,k}) + \dot{Z}_k + c_{P,k}\dot{E}_{D,k} \tag{8.23b}$$

In each of Equations 8.23, the last term on the right involves the rate of exergy destruction. As discussed next, these terms provide measures of the cost of exergy destruction, at least approximately.

Assuming that the product ($\dot{E}_{P,k}$) is fixed and that the unit cost of fuel ($c_{F,k}$) of the kth component is independent of the exergy destruction, we can define the cost of exergy destruction by the last term of Equation 8.23a as

$$\dot{C}_{D,k} = c_{F,k}\dot{E}_{D,k} \qquad (\dot{E}_{P,k} \text{ fixed}) \tag{8.24a}$$

As the fuel rate ($\dot{E}_{F,k}$) must account for the fixed product rate ($\dot{E}_{P,k}$) *and* the rate of exergy destruction ($\dot{E}_{D,k}$), we may interpret $\dot{C}_{D,k}$ in Equation 8.24a as the cost rate of the additional fuel that must be supplied to the kth component (above the rate needed for the product) to cover the rate of exergy destruction.

Alternatively, assuming that the fuel ($\dot{E}_{F,k}$) is fixed and that the unit cost of product ($c_{P,k}$) of the kth component is independent of the exergy destruction, we can define the cost of exergy destruction by the last term of Equation 8.23b as

$$\dot{C}_{D,k} = c_{P,k}\dot{E}_{D,k} \qquad (\dot{E}_{F,k} \text{ fixed}) \tag{8.24b}$$

When the fuel ($\dot{E}_{F,k}$) is fixed, the exergy destruction ($\dot{E}_{D,k}$) reduces the product of the kth component ($\dot{E}_{P,k}$), and so Equation 8.24b can be interpreted as the monetary loss associated with the loss of product.

In practice neither of the sets of assumptions used to define $\dot{C}_{D,k}$, namely Equations 8.24a or 8.24b, is strictly satisfied. Accordingly, these equations represent only plausible *approximations of the average costs* associated with exergy destruction in the kth component. For most applications, Equation 8.24a gives a lower estimate and Equation 8.24b gives a higher estimate with the actual exergy destruction cost being somewhere between the two. In the following, we use Equation 8.24a to estimate the cost of exergy destruction. Let us consider next the rationale for this practice.

In a well-designed system, the exergy destruction in the kth component directly affects the capital investment for the same component and, in some cases, indirectly affects the capital investment and the fuel costs of other components. For most well-designed components, as the exergy destruction decreases or as the efficiency increases, the term $\dot{C}_{D,k}$ decreases, but the capital investment $\dot{Z}^{CI}$ increases. The design optimization of a single component in *isolation,* discussed in Section 9.2, consists of finding the appropriate trade-offs between $\dot{C}_{D,k}$ and $\dot{Z}_k$ that minimize the unit cost of the product generated in the same component. At the *cost optimal point,* the ratio $\dot{C}_{D,k}/\dot{Z}_k$ remains approximately constant (see Equation 9.16). Accordingly, the lower the cost per exergy unit used to calculate $\dot{C}_{D,k}$, the lower the cost-optimal value of $\dot{Z}_k$. Thus, by using Equation 8.24a to estimate the cost of exergy destruction, $(\dot{C}_{D,k})$, we take a prudent approach with respect to the required capital investment costs. This philosophy is consistent with common practice in the design of industrial systems.

Reconsidering Equations 8.23, observe that the terms in parenthesis on the right side involve the rate of exergy loss from the kth component $\dot{E}_{L,k}$ and the associated cost rate $\dot{C}_{L,k}$. As noted in Section 8.1.4, the simplest approach to costing an exergy loss is to use Equation 8.20: $\dot{C}_{L,k} = 0$. Since this assumption is consistent with the purpose of evaluating and optimizing the design of a system, we will assume that Equation 8.20 applies in the derivation of the remaining thermoeconomic variables to be considered in this section. In passing, we may also note that use of Equation 8.21a in Equation 8.23a or use of Equation 8.21b in Equation 8.23b would result in a zero value for the term in parenthesis, and thus no exergy loss cost would be charged to the average unit cost of the product of the kth component $c_{P,k}$.

The Effects of Exergy Destruction and Exergy Loss on Efficiency and Costs.

Before continuing with the discussion of the remaining two thermoeconomic variables (r_k and f_k), we consider the effects of changes in the exergy destruction or exergy loss in one component on the exergy input to the *overall system*. This issue is important to both the design of new systems and the performance improvement of existing systems. For simplicity, we assume here that the product of the overall system and the exergetic efficiencies of the remaining components remain constant as the exergy destruction and/or the exergy loss in a component vary.

Let us begin by considering the system shown in Figure 8.5a. The system consists of a series of three components in which the product of each of the first two components is the fuel of the following component. We keep the product of the third component constant ($\dot{E}_{P,\mathrm{III}}$ = constant) and assume that the sum of exergy destruction and exergy loss in this component increases:

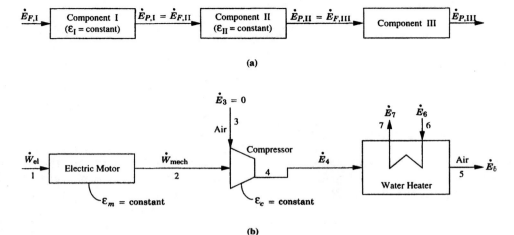

Figure 8.5 Systems in which the product of one component is the fuel of the next component.

$\Delta \dot{E}_{D,III} + \Delta \dot{E}_{L,III} = 10^m$ kW, where m is a positive integer. This results in the following changes of the exergy streams in Figure 8.5a: $\Delta \dot{E}_{F,III} = \Delta \dot{E}_{P,II} = 10^m$ kW; $\Delta \dot{E}_{F,II} = \Delta \dot{E}_{P,I} = 10^m$ kW$/\epsilon_{II}$; and $\Delta \dot{E}_{F,I} = 10^m$ kW$/(\epsilon_I \epsilon_{II})$. Thus, because $\epsilon_I < 1$ and $\epsilon_{II} < 1$, the additional fuel to the total system ($\Delta \dot{E}_{F,I}$) that is required to cover the increase in exergy destruction and/or exergy loss in the third component is larger than the sum ($\Delta \dot{E}_{D,III} + \Delta \dot{E}_{L,III}$).

By formulating cost rate balances for components I and II, using Equation 8.20, and neglecting the $\dot{Z}$ terms, we obtain

$$c_{P,I}^0 \dot{E}_{P,I} = c_{F,I}^0 \dot{E}_{F,I} \quad \text{or} \quad c_{P,I}^0 = \frac{c_{F,I}^0}{\epsilon_I}$$

$$c_{P,II}^0 \dot{E}_{P,II} = c_{F,II}^0 \dot{E}_{F,II} \quad \text{or} \quad c_{P,II}^0 = \frac{c_{F,II}^0}{\epsilon_{II}}$$

where superscript 0 indicates that only fuel costs are considered. Recognizing that $c_{P,I} = c_{F,II}$, these equations reduce to

$$c_{F,III}^0 = c_{P,II}^0 = \frac{c_{F,I}^0}{\epsilon_I \epsilon_{II}}$$

Combining this equation with the relation

$$\Delta \dot{E}_{F,I} = \frac{\Delta \dot{E}_{D,III} + \Delta \dot{E}_{L,III}}{\epsilon_I \epsilon_{II}}$$

we obtain

$$\frac{c_{F,III}^0}{c_{F,I}^0} = \frac{\Delta \dot{E}_{F,I}}{\Delta \dot{E}_{D,III} + \Delta \dot{E}_{L,III}}$$

or

$$c_{F,I}^0 \, \Delta \dot{E}_{F,I} = c_{F,III}^0 (\Delta \dot{E}_{D,III} + \Delta \dot{E}_{L,III}) \qquad (8.24c)$$

This equation states that the cost of the additional fuel to the overall system is equal to the increase in the cost of exergy destruction and exergy loss in the third component, as expressed by the right side of the last equation. Note that the cost of exergy destruction and exergy loss in the overall system cannot be calculated by adding the costs of exergy destruction and exergy loss in the components because the c_F value for each component already contains information related to the exergy destruction and exergy loss in all upstream components. For further discussion, see Section 8.3.2.

A system corresponding to the arrangement of Figure 8.5a is presented in Figure 8.5b. The system consists of an electric motor, an air compressor, and a water heater. For simplicity the exergy flow rates $\dot{E}_7 - \dot{E}_6$ and $\dot{E}_5$, the pressure p_5, and the exergetic efficiencies ϵ_m and ϵ_c of the electric motor and compressor, respectively, are each fixed. Then, an increase in the exergy destruction of 10^m kW caused by an increase in the pressure drop in the water heater results in an increase in the required electric power for the motor by 10^m kW/$(\epsilon_c \epsilon_m)$. If the 10^m kW increase of exergy destruction would occur in the air compressor, however, the effect on the electric power would be smaller: 10^m kW/ϵ_m.

From these examples, we conclude that the economic importance of exergy destruction and exergy loss in a component depends on the *relative position* of the component with respect to fuel and product streams of the overall system: A change in the exergy destruction or exergy loss in a component closer to a product stream generally has a *greater impact* on the fuel supplied to the overall system than a change in the exergy destruction (or exergy loss) of the same magnitude in a component closer to the component where the fuel to the overall system is supplied. That is, to provide a component where incremental exergy destruction and/or exergy loss occur with the exergy it requires, additional exergy is generally destroyed in the upstream components, and so the exergy supplied to the overall system increases disproportionately.[3]

This conclusion applies to any system with any arrangement of components. In a steam power system, for instance, avoidable exergy destruction (or exergy loss) in the steam turbine has a larger impact on the system efficiency and cost of electricity than does avoidable exergy destruction (or exergy loss) of the same magnitude in the boiler.

---

[3]For complex systems the approach of Reference 8 may be used to develop algebraic relations expressing the interdependence of efficiencies of various components.

8.2.2 Relative Cost Difference

The relative cost difference r_k for the kth component is defined by

$$r_k = \frac{c_{P,k} - c_{F,k}}{c_{F,k}} \tag{8.25}$$

This variable expresses the relative increase in the average cost per exergy unit between fuel and product of the component. The relative cost difference is a useful variable for evaluating and optimizing a system component. In an iterative cost optimization of a system, for example, if the cost of fuel of a major component changes from one iteration to the next, the objective of the cost optimization of the component should be to minimize the relative cost difference instead of minimizing the cost per exergy unit of the product for this component (see Section 9.6).

With Equations 8.2, 8.20, and 8.23a, Equation 8.25 becomes

$$r_k = \frac{c_{F,k}(\dot{E}_{D,k} + \dot{E}_{L,k}) + (\dot{Z}_k^{CI} + \dot{Z}_k^{OM})}{c_{F,k}\dot{E}_{P,k}} \tag{8.26}$$

Equation 8.26 reveals the *real cost sources* associated with the kth component. These sources, which cause an increase in the cost per exergy unit between fuel and product, are the cost rates associated with

- Capital investment $\dot{Z}_k^{CI}$
- Operating and maintenance $\dot{Z}_k^{OM}$
- Exergy destruction $c_{F,k}\dot{E}_{D,k}$
- Exergy loss $c_{F,k}\dot{E}_{L,k}$

Using the exergetic efficiency of the kth component,

$$\epsilon_k = \frac{\dot{E}_{P,k}}{\dot{E}_{F,k}} = 1 - \frac{\dot{E}_{D,k} + \dot{E}_{L,k}}{\dot{E}_{F,k}}$$

Equation 8.26 may be written as

$$r_k = \frac{1 - \epsilon_k}{\epsilon_k} + \frac{\dot{Z}_k^{CI} + \dot{Z}_k^{OM}}{c_{F,k}\dot{E}_{P,k}} \tag{8.27}$$

The relative cost difference can be expressed alternatively using Equations 8.2, 8.20, and 8.23b. This is left as an exercise. The expressions given by

Equations 8.26 and 8.27 are preferred, however, since they are in accord with the evaluation of the exergy destruction cost and the exergy loss cost discussed in the previous section.

8.2.3 Exergoeconomic Factor

As Equations 8.26 and 8.27 indicate, the cost sources in a component may be grouped in two categories. The first consists of non-exergy-related costs (capital investment and operating and maintenance expenses), while the second category consists of exergy destruction and exergy loss. In evaluating the performance of a component, we want to know the relative significance of each category. This is provided by the *exergoeconomic factor* f_k defined for component k by

$$f_k = \frac{\dot{Z}_k}{\dot{Z}_k + c_{F,k}(\dot{E}_{D,k} + \dot{E}_{L,k})} \qquad (8.28)$$

where, as before, $\dot{Z}_k = \dot{Z}_k^{\text{CI}} + \dot{Z}_k^{\text{OM}}$. The total cost rate causing the increase in the unit cost from fuel to product is given by the denominator in Equation 8.28. Accordingly, the exergoeconomic factor expresses as a ratio the contribution of the non-exergy-related cost to the total cost increase. A low value of the exergoeconomic factor calculated for a major component suggests that cost savings in the entire system might be achieved by improving the component efficiency (reducing the exergy destruction) even if the capital investment for this component will increase. On the other hand, a high value of this factor suggests a decrease in the investment costs of this component at the expense of its exergetic efficiency. Typical values of the exergoeconomic factor depend on the component type. For instance, this value is *typically* lower than 55% for heat exchangers, between 35 and 75% for compressors and turbines, and above 70% for pumps.

8.3 THERMOECONOMIC EVALUATION

This section illustrates the use of the thermoeconomic variables introduced thus far for the evaluation of thermal systems. First, the design evaluation of a new system is presented and then the performance evaluation of an existing system is discussed. The difference between these two cases is in the way that investment costs are treated: In the second case, all charges directly associated with the investment costs of *existing* equipment are *sunk costs* (Section 7.3) and thus are not considered in the analysis.

8.3.1 Design Evaluation

A detailed thermoeconomic evaluation of the design of a thermal system is based on a set of variables calculated for each component of the system. Thus, for the kth component we calculate the

- Exergetic efficiency ϵ_k
- Rates of exergy destruction $\dot{E}_{D,k}$ and exergy loss $\dot{E}_{L,k}$
- Exergy destruction ratio $y_{D,k}$ and exergy loss ratio $y_{L,k}$
- Cost rates associated with capital investment $\dot{Z}_k^{CI}$, operating and maintenance expenses $\dot{Z}_k^{OM}$, and their sum $\dot{Z}_k$
- Cost rate of exergy destruction $\dot{C}_{D,k}$
- Relative cost difference r_k
- Exergoeconomic factor f_k

To improve the cost effectiveness of a thermal system consisting of several components, we recommend the following methodology:

1. Rank the components in descending order of cost importance using the sum $\dot{Z}_k + \dot{C}_{D,k}$
2. Consider design changes initially for the components for which the value of this sum is high.
3. Pay particular attention to components with a high relative cost difference r_k, especially when the cost rates $\dot{Z}_k$ and $\dot{C}_{D,k}$ are high.
4. Use the exergoeconomic factor f_k to identify the major cost source (capital investment or cost of exergy destruction).
   a. If the f_k value is high, investigate whether it is cost effective to reduce the capital investment for the kth component at the expense of the component efficiency.
   b. If the f_k value is low, try to improve the component efficiency by increasing the capital investment.
5. Eliminate any subprocesses that increase the exergy destruction or exergy loss without contributing to the reduction of capital investment or of fuel costs for other components.
6. Consider improving the exergetic efficiency of a component if it has a relatively low exergetic efficiency or relatively large values of the rate of exergy destruction, the exergy destruction ratio, or the exergy loss ratio.

When applying this methodology, recognize that the values of all thermoeconomic variables depend on the component types: heat exchanger, compressor, turbine, pump, chemical reactor, and so forth. Accordingly, whether a particular value is judged to be *high* or *low* can be determined only with reference to a particular class of components. *Typical* ranges of the exergoeconomic factor for some component types are given at the close of Section 8.2.3.

The foregoing methodology for improving the cost effectiveness of a thermal system is illustrated in Example 8.2 for the cogeneration system case study. The cogeneration system provides *only an elementary example* to demonstrate the application of thermoeconomic techniques for evaluating and optimizing the design of a system. These techniques become particularly powerful and effective in more complex systems where conventional techniques do not suffice (Sections 1.6.3 and 9.6).

Example 8.2   Using data from Example 8.1 for the cogeneration system case study, conduct a detailed thermoeconomic evaluation of the system, identify the effects of the design variables on costs, and suggest values of the design variables that would make the overall system more cost effective than the case of Example 8.1.

Solution

MODEL

1. The following decision variables are used for the cogeneration system: The pressure ratio p_2/p_1, the isentropic efficiencies of the air compressor η_{sc} and gas turbine η_{st}, the temperature T_3 of the preheated air at the outlet of the air preheater, and the combustion gas temperature T_4 at the inlet to the gas turbine (see Section 2.5 for discussion).

2. The four elements of the model used in the solution to Example 8.1 apply.

3. The thermodynamic performance models for the components are given in Section 2.5.

4. We assume that an experienced cost engineer is available for consulting; and the engineer has explained the *qualitative* effects of some design parameters on the cost of purchasing each system component. For example, we know that the costs of purchasing the gas turbine and the air compressor increase with increasing pressure ratio p_2/p_1 and increasing values of the respective isentropic efficiencies; moreover, the costs of purchasing the gas turbine and the combustion chamber increase exponentially with increasing temperature T_4.[4]

---

[4]A high T_4 value requires application of special alloys and advanced internal cooling techniques of the turbine blades and vanes.

5. The *quantitative* effects of the design parameters on the costs of purchasing each component of the cogeneration system are provided by the cost functions of Appendix B. In practice, such cost functions are generally not available, and the services of a cost engineer would be required: After each design modification, the necessary thermodynamic data would be submitted to the cost engineer to determine the new purchased-equipment costs. In this example, and in all the following ones using the cogeneration system case study, we assume that the answers of the cost engineer are identical to the answers obtained from the cost functions of Appendix B.

ANALYSIS The accompanying table (on page 442) summarizes the thermoeconomic variables calculated for each component of the cogeneration system using the data from Example 8.1 and the definitions from Tables 3.3 and 8.2. The variables include the exergetic efficiency ϵ, rate of exergy destruction $\dot{E}_D$, exergy destruction ratio $y_D = \dot{E}_D/\dot{E}_{10}$, average costs per unit of fuel exergy c_F and product exergy c_P, cost rate of exergy destruction $\dot{C}_D$, investment and O&M cost rate $\dot{Z}$, relative cost difference r, and exergoeconomic factor f. In accord with the presented methodology, the components are listed in order of descending value of the sum $\dot{C}_D + \dot{Z}$. The analysis of the data presented in the table leads to the following conclusions:

The combustion chamber, the gas turbine, and the air compressor have the highest values of the sum $\dot{Z} + \dot{C}_D$ and are, therefore, the most important components from the thermoeconomic viewpoint. The low value of the variable f for the combustion chamber shows that the costs associated with the combustion chamber are almost exclusively due to exergy destruction. As discussed in Section 3.5.4, a part of the exergy destruction in a combustion chamber can be avoided by preheating the reactants and by reducing the heat loss and the excess air, but this usually leads only to a small reduction in the exergy destruction. For simplicity, we assume that the heat loss cannot be further reduced. The excess air is determined by the desired temperature T_4 at the inlet to the gas turbine. The temperature T_4 is a key design variable for it affects both the performance of the entire system (exergy destruction in the combustion chamber, gas turbine, air preheater, and HRSG, and exergy loss associated with stream 7) and the investment costs of the components.

An increase in the heat transfer rate in the air preheater achieved through an increase in temperature T_3 also results in a decrease of the exergy destruction in the combustion chamber. Thus, the temperature T_3 is also a key design variable because, in addition to the combustion chamber, it affects the exergy loss associated with stream 7 as well as the performance and investment costs of the air preheater and the heat-recovery steam generator. Holding all other decision variables constant, the higher the temperature T_3, the smaller the average temperature difference in the air preheater and the heat-recovery steam generator. A decrease in the average temperature difference in these

Component	ϵ (%)	$\dot{E}_D$ (MW)	y_D (%)	c_F ($/GJ)	c_P ($/GJ)	$\dot{C}_D$ ($/h)	$\dot{Z}$ ($/h)	$\dot{C}_D + \dot{Z}$ ($/h)	r (%)	f (%)
Combustion chamber	80.37	25.48	29.98	11.45	14.51	1050	68	1118	26.7	6.1
Gas turbine	95.20	3.01	3.54	14.51	18.76	157	753	910	29.2	82.7
Air compressor	92.84	2.12	2.50	18.76	27.80	143	753	896	48.2	84.0
HRSG	67.17	6.23	7.33	14.51	27.36	326	264	590	88.5	44.8
Air preheater	84.55	2.63	3.09	14.51	20.81	137	189	326	43.4	57.9

heat exchangers results in an increase in both the exergetic efficiency and the capital investment for each heat exchanger.

To summarize: By considering measures for reducing the high cost rate associated with the exergy destruction in the combustion chamber of the cogeneration system, two key design variables have been identified, the temperatures T_3 and T_4. An increase in these temperatures reduces the $\dot{C}_D$ value for the combustion chamber and other components (Section 9.5.1) but increases their capital investment costs.

Turning next to the gas turbine, which has the second highest value of the sum $\dot{Z} + \dot{C}_D$, the relatively large value of factor f suggests that the capital investment and O&M costs dominate. According to assumption 4 of the cost model, the capital investment costs of the gas turbine depend on temperature T_4, pressure ratio p_4/p_5, and isentropic efficiency η_{st}. To reduce the high $\dot{Z}$ value associated with the gas turbine, we should consider a reduction in the value of at least one of these variables.

The air compressor has the highest f value and the second highest relative cost difference r among all components. Thus, we would expect the cost effectiveness of the entire system to improve if the $\dot{Z}$ value for the air compressor is reduced. This may be achieved by reducing the pressure ratio p_2/p_1 and/or the isentropic efficiency η_{sc}.

The heat recovery steam generator has the lowest exergetic efficiency and the highest r value among all the components. As the f value indicates, almost 45% of the relative cost difference is caused by the $\dot{Z}$ value in this component, with the remaining 55% caused by exergy destruction. Thus, we might conclude that a decrease of the exergy destruction in the HRSG could be cost effective for the entire system even if this would increase the investment costs associated with this component. The exergy destruction in the HRSG can be reduced by decreasing the values of T_6 and T_7. A decrease in the value of T_7 also results in a decrease in the exergy loss from the total system. In terms of the decision variables, temperatures T_6 and T_7 may be reduced by increasing T_3 and/or decreasing T_4 at fixed values of the remaining decision variables.

The relatively high value of f in the air preheater suggests a reduction in the investment costs of this component. This can be achieved by decreasing T_3. It should be noted, however, that changes suggested by the evaluation of this component should only be considered if they do not contradict changes suggested by components with a larger value of $\dot{C}_D + \dot{Z}$.

Summarizing the foregoing conclusions, the following changes in the design variables are expected to improve the cost effectiveness of the system:

- Increase the value of T_3 as suggested by the evaluation of the combustion chamber and HRSG.
- Decrease the pressure ratio p_2/p_1 (and thus p_4/p_5) and the isentropic efficiencies η_{sc} and η_{st}, as suggested by the evaluation of the air compressor and gas turbine.

- Maintain T_4 fixed, since we get contradictory indications from the evaluations of the combustion chamber on one side and the gas turbine and HRSG on the other side.

To illustrate the effect of the suggested changes in the decision variables on the overall costs, we use the following new values of the design variables, giving the Example 8.1 values in parenthesis: $T_3 = 870$ K (850 K); $T_4 = 1520$ K (1520 K); $p_2/p_1 = 9$ (10); $\eta_{sc} = 0.85$ (0.86); and $\eta_{st} = 0.85$ (0.86). As a result of these changes, the cost flow rate $\dot{C}_{P,tot}$ (Equation 8.1), which represents the objective function in the optimization procedure, is reduced from \$3617/h to \$3354/h. The details of this calculation are left as an exercise.

COMMENTS

1. There can be no assurance that the changes in the values of the design variables suggested by the thermoeconomic methodology will necessarily result in an improvement in the cost effectiveness of the entire system. The methodology, however, does provide a plausible exploratory approach for improving the cost effectiveness of a thermal system.
2. After becoming more familiar with the behavior of the cogeneration system, the reader will recognize that temperature T_4 should have been decreased to improve the cost effectiveness of the cogeneration system. Some iterations are always necessary to establish an improved system design.
3. A conventional system evaluation based on energy balances and purchased-equipment cost calculations cannot provide the above information with an equal degree of confidence. Compared with a thermoeconomic evaluation, many more trial-and-error attempts typically would be required with a conventional optimization approach to a complex system to achieve comparable improvements in the system design.

8.3.2 Performance Evaluation

The evaluation of the performance of an *existing system* is conducted in a parallel manner using the same thermoeconomic variables as in the design evaluation of a new system. Since the capital investments ($\dot{Z}^{CI}$) of all existing components represent *sunk costs*, capital investment is ignored in such evaluations. Moreover, we also elect for simplicity to neglect the effect of the operating and maintenance costs. Accordingly, only fuel-related costs are considered here.

When calculated on such a basis, the values of the thermoeconomic variables (see variables of Example 8.3) assist with a range of decisions related to the performance of an existing system. For example, selected variables help to understand the effects of a malfunction in a component on the performance of the other components and the total system. With the aid of on-line performance evaluation software including reconciliation of measured data, the actual values of the thermoeconomic variables can be compared with design (target) values provided by simulation software. The comparison of actual and target costs per exergy unit allows an early detection of component malfunctions and a rapid identification of the malfunction source. Moreover, exergy stream cost data can be used to decide whether a malfunctioning component should be replaced. The thermoeconomic variables also illustrate the flow of fuel costs in the system and the cost formation process as the exergy of the fuel to the overall system is converted into exergy of the product streams.

Let us now develop expressions for the thermoeconomic variables neglecting capital investment costs and the operating and maintenance expenses. The use of these variables is illustrated in Example 8.3. For present purposes, the cost rate balance for the kth component, Equation 8.4a, appears as

$$\sum_e \dot{C}^0_{e,k} + \dot{C}^0_{w,k} = \dot{C}^0_{q,k} + \sum_i \dot{C}^0_{i,k} \tag{8.29}$$

where, as in Section 8.2.1, superscript 0 indicates that only fuel costs are considered. The cost rate equation for each exergy stream takes the form

$$\dot{C}^0_j = c^0_j \dot{E}_j \tag{8.30}$$

The variables $\dot{C}^0_j$ assist in understanding the flow of fuel-related costs within the system.

With $\dot{Z}_k = 0$, Equation 8.20, and the definition of the exergetic efficiency ϵ_k (Equation 3.29), Equation 8.19 reduces to give

$$c^0_{P,k} = \frac{c^0_{F,k}}{\epsilon_k} \tag{8.31}$$

As $\epsilon_k < 1$, this equation shows that, even when only fuel-related costs are considered in the kth component, the average cost per exergy unit of product ($c^0_{P,k}$) is always greater than the average cost per exergy unit of fuel ($c^0_{F,k}$).

When non-exergy-related costs are ignored, the exergoeconomic factor vanishes. As Equation 8.27 shows, the relative cost difference depends only on the exergetic efficiency and thus does not provide additional information. The cost rate of exergy destruction in component k, denoted by $\dot{C}^0_{D,k}$, is

$$\dot{C}^0_{D,k} = c^0_{F,k} \dot{E}_{D,k} \tag{8.32}$$

The value of $c^0_{F,k}$ is determined by the exergy destruction and exergy loss in all upstream components interconnected with the kth component, and the cost of the fuel supplied to the overall system. Equation 8.32 shows that the variable $c_{F,k}^0$ is a measure of the *cost importance* of a unit of exergy destruction within the kth component when non-exergy-related costs are ignored. Since only fuel-related costs are considered, $c_{F,k}^0$ is also a measure of the *thermodynamic importance* of the exergy destruction within the kth component. The value of $\dot{C}_{D,k}^0$ is due to the exergy destruction within component k but is also affected by the variables determining $c_{F,k}^0$.

Recalling the discussion of Figure 8.5*a* in Section 8.2.1 for illustration, we can say that 10^m kW of exergy destruction within component III is *thermodynamically* and, of course, economically more important than the same exergy destruction rate within component II. Using Equation 8.24c, this exergy destruction rate is actually $c_{F,\text{III}}^0/c_{F,\text{II}}^0$ times more important. Accordingly, in trying to improve the thermodynamic performance of a system, we should initially focus on the efficient operation of components with the highest $c_{F,k}^0$ values.

As illustrated in the example of Section 8.2.1, the $c_{F,k}$ value represents the average cost at which each exergy unit of the fuel for the kth component was provided in upstream components. Therefore, the $c_{F,k}$ value is affected by the interdependence of the kth component with all upstream components. Any malfunction in an upstream component interconnected with the kth component will cause an increase in the $c_{F,k}$ value.

Denoting the cost per exergy unit of fuel provided to the overall system by $c_{F,\text{tot}}$, we can calculate the ratios $c_{F,k}^0/c_{F,\text{tot}}$, $\dot{C}_j^0/c_{F,\text{tot}}$ which are independent of the cost of fuel supplied to the total system. The dimensionless variable $c_{F,k}^0/c_{F,\text{tot}}$ is greater or equal to unity and expresses how many units of fuel exergy must be supplied to the overall system to provide the kth component with one unit of fuel exergy. The variable $\dot{C}_j^0/c_{F,\text{tot}}$, which has exergy rate units, is known as the *exergetic cost* of the jth stream [9, 10]. The cost balance, Equation 8.29, shows that in each component the sum of the exergetic costs associated with the inlet flows is equal to the sum of the exergetic costs associated with the outlet flows. The *unit of exergetic cost* $c_j^0/c_{F,\text{tot}}$ shows how many units of fuel exergy must be supplied to the overall system to generate one unit of exergy associated with the jth stream.

Example 8.3   Using data from Example 8.1, and considering only fuel-related costs for the cogeneration system case study, calculate (a) the levelized cost rate $\dot{C}^0$, the levelized cost per exergy unit c^0, and the ratios $\dot{C}^0/c_{F,\text{tot}}$ and $c^0/c_{F,\text{tot}}$ associated with each stream of the system, and (b) the variables c_F^0, c_P^0, $\dot{C}_D^0$, and $c_F^0/c_{F,\text{tot}}$ for each system component. Comment on the calculated values.

Solution

ASSUMPTIONS

1. The costs associated with capital investment and operation and maintenance are neglected.
2. All remaining assumptions of Example 8.1 apply.

ANALYSIS

(a) Referring to the solution of Example 8.1, the linear system of Equations (1)–(12) remains valid, except that now we neglect the costs associated with capital investment and operation and maintenance. That is, we set

$$\dot{Z}_{ac} = \dot{Z}_{ph} = \dot{Z}_{cc} = \dot{Z}_{gt} = \dot{Z}_{hrsg} = 0$$

The values of the levelized cost rates $\dot{C}_1^0$ through $\dot{C}_{12}^0$ calculated from the solution of the resulting system of equations are given in the following table:

Stream No.	$\dot{E}$ (MW)	$\dot{C}^0$ ($/h)	c^0 ($/GJ)	$\dot{C}^0/c_{F,tot}$ (MW)	$c^0/c_{F,tot}$
1	0	0	0.00	0.000	0.000
2	27.538	818	8.26	49.719	1.808
3	41.938	1265	8.39	76.895	1.835
4	101.454	2663	7.29	161.889	1.596
5	38.782	1018	7.29	61.884	1.596
6	21.752	571	7.29	34.709	1.596
7	2.773	73	7.29	4.424	1.596
8	0.062	0	0.00	0.000	0.000
9	12.810	498	10.80	30.285	2.364
10	84.994	1398	4.57	84.994	1.000
11	29.662	818	7.66	49.719	1.676
12	30.000	827	7.66	50.286	1.676

The values of $\dot{C}_j^0$ given in the table illustrate the flow of fuel-related levelized costs within the system. In particular, note that the cost rate associated with the fuel supply to the overall system ($\dot{C}_{10} = \$1398/h$) is apportioned between the three streams exiting the system: net power, steam, and effluent. That is, $\dot{C}_{12}^0 = \$827/h$, $\dot{C}_9^0 = \$498/h$, and $\dot{C}_7^0 = \$73/h$. When only fuel-related costs are to be considered, this information can be used for costing the two useful output streams: net power and electricity. Note that before we calculate the final costs at which power and electricity are generated in this system, the cost rate $\dot{C}_7^0 = \$73/h$ must be apportioned between net power and electricity, as was done in Example 8.1, giving $\dot{C}_{12}^{0''} = \$878/h$ and $\dot{C}_{9}^{0''} = \$520/h$.

The levelized costs per exergy unit c_1^0 through c_{12}^0 are calculated from $c_j^0 = \dot{C}_j^0/\dot{E}_j$ and are also listed in this table. The costs per exergy unit of net power and stream are $c_{12}^0 = \$7.66/GJ$ and $c_9^0 = \$10.80/GJ$. After apportioning the cost rate $\dot{C}_7^0$ between the two product streams, we obtain $c_{12}^{0''} = \$8.13/GJ$ and $c_9^{0''} = \$11.28/GJ$.

The last two columns of this table help us consider the results shown in the first two columns from a different perspective: The *exergetic cost* $\dot{C}_j^0/c_{F,tot}$, expressed in megawatts, gives the exergy feed required in a system to generate the jth exergy rate ($\dot{E}_j$) exiting this system. Thus, 50.286, 30.285, and 4.424 MW must be supplied to the total system to generate streams 12 (net power), 9 (steam), and 7 (effluent), respectively. The total exergy input to the system ($\dot{E}_{10}$) is equal to the sum of these numbers. The dimensionless ratio $c_j^0/c_{F,tot}$, shown in the last column of the table, accounts for the units of feed exergy to the total system that are required to generate one unit of exergy associated with the jth stream. Streams 9, 3, 2, 11, and 12 have the highest values of this ratio and are, therefore, the most expensive ones to be generated on a per unit of exergy basis.

(b) The following table provides values of selected thermoeconomic variables for each component calculated as in Example 8.2: the average cost per unit of fuel exergy c_F^0, the average cost per unit of product exergy c_P^0, the cost rate of exergy destruction $\dot{C}_D^0$, and the ratio $c_F^0/c_{F,tot}$.

Component	c_F^0 ($/GJ)	c_P^0 ($/GJ)	$\dot{C}_D^0$ ($/h)	$\dfrac{c_F^0}{c_{F,tot}}$
Air compressor	7.66	8.25	59	1.676
Air preheater	7.29	8.62	69	1.596
Combustion chamber	5.83	7.29	534	1.276
Gas turbine	7.29	7.66	79	1.596
HRSG	7.29	10.85	164	1.596
Total system	4.57	9.08	649	1.000

The c_F^0 column of the table shows that the combustion chamber has the lowest fuel cost and the air compressor the highest fuel cost. Accordingly, the unit exergy destruction in the air compressor costs more than in any other system component. Thus, measures to improve the thermodynamic performance of the cogeneration system should initially consider improving the efficiency of the air compressor. In the gas turbine, air preheater, and HRSG, each unit of fuel exergy is provided at the same average cost ($7.29/GJ). The HRSG and the air preheater have the highest unit cost of product c_P^0. Note that this is a different ranking than the one obtained in Example 8.2, where non-exergy-related costs are included in the calculations. In Example 8.2, the air compressor and the HRSG have the highest unit costs c_P^0. This illustrates that the conclusions from a thermoeconomic evaluation may be different when the *performance of an existing system* is the objective than when the *design of a new system* is the objective.

The $\dot{C}_D^0$ values show how much the inefficiencies in each component cost the system operator when only fuel costs are considered in the cogeneration system. The highest costs of exergy destruction occur in the combustion chamber and the HRSG. The cost ratio $c_F^0/c_{F,\text{tot}}$ shown in the last column of the table accounts for the units of methane exergy (stream 10) that must be supplied to the total system to provide the respective component with one unit of fuel exergy. Monitoring the ratios $c_F^0/c_{F,\text{tot}}$ and $c_j^0/c_{F,\text{tot}}$ with the aid of on-line evaluation software and comparing the actual values of these variables with design (target) values allows an early detection of component malfunctions and a rapid identification of the malfunction source.

8.4 ADDITIONAL COSTING CONSIDERATIONS

We have considered thus far the *total exergy* associated with material streams entering or exiting a component. In some systems, however, the *chemical* and *physical exergy* of streams (Section 3.1.2) might be supplied or generated at different unit costs. In such cases, the accuracy of the thermoeconomic analysis may be improved by taking these differing unit costs into consideration. For similar reasons, we might identify two subcomponents of the physical exergy associated with streams of matter, the *thermal* and *mechanical exergy,* and cost them separately. These cases and some other exergy stream-costing concepts are briefly discussed below. For simplicity in these discussions, we neglect the kinetic and the potential exergy. For more details, see References 2, 7, and 11–14.

8.4.1 Costing Chemical and Physical Exergy

Referring to Section 3.1.2, the total exergy transfer rate associated with a stream of matter is the sum of four contributions: physical, kinetic, potential, and chemical exergy. That is

$$\dot{E}_j = \dot{E}_j^{\text{PH}} + \dot{E}_j^{\text{KN}} + \dot{E}_j^{\text{PT}} + \dot{E}_j^{\text{CH}} \tag{3.1}$$

Neglecting the kinetic and potential contributions, this becomes simply

$$\dot{E}_j = \dot{E}_j^{\text{PH}} + \dot{E}_j^{\text{CH}}$$

Denoting the average costs per unit of physical and chemical exergy by c_j^{PH} and c_j^{CH}, respectively, the cost rate associated with stream j is then

$$\dot{C}_j = c_j \dot{E}_j = c_j^{CH} \dot{E}_j^{CH} + c_j^{PH} \dot{E}_j^{PH} \tag{8.33}$$

The term $\dot{C}_j$ of this equation represents the cost rate associated with the total exergy of stream j, whereas the two terms on the right side denote the cost rates associated with chemical and physical exergy. Owing mainly to the different auxiliary relations that are applicable, the value of the average cost per total exergy unit (c_j) calculated from Equation 8.33 is, in general, different than the value calculated when the total exergy is not split into its components (Section 8.1).

The cost rate balance (Equation 8.4b) now takes the form

$$\sum_e (c_e^{CH} \dot{E}_e^{CH} + c_e^{PH} \dot{E}_e^{PH})_k + c_{w,k} \dot{W}_k$$
$$= c_{q,k} \dot{E}_{q,k} + \sum_i (c_i^{CH} \dot{E}_i^{CH} + c_i^{PH} \dot{E}_i^{PH})_k + \dot{Z}_k \tag{8.34}$$

When the costs of physical and chemical exergy are considered separately, the auxiliary relations required to solve for the unknowns in a thermoeconomic analysis differ from those considered in Section 8.1 and tabulated in Table 8.2. The appropriate auxiliary relations for selected components are provided in the second to last row of Table 8.3, which is closely patterned after Table 8.2. The last row of Table 8.3 shows the variable to be calculated from the cost balance (Equation 8.34) when we distinguish between chemical and physical exergy.

Let us consider the entries of Table 8.3 in some detail. First note that the cost per unit of chemical exergy of a stream remains constant when its chemical composition does not change. In Table 8.3, this rule is applied to the working fluids in a compressor, turbine, heat exchanger, and boiler. In the case of complete combustion (e.g., in a combustion chamber or a boiler), a zero cost is assigned to the chemical exergy associated with the exiting combustion products because this exergy cannot be retrieved by practical means. If the combustion is incomplete, however, the cost per unit of chemical exergy of the exiting stream that contains the incomplete reaction products is set equal to the cost per unit of fuel chemical exergy. Thus, in Table 8.3 we set $c_3^{CH} = c_1^{CH}$ for the case of incomplete combustion in a combustion chamber or boiler. In the same table we set $c_3^{CH} = c_3^{PH}$ for a gasifier because the chemical exergy and the physical exergy of the exiting stream are each generated simultaneously in the same process.

Example 8.4   Using data from Tables 3.1, 7.6, and 7.15 for the cogeneration system case study, and accounting separately for the costs of physical and

Table 8.3 Auxiliary thermoeconomic relations for selected components at steady-state operation when physical and chemical exergy are considered separately[a]

Component	Compressor, Pump, or Fan	Turbine or Expander	Heat Exchanger[b]	Mixing Unit	Gasifier or Combustion Chamber	Boiler
Schematic						
Auxiliary thermoeconomic relations	$c_2^{CH} = c_1^{CH}$	$c_2^{PH} = c_3^{PH} = c_1^{PH}$ $c_2^{CH} = c_3^{CH} = c_1^{CH}$	$c_4^{PH} = c_3^{PH}$ $c_4^{CH} = c_3^{CH}$ $c_2^{CH} = c_1^{CH}$	$c_3^{CH} = \dfrac{\dot{C}_1^{CH} + \dot{C}_2^{CH}}{\dot{E}_3^{CH}}$	$c_3^{CH} = c_3^{PH}$ (gasifier) $c_3^{CH} = c_1^{CH}$ (incomplete combustion) $c_3^{CH} = 0$ (complete combustion)	$c_6^{CH} = c_5^{CH},\ c_8^{CH} = c_7^{CH}$ $c_3^{CH} = c_1^{CH},\ c_4^{CH} = 0$ $\dfrac{\dot{C}_6^{PH} - \dot{C}_5^{PH}}{\dot{E}_6^{PH} - \dot{E}_5^{PH}} = \dfrac{\dot{C}_8^{PH} - \dot{C}_7^{PH}}{\dot{E}_8^{PH} - \dot{E}_7^{PH}}$ For c_3^{PH} and c_4^{PH} see Section 8.1.4 and Equations 8.14
Variable calculated from cost balance	c_2^{PH}	c_w	c_2^{PH}	c_3^{PH}	c_3^{PH}	c_6^{PH} or c_8^{PH}

[a]The cost rates $\dot{C}_F$ and $\dot{C}_P$ for these components are defined in Table 8.2.

[b]These relations assume that the purpose of the heat exchanger is to heat the cold stream ($T_1 \geq T_0$). If the purpose of the heat exchanger is to provide cooling ($T_3 \leq T_0$), then the following relations should be used: $\dot{C}_P = \dot{C}_4 - \dot{C}_3$; $\dot{C}_F = \dot{C}_1 - \dot{C}_2$; $c_2^{PH} = c_1^{PH}$; $c_2^{CH} = c_1^{CH}$, and $c_4^{CH} = c_3^{CH}$. The variable c_4^{PH} is calculated from the cost balance.

chemical exergy, (a) calculate the levelized cost rates and the levelized costs per unit of exergy associated with all streams in the system and (b) compare with the results of Example 8.1.

Solution

MODEL

1. The four elements of the model used in the solution to Example 8.1 apply.
2. The fuel to the total system (stream 10) is supplied from a pipeline operating at a pressure higher than p_{10} and is throttled to pressure p_{10} before being introduced to the combustion chamber. According to the purchasing agreement, the pipeline operators charge, in effect, only for the chemical exergy of the fuel, and thus a zero cost is assigned to the physical exergy of stream 10.

ANALYSIS

(a) Paralleling the analysis of Example 8.1, we formulate the cost rate balances (Equation 8.34) and the auxiliary relations (Table 8.3) for each component of the cogeneration system. By solving the resulting linear system of equations, we obtain the values of the thermoeconomic variables summarized in the following table. The table gives the levelized cost rates ($\dot{C}$) and levelized costs per exergy unit (c) associated with physical (superscript PH), chemical (superscript CH), and total (no superscript) exergy for each stream of the cogeneration system.

Stream No.	$\dot{C}^{PH}$ ($/h)	$\dot{C}^{CH}$ ($/h)	$\dot{C}$ ($/h)	c^{PH} ($/GJ)	c^{CH} ($/GJ)	c ($/GJ)
1	0	0	0	0	0	0
2	2767	0	2767	27.91	0	27.95
3	3852	0	3852	25.51	0	25.54
4	5318	0	5318	14.61	0	14.56
5	2021	0	2021	14.61	0	14.48
6	1125	0	1125	14.61	0	14.37
7	127	0	127	14.61	0	12.68
8	0	0	0	0	0	0
9	1263	0	1263	27.45	0	27.38
10	0	1398	1398	0	4.60	4.57
11	—	—	2014	—	—	18.86
12	—	—	2037	—	—	18.86

(b) Comparing the values determined in this example for $\dot{C}$ and c with those obtained in Example 8.1, we note that for each stream the difference is small, almost negligible for practical purposes. This is due to the fact that the chemical exergy of the fuel supplied to the total system is used in a complete combustion process immediately after the fuel enters the system. In different systems, where the chemical exergy of the fuel provided to the overall system is first processed without being used entirely in a combustion process (e.g., in a gasifier [7] or a steam reformer), the calculated values of $\dot{C}$ and c may differ significantly when the costs of physical and chemical exergy are considered separately from the calculated values of $\dot{C}$ and c using only total exergy transfer rates.

In this section, we have represented the total exergy in terms of physical and chemical exergy contributions and then assigned unit costs to each of the contributions. In some applications it is also possible and appropriate to further divide each of the chemical and physical contributions into terms of subcontributions that are costed individually. These types of costing are considered in Sections 8.4.2 and 8.4.3. Note that the separate consideration of chemical and physical exergy in costing applications is a precondition for costing reactive and nonreactive exergy (Section 8.4.2) and/or mechanical and thermal exergy (Section 8.4.3).

8.4.2 Costing Reactive and Nonreactive Exergy

While discussing chemical exergy in Section 3.1.4, we note that the term chemical exergy does not necessarily imply a chemical reaction. This is exemplified by Equation 3.15 where the term chemical exergy refers to the work of an expansion process that involves no chemical reaction. In general, however, the chemical exergy concept involves both chemical reaction and nonreactive processes such as expansion, compression, mixing, and separation. The portion of the chemical exergy associated with nonreactive effects generally cannot be used in actual processes and, therefore, for most thermal systems has no economic importance. In Table 8.3, for example, note that the combustion products exiting a combustion chamber after complete combustion possess chemical exergy that is regarded to have no economic value.

With the exception of separation processes (water desalination, air separation, distillation, etc.) where the objective is to increase the nonreactive part of chemical exergy, the real economic value of chemical exergy is associated with chemical reactions. Thus, in certain thermoeconomic evaluations it is appropriate to split the chemical exergy of stream j into two components, the *reactive exergy* $\dot{E}_j^R$ and *nonreactive exergy* $\dot{E}_j^N$, which are associated with chemical and nonchemical effects, respectively:

$$\dot{E}_j^{CH} = \dot{E}_j^R + \dot{E}_j^N \qquad (8.35)$$

In Reference 11 the two components of chemical exergy are called *reaction exergy* and *environmental exergy,* respectively.

Before proceeding with the costing of the reactive and nonreactive components, the calculation of reactive and nonreactive exergy is discussed. We begin by noting that the reactive exergy of a *single substance present* in the standard reference exergy environment (Section 3.4.1) is zero. Thus, for reference substances (O_2, N_2, CO_2, H_2O, etc.), the nonreactive exergy is identical to the chemical exergy. For single substances *not present* in the reference environment, the hypothetical system of Figure 8.6 is used to determine the reactive exergy: We think of the nonreference substance reacting with reference substances (e.g., atmospheric air) in a reversible, isothermal, and isobaric reaction to form the reaction products. Both the reactants and the reaction products are supplied to and removed from the system at temperature T_0 and pressure p_0. Heat transfer occurs only with the environment. Figure 8.6 shows the case of a fuel reacting with atmospheric air.

Using Figure 8.6, the molar reactive exergy is evaluated as the work per kmol of fuel obtained from the reaction chamber in the absence of internal irreversibilities, $(\dot{W}/\dot{n}_{fuel})_{rev}$, which equals $-\Delta G$ for the complete chemical reaction. In the case of a pure hydrocarbon fuel C_aH_b, for example, the molar reactive exergy of the fuel is

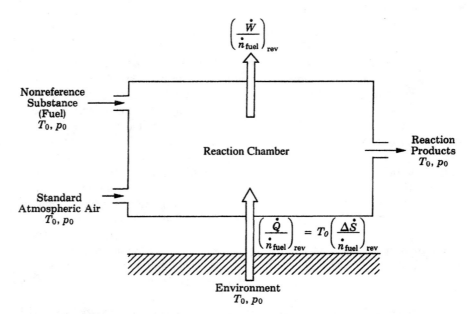

Figure 8.6 System used to calculate the reactive exergy of a nonreference substance. Assumptions: reversible, isothermal, and isobaric reaction at T_0 and p_0. Heat transfer occurs only with the environment.

$$\overline{e}_F^R = \overline{g}_F + 4.76 \left(a + \frac{b}{4} \right) \overline{g}_{air} - a\overline{g}_{CO_2} - \frac{b}{2} \overline{g}_{H_2O}$$

$$- 3.76 \left(a + \frac{b}{4} \right) \overline{g}_{N_2} \tag{8.36}$$

In this equation we assume for simplicity that air is a mixture of only O_2 (21 vol %) and N_2 (79 vol %). The molar Gibbs functions of CO_2, H_2O, and N_2 in the combustion products are calculated at T_0 and at the corresponding partial pressure.

The differences in the models used for calculating chemical exergy (Figure 3.5 and Equation 3.17b) and reactive exergy (Figure 8.6 and Equation 8.36) lie in the oxidant (pure oxygen and atmospheric air, respectively) and in the fact that in the reactive exergy calculation the reaction products leave the reaction system as a mixture, whereas in the chemical exergy calculation the reaction products are removed separately (unmixed) from the system. The reactive exergy is independent of the model used to calculate the standard chemical exergy values (Section 3.4.1)

The molar reactive exergy of a *gas mixture* may be calculated from

$$\overline{e}^R = \sum_j x_j \overline{e}_j^R \tag{8.37}$$

where x_j is the mole fraction of the jth gas in the mixture.

The nonreactive exergy is the difference between chemical and reactive exergy

$$\overline{e}^N = \overline{e}^{CH} - \overline{e}^R \tag{8.38}$$

After this review of the procedures used for calculating the reactive and nonreactive exergy, we proceed with the costing of these exergy components. Both the reactive exergy and the associated costs are conserved in all components where no chemical reaction occurs. At steady state, we write

$$\sum_i \dot{E}_i^R = \sum_e \dot{E}_e^R \tag{8.39}$$

and

$$\sum_i c_i^R \dot{E}_i^R = \sum_e c_e^R \dot{E}_e^R \tag{8.40}$$

Here, the subscripts i and e refer to the material streams at the inlet and outlet of the component, respectively. The cost rate associated with the chemical exergy of stream j would be written as

$$\dot{C}_j^{CH} = \dot{C}_j^R + \dot{C}_j^N \qquad (8.41a)$$

or

$$c_j^{CH}\dot{E}_j^{CH} = c_j^R\dot{E}_j^R + c_j^N\dot{E}_j^N \qquad (8.41b)$$

When the reactive exergy represents the only cost-relevant component of the chemical exergy, Equations 8.41 reduce to

$$\dot{C}_j^{CH} = c_j^{CH}\dot{E}_j^{CH} = \dot{C}_j^R = c_j^R\dot{E}_j^R \qquad (8.42)$$

In this case, all costs calculated in the thermoeconomic analysis are independent of the model used to calculate the standard chemical exergy values.

Finally, the introduction of reactive exergy allows for a more rational definition of the exergetic efficiency for a mixing or separation process. For the mixing process shown in Tables 3.3 and 8.2, for instance, we obtain

$$\epsilon = \frac{\dot{E}_3 - \dot{E}_3^R}{(\dot{E}_1 - \dot{E}_1^R) + (\dot{E}_2 - \dot{E}_2^R)} \qquad (8.43)$$

This definition considers only the nonchemical effects associated with the mixing process and makes the exergetic efficiency independent of the potential for reaction associated with the streams being mixed.

8.4.3 Costing Mechanical and Thermal Exergy

For ideal gases and incompressible liquids, the physical exergy can be expressed unambiguously in terms of *mechanical exergy* and *thermal exergy,* which are associated with the pressure and temperature of the substance, respectively:

$$\dot{E}^{PH} = \dot{E}^M + \dot{E}^T \qquad (8.44)$$

This is exemplified by Equation 3.14 giving the physical exergy associated with an ideal-gas stream for the case of a constant specific heat ratio. The equation consists of an expression involving pressure p but not temperature, and an expression involving temperature T but not pressure. These may be identified, respectively, as the mechanical and thermal exergy. Although splitting the physical exergy into these components is not *generally* possible, plausible means for circumventing this limitation have been suggested [12]. Applications are discussed in the literature [2, 12, 13].

In thermoeconomics, it is meaningful to distinguish between the mechanical and thermal contributions only in cases where each is supplied to a system at significantly different costs. The cost rate associated with the physical exergy of stream j is then

$$\dot{C}_j^{PH} = c_j^{PH}\dot{E}_j^{PH} = c_j^M\dot{E}_j^M + c_j^T\dot{E}_j^T = \dot{C}_j^M + \dot{C}_j^T \tag{8.45}$$

Compared with the case where the physical exergy is not split into these components, the introduction of mechanical and thermal exergy increases the number of auxiliary relations required to solve for the unknown variables c_j^M and c_j^T. For example, in formulating the additional auxiliary relations for the components shown in Table 8.3, we note that the auxiliary relations given in this table for the costs per unit of physical exergy apply now to the costs per unit of thermal exergy. To formulate auxiliary relations for mechanical exergy, we must consider that the mechanical exergy in a process changes as a result of friction, compression, expansion, and phase change. Thus, the following additional auxiliary relations are used for mechanical exergy:

1. In processes involving *removal* of mechanical exergy from the working fluid (e.g., through expansion and/ or friction or through condensation at a pressure greater than p_0), the appropriate thermoeconomic relation is

$$c_e^M = c_i^M \tag{8.46}$$

   where the subscripts i and e refer to the inlet and outlet state of the working fluid, respectively.
2. For a compression or vaporization process, where both mechanical and thermal exergy are *supplied* to the working fluid, we write

$$\frac{\dot{C}_e^T - \dot{C}_i^T}{\dot{E}_e^T - \dot{E}_i^T} = \frac{\dot{C}_e^M - \dot{C}_i^M}{\dot{E}_e^M - \dot{E}_i^M} \tag{8.47}$$

   This equation implies that the *addition* of mechanical and thermal exergy to the working fluid occurs at the same cost per unit of supplied exergy.
3. When a working fluid condenses at a pressure lower than the pressure p_0 of the environment, mechanical exergy is *added* to it. This case is directly related to the costing of condensers. A simple approach is to set $c_e^M = c_i^M$ and $c_e^T = c_i^T$ and apportion all remaining costs associated with the condenser to all components served by it.[5] This apportioning is done according to the ratio $\Delta\dot{S}_k/\Sigma_k\,\Delta\dot{S}_k$ where $\Delta\dot{S}_k$ refers to the entropy

---

[5]In a vapor power cycle, for example, the condenser serves *all* other components.

increase of the working fluid $(\dot{S}_e - \dot{S}_i)_k$ in the kth component served by the condenser.[6]

The distinction between mechanical and thermal exergy permits the definition of a more rational exergetic efficiency for some components. For example, if the purpose of the heat exchanger of Table 8.2 is to heat the cold stream and no phase change occurs in that stream, we write

$$\epsilon = \frac{\dot{E}_2^T - \dot{E}_1^T}{(\dot{E}_3 - \dot{E}_4) + (\dot{E}_1^M - \dot{E}_2^M) + (\dot{E}_1^{CH} - \dot{E}_2^{CH})} \tag{8.48}$$

In the above definition of the heat exchanger exergetic efficiency, all pressure drops are considered on the fuel side.

The improvements in the accuracy of the exergetic and thermoeconomic evaluations obtained through splitting of the physical exergy into its components are often slight and do not always justify the increase in the complexity of the calculations.

8.4.4 Diagram of Unit Cost Versus Specific Exergy

To understand the cost formation process and the cost flow between components, for each stream of matter it is necessary to account for every addition and removal of exergy and for the cost at which each addition or removal of exergy occurs. Using this information we may plot for each stream and for the xth form of exergy (total, chemical, physical, reactive, nonreactive, mechanical, or thermal exergy) the cost per unit of exergy as a function of the specific exergy, $c^x = f(e^x)$. Such a diagram illustrates for each stream of matter the history of the exergy and cost of the xth exergy form. With this approach, the need for auxiliary thermoeconomic relations is eliminated but at the expense of increasing the required computations. For further discussion see References 2, 13, and 14.

8.5 CLOSURE

A thermoeconomic (exergoeconomic) analysis combines thermodynamic and economic analyses at the component level. Costing in a thermoeconomic analysis is based on exergy, which in most cases is the only rational basis for assigning costs to the exergy streams of thermal systems.

---

[6]The rationale for this cost-apportioning approach is that in a thermodynamic cycle the entropy rate of the working fluid must be decreased in the condenser by the amount $\Sigma_k \Delta\dot{S}_k$. This represents the sum of the rates of entropy addition to the working fluid in the remaining components.

A thermoeconomic analysis identifies the real cost sources in a thermal system. A comparison between investment and O&M costs and costs associated with exergy destruction for major components is useful in developing design changes that improve the cost effectiveness of the entire system. In addition, a thermoeconomic analysis can be used to understand the cost formation process within a system and to calculate separately the cost at which each product stream is generated.

A thermoeconomic analysis is also very useful, particularly for complex thermal systems, in optimizing the entire system or specific variables in a single component. Applications of thermoeconomics to optimization are discussed in the next chapter.

REFERENCES

1. G. Tsatsaronis, Combination of exergetic and economic analysis in energy-conversion processes, in A. Reis, et al., eds., *Energy Economics and Management in Industry,* Proceedings of the European Congress, Algarve, Portugal, April 2–5, 1984, Permagon Press, Oxford, England, Vol. 1, pp. 151–157.

2. G. Tsatsaronis, Thermoeconomic analysis and optimization of energy systems, *Prog. Energy Combustion Sci.,* Vol. 19, 1993, pp. 227–257.

3. E. P. DeGarmo, W. G. Sullivan, and J. A. Bontadelli, *Engineering Economy,* 9th ed., Macmillan, New York, 1993.

4. R. A. Gaggioli, Second law analysis for process and energy engineering, in R. A. Gaggioli, ed., *Efficiency and Costing,* ACS Symposium Series, Vol. 235, American Chemical Society, Washington, D.C., 1983, pp. 1–50.

5. G. Tsatsaronis and M. Winhold, *Thermoeconomic Analysis of Power Plants,* EPRI AP-3651, RP 2029-8, Final Report, Electric Power Research Institute, Palo Alto, CA, August 1984.

6. G. Tsatsaronis and M. Winhold, Exergoeconomic analysis and evaluation of energy conversion plants. Part I—a new general methodology; Part II—analysis of a coal-fired steam power plant, *Energy—Int. J.,* Vol. 10, No. 1, 1985, pp. 69–94.

7. G. Tsatsaronis, M. Winhold, and C. G. Stojanoff, *Thermoeconomic Analysis of a Gasification-combined-cycle Power Plant,* EPRI AP-4734, RP2029-8, Final Report, Electric Power Research Institute, Palo Alto, Ca, August 1986.

8. A. Valero, D. Wimmert, and C. Torres, SYMBCOST: A program for symbolic computation of exergoeconomics cost parameters, in G. Tsatsaronis, et al., eds., *Computer-Aided Energy-Systems Analysis* AES-Vol. 21, American Society of Mechanical Engineers, New York, 1991, pp. 13–21.

9. A. Valero, M. Lozano, and C. Torres, A general theory of exergy savings—Parts I–III, in R. A. Gaggioli, ed., *Computer-Aided Engineering of Energy Systems,* AES-Vol. 2–3, American Society of Mechanical Engineers, New York, 1986, pp. 1–121.

10. A. Valero, M. Lozano, L. Serra, and C. Torres, Application of the exergetic cost theory to the CGAM problem, *Energy—Int. J.* Vol. 19, 1994, pp. 365–381.

11. G. Tsatsaronis, J. J. Pisa, and L. M. Gallego, Chemical exergy in exergoeconomics, in R. Cai, and M. J. Moran, eds., *Thermodynamic Analysis and Improvement of Energy Systems,* Proceedings of the International Symposium, Beijing, China, June 5–8, 1989, Pergamon Press, Oxford, 1989, pp. 195–200.

12. G. Tsatsaronis, J. J. Pisa, and L. Lin, The effect of assumptions on the detailed exergoeconomic analysis of a steam power plant design configuration; Part I: theoretical development; Part II: results and discussion, in S. Stecco and M. J. Moran, eds., *A Future for Energy,* Proceedings of the Florence World Energy Research Symposium, Florence, Italy, May 28–June 1, 1990, pp. 771–792.

13. G. Tsatsaronis and J. Pisa, Exergoeconomic evaluation and optimization of energy systems—application to the CGAM problem, *Energy—Int. J.,* Vol. 19, 1994, pp. 287–321.

14. G. Tsatsaronis, L. Lin, and J. Pisa, Exergy costing in exergoeconomics, *J. Energy Resources Tech.,* Vol. 115, 1993, pp. 9–16.

PROBLEMS

8.1 Repeat the calculations of Examples 8.1 and 8.2 using the levelized annual costs in current dollars for a levelization time period of 20 years (Table 7.15) instead of 10 years. All other assumptions and data of Examples 8.1 and 8.2 remain unchanged. Compare the new results with those of Examples 8.1 and 8.2. What *qualitative* changes of the design variables are recommended from this analysis?

8.2 Consider an overall system consisting of the gasification and cogeneration systems shown in Figure P8.2. The numbers given for each stream represent exergy flows. The levelized cost per exergy unit of stream 1 is $c_1 = \$3.0/GJ$. The cost associated with stream 7 (combustion air and water) is zero. The levelized cost rates associated with capital investment and operating and maintenance expenses are

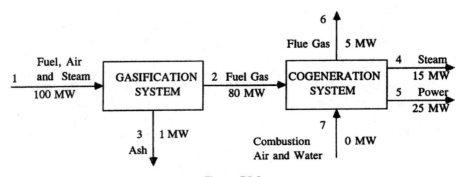

Figure P8.2

$$\text{Gasification system} \quad \dot{Z}_{\text{gas}} = \$1.0/\text{s}$$

$$\text{Cogeneration system} \quad \dot{Z}_{\text{cogen}} = \$0.5/\text{s}$$

The flue gas (stream 6) and ash (stream 3) are discharged to the environment without any additional expenses.

Calculate: (a) the exergy destruction rates and the exergy destruction ratios for the gasification system and the cogeneration system; (b) the exergetic efficiencies of the gasification system, the cogeneration system, and the overall system; (c) the cost rates and the costs per exergy unit associated with streams 2 (fuel gas), 4 (steam), and 5 (electricity).

8.3   Figure P8.3 shows an air compressor with intercooling. The numbers given for each stream represent exergy flows. All costs are levelized costs. The cost per exergy unit of stream 2 is $c_2 = \$10.0/\text{GJ}$. The costs associated with streams 1 (air) and 5 (feedwater inlet) are zero. The cost rates associated with capital investment and operating and maintenance expenses are

$$\text{Compressor} \quad \dot{Z}_c = \$0.01/\text{s}$$

$$\text{Feedwater heater} \quad \dot{Z}_{\text{fwh}} = \$0.02/\text{s}$$

Calculate: (a) the exergy destruction rates and the exergy destruction ratios for the compressor and the feedwater heater; (b) the exergetic efficiencies of the compressor, the feedwater heater, and the overall system shown in Figure P8.3; (c) the cost rate and the cost per exergy unit associated with streams 3 (compressed air), 4 (cooled compressed air), and 6 (preheated feedwater).

8.4   To consider the effect of aggregation level (Section 8.1.2) for the overall system shown in Figure P8.3, formulate a cost balance for the sys-

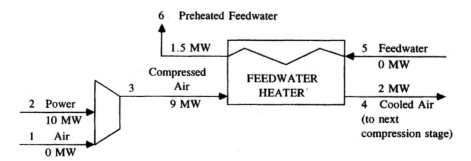

Figure P8.3

tem without considering separately the air compressor and the feedwater heater. Develop the required auxiliary relation and calculate the cost rates and the costs per exergy unit associated with cooled compressed air and preheated feedwater. Discuss.

8.5   Using Equations 8.20 and 8.23b, develop an expression for the relative cost difference r_k. List all assumptions made for this development.

8.6   For the system of Figure P8.2, (a) calculate all thermoeconomic variables associated with the gasification system and the cogeneration system, and (b) discuss options for improving the cost effectiveness of the overall system.

8.7   For the system of Figure P8.3, (a) calculate all thermoeconomic variables associated with the air compressor and the feedwater heater, and (b) discuss options for improving the cost effectiveness of the overall system.

8.8   For the cogeneration system of Example 8.1, explain why the cost per exergy unit of steam is *higher* than the cost per exergy unit of electricity generated by the system. Is this a general result that would apply to both a new system design *and* an existing system (where only fuel costs are considered) if both systems have the configuration of Figure 1.7? Should the owner of the cogeneration system try to *sell* the two product streams at prices equal to the calculated costs?

8.9   In the cogeneration system of Figure 8.2, the cost per exergy unit of steam is *lower* than the cost per exergy unit of electricity generated by the system. Answer the same questions as in Problem 8.8. If the answers are different than in Problem 8.8, discuss the effects causing these differences.

8.10   A high-pressure (HP) boiler and a low-pressure (LP) boiler will be added to the steam-generating system of a chemical plant. Both boilers use the same fuel and have about the same heat loss rate $\dot{Q}$. The average combustion temperature is higher in the HP than in the LP boiler. Should the design engineer spend the same amount of money to insulate each boiler? Justify your answer.

8.11   For the system shown in Figure 8.5b calculate for each stream the exergy flow rate and the cost per exergy unit ignoring heat loss. Assume $\dot{W}_{el} = 100$ kW, $c_1 = 5$ ¢/kWh, $\epsilon_m = 95\%$, $\varepsilon_c = 70\%$, $\dot{E}_5 = 30$ kW, $\dot{E}_6 = 5$ kW and $c_6 = \$2.0/GJ$. The exergetic efficiency of the water heater is 30%. The contributions of the investment and O&M costs may be neglected. What values would you calculate for c_5 and c_7 if only one cost balance for the overall system would be formulated?

8.12   Calculate the reactive exergy of streams 3, 4, and 10 in Figure 1.7.

THERMOECONOMIC OPTIMIZATION

As discussed in Section 1.2 the term *optimal* can take on many guises depending on the application. In this chapter, unless otherwise indicated, *optimization* means the modification of the structure *and* the design parameters of a system to *minimize the total levelized cost of the system products* under *boundary conditions* associated with available materials, financial resources, protection of the environment, and governmental regulation, together with the safety, operability, reliability, availability, and maintainability of the system.

A *thermodynamic* optimization aims at minimizing the thermodynamic inefficiencies: exergy destruction and exergy loss (Sections 3.5 and 6.2). The objective of a *thermoeconomic* optimization, however, is to *minimize costs,* including costs owing to thermodynamic inefficiencies. A truly optimized system is one for which the magnitude of *every* significant thermodynamic inefficiency is justified by considerations related to costs or is imposed by at least one of the above boundary conditions.

This chapter presents thermoeconomic approaches to optimization involving the calculation of the cost-optimal exergetic efficiency for *single* system components and the *iterative optimization* of complex systems. The optimization of heat exchanger networks and some conventional mathematical optimization techniques are also discussed in Sections 9.3 and 9.4, respectively. The results from the optimization of the cogeneration system case study are presented in Section 9.5.

9.1 INTRODUCTION TO OPTIMIZATION

Engineers involved in thermal system design are required to answer questions such as: What processes or equipment items should be selected and how should they be arranged? What is the preferred size of a component or group of components? Should some equipment items be used in parallel (parallel trains) in specific processes to increase the overall system availability? What is the best temperature, pressure, flow rate, and chemical composition of each stream in the system? To answer these questions, engineers need to *formulate an appropriate optimization problem.* Appropriate problem formulation is usually the most important and sometimes the most difficult step of a successful optimization study. In the following we discuss the essential features of optimization problems. More details are given in the literature [1–3].

System Boundaries. The first step in an optimization study is to define clearly the boundaries of the system to be optimized. All the subsystems that significantly affect the performance of the system under study should be included in the optimization problem. When dealing with complex systems, it is often desirable to break them down into smaller subsystems that can be optimized individually. In such cases, the selection of the subsystem boundaries is very important because the optimization results may be affected by this selection. This issue is part of the *suboptimization* considerations discussed later in the present section.

Optimization Criteria. The selection of criteria on the basis of which the system design will be evaluated and optimized is a key element in formulating an optimization problem. Optimization criteria may be *economic* (total capital investment, total annual levelized costs, annual levelized net profit, return on investment, or any of the profitability evaluation criteria discussed in Sections 7.5.1 and 7.5.2), *technological* (thermodynamic efficiency, production time, production rate, reliability, total weight, etc.), and *environmental* (e.g., rates of emitted pollutants). An optimized design is characterized by a minimum or maximum value, as appropriate, for each selected criterion.

In practice, it is usually desirable to develop a design that is "best" with respect to more than one criterion. As these criteria usually compete with each other, it is impossible to find a solution that, for example, simultaneously minimizes costs and environmental impact while maximizing efficiency and reliability. Advanced techniques for solving certain optimization problems with multiple competing criteria are discussed in the literature [4–6] but are not covered in this book. Here, only one primary criterion is used as an optimization performance measure—the total annual levelized cost of the system products. All other criteria (return on investment, reliability, environmental performance, etc.) are treated as *problem constraints* or *parameters.*

Variables. Another essential element in formulating an optimization problem is the selection of the *independent variables* that adequately characterize the possible design options. In selecting these variables, it is necessary to (a) include all the important variables that affect the performance and cost effectiveness of the system, (b) not include fine details or variables of minor importance, and (c) distinguish among independent variables whose values are amenable to change: the *decision variables* and the *parameters* whose values are *fixed* by the particular application. In optimization studies, only the decision variables may be varied; the parameters are independent variables that are *each given one specific and unchanging value* in any particular model statement. The variables whose values are calculated from the independent variables using the mathematical model are the *dependent variables*. The decision variables, the parameters, and the dependent variables for the cogeneration system case study are given in Section 2.5.

In a *preliminary* design, for example, it is generally not necessary to consider the details of the design of each system component. A compressor may be aptly characterized by just the pressure ratio, the volumetric flow, and the isentropic or polytropic efficiency. For a heat exchanger, it may be sufficient to use the heat transfer surface area and the pressure drop on each side. Detailed component design variables would normally be considered in a separate design study involving each system component individually.

In some thermal systems, there are decision variables that affect the product cost without influencing the thermodynamic performance (efficiency) of the system. As a simplification in practice, these variables often may be considered separately and optimized *after* the decision variables affecting both cost *and* efficiency have been optimized.

The Mathematical Model. A mathematical model is a description in terms of mathematical relations, invariably involving some idealization, of the functions of a physical system [3]. The mathematical model describes the manner in which all problem variables are related and the way in which the independent variables affect the performance criterion. The mathematical model for an optimization problem consists of

- An objective function to be maximized or minimized
- Equality constraints
- Inequality constraints

The *objective function* expresses the optimization criterion as a function of the dependent and independent variables. In the cogeneration system case study, for example, the objective function obtained from Equation 8.1 is

$$\text{Minimize } \dot{C}_{P,\text{tot}} = \dot{C}_{F,\text{tot}} + \dot{Z}_{tot}^{\text{CI}} + \dot{Z}_{\text{tot}}^{\text{OM}} \qquad (9.1)$$

The variables $\dot{C}_{F,\text{tot}}$, $\dot{Z}_{\text{tot}}^{\text{CI}}$, and $\dot{Z}_{\text{tot}}^{\text{OM}}$ are functions of the decision variables for this problem. The functions are given in Section 2.5 and in Appendix B, which present the thermodynamic model and the economic model, respectively, of the cogeneration system. In this application we minimize the total cost rate associated with the product $\dot{C}_{P,\text{tot}}$ instead of the cost per unit of product exergy c_p. This can be done since the product $\dot{E}_P$ is constant: The generated net power is specified as 30 MW, and the steam supplied by the system is specified as saturated vapor at 20 bars and a mass flow rate of 14 kg/s.

The *equality* and *inequality constraints* are provided by appropriate thermodynamic and cost models as well as by appropriate boundary conditions. These models generally include material and energy balance equations for each component, relations associated with the engineering design such as maximum or minimum values for temperatures and pressures, the performance and the cost of each component as well as physical and chemical properties of the substances involved. The models also contain equations and inequalities that specify the allowable operating ranges, the maximum or minimum performance requirements, and the bounds on the availability of resources.

The development of a mathematical model requires a good understanding of the system and is normally a time-consuming activity. In modeling an optimization problem, the most common mistake is to omit something significant. A discussion of how to reduce such omissions by systematically checking the model before trying to compute with it is given in Reference 3. Process design simulations and flowsheeting software (Section 1.6.3) are also very useful for modeling purposes.

Suboptimization. Suboptimization is usually applied to complex thermal systems, particularly when the optimization of the entire system may not be feasible owing to complexity. Suboptimization is the optimization of one part of a problem or of a subsystem, ignoring some variables that affect the objective function or other subsystems. Suboptimization is useful when neither the problem formulation nor the available optimization techniques allow a solution to the entire problem. In practice, suboptimization of a system may also be necessary because of economic and practical considerations such as limitations on time or manpower. However, suboptimization of all subsystems separately does not necessarily ensure optimization of the overall system. An alternative to suboptimization is provided by the iterative thermoeconomic optimization approach discussed in Section 9.6.

9.2 COST-OPTIMAL EXERGETIC EFFICIENCY FOR AN ISOLATED SYSTEM COMPONENT

Several mathematical approaches may be applied to optimize the design of a single system component in *isolation* from the remaining system components. Some of these approaches are mentioned in Section 9.4. The present section involves a *thermoeconomic* approach that illustrates clearly the connections between thermodynamics and economics. With this approach, the cost-optimal exergetic efficiency is obtained for a component isolated from the remaining system components. The same approach can also be used for suboptimizing the thermoeconomically major components of a complex system (see Section 9.6). The approach is based on assumptions A1 through A4, expressed analytically by Equations 9.2–9.6.

Assumption A1. The exergy flow rate of the product $\dot{E}_P$ and the unit cost of the fuel c_F remain constant for the kth component to be optimized:

$$\dot{E}_{P,k} = \text{constant} \tag{9.2}$$

$$c_{F,k} = \text{constant} \tag{9.3}$$

These equations, which represent constraints of the optimization problem, define mathematically what is meant by isolation in the present context.

Assumption A2. For every system component, we expect the investment costs to increase with increasing capacity and increasing exergetic efficiency of the component. Here we assume that for the kth component the total capital investment (TCI_k) can be represented, at least approximately, by the following relation [7–9]:

$$\text{TCI}_k = B_k \left(\frac{\varepsilon_k}{1 - \varepsilon_k}\right)^{n_k} \dot{E}_{P,k}^{m_k} \tag{9.4}$$

where $\dot{E}_{P,k}$ is the exergy rate of the product for the kth component and ε_k is the component's exergetic efficiency. The term $[\varepsilon_k/(1 - \varepsilon_k)]^{n_k}$ expresses the effect of efficiency (thermodynamic performance), while the term $\dot{E}_{P,k}^{m_k}$ expresses the effect of capacity (component size) on the value of TCI_k. Equation 9.4 is assumed to be valid within a certain range of the design options or design conditions for the kth component. Within this range, the parameter B_k and the exponents n_k and m_k are constant. However, as in end-of-chapter Problem 9.3, B_k can depend on thermodynamic variables. To apply the present approach iteratively in such a case, the current values of the underlying thermodynamic variables are used in each iteration.

The total capital investment associated with the kth component may be estimated from the purchased-equipment cost of the same component using the factors discussed in Section 7.1 (see, e.g., Equations 7.17). When the purchased-equipment costs for different design conditions of the kth component are known, the method of least squares can be used to calculate the values of the constants B_k, n_k, and m_k in Equation 9.4 [7]. The application of the least-squares method is simplified if the value of the capacity exponent m_k can be assumed equal to the scaling exponent α (Table 7.3) for the respective equipment item, but this is not always appropriate.

The discussion thus far refers only to components for which a meaningful exergetic efficiency may be defined (see Section 3.5.3). If no such efficiency can be formulated for a component (e.g., for a throttling valve or a cooler), this component may be optimized in isolation by minimizing the sum of its investment and operating and maintenance costs. However, a preferred approach is to optimize such a component *together* with the component(s) it serves.

Assumption A3. Usually a part of the operating and maintenance (O&M) costs depends on the total investment costs and another part on the actual production rate. Here we assume that the *annual* levelized operating and maintenance costs attributed to the kth system component may be represented, at least approximately, by [8, 9]

$$Z_k^{\text{OM}} = \gamma_k(\text{TCI}_k) + \omega_k \tau \dot{E}_{P,k} + R_k \tag{9.5}$$

In this equation, γ_k is a coefficient that accounts for the part of the fixed operating and maintenance costs depending on the total capital investment associated with the kth component[1]; ω_k is a constant that accounts for the variable operating and maintenance costs associated with the kth component, and denotes the O&M cost per unit of product exergy; τ is the average annual time of plant operation at the nominal load; and R_k includes all the remaining operating and maintenance costs that are independent of the total capital investment and the exergy of the product.

Normally the annual operating and maintenance costs are calculated for the *entire* system (see, e.g., Table 7.15 for the cogeneration system case study). In the absence of other information, the purchased-equipment cost for each component may be used as a weighting factor to apportion the operating and maintenance costs of the entire system among its components.

Assumption A4. The economic analysis of the system being considered (see, e.g., Table 7.12 for the cogeneration system of Figure 1.7) is simplified

---

[1]In large conventional electric power plants, for example, an average value for the coefficient γ_k of 0.015 × CELF may be assumed for all plant components; CELF is the constant-escalation levelization factor, Equation 7.34. For relatively small thermal systems, the coefficient γ_k could be as high as 0.10.

by neglecting the effects of financing, inflation, taxes, insurance, and construction time and by considering the startup costs, working capital and the costs of licensing, research, and development together with the total capital investment. The annual carrying charge associated with the kth component is then obtained by multiplying the total capital investment for this component (TCI_k) by the capital recovery factor β, given by Equation 7.28, as follows:

$$Z_k^{\mathrm{CI}} = \beta(\mathrm{TCI}_k) \tag{9.6}$$

Assumptions A1 through A4 (Equations 9.2 through 9.6) form the *cost model*. The total annual levelized costs excluding fuel costs associated with the kth component are obtained by combining Equations 9.5 and 9.6

$$Z_k = Z_k^{\mathrm{CI}} + Z_k^{\mathrm{OM}} = (\beta + \gamma_k)(\mathrm{TCI}_k) + \omega_k \tau \dot{E}_{P,k} + R_k \tag{9.7}$$

The corresponding cost rate $\dot{Z}_k$ (Equation 8.2) is obtained by dividing Equation 9.7 by τ:

$$\dot{Z}_k = \frac{\beta + \gamma_k}{\tau}(\mathrm{TCI}_k) + \omega_k \dot{E}_{P,k} + \frac{R_k}{\tau} \tag{9.8a}$$

Then with Equation 9.4

$$\dot{Z}_k = \frac{(\beta + \gamma_k)B_k}{\tau}\left(\frac{\varepsilon_k}{1 - \varepsilon_k}\right)^{n_k} \dot{E}_{P,k}^{m_k} + \omega_k \dot{E}_{P,k} + \frac{R_k}{\tau} \tag{9.8b}$$

The objective function to be minimized expresses the cost per exergy unit of product for the kth component. Accordingly, with Equations 8.19 and 8.20, we write

$$\mathrm{Minimize}\ c_{P,k} = \frac{c_{F,k}\dot{E}_{F,k} + \dot{Z}_k}{\dot{E}_{P,k}} \tag{9.9}$$

With Equations 3.29 and 9.8b, this objective function may be expressed as

$$\mathrm{Minimize}\ c_{P,k} = \frac{c_{F,k}}{\varepsilon_k} + \frac{(\beta + \gamma_k)B_k}{\tau \dot{E}_{P,k}^{1-m_k}}\left(\frac{\varepsilon_k}{1 - \varepsilon_k}\right)^{n_k} + \omega_k + \frac{R_k}{\tau \dot{E}_{P,k}} \tag{9.10}$$

The values of parameters β, γ_k, B_k, τ, ω_k, and R_k remain constant during the optimization process, and so $c_{P,k}$ varies only with ε_k. Thus, the optimization problem reduces to the minimization of Equation 9.10 subject to the constraints expressed by Equations 9.2 and 9.3. The minimum cost per exergy unit of product is obtained by differentiating Equation 9.10 and setting the derivative to zero:

$$\frac{dc_{P,k}}{d\varepsilon_k} = 0$$

The resulting cost-optimal (superscript OPT) exergetic efficiency is

$$\varepsilon_k^{OPT} = \frac{1}{1 + F_k} \tag{9.11}$$

where

$$F_k = \left(\frac{(\beta + \gamma_k)B_k n_k}{\tau c_{F,k} \dot{E}_{P,k}^{1-m_k}} \right)^{1/(n_k+1)} \tag{9.12}$$

Equations 9.11 and 9.12 show that the cost-optimal exergetic efficiency increases with increasing cost per exergy unit of fuel $c_{F,k}$, increasing annual number of hours of system operation τ, decreasing capital recovery factor β, decreasing fixed O&M cost factor γ_k, and decreasing cost exponent n_k.

Equation 9.11 may be rewritten as

$$F_k = \frac{1 - \varepsilon_k^{OPT}}{\varepsilon_k^{OPT}} \tag{9.13a}$$

or with Equations 3.28 and 3.29 as

$$F_k = \left(\frac{\dot{E}_{D,k} + \dot{E}_{L,k}}{\dot{E}_{P,k}} \right)^{OPT} \tag{9.13b}$$

Since the exergy rate of the product is assumed constant during optimization, the cost optimal value of the sum $(\dot{E}_{D,k} + \dot{E}_{L,k})$ is given by

$$(\dot{E}_{D,k} + \dot{E}_{L,k})^{OPT} = \dot{E}_{P,k} F_k = \dot{E}_{P,k} \left(\frac{1 - \varepsilon_k^{OPT}}{\varepsilon_k^{OPT}} \right) \tag{9.14}$$

At this point, a simplification of assumption A3 allows some additional results to be obtained: In Equation 9.5 (and in Equations 9.7, 9.8, and 9.10) we may neglect the last two terms on the right side referring to certain portions of the O&M costs since these costs are often small compared with the remaining term(s) on the same side of the respective equations. With this simplification and with Equation 3.29, Equation 9.10 can be expressed in terms of $(\dot{E}_{D,k} + \dot{E}_{L,k})$ as

$$\text{Minimize } c_{P,k} = c_{F,k}\left(1 + \frac{\dot{E}_{D,k} + \dot{E}_{L,k}}{\dot{E}_{P,k}}\right) + \frac{(\beta + \gamma_k)B_k}{\tau\dot{E}_{P,k}^{1-m_k}}\left(\frac{\dot{E}_{P,k}}{\dot{E}_{D,k} + \dot{E}_{L,k}}\right)^{n_k}$$

(9.15)

By differentiating Equation 9.15 with respect to $(\dot{E}_{D,k} + \dot{E}_{L,k})$ and setting the derivative to zero, we obtain after some reduction the following relation between the cost-optimal values of the cost rates expressed by $c_{F,k}(\dot{E}_{D,k} + \dot{E}_{L,k})$ and $\dot{Z}_k$:

$$n_k = \frac{c_{F,k}(\dot{E}_{D,k} + \dot{E}_{L,k})^{\text{OPT}}}{\dot{Z}_k^{\text{OPT}}}$$

(9.16)

Thus, under assumptions A1, A2, A4, and the simplified assumption A3, when the kth component is *optimized in isolation,* the cost exponent n_k in Equations 9.4, 9.8b, 9.10, and 9.12 expresses the ratio between the cost-optimal rate associated with exergy destruction and exergy loss and the cost-optimal rate associated with capital investment.

With Equations 9.14 and 9.16, we obtain the following expressions for the cost-optimal values of the non-fuel-related cost rate $\dot{Z}_k$, the relative cost difference r_k, Equation 8.27, and the exergoeconomic factor f_k, Equation 8.28:

$$\dot{Z}_k^{\text{OPT}} = c_{F,k}\dot{E}_{P,k}\frac{F_k}{n_k}$$

(9.17)

$$r_k^{\text{OPT}} = \frac{n_k + 1}{n_k}F_k$$

(9.18)

$$f_k^{\text{OPT}} = \frac{1}{1 + n_k}$$

(9.19)

The minimization of the product cost for a single component is illustrated in Figure 9.1, which in particular might be a representation of Equation 9.10

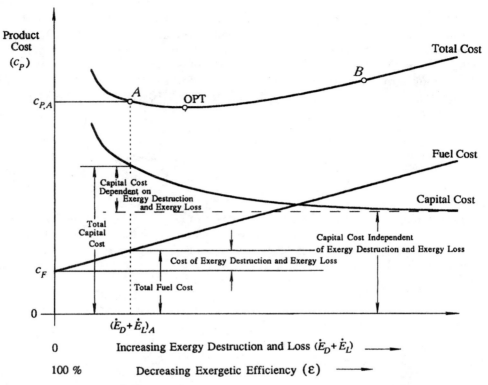

Figure 9.1 Schematic of the contributions of fuel cost and capital cost to the total product cost, as a function of the exergetic efficiency and the sum of exergy destruction and exergy loss.

or 9.15. This figure shows the effects of fuel cost and capital cost on the average cost per exergy unit of product c_P, as functions of the exergetic efficiency ε, and the sum of exergy destruction and exergy loss $\dot{E}_D + \dot{E}_L$. The abscissa is linear with respect to $\dot{E}_D + \dot{E}_L$ and nonlinear with respect to ε. The total cost is the sum of capital and fuel costs. For a design point indicated by A, the figure shows the parts of capital cost and fuel cost that depend on $\dot{E}_D + \dot{E}_L$. The average cost per exergy unit of fuel c_F is taken as constant.

Point A in Figure 9.1 is characterized by capital costs that are too high, whereas the fuel cost at point B is too high. The optimal design point is denoted by OPT. The total cost curve is usually flat around the optimal point. Therefore, several points around the optimal point may be regarded as nearly optimal. For simplicity, only capital investment and fuel costs have been considered in Figure 9.1. However, practically the same curves would be obtained qualitatively if we would include the contribution of the fixed O&M costs together with the investment costs and the variable O&M costs together with the fuel costs.

The procedure for optimizing a single component in isolation as presented in this section may also be applied to a *group* of components by appropriately adjusting the definition of fuel and product and the cost calculations (Equations 9.4 and 9.5). However, the larger the number of components considered in a group, the more difficult it may be to develop an acceptable cost equation such as Equation 9.4. The O&M costs can always be adjusted to fit Equation 9.5.

To use this optimization approach we must be able to express the total capital investment of a system component as a function of the exergetic efficiency and the capacity through a relation similar to Equation 9.4. This may not be possible for all components for which a meaningful exergetic efficiency is defined. The total capital investment associated with a heat exchanger, for example, cannot always be represented as in Equation 9.4. A single heat exchanger, however, may be optimized in isolation by determining the optimal minimum temperature difference.

Assumption A1 expressed by Equations 9.2 and 9.3 is, mathematically speaking, not fulfilled for components of complex plants because when several system variables are varied simultaneously, the variables $\dot{E}_{P,k}$ and $c_{F,k}$ generally do not remain constant for the kth component. Thus, when applied to a system component, the single-component optimization expressed by Equations 9.11, 9.12, and 9.14–9.19 may represent only a plausible *approximation*.

In thermal design, we are interested in optimizing the overall system. The conditions that optimize a single component in isolation usually do not correspond to an overall system optimization. Therefore, a single-component optimization generally has only limited usefulness. One of the most appropriate applications, however, may occur during the preliminary optimization phase, but then only to those components whose costs dominate: the components with the highest values for the variables $\dot{Z}_k$ and $\dot{C}_{D,k}$. The cost optimization of an overall thermal system is illustrated in Section 9.5 by application to the cogeneration system case study. In Section 9.3 we consider the optimization of heat exchanger networks—an important special class of thermal systems often encountered in practice.

9.3 OPTIMIZATION OF HEAT EXCHANGER NETWORKS

Heat exchanger networks (HENs) can be found in most complex thermal systems. The *typical* heat exchanger network design problem has several features related to a set of *process streams:* (i) An inlet temperature, a desired outlet temperature, a mass flow rate, and an average specific heat (c_p) are specified for each stream. (ii) Available *utility streams* such as steam and cooling water are also identified. (iii) Costing and thermal data are generally known, as for example, the cost of a heat exchanger as a function of the heat transfer area, utility costs, and appropriate heat transfer coefficients.

The optimization of a HEN involves finding the design configuration that minimizes the total annual levelized cost associated with the network. For optimization purposes it is assumed that the HEN consists of counterflow heat exchangers with negligible heat transfer to the surroundings.

Owing to the combinatorial nature of design (Section 1.5.3), a very large number of alternative HEN configurations can exist even for a relatively small number of process streams. In a problem that involves N_h *hot streams* and N_c *cold streams,* the total number of possible HEN configurations N_{hen} is

$$N_{hen} = (N_h \times N_c)! \tag{9.20}$$

This equation shows that N_{hen} increases very rapidly as the number of process streams increases. For example, there are 24 possible HEN configurations for a problem involving two hot and two cold streams. However, only a few of them are technically or economically feasible. For example, configurations requiring heat transfer from a cold to a hot stream can be eliminated on second-law grounds. Other configurations can be eliminated on the basis of cost. Still, an approach based on the complete enumeration and evaluation of all viable alternatives is time consuming at best, and may not be practical.

To avoid the need to evaluate all possibilities, plausible exploratory methods have been developed to assist the design of HENs. A review of such methods is given in Reference 10. In spite of the amount of work in this area, most available methods still have a number of limitations. Details about the various methods may be found in the literature [10–20]. Section 9.3 is based on the pinch analysis method, one of the most effective and easy to implement methods. In Section 9.3.1, we consider only two process streams. A problem with a larger number of streams is discussed in Sections 9.3.2 through 9.3.7.

9.3.1   Temperature–Enthalpy Rate Difference Diagram

Temperature–enthalpy difference diagrams play a central role in the analysis of HENs. Figure 9.2 provides such a diagram for the case of two streams, a hot stream (HS) and a cold stream (CS). The hot stream must be cooled from the inlet temperature T_{hi} to the outlet temperature T_{he}, whereas the cold stream must be heated from the inlet temperature T_{ci} to the outlet temperature T_{ce}. The direction of these processes is indicated by the arrows in the figure. All stream temperatures and mass flow rates are assumed to be fixed. The heat transfer rates are

$$\dot{Q}_h = \dot{m}_h(h_{he} - h_{hi}) \equiv (\Delta \dot{H})_h \tag{9.21}$$

$$\dot{Q}_c = \dot{m}_c(h_{ce} - h_{ci}) \equiv (\Delta \dot{H})_c \tag{9.22}$$

for the hot and cold stream, respectively. For cooling and heating purposes, a *cold utility* (e.g., cooling water) and a *hot utility* (e.g., steam) are available with the respective heat transfer rates (*utility loads*) of $\dot{Q}_{cu}$ and $\dot{Q}_{hu}$.

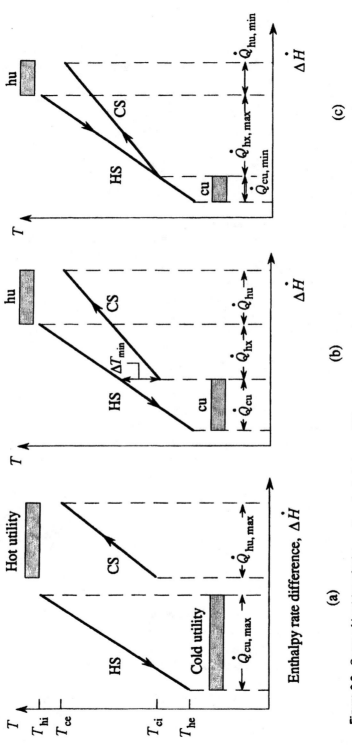

Figure 9.2 Cases of heat transfer integration of a hot stream (HS), a cold stream (CS), a hot utility (hu), and a cold utility (cu); hx = heat exchanger.

In Figure 9.2*a*, both the hot-stream cooling and the cold-stream heating requirements are *exclusively met* by the cold and the hot utilities. This arrangement establishes the *maximum* utility loads ($\dot{Q}_{\text{hu,max}}$ and $\dot{Q}_{\text{cu,max}}$). Among all design alternatives, this case is characterized by the highest utility costs and the lowest capital investment costs. When utility costs, capital investment costs, and the temperatures of the two streams are considered, we will usually find that the arrangement of Figure 9.2*a* is not cost optimal, however.

This can be seen by shifting the cold stream parallel to the enthalpy rate difference axis as shown in Figure 9.2*b*. In the arrangement of this figure, a part of the cold stream heating is done by directly cooling the hot stream. The enthalpy rate difference $\Delta\dot{H}$ in the overlapping interval corresponds to the heat transfer $\dot{Q}_{\text{hx}}$ from the hot to the cold stream (the *heat exchanger load*) that could be accomplished in a heat exchanger. The location with the minimum temperature difference ΔT_{min} in the heat exchanger is called the *pinch point* or simply *pinch*. The temperature difference at the pinch is a measure of the thermodynamic forces driving the heat transfer process in the heat exchanger and, thus, of the required heat transfer area A_{hx}. A parallel shifting of the cold stream shows that the temperature difference at the pinch is also a measure of the utility loads: The smaller ΔT_{min}, the larger $\dot{Q}_{\text{hx}}$ and A_{hx} and the smaller $\dot{Q}_{\text{hu}}$ and $\dot{Q}_{\text{cu}}$. Thus, with decreasing ΔT_{min}, the fuel and operating costs associated with the hot and cold utilities decrease, whereas the total capital cost generally increases. The optimization problem is to find the value of ΔT_{min} that minimizes the sum of the capital cost and the fuel and operating costs for the overall system.

Figure 9.2*c* shows the thermodynamically best arrangement, characterized by a zero pinch: $\Delta T_{\text{min}} = 0$ and the maximum value of $\dot{Q}_{\text{hx}}$. This diagram establishes the *minimum* utility loads ($\dot{Q}_{\text{hu,min}}$ and $\dot{Q}_{\text{cu,min}}$) for the given task. The cost optimal values of $\dot{Q}_{\text{hu}}$ and $\dot{Q}_{\text{cu}}$ are always larger than the minimum values shown in Figure 9.2*c* and are usually smaller than the maximum values indicated in Figure 9.2*a*.

In the case of only two streams of matter, the pinch position in the heat exchanger depends on the *heat capacity rates* ($\dot{m}c_p$) of the streams and on whether there is a phase change. In the absence of a phase change, there are three cases:

- $\dot{m}_h c_{\text{ph}} > \dot{m}_c c_{\text{pc}}$: The pinch occurs at the heat exchanger end determined by T_{hi} and the heat exchanger outlet temperature of the cold stream.
- $\dot{m}_h c_{\text{ph}} < \dot{m}_c c_{\text{pc}}$: The pinch occurs at the heat exchanger end determined by T_{ci} and the heat exchanger outlet temperature of the hot stream (Figure 9.2*b*)
- $\dot{m}_h c_{\text{ph}} = \dot{m}_c c_{\text{pc}}$: The profiles in the temperature–enthalpy difference diagram are parallel (balanced heat exchanger, Section 6.2.4). The temperature difference between the streams is constant and no pinch is formed.

When there is a phase change in at least one of the streams, the pinch may appear at either end or anywhere between the two ends of the heat exchanger.

9.3.2 Composite Curves and Process Pinch

So far we have considered only the case of one hot stream and one cold stream. Most HENs have more than one stream in each category. For such applications it is necessary to combine the temperature characteristics of all hot streams into a single *hot composite curve* and to combine those of all the cold streams into a single *cold composite curve*. When both curves are plotted in a temperature–enthalpy rate difference diagram, the *process pinch* can be identified: the location with the minimum temperature difference ΔT_{min} between the two curves. The construction of the hot and cold composite curves is illustrated with the aid of the following example adapted from Reference 18.

Example 9.1 For the system consisting of two hot and two cold streams with the data given in the following table, draw the hot and cold composite curves.

Stream Number	Inlet Temperature T_i (K)	Outlet Temperature T_e (K)	Heat Capacity Rate $\dot{m}c_p$ (kW/K)
1 (h)	400	310	2.0
2 (c)	300	390	1.8
3 (c)	330	370	4.0
4 (h)	450	350	1.0

Solution

MODEL

1. All heat transfers are accomplished in counterflow heat exchangers.
2. The enthalpy of each stream depends only on temperature.[2]
3. Heat transfer to the surroundings is negligible.
4. All changes of kinetic and potential energy are negligible.
5. The average specific heat of each stream of matter is constant.

---

[2]The effect of pressure drop is neglected in pinch analysis. This represents a shortcoming of the method.

ANALYSIS Figure E.9.1 shows the heating or cooling process for each stream on a temperature–enthalpy rate difference diagram. This representation allows the identification of three temperature intervals to be considered in drawing the composite curves for the hot and the cold streams.

For the hot streams 1 and 4, the first temperature interval is determined through the temperatures T_{4i} and T_{1i}: $\Delta T_{hI} = T_{4i} - T_{1i} = (450 - 400)$ K. Only stream 4 exists in this interval. The enthalpy rate difference in this interval is

$$\Delta \dot{H}_{hI} = \dot{m}_4 c_{p4} \Delta T_{hI} = 1.0 \text{ kW/K} \times (450 - 400) \text{ K} = 50 \text{ kW}$$

The second temperature interval for the hot streams is given by $\Delta T_{hII} = T_{1i} - T_{4e} = (400 - 350)$ K. Since both streams 1 and 4 exist in this interval, the corresponding enthalpy rate difference is

$$\Delta \dot{H}_{hII} = (\dot{m}_1 c_{p1} + \dot{m}_4 c_{p4})\, \Delta T_{hII} = (2.0 + 1.0) \text{ kW/K}$$

$$\times (400 - 350) \text{ K} = 150 \text{ kW}$$

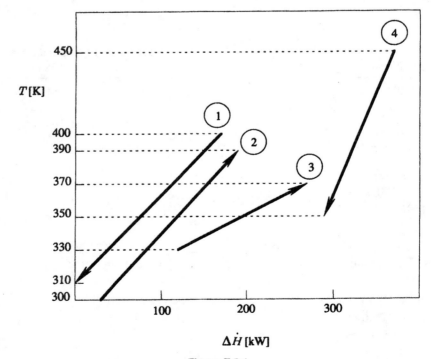

Figure E.9.1

Similarly, the third and last temperature interval for the hot streams is $\Delta T_{hIII} = T_{4e} - T_{1e} = (350 - 310)$ K with the corresponding enthalpy rate difference $\Delta \dot{H}_{hIII} = 80$ kW. The data calculated for each temperature interval of the hot streams are used to draw the hot composite curve in Figure 9.3. This curve covers the temperature range from 310 K (T_{1e}) to 450 K (T_{4i}). The total heat transfer rate available from the two hot streams is 280 kW.

The cold composite curve is determined similarly by considering the cold streams: For the cold streams 2 and 3, we identify the temperature intervals $\Delta T_{cI} = T_{3i} - T_{2i} = 330 - 300 = 30$ K, $\Delta T_{cII} = T_{3e} - T_{3i} = 370 - 330 = 40$ K and $\Delta T_{cIII} = T_{2e} - T_{3e} = 390 - 370 = 20$ K. The respective enthalpy rate differences are $\Delta \dot{H}_{cI} = 1.8 \times 3.0 = 54$ kW, $\Delta \dot{H}_{cII} = (1.8 + 4.0) \times 40 = 232$ kW, and $\Delta \dot{H}_{cIII} = 1.8 \times 20 = 36$ kW. Note that in the second temperature interval both streams 2 and 3 exist. This information is used to draw the cold composite curve in Figure 9.3. This curve covers the temperature range between 300 K (T_{2i}) and 390 K (T_{2e}). The total heat transfer rate required for heating the cold streams is 322 kW.

The procedure used in Example 9.1 to evaluate the enthalpy rate differences can be formulated generally as follows: The change in the enthalpy rate $\Delta \dot{H}_i$ within a given temperature interval ΔT_i is calculated from

$$\Delta \dot{H}_i = \sum_j (\dot{m}c_p)_j \, \Delta T_i \qquad (9.23)$$

where $\Delta \dot{H}_i$ denotes the enthalpy change for the hot (or cold) composite curve and subscript j refers to all hot (or cold) streams that exist in the given temperature interval ΔT_i.

Referring to Figure 9.3, the hot and the cold composite curves can be shifted horizontally in the $T - \Delta \dot{H}$ diagram until a *specified* minimum temperature difference between the two curves is obtained. A value of $\Delta T_{min} = 10$ K is used for the system of Example 9.1 and is shown in Figure 9.3. In this figure the pinch occurs at a temperature of 330 K for the cold composite curve and 340 K for the hot composite curve. In subsequent discussions these temperatures are denoted by $T_{cc,pinch}$ and $T_{hc,pinch}$, respectively. The seven enthalpy intervals $\Delta \dot{H}_1$ through $\Delta \dot{H}_7$ shown in Figure 9.3 are used to estimate the total heat transfer surface area required for this HEN, as discussed in Section 9.3.6.

The following conclusions can be drawn from the pinch analysis using the composite curves of Figure 9.3:

- The minimum hot utility load ($\dot{Q}_{hu,min}$) is determined graphically as the horizontal distance between the high-temperature ends of the composite

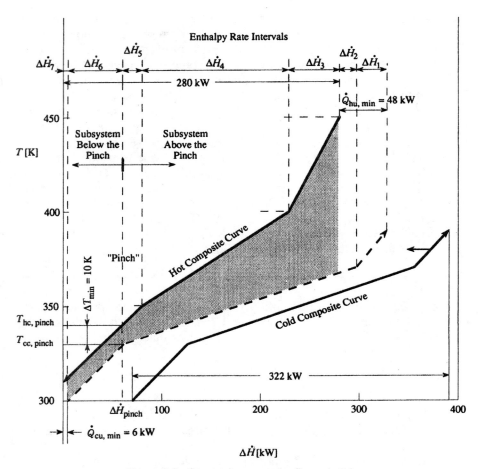

Figure 9.3   Composite curves for Example 9.1

curves. Similarly, the minimum cold utility load ($\dot{Q}_{cu,min}$) is calculated as the horizontal distance between the low-temperature ends of the composite curves. At $\Delta T_{min} = 10$ K, we read from Figure 9.3 $\dot{Q}_{hu,min} = 48$ kW and $\dot{Q}_{cu,min} = 6$ kW. These values represent the thermodynamic optimum (minimal use of the utilities) for this system when $\Delta T_{min} = 10$ K. If the hot utility supplies more energy than 48 kW, $\dot{Q}_{hu} = 48$ kW $+ \dot{Q}_e$ where $\dot{Q}_e$ denotes an excess heat transfer, the energy rate $\dot{Q}_e$ is transferred through the system without use and is finally rejected in the cold utility: $\dot{Q}_{cu} = 6$ kW $+ \dot{Q}_e$. Thus, an increase in the minimum heating load $\dot{Q}_{hu,min}$ shifts the cold composite curve to the right and results in an equal increase in the cooling load $\dot{Q}_{cu,min}$.

- The smaller the selected value for ΔT_{min}, the smaller the requirements from the hot and cold utilities.

- For the special case considered, the pinch is determined by the inlet temperature of the cold stream 3 (330 K) and the heat capacity rates $\dot{m}_1 c_{p1}$ and $\dot{m}_2 c_{p2}$. If it is possible to change the pinch position by modifying these parameters, the overall energetic integration of the hot and cold streams may be improved by further reducing the required utilities for the same value of $\Delta T_{\min}$.

9.3.3 Maximum Energy Recovery

The composite curves and knowledge of the pinch position are very useful in a HEN analysis. The pinch position is determined by the temperatures $T_{\text{hc,pinch}}$ and $T_{\text{cc,pinch}}$ for the hot and cold composite curves, respectively. The respective points on the composite curves have the same $\Delta \dot{H}$ value, denoted by $\Delta \dot{H}_{\text{pinch}}$. In Figure 9.3, the value of $\Delta \dot{H}_{\text{pinch}}$ is 60 kW. At the process pinch, the entire system can be divided into two subsystems, each of which is in energy balance with its respective utility. Above the pinch, that is, above $\Delta \dot{H}_{\text{pinch}}$, only a hot utility is required: The subsystem represents a *net heat sink*. The use of a cold utility in this region would only increase the hot utility demand. Similarly, below the pinch, that is, below $\Delta \dot{H}_{\text{pinch}}$, only a cold utility is required: The subsystem represents a *net heat source*. If heat is transferred from a hot stream above the pinch to a cold stream below the pinch (heat transfer *across* the pinch), the energy balances for the two subsystems require that the loads of both the hot and the cold utilities be increased by the amount transferred. A design with no heat transfer across the pinch gives the *maximum energy recovery* (MER) possible [14]. For a specified value of $\Delta T_{\min}$, this term implies that the maximum energy of the hot streams is transferred to (recovered by) the cold streams so that the utility loads are minimized.

From the discussion thus far, the following rules are obtained for the *thermodynamic optimization* of a HEN:

- Do not transfer heat across the pinch.
- Use a hot utility only above the pinch.
- Use a cold utility only below the pinch.

9.3.4 Calculation of Utility Loads

The graphical approach for determining the minimum hot and cold utility loads at various values of $\Delta T_{\min}$ discussed in Section 9.3.2 is generally not accurate. These loads may be calculated accurately with the aid of the following algorithm based on temperature–interval analysis:

1. Specify a value for $\Delta T_{\min}$ at the process pinch (minimum driving force between the hot and cold streams).

2. Reduce the temperatures of all hot streams by $\Delta T_{min}/2$ and increase the temperatures of all cold streams by $\Delta T_{min}/2$. This is conveniently shown in a temperature–interval diagram (see Figure 9.4 for the streams of Example 9.1). This step identifies all temperature intervals determined by the new inlet and outlet temperatures of all hot and cold streams (five intervals in Figure 9.4).

3. Starting with the highest-temperature interval calculate the net heat transfer for each interval (subscript i) separately using the following equation:

$$\dot{Q}_i = \left[\overset{\text{cold}}{\sum} (\dot{m}c_p)_c - \overset{\text{hot}}{\sum} (\dot{m}c_p)_h \right] \Delta T_i \qquad (9.24)$$

A negative value of $\dot{Q}_i$ indicates a *surplus* of energy available from the hot streams in this interval. This surplus may be used to heat cold streams at lower-temperature intervals or (below the pinch) the surplus may be transferred to a cold utility. A positive value for $\dot{Q}_i$ means that an energy *deficit* exists in this interval. This deficit can be covered from

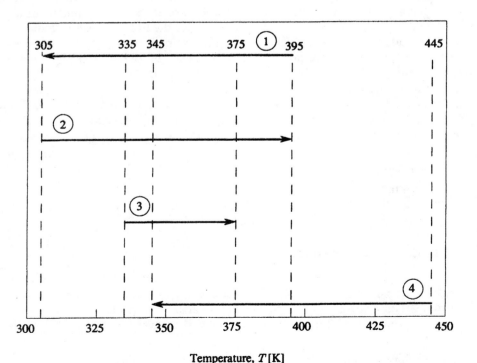

Figure 9.4 Temperature intervals for Example 9.1.

a hot utility or from energy available at higher-temperature intervals. Figure 9.5 shows the values of $\dot{Q}_i$ for the five intervals of Figure 9.4. The first three columns of Figure 9.5 form a so-called *cascade diagram*. This diagram shows how the energy cascades through the temperature intervals.

4. Calculate a *cumulative net heat transfer* $(\Delta \dot{Q}_i)$ for each interval starting with the highest temperature interval:

$$\Delta \dot{Q}_i = \sum_{j=1}^{i} \dot{Q}_j \qquad (9.25)$$

The values of $\Delta \dot{Q}_i$ for Example 9.1 are given in Figure 9.5. The cumulative net heat transfer is calculated for an interval i and, if the value is negative, represents the possible energy transfer to the next lower interval $i + 1$. Therefore, the $\Delta \dot{Q}_i$ values are assigned to the temperature at which the energy transfer takes place: the lowest temperature of the

Temperature Interval Number (i)	Adjusted Temperature [K]	$\dot{Q}_i$ [kW]	$\Delta \dot{Q}_i$ [kW]	$\Delta \dot{Q}_i^*$ [kW]
	445		0	- 48
1		- 50		
	395		- 50	- 98
2		- 24		
	375		- 74	- 122
3		+ 84		
	345		+ 10	- 38
4		+ 38		
	335		+ 48	0
5		- 6		
	305		+ 42	- 6

Figure 9.5 Cascade diagram and calculation of utility loads for Example 9.1.

ith interval. The largest $\Delta\dot{Q}_i$ value determines the *minimum hot utility load* ($\dot{Q}_{hu,min} = \Delta\dot{Q}_{i,max}$) for the selected value of ΔT_{min}. From Figure 9.5, we see that this occurs in the fourth temperature interval. Thus, $\dot{Q}_{hu,min} = 48$ kW.

5. Calculate a *modified* cumulative net heat transfer ($\Delta\dot{Q}_i^*$) using

$$\Delta\dot{Q}_i^* = \Delta\dot{Q}_i - \dot{Q}_{hu,min} \tag{9.26}$$

The $\Delta\dot{Q}_i^*$ values are also assigned to the lowest temperature of the ith interval. The temperature at which $\Delta\dot{Q}_i^*$ vanishes is called *pinch temperature* T_{pinch} because it identifies the position of the pinch. The temperature of the hot streams at the pinch is

$$T_{hc,pinch} = T_{pinch} + \Delta T_{min}/2 \tag{9.27}$$

while the temperature of the cold streams at the pinch is

$$T_{cc,pinch} = T_{pinch} - \Delta T_{min}/2 \tag{9.28}$$

As Figure 9.5 shows, T_{pinch} is equal to 335 K for the system of Example 9.1. With Equations 9.27 and 9.28 we obtained for this example the values determined in Section 9.3.2: $T_{hc,pinch} = 340$ K and $T_{cc,pinch} = 330$ K. The *minimum cold utility load* ($\dot{Q}_{cu,min}$) for the selected value of ΔT_{min} is equal to the value of $\Delta\dot{Q}_i^*$ for the lowest process temperature. From Figure 9.5 we obtain $\dot{Q}_{cu,min} = -6$ kW.

This algorithm provides the same information as the composite curves. It identifies the pinch and calculates the minimum utility loads. In addition to the accurate calculation of these loads, the advantages of the algorithm are that it can be easily implemented on a computer and can be adapted to cover cases where the assumption of constant specific heats is invalid or where the allowed ΔT_{min} value depends on the streams matched [14].

9.3.5 Grand Composite Curve

The values of $\Delta\dot{Q}_i^*$ can be used to draw the *grand composite curve* in a plot of temperature versus $-\Delta\dot{Q}^*$. The grand composite curve shows the cumulative surplus or deficit of energy for each temperature. The temperatures used in this diagram are *not the original* ones but the temperatures shifted by $\Delta T_{min}/2$. The grand composite curve can also be constructed directly from the hot and cold composite curves by starting at the pinch condition and requiring a zero heat flow at the average of the hot and cold pinch temperatures. Figure 9.6 shows the grand composite curve for the streams of Example 9.1.

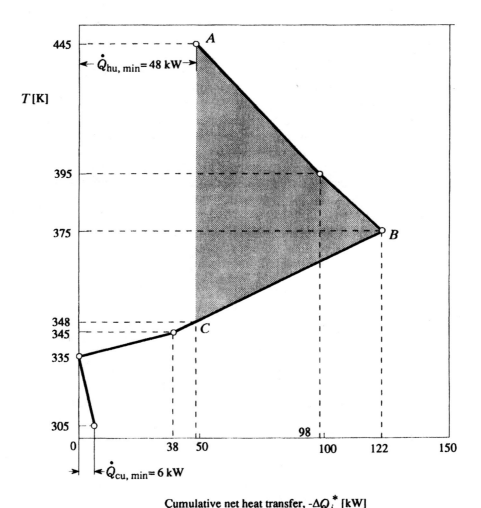

Figure 9.6   Grand composite curve for Example 9.1.

The grand composite curve indicates the quality and type of the required utilities. For example, since the heating requirements between 348 and 445 K (shaded area in Figure 9.6) can be covered internally through cooling of the hot streams, the hot utility does not have to be supplied at the highest process temperature (450 K) but at a lower temperature. To satisfy the requirement of $\Delta T_{min} = 10$ K, it is sufficient for the hot utility to be available at a minimum temperature of 353 K, which is the sum of the shifted temperature 348 K and $\Delta T_{min}/2$. This minimum temperature determines the *quality* of the required utility.

The slope of the grand composite curve in the area where external heating is required (335–348 K) is relatively small. This information should lead us to use a hot utility with a relatively small slope in a T vs. $\Delta \dot{H}$ plot (e.g., steam that is condensed) to reduce the exergy destruction during the heat transfer from the hot utility. If, however, the slope of the grand composite curve in the temperature range requiring the supply of a hot utility would be steep, the exhaust gas from a combustion process, if available, would represent a better match. Thus, the slope of the grand composite curve assists in the selection of the appropriate *type* of utility.

Finally, the grand composite curve indicates the exergy destruction that is avoidable and unavoidable in the HEN design. The exergy destruction associated with the heat transfer processes corresponding to the area ABC in Figure 9.6 is unavoidable for the specified value of ΔT_{min}, whereas the exergy destruction associated with the heat transfer processes from the hot utility and to the cold utility depends on the matching of the temperature profiles between the streams and utilities.

9.3.6 Estimation of the Required Total Heat Transfer Surface Area

The total heat transfer surface area required for the entire HEN at a specified ΔT_{min} at the process pinch can be estimated before the actual design of the HEN is attempted (Section 9.3.7) by assuming that the temperature differences between the hot and cold streams are used in a *thermodynamically optimal*[3] way in counterflow heat exchangers as explained next: The hot and cold composite curves (including the utilities) are divided into *enthalpy rate intervals* as shown in Figure 9.3 (seven intervals $\Delta \dot{H}_1$ through $\Delta \dot{H}_7$) for the curves of Example 9.1. Within each enthalpy rate interval (subscript j) there is an associated heat transfer rate: $\dot{Q}_j = \Delta \dot{H}_j$ between the hot and cold composite curves. The heat transfer is assumed to be accomplished in a counterflow heat exchanger at a log-mean temperature difference $\Delta T_{lm,j}$ calculated as follows:

$$\Delta T_{lm,j} = \frac{\Delta T_{gr,j} - \Delta T_{sm,j}}{\ln \dfrac{\Delta T_{gr,j}}{\Delta T_{sm,j}}} \tag{9.29}$$

where subscripts gr and sm indicate the larger and smaller temperature difference, respectively, at the two ends of the heat exchanger.

---

[3]The term *thermodynamically optimal* means here that a complete matching between hot and cold streams is achieved in each enthalpy rate interval, but without consideration of the number of heat exchangers that would be required to accomplish this matching.

Assuming a constant overall heat transfer coefficient U_j (Equation 6.1) for the jth interval, the total heat transfer area may be calculated from

$$A_{\text{hen}} = \sum_j \frac{\dot{Q}_j}{U_j \, \Delta T_{\text{lm},j}} \tag{9.30}$$

When there are multiple streams in an interval, a more accurate relation for A_{hen} is obtained from

$$A_{\text{hen}} = \sum_j \frac{1}{\Delta T_{\text{lm},j}} \left(\sum_i^{\text{hot}} \frac{\dot{Q}_{ji}}{h_i} + \sum_k^{\text{cold}} \frac{\dot{Q}_{jk}}{h_k} \right) \tag{9.31}$$

where h_i and h_k are the individual convective heat transfer coefficients (Section 4.3) for the ith hot stream and the kth cold stream, respectively. $\dot{Q}_{ji}$ is the rate of heat transfer from the ith hot stream within the jth enthalpy rate interval. Similarly, $\dot{Q}_{jk}$ is the rate of heat transfer to the kth cold stream within the jth enthalpy rate interval. The thermal resistances of the walls separating the hot and cold fluids are neglected in Equation 9.31.

The approximate procedures discussed in this section do not necessarily provide the same results as would be obtained from a detailed design of the HEN. However, Equation 9.30 or 9.31 gives a reasonable estimate of the total area required for a HEN without first designing the HEN. These equations are particularly useful during the preliminary design stage when quick estimates of the capital cost associated with the HEN are required. The detailed design of a HEN is discussed in Section 9.3.7. The following example adapted from Reference 18 illustrates the estimation of the total heat transfer surface area.

Example 9.2 For the streams and data of Example 9.1, estimate the total heat transfer area if the individual convective heat transfer coefficients for the hot streams are $h_1 = 500$ W/m^2 K and $h_4 = 800$ W/m^2 K and for the cold streams are $h_2 = 750$ W/m^2 K and $h_3 = 600$ W/m^2 K. The hot and cold utilities are available at 500 and 298 K, with heat transfer coefficients of 5000 and 850 W/m^2 K, respectively.

Solution

MODEL

1. The assumptions of Example 9.1 apply.

| Enthalpy Rate Interval Number (j) | Enthalpy Rate Difference ΔH_j (kW) | Temperature Intervals (K) | | Log-Mean Temperature Difference $\Delta T_{\mathrm{lm},j}$ (K) | Heat Transfer Rates (kW) | | | | | |
| | | Hot Side | Cold Side | | Streams | | | | Utilities | |
					1	2	3	4	hu	cu
1	36	500–500	370.0–390.0	119.7		36.0			36.0	
2	12	500–500	367.9–370.0	131.0		3.7	8.3		12.0	
3	50	450–400	359.3–367.9	59.0		15.5	34.5	50.0		
4	150	400–350	333.4–359.3	26.8	100.0	46.5	103.5	50.0		
5	20	350–340	330.0–333.4	13.0	20.0	6.2	13.8			
6	54	340–313	300.0–330.0	11.4	54.0	54.0				
7	6	313–310	298.0–298.0	13.4	6.0					6.0

2. In each enthalpy rate interval of the composite curves, Figure 9.3, the temperature differences between the hot and cold streams are used in a thermodynamically optimal way in counterflow heat exchangers.

ANALYSIS   Referring to Figure 9.3, the first enthalpy interval, $\Delta \dot{H}_1$, corresponds to the temperature interval 370–390 K of the cold composite curve. The only heat transfer in this interval is from the hot utility to stream 2: $\Delta \dot{H}_1 = \dot{Q}_{12} = 1.8$ kW/K $\times$ (390 $-$ 370)K = 36.0 kW. The log-mean temperature difference in this interval is

$$\Delta T_{\text{lm,1}} = \frac{(500 - 370) - (500 - 390)}{\ln \left(\dfrac{500 - 370}{500 - 390} \right)} = 119.72 \text{ K}$$

The second enthalpy interval of Figure 9.3, $\Delta \dot{H}_2$, corresponds to the inlet temperature of the fourth stream (hot) and the outlet temperature of the third stream (cold). The heat transfer from the hot utility in this interval is the difference in heat transfer rates: $\Delta \dot{H}_2 = \dot{Q}_{\text{hu,min}} - \Delta \dot{H}_1 = 48.0 - 36.0 = 12.0$ kW. The temperature difference for the two cold streams appearing in this interval is $\Delta \dot{H}_2 / (\dot{m}_2 c_{p2} + \dot{m}_3 c_{p3}) = 12.0/(1.8 + 4.0) = 2.069$ K. Thus, in the second interval the inlet temperature of the cold streams is (370.0 $-$ 2.069) K = 367.931 K. The log-mean temperature difference for the second interval is

$$\Delta T_{\text{lm,2}} = \frac{(500 - 367.931) - (500 - 370)}{\ln \left(\dfrac{500 - 367.931}{500 - 370} \right)} = 131.03 \text{ K}$$

In the second interval, the heat transfer rates supplied to cold streams 2 and 3 are $\dot{Q}_{22} = 1.8 \times 2.069 = 3.724$ kW and $\dot{Q}_{23} = 4.0 \times 2.069 = 8.276$ kW, respectively.

The temperature intervals and the heat transfer rates for the remaining enthalpy rate intervals are determined similarly. The results are summarized in the accompanying table, in which some numbers are rounded.

The total heat transfer surface is then calculated from Equation 9.31

$$A_{\text{hen}} = \frac{1}{119.7} \left(\frac{36.0}{5000} + \frac{36.0}{750} \right) + \frac{1}{131.0} \left(\frac{12.0}{5000} + \frac{3.7}{750} + \frac{8.3}{600} \right)$$

$$+ \frac{1}{59.0} \left(\frac{50.0}{800} + \frac{15.5}{75} + \frac{34.5}{600} \right)$$

$$+ \frac{1}{26.8} \left(\frac{100.0}{500} + \frac{50.0}{800} + \frac{46.5}{750} + \frac{103.5}{600} \right)$$

$$+ \frac{1}{13.0} \left(\frac{20.0}{500} + \frac{6.2}{750} + \frac{13.8}{600} \right)$$

$$+ \frac{1}{11.4} \left(\frac{54.0}{500} + \frac{54.0}{750} \right) + \frac{1}{13.4} \left(\frac{6.0}{500} + \frac{6.0}{850} \right)$$

or $A_{hen} = 44.2 \text{ m}^2$.

COMMENTS

1. In this example the hot utility is assumed to be available at 500 K. However, in Section 9.3.5 we determined that the temperature of the hot utility is required only to be greater than or equal to 353 K. If a hot-utility temperature lower than 500 K would be considered, the total heat transfer area would increase.
2. If a cold-utility temperature lower than 298 K would be considered, the total heat transfer area would decrease.

9.3.7 HEN Design

Thus far we have presented a simple procedure for (a) calculating the minimum heating and cooling requirements for a HEN and (b) estimating the total heat exchanger area required. These calculations are possible without even developing a flow sheet for the HEN and are therefore very useful for screening various options.

In designing a HEN, it is also useful to know the *minimum number of heat exchangers required* ($N_{hx,min}$). This number, which includes the number of heat exchangers in which a hot or cold utility is used (utility heaters and coolers), is calculated from [11]

$$N_{hx,min} = N_h + N_c + N_{hu} + N_{cu} - 1 \qquad (9.32)$$

where N_h and N_c denote the number of hot and cold streams, respectively, and N_{hu} and N_{cu} denote the number of hot and cold utilities, respectively. The minimum number of heat exchangers required for the streams of Example 9.1 is five.

Since the number of heat exchangers required is always less than or equal to the number of heat exchangers in a maximum energy recovery (MER) network (Section 9.3.3), it is convenient to develop initially a HEN characterized by maximum energy recovery for a specified ΔT_{min} value at the pinch and then modify the HEN design, if necessary, to increase its cost effectiveness. In the present discussion, we illustrate this procedure for the case of

Example 9.1 and present some general guidelines for HEN design. More details may be found in the literature [14, 15, 17, 18].

MER Design. The design of a HEN for maximum energy recovery should be considered in two parts: First, a network for the streams above the pinch is designed and then another for the streams below the pinch. In both cases we start the design *at* the pinch and move away. This approach ensures that no heat is transferred across the pinch.

Several rules have been proposed for guiding the development of a HEN with maximum energy recovery, together with a methodology for their application. The rules, which include both heuristic rules [14, 17] and rules based explicitly on exergy considerations [20], are considered next.

The matching of streams above and below the pinch is based on their temperatures and heat loads. The following rules, which restate or complement the design guidelines presented in Section 3.6, may be used to improve this matching and reduce the number of iterations required in a subsequent step to obtain the cost-optimal HEN design.

- Do not use a cold utility above the pinch or a hot utility below the pinch.
- Avoid the use of excessively large or excessively small temperature differences (thermodynamic driving forces).
- Minimize the mixing of streams with differences in temperature, pressure, or chemical composition.
- Minimize the use of intermediate heat transfer fluids when exchanging heat between two streams.
- Try to match streams where the final temperature of one is close to the initial temperature of the other.
- Try to match streams with similar heat capacity rates ($\dot{m}c_p$). If there is a significant difference between the heat capacities of the hot and cold streams, consider splitting the stream with the larger heat capacity.
- Maximize the load of heat exchangers.
- Ensure that the minimum temperature difference in all heat exchangers is equal to or larger than the selected ΔT_{min} at the pinch. This condition is met when the following relations are satisfied at the pinch by the streams present there:
  Above the pinch:

$$(N_h)_{\text{pinch}} \leq (N_c)_{\text{pinch}} \tag{9.33}$$

and

$$\dot{m}_h c_{\text{ph}} \leq \dot{m}_c c_{\text{pc}} \tag{9.34}$$

Below the pinch:

$$(N_h)_{\text{pinch}} \geq (N_c)_{\text{pinch}} \tag{9.35}$$

and

$$\dot{m}_h c_{\text{ph}} \geq \dot{m}_c c_{\text{pc}} \tag{9.36}$$

Here $(N_h)_{\text{pinch}}$ and $(N_c)_{\text{pinch}}$ denote the number of hot and cold streams, respectively, at the pinch. Equations 9.34 and 9.36 refer to the matching between one hot and one cold stream at the pinch. Equations 9.33 through 9.36 must be satisfied simultaneously. If Equation 9.33 is not satisfied, a cold stream must be split. When Equation 9.34 cannot be satisfied for all stream matchings above the pinch, a stream (normally a hot stream) must be split. If Equation 9.35 is not satisfied, a hot stream must be split. When Equation 9.36 cannot be satisfied for all stream matchings below the pinch, a stream (normally a cold stream) must be split.

- Consider additional factors as appropriate for the case being considered (e.g., the distances between the streams of the actual system, compatibility of the substances in the streams to be matched, use of existing equipment, and HEN control).

Continuing with the case of Example 9.1, let us develop a MER design for the system consisting of two hot and two cold streams. In Figure 9.3 we established that $T_{\text{hc,pinch}} = 340$ K and $T_{\text{cc,pinch}} = 330$ K. In the following we discuss initially the design of the subsystem below the pinch and then the design of the subsystem above the pinch.

Below the pinch there is only one hot stream (stream 1) and one cold stream (stream 2). Thus, $(N_h)_{\text{pinch}} = (N_c)_{\text{pinch}}$, and Equation 9.35 is satisfied. Also Equation 9.36 is satisfied since $\dot{m}_1 c_{p1} = 2.0$ kW/K and $\dot{m}_2 c_{p2} = 1.8$ kW/K. Below the pinch the maximum heat transfer rate from stream 1 is 2.0 kW/K $\times$ (340 − 310) K = 60 kW and the maximum heat transfer rate to stream 2 is 1.8 kW/K $\times$ (330 − 300) K = 54 kW. The smaller value determines the load of the heat exchanger between streams 1 and 2 (heat exchanger I): $\dot{Q}_{\text{I}} = 54$ kW. In this heat exchanger, stream 1 is cooled from 340 to 313 K. The further cooling of stream 1 to 310 K occurs in the cooler (cu), the load of which is 6 kW.

Above the pinch all four streams exist. At the pinch there is one hot stream (stream 1) and two cold streams (streams 2 and 3). Thus, Equation 9.33 is satisfied. Equation 9.34 dictates that we exclude the matching of stream 1 ($\dot{m}_1 c_{p1} = 2.0$ kW/K) with stream 2 ($\dot{m}_2 c_{p2} = 1.0$ kW/K). For Equation 9.34 to be satisfied, hot stream 1 must be matched with cold stream 3 ($\dot{m}_3 c_{p3} = 4.0$ kW/K). Above the pinch, the maximum heat transfer rate from stream 1 is 2.0 kW/K $\times$ (400 − 340) K = 120 kW and to stream 3 is 4.0 kW/K $\times$

$(370 - 330)$ K $= 160$ kW. Thus, the load of heat exchanger II between streams 1 and 3 is 120 kW. Similarly, we determine the load of heat exchanger III between streams 4 and 2 to be 100 kW. For each of the cold streams 3 and 2 we need to use a utility heater denoted by hu-a and hu-b with a load of 40 and 8 kW, respectively, to bring these streams to their final temperatures. All in all, six heat exchangers are required for the MER design of the system of Example 9.1: Three stream to stream heat exchangers (I, II and III), two utility heaters (hu-a and hu-b), and one utility cooler (cu).

The MER network obtained thus far is shown in Figure 9.7. A *grid representation* [19] is used in this figure in addition to the flow sheet representation shown in the left upper corner. In the grid representation, the hot streams run from right to left at the top of the diagram and the cold streams run countercurrent at the bottom of the figure. A block identifying each stream by number, 1–4, is located at the stream inlet. A heat exchanger is shown as two circles connected by a vertical line with the load written beneath the lower circle.

With a grid representation, a HEN design can be developed using any sequence of hot and cold stream matches and any design change is possible *without reordering or rerouting the streams*. When the pinch temperatures are marked in such a diagram, heat exchangers transferring heat across the pinch are easily identified.

Cost-Optimal HEN Design. The number of heat exchangers required for the overall system is always less than or equal to that for a MER network. Moreover, the number of heat exchangers should be kept as small as possible because of technical and economic considerations. A MER design satisfies the minimum utility requirements, but usually leads to *loops* across the pinch. Loops are multiple links between streams that had to be established to prevent heat from crossing the pinch. Whenever a loop exists, a path on the grid diagram can be traced that starts at one heat exchanger and returns to the same exchanger (see dotted line in Figure 9.7). The existence of a loop implies that there is an extra heat exchanger in the network—that is, the number of heat exchangers is greater than the minimum number given by Equation 9.32. Indeed, the six heat exchangers shown in Figure 9.7 are one more than the minimum number of five calculated at the beginning of this section. As Figure 9.7 corresponds to a MER design, steps taken to reduce the number of heat exchangers increase the load of each utility. Thus there is a trade-off: We reduce the number of heat exchangers to reduce the capital costs, but this results in an increase in the utility costs.

To reduce the number of heat exchangers, we focus on each loop appearing in the MER design and *break* the loop by removing the heat exchanger with the smallest load in the path and allowing some heat transfer across the pinch.[4]

---

[4]If more than one loop exists, we first break the loop including the heat exchanger with the smallest heat load.

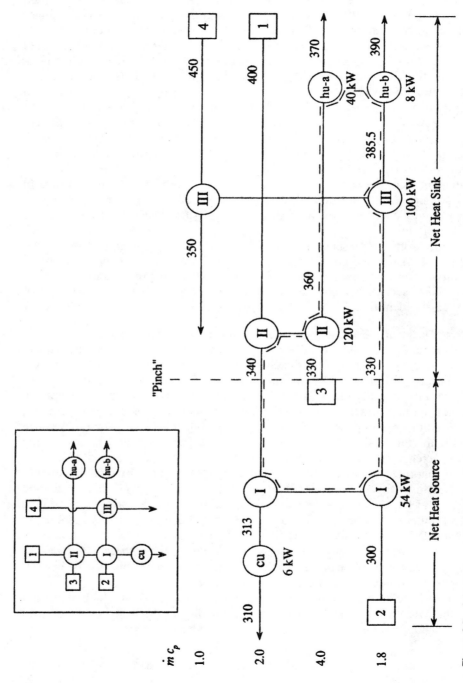

Figure 9.7 Grid representation of the entire heat exchanger network for Example 9.1. A flow sheet representation is shown in the upper left corner. The dashed line identifies a loop.

494

This procedure, called *network relaxation,* or *energy relaxation,* normally introduces at least one violation of the specified value for ΔT_{min}. The ΔT_{min} value is restored by shifting energy along a *path* that connects a heater and a cooler in a network. This shifting increases the utilities load.

Figure 9.8 shows the grid and flow sheet representations of the network for Example 9.1 after the loop shown in Figure 9.7 is eliminated. Comparing Figure 9.7 to 9.8, we see that the utility heater whose load is 8 kW has been removed from the network. The design of Figure 9.8 includes just five heat exchangers: I, II, III, plus one utility heater, and one utility cooler. An analysis of the network shows that to keep the minimum temperature difference of $\Delta T_{min} = 10$ K after this network relaxation step, the load of each utility must be increased by 0.8 kW. The dotted line in Figure 9.8 shows the path along which energy is shifted from the hot utility to the cold utility. Detailed procedures for developing new network designs and for obtaining the design of Figure 9.8 from that of Figure 9.7 are discussed in the literature [14, 17, 18].

The cost optimization of a HEN consists in finding the design that has the smallest total annual cost. Accordingly, the determination of the more cost-effective design between the two designs shown in Figures 9.7 and 9.8 depends on the economic parameters and assumptions (see, e.g., Table 7.9) used in the optimization. For brevity in the present discussion this important practical step is omitted.

The pinch method for the design of HENs has some significant limitations. A limitation noted in the discussion of Example 9.1 is that the effect of pressure drop is neglected. Additionally, the method requires knowledge of the mass flow rate, the average specific heat, and the inlet and outlet temperatures of each stream. In designing a thermal system, however, the temperatures and mass flow rates of streams are often treated as decision variables, and thus the values of these variables would be unknown at the outset of the design process. Accordingly, there is some circular analysis inherent in the pinch method. This is resolved in practice by making assumptions about the unknown quantities and relaxing the assumptions selectively as the design evolves iteratively.

9.3.8 Integration of a HEN with Other Components

This section briefly discusses the integration of a HEN with a power cycle, a heat pump, or a distillation column. The question to be answered here is how should these components be placed relative to the HEN pinch to make the integration thermodynamically advantageous compared with the stand-alone option. Additional details are given in References 14, 21, 22.

Power cycles receive an energy rate $\dot{Q}_{in}$ at an average high temperature T_{in}, convert part of it to power and reject $\dot{Q}_{out}$ at an average lower temperature T_{out}. The exergy rate associated with $\dot{Q}_{out}$ represents an exergy loss for stand-alone power cycles. If a power cycle is integrated with a HEN so that $T_{out} > T_{hc,pinch}$, however, the exergy rate associated with $\dot{Q}_{out}$ may be used in the network to reduce the hot utility load. Similarly, if a power cycle is installed

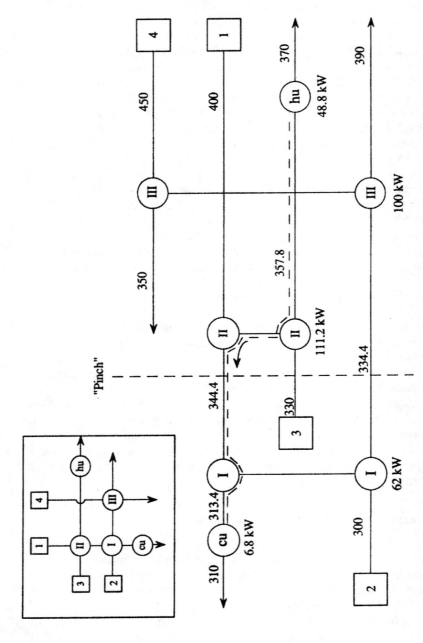

Figure 9.8 Grid representation of the entire heat exchanger network for Example 9.1, after energy relaxation. A flow sheet representation is shown in the upper left corner. The dashed line indicates a path along which energy is shifted.

below the pinch ($T_{hc,pinch} > T_{in}$), the cycle may use part of the exergy that would be discarded to a cold utility and, thus, reduce the total exergy rejected in the cold utility. However, if $\dot{Q}_{in}$ is supplied above the pinch and $\dot{Q}_{out}$ is rejected below the pinch, nothing is gained from the integration because, in effect, $\dot{Q}_{out}$ is transferred across the pinch and the cold utility load is increased by this amount. Hence, *a power cycle should be placed either above or below the pinch but not across the pinch.*

Power cycles requiring high input temperatures, such as gas turbines, will normally be integrated above the process pinch. Rankine cycles may be placed either above or below, depending on the working temperatures and the pinch temperature. As exhaust gases are invariably discarded below the pinch, it is impossible to gain the advantage of HEN integration with a power cycle for the *entire* energy transfer available from these gases.

Turning next to the case of *heat pumps,* such devices *should be placed across the pinch* so that energy is transferred from a hot stream below the pinch to a cold stream above the pinch. This reduces both the hot and cold utility loads. Placing a heat pump entirely above the pinch does not provide any thermodynamic or economic benefit compared with the stand-alone option because power would substitute for utility energy. Placing a heat pump below the pinch leads to a worse situation: The net overall effect is that the power supplied to the heat pump is converted into an additional load for the cold utility. The use of a heat pump in a HEN should be considered, for example, when the slopes of both the heating and the cooling requirements in the grand composite curve are small, and the temperature difference between these requirements can be efficiently covered by a heat pump.

Distillation columns are major exergy consumers in chemical plants: An energy rate $\dot{Q}_{in}$ is supplied in a reboiler and an energy rate $\dot{Q}_{out}$ is removed in a condenser. *Distillation columns should be placed entirely either above or below the pinch of a HEN.* If a distillation column falls across the pinch, the result is thermodynamically or economically no better than if the column would be installed as a stand-alone unit. The temperature at which $\dot{Q}_{in}$ is supplied depends on the operating pressure of the distillation column. Therefore, if a distillation column falls across the pinch in a specific design configuration, it might be possible to shift the column operation entirely above the pinch by raising the column pressure or to shift the operation entirely below the pinch by lowering the column pressure.

9.3.9 Closure

This section has provided an introduction to the pinch method for the design and optimization of heat exchanger networks. Although a seemingly formal approach, well laden with rules, pinch analysis is fundamentally an embodiment of the main theme of this book: Use second-law reasoning in thermal system design. Pinch analysis is an offshoot of exergy analysis, but with calculations being conveniently replaced by primarily graphical means [15, 23, 24].

Pinch analysis can be effectively used for the evaluation and *thermodynamic* optimization of a HEN and its integration with the components discussed in Section 9.3.8. Other processes, particularly when involving chemical reactions, cannot be evaluated using pinch analysis as effectively (if at all) as using exergy analysis. In addition, because of the limitations discussed at the end of Section 9.3.7, pinch analysis might not detect some opportunities for improving the design of a system. Such opportunities might be brought to light by an exergy analysis, however, which is a more general and powerful tool [24]. Still, pinch analysis and exergy analysis should be considered as complementary and not competing methods for the design and optimization of thermal systems.

In Section 9.4 we survey some commonly used analytical and numerical optimization techniques. Then in Section 9.5 we discuss the optimization of the cogeneration system. In Section 9.6 we present exergy-related design considerations applicable to complex thermal systems.

9.4   ANALYTICAL AND NUMERICAL OPTIMIZATION TECHNIQUES

Since there is no general mathematical method for conducting the search for an optimal value, many methods have been developed for the efficient solution of particular classes of problems. The optimization methods can be categorized according to the nature of the objective function, the constraints, and the decision variables involved: The objective function may (a) contain only a single decision variable or many decision variables, (b) be continuous or contain discontinuities, and (c) be linear or nonlinear. The constraints of the optimization problem may be expressed as linear or nonlinear equations (or inequalities). The decision variables may be continuous, integers, or a combination. In the following subsections some of the more commonly used techniques applicable to these types of optimization problems are briefly discussed. Section 9.4.1 refers to unconstrained and constrained functions of a single variable. Section 9.4.2 deals with unconstrained nonlinear objective functions. Sections 9.4.3 and 9.4.4 refer to constrained multivariable optimization problems. Specific optimization techniques are briefly discussed in Section 9.4.5. The classification of optimization problems we provide in Sections 9.4.1 through 9.4.5 is not unique. The same methods may be grouped in different ways. Further discussion is provided in References 1–3, 25, 26, and the current literature.

In applying numerical optimization methods, we must consider the tradeoffs between the complexity of the numerical procedure and the number of objective-function evaluations needed. For example, in some problems a simulation may be required to generate the function values. In other cases we might not have a model of the system to be optimized and be required to operate a similar existing system at various input levels to evaluate the output values. The generation of a new value for the objective function in such circumstances may be very costly. The efficiency of the numerical method becomes then a key criterion in selecting the appropriate optimization method.

9.4.1 Functions of a Single Variable

This is the most elementary type of optimization problem. The techniques applicable to this type of problem are also generally important because some techniques applicable to multivariable functions involve repeated use of a single-variable search.

Graphical Methods. The graphical procedure where a number of values of the objective function are calculated and then plotted is perhaps the most elementary method of finding an *extremum* (maximum or minimum) point of an objective function. Depending on the required accuracy, however, another more detailed graph might have to be drawn in the region near the extremum.

Indirect Methods. As developed in elementary calculus, a necessary condition for an extremum point of a differentiable objective function is that the point be a *stationary point*: The first derivative of the objective function vanishes at that point. If the second derivative is positive, the stationary point is a *local minimum*. If the second derivative is negative, the point is a *local maximum*. An *inflection point,* or *saddle point,* is a stationary point that does not correspond to a local extremum (e.g., when the second derivative is zero and the third derivative is nonzero). If a function is bounded in an interval (*constrained function*), the *boundary points* also can qualify as local optima. The *global extremum* is identified among the local extrema by computing the value of the objective function at the local extrema and selecting the best value.

The methods for determining an extremum by using *derivatives* and values of the objective function are called *indirect methods.* Such methods include *Newton's method,* the *finite difference approximation of Newton's method* (or *quasi-Newton method*), and the *Secant method* [2].

Direct Methods. Computers have enabled the application of *direct methods,* that is, methods that search for an extremum by directly comparing function values at a sequence of trial points without involving derivatives. The direct numerical methods generally can more easily treat problems involving objective functions with discontinuities, points of inflection, and boundary points.

The *region elimination methods* are direct search methods that enable the exclusion of a portion of the decision-variable range from consideration in each successive stage of the search for an extremum. The search terminates when the remaining subinterval for the decision variable is sufficiently small. These methods include the *two-point equal interval search,* the *method of bisecting* (or *dichotomous search*), the *Fibonacci method,* and the *golden section method* [1, 2, 25]. The last two methods are considered the most efficient.

The region elimination methods require only a simple comparison of objective-function values at two trial points. The *point estimation methods,* or *polynomial approximation methods,* involve, in addition to the function values, the magnitude of the difference between the function values. The point estimation methods usually involve a quadratic or cubic approximation of the objective function. *Powell's method,* which uses a successive quadratic estimation, is considered the most efficient algorithm among the point estimation methods.

Point estimation methods converge more rapidly than region elimination methods for well-behaved unimodal functions.[5] However, for strongly skewed or multimodal functions the golden section search is more reliable. A Powell-type successive quadratic estimation search is recommended for general use [1]. If this search encounters difficulties in the course of iteration, the optimization procedure should continue with a golden section search.

9.4.2 Unconstrained Multivariable Optimization

The *unconstrained* nonlinear programming methods used for multivariable optimization are iterative procedures in which the following two steps are repeated: (a) starting from a given point choose a search direction and (b) minimize or maximize in that direction to find a new point. These methods mainly differ in how they generate the search directions.

Direct methods in this category include the *random search,* the *grid search,* the *univariate search,* the *sequential Simplex method,* the *Hooke–Jeeves pattern search method,* and *Powell's conjugate direction method* [1, 2]. These methods are relatively simple to understand and execute. However, they are not as efficient and robust as many of the indirect methods.

The indirect methods include the *steepest–descent/ascent gradient method* (or *Cauchy's method*), the *conjugate gradient methods, Newton's method, Marquardt's method,* the *secant methods,* and the *Broyden–Fletcher–Goldfarb–Shanno (BFGS) method* [1, 2]. Various computer codes used for unconstrained optimization are listed in Reference 2.

9.4.3 Linear Programming Techniques

The term *linear programming* refers to the optimization procedure applied to problems in which both the objective function and the constraints are linear. The optimal solution of a linear programming problem lies on some constraint or at the intersection of more than one constraint and not in the interior of the convex region where the inequality constraints are satisfied.

Linear programming techniques are widely used because of the availability of commercial software for solving large problems. Problems involving two

---

[5]A function is unimodal in an interval if and only if it is monotonic on either side of the single optimal point in that interval.

variables may be solved graphically. Problems involving more than two variables are solved by the *linear programming simplex method,*[6] the *revised simplex method,* or the *Karmarkar algorithm* [1, 22, 25, 26]. A list of linear programming computer codes is given in Reference 2.

9.4.4 Nonlinear Programming with Constraints

Problems with a nonlinear objective function and/or constraints are the most common type encountered in thermal design optimization. In a problem in which there are n independent variables and m equality constraints, we might attempt initially to solve each of these constraints explicitly for a set of $n - m$ variables, and then express the objective function in terms of these variables. If all the equality constraints can be removed, and there are no inequality constraints, the objective function might then be differentiated with respect to each of the remaining $n - m$ variables and the derivatives set equal to zero. Few practical problems may by solved using this simple method, however, and additional approaches are required.

The five major approaches for solving nonlinear programming problems with constraints are the Lagrange multiplier method, iterative quadratic programming methods, iterative linearization methods, penalty function methods, and direct-search methods [1–3, 25, 26]. In the *Lagrange multiplier method,* the Lagrange multipliers are introduced and an augmented objective function (the Lagrangian) is defined (see Appendix A for an illustration). The term *quadratic programming* refers to a procedure that minimizes a quadratic function subject to constraints of the linear inequality and/or equality type. The *linearization methods* solve the general nonlinear programming problem by linearizing the problem and successively applying linear programming techniques. The *penalty function methods* are based on the transformation of the constrained problem into a problem in which a single unconstrained function is minimized. One of the most effective *direct-search methods* is *Box's complex method,* which is an adaptation of the simplex direct-search method (Section 9.4.2). The quadratic programming methods, the linearization methods, and the penalty function methods are better for general problems than the Lagrange and direct-search methods. Computer codes for nonlinear programming with constraints are discussed in References 1–3.

9.4.5 Other Techniques

Dynamic Programming. Dynamic programming procedures are applied in a multistage manner even if the system under consideration is not itself comprised of separate subsystems. Dynamic programming methods decompose a multivariate interconnected optimization problem into a sequence of subproblems that can be solved serially. Each of the subproblems may contain one

---

[6]This method has no relationship to the simplex method noted in Section 9.4.2.

or a few decision variables. More details are given in References 2, 25, and 26.

Integer Programming. Some optimization problems require that some or all of the decision variables must be integer valued. The term *pure integer programming* refers to the class of problems where all of the decision variables are restricted to be integers. In *mixed integer programming* some of the variables are restricted to be integers while others may assume continuous (fractional) values. Linear and nonlinear integer programming is discussed in References 1, 2, and 25.

Geometric Programming. *Geometric programming* refers to constrained optimization problems involving a special class of polynomial functions as the objective function and the constraints of the problem. Geometric programming methods are discussed in References 1, 25, and 26.

Variational Calculus. *Variational calculus* is devoted to finding the *function* that maximizes or minimizes the value of an integral. An example is the problem of determining an optimal temperature profile in a reactor. Another example is presented in Section 5.1. Some principles of variational calculus are discussed in Appendix A. Additional details are given in References 25 and 26.

9.5 DESIGN OPTIMIZATION FOR THE COGENERATION SYSTEM CASE STUDY

In this section we bring closure to the discussion of the cogeneration system case study initiated with the primitive problem statement of Section 1.3.2 and considered several times subsequently. Referring to Figure 1.2, the life-cycle design flow chart, the present results would evolve during the detailed design stage, and would contribute significantly to the development of the final flow sheets for the system. Though important from an engineering perspective, the remaining aspects of the design flow chart are not covered here, however. The interested reader should consult more specialized textbooks.

9.5.1 Preliminaries

Two types of optimization are considered here: thermodynamic and economic. The objective of the thermodynamic optimization is to *maximize the exergetic efficiency* of the cogeneration system. In the economic optimization, the objective is to *minimize the levelized costs* of the product streams (electricity and steam) generated by the cogeneration system. The thermodynamic optimization is based on the thermodynamic model of Section 2.5, whereas the economic optimization employs both the thermodynamic model and the cost model of Appendix B. Box's complex method[7] (Section 9.4.4) and the corresponding computer program from Reference 27 were used to calculate the results presented in Sections 9.5.2 and 9.5.3.

When the cost model of Appendix B is used together with the thermodynamic model of Section 2.5, complex interactions are observed among the various thermodynamic and cost variables. Still, with reference to Figure 1.7, some general relationships can be identified: The thermodynamic performance of the air compressor affects the performance and costs of the air preheater through the temperature T_2. In turn, the preheater performance affects the performance and costs of the combustion chamber and the heat recovery steam generator through the temperatures T_3 and T_6. Furthermore, the performance of the combustion chamber and gas turbine affect (a) the performance and costs of the air preheater and the heat recovery steam generator through the temperature T_5 and (b) the costs of stream 2 and, indirectly, the costs of the remaining streams through the power to the compressor $\dot{W}_{11}$.

Before proceeding with the thermodynamic or economic optimization of an overall system, the design engineer may use the thermodynamic model to study the effect on the overall exergetic efficiency (ε_{tot}) of a change in some of the decision variables, while keeping all other variables constant. The results of such a parametric study for the cogeneration system of Figure 1.7 may be summarized as follows:

- ε_{tot} *increases* monotonically with increasing isentropic compressor efficiency η_{sc}, and with increasing isentropic turbine efficiency η_{st}.
- ε_{tot} *increases* as the compressor pressure ratio p_2/p_1 increases, passes through a maximum, and then decreases. The pressure ratio at which the maximum value of ε_{tot} occurs increases with increasing T_4, the temperature of the gases entering the gas turbine.
- ε_{tot} *decreases* as T_4 increases. However, if the temperature of the air exiting the preheater, T_3, is appropriately adjusted each time the temperature T_4 is changed, ε_{tot} *increases* as T_4 and T_3 increase.

9.5.2 Thermodynamic Optimization

Table 9.1 gives the values of the decision variables and selected parameters for the thermodynamically optimal (TO) design. For comparison, values for the base design (Tables 1.2, 3.1, and 3.2) and the cost-optimal (CO) design are also presented. As might have been expected, the thermodynamic optimum is obtained at the maximum values of the decision variables η_{sc}, η_{st}, and T_4 and the minimum values of $\dot{m}_1$, $\dot{m}_{10}$, and ΔT_{min} in the heat recovery steam generator.

---

[7]A direct-search method, such as Box's complex method, should be used cautiously, however, as it is possible for the method to identify a local optimum rather than the global optimum.

Table 9.1 Values of the decision variables and selected parameters for the base design, thermodynamically optimal (TO) design, and cost-optimal (CO) design[a]

Parameter	Base Design	TO Design	CO Design
Compressor pressure ratio, p_2/p_1	10.0	16.0	6.0
Compressor isentropic efficiency, η_{cs} (%)	86.0	88.0[b]	81.1
Turbine isentropic efficiency, η_{ts} (%)	86.0	90.0[b]	84.7
Air preheater outlet temperature, T_3 (K)	850.0	792.4	903.0
Combustion products temperature, T_4 (K)	1520.0	1550.0	1462.9
Pinch temperature difference in the HRSG, $\Delta T_{min,hrsg}$ (K)	40.2	15.0	50.8
Air mass flow rate, $\dot{m}_1$ (kg/s)	91.28	74.23	122.20
Methane mass flow rate, $\dot{m}_{10}$ (kg/s)	1.64	1.51	1.83

[a]The optimal values of the decision variables given for the TO and CO designs are not unique. The same values of the maximum exergetic efficiency and the minimum overall cost rate may be obtained through other combinations of the values of the decision variables. In addition, many different sets of the decision variables values lead to nearly optimal values of the objective functions.

[b]For the TO design, the purchased-equipment cost calculations using the functions from Appendix B are based on $\eta_{cs} = 87.97\%$ and $\eta_{ts} = 89.99\%$.

In Table 9.2, the values of three important exergy-related variables introduced in Chapter 3 are listed for the overall cogeneration system and each of the system components. These are the rate of exergy destruction $\dot{E}_{D,k}$, exergy destruction ratio $\dot{E}_{D,k}/\dot{E}_{F,tot}$, and exergetic efficiency ε_k. As in Table 9.1, values for the base design and the cost-optimal design are provided for comparison with the thermodynamically optimal values.

From Table 9.2 we see that the overall exergetic efficiency of the TO design is 54.4%, whereas the corresponding values for the other two cases are significantly lower. Still, the exergetic efficiencies of the components are not invariably higher in the TO design, indicating again that the optimal performance of components does not necessarily correspond to the optimal performance of an overall system, and conversely. Table 9.2 also shows that the component values for the exergy destruction rate and exergy destruction ratio are generally, but not invariably, lower in the TO design than in the other two designs. The values of these exergy-destruction variables for the overall system are, of course, lower in the TO design than in the other two designs.

The constraints on the values of the decision variables η_{sc}, η_{st}, p_2/p_1, T_3, and T_4 limit the maximum value of ε_{tot} that can be obtained in practice. For example, when the thermodynamically optimal values are used for the remaining variables, the maximum value of ε_{tot} would be obtained for p_2/p_1 greater than 16, which according to the thermodynamic model of Section 2.5 exceeds the maximum allowed value for p_2/p_1. Thus, for this cogeneration system the thermodynamic optimum is obtained at a boundary point not only with respect to η_{sc}, η_{st}, and T_4 but also with respect to the pressure ratio p_2/p_1.

Table 9.2 Exergy destruction $\dot{E}_D$, exergy destruction ratio $\dot{E}_D/\dot{E}_{F,tot}$, and exergetic efficiency ε for the kth component of the cogeneration system for the base design and the TO and CO designs

Component	Base Design[a]			TO Design			CO Design		
	$\dot{E}_{D,k}$ (MW)	$\dfrac{\dot{E}_{D,k}}{\dot{E}_{F,tot}}$ (%)	ε (%)	$\dot{E}_{D,k}$ (MW)	$\dfrac{\dot{E}_{D,k}}{\dot{E}_{F,tot}}$ (%)	ε (%)	$\dot{E}_{D,k}$ (MW)	$\dfrac{\dot{E}_{D,k}}{\dot{E}_{F,tot}}$ (%)	ε (%)
Combustion chamber	25.75	30.20	79.8	23.81	30.32	78.8	28.70	30.48	80.8
Heat-recovery steam generator	6.23	7.31	67.2	5.58	7.10	69.6	6.33	6.67	66.8
Gas turbine	3.01	3.53	95.2	2.29	2.91	96.4	3.16	3.33	95.0
Air preheater	2.63	3.08	84.6	1.08	1.38	82.8	5.31	5.60	84.3
Air compressor	2.12	2.49	92.8	1.66	2.11	94.6	3.36	3.54	89.0
Total system	39.74	46.61[b]	50.1	34.42	43.83[b]	54.4	46.86	49.63[b]	45.1

[a]In developing Tables 9.1, 9.2, and 9.3, the fuel is assumed to enter the system at a pressure $p_{10} = 40$ bars instead of the nominal 12-bar value used in Table 3.2. This results in some minor differences between the entries of Tables 3.2 and 9.2. The higher pressure is used in the present evaluations because the pressure reduction unit receiving methane from the high-pressure supply line at 40 bars is considered together with the combustion chamber. This was done to eliminate the effect of a varying methane supply pressure on the optimization results.

[b]An additional 3.25, 1.74, and 5.28% of the exergy supplied by the fuel to the total system is carried out of the system at state 7, respectively; this represents an exergy loss for the total system.

9.5.3 Economic Optimization

Table 9.3 provides the costs obtained in the economic optimization. Also presented in this table for comparison purposes are the costs calculated for the base design and the costs determined using the results of the thermodynamically optimal design. The table presents the total cost flow rate (Section 8.1) in $/h, the cost of electricity in ¢/kWh, and the cost of steam in ¢/kg.

Comparing the cost data of the cost-optimal and thermodynamically-optimal designs shows striking differences: The total cost flow rate and the cost of electricity are significantly higher in the TO design than in the CO design. These differences result mainly from the investment costs of the air compressor and the gas turbine that are very high in the TO design.

9.5.4 Closure

In the design of a thermal system, the total cost associated with the thermodynamically optimal design is always higher than for the cost-optimal design and, as in the present application, sometimes significantly higher. Accordingly, studies focusing only on the thermodynamically optimal performance for the design of a new system can lead to gross misevaluations and skewed decision making.

The economic optimization results discussed here are obtained without applying any of the *thermoeconomic techniques* discussed in Sections 8.3, 9.2, and 9.3 because (i) the present system is simple, (ii) its structure was specified (Section 1.5.4), and (iii) both the thermodynamic and the cost models are complete and allow the use of a numerical optimization technique. The next section discusses an iterative thermoeconomic optimization technique that may be used when one or more of these conditions is not satisfied.

9.6 THERMOECONOMIC OPTIMIZATION OF COMPLEX SYSTEMS

Complex thermal systems cannot always be optimized using the mathematical techniques discussed in Section 9.4 and illustrated by the results presented in Section 9.5. The reasons include incomplete models, system complexity, and structural changes:

Table 9.3 Calculated costs for the base design, thermodynamically optimal (TO) design, and cost-optimal (CO) design

Parameter	Base Design	TO Design	CO Design
Total cost flow rate ($/h)	3617	9089	2870
Cost of electricity (¢/kWh)	7.41	21.34	5.55
Cost of steam (¢/kg)	2.77	5.33	2.39

- Some of the input data and functions required for the thermodynamic and, particularly, the economic model might not be available or might not be in the required form. For example, it is not always possible to express the purchased-equipment costs as a function of the appropriate thermodynamic decision variables.
- Even if all the required information is available, the complexity of the system might not allow a satisfactory mathematical model to be formulated and solved in a reasonable time.
- The analytical and numerical optimization techniques are applied to a *specified structure* of the thermal system. However, a significant decrease in the product costs may be achieved through *changes in the structure* of the system. It is not always practical to develop a mathematical model for *every* promising design configuration of a system. More importantly, analytical and numerical optimization techniques cannot suggest structural changes that have the potential of improving the cost effectiveness.

The usual approach to the optimization of such complex systems is to iteratively optimize subsystems and/or ignore the influence of some structural changes and decision variables. An alternative to this approach is an *iterative thermoeconomic optimization technique* that consists of the following seven steps:

1. In the first step a workable design is developed. The guidelines presented in Table 1.1 and Sections 3.6 and 9.3.7 may assist in developing a workable design that is relatively close to the optimal design. The use of these guidelines can reduce, therefore, the total number of iterations required.

2. A detailed thermoeconomic analysis and evaluation (Sections 3.5 and 8.3) and a pinch analysis (Section 9.3) are conducted for the design configuration developed in the previous step. The results are used, as discussed in Sections 3.6, 8.3, and 9.3, to determine design changes that are expected to improve the design being considered. In this step, and in steps 3–5, we consider only changes in the decision variables that affect *both* the exergetic efficiency and the investment costs. The remaining decision variables are optimized in step 6. This reduces the number of variables that must be optimized simultaneously.

3. If the system has one or two components for which the sum of the cost rates $(\dot{Z}_k + \dot{C}_{D,k})$ is *significantly* higher than the same sum for the remaining components, the designs of these components are modified to approach their corresponding cost-optimal exergetic efficiency, given by Equations 9.11 and 9.12. To apply these equations, we assume that the costs per exergy unit remain constant for all inlet streams. Examples of major components whose designs may be approximately optimized *in isolation* include the steam reformer of a chemical plant and the gasifier and gas turbine system of an integrated gasification-combined-cycle power plant. This step is meaningful

only for components where *each* of the terms $\dot{Z}_k$ and $\dot{C}_{D,k}$ has a significant contribution to the costs associated with the respective component. If not, step 3 should either be omitted or, preferably, replaced by an efficiency maximization procedure when $\dot{C}_{D,k}$ is the dominating cost rate, or an investment cost minimization procedure when $\dot{Z}_k$ is the dominating cost rate.

As shown in the table for Example 8.2, for example, this step should not be applied to the combustion chamber of the cogeneration system sample problem because the sum $\dot{Z}_{cc} + \dot{C}_{D,cc}$ is not considerably higher than the same sum for the other system components. Also, in Table 9.2 we note that the exergetic efficiency of the combustion chamber in the CO design is higher than in the TO design and in the base design. Although the exergetic efficiency in the CO design is not maximized, it still has a relatively high value dictated by the dominating cost rate associated with exergy destruction ($\dot{C}_{D,cc}$).

An alternative to using Equations 9.11 and 9.12 is to optimize one or two decision variables for each of the most important components by graphically minimizing the relative cost difference (Equation 8.25) calculated for the components at various values of the corresponding decision variable(s). In this case it is unnecessary to assume that the costs per exergy unit remain constant for each inlet stream. This approach is used in References 28 and 29 to optimize the gasification temperature in an integrated gasification–combined-cycle power plant.

4. For the remaining components, particularly the ones having a relatively high value of the sum $\dot{Z}_k + \dot{C}_{D,k}$, the relative deviations of the actual values from the cost-optimal values for the exergetic efficiency and relative cost difference are calculated:

$$\Delta\varepsilon_k = \frac{\varepsilon_k - \varepsilon_k^{\text{OPT}}}{\varepsilon_k^{\text{OPT}}} \times 100 \tag{9.37}$$

$$\Delta r_k = \frac{r_k - r_k^{\text{OPT}}}{r_k^{\text{OPT}}} \times 100 \tag{9.38}$$

Here ε_k and r_k are the actual values and $\varepsilon_k^{\text{OPT}}$ and r_k^{OPT} are the cost-optimal values for the exergetic efficiency and the relative cost difference, respectively. The cost-optimal values are calculated using Equations 9.11, 9.12, and 9.18. The design of these remaining thermoeconomically important components is modified to reduce the values of $\Delta\varepsilon_k$ and Δr_k. If conflicting design changes are suggested from the evaluation of different components, the design changes for the components with the higher $\dot{Z}_k + \dot{C}_{D,k}$ values prevail. When this step cannot be implemented because Equation 9.4 is unavailable, the approach of Section 8.3.1 is invoked to suggest design changes that would improve the cost effectiveness of the overall system.

5. Based on the results from steps 2, 3, and 4, a new design is developed and the value of the objective function for this design is calculated. If in

comparison with the previous design this value has been improved, we may decide to proceed with another iteration that involves steps 2–5. If, however, the value of the objective function is not better in the new design than in the previous one, we may either revise some design changes and repeat steps 2–5 or proceed with step 6.

6. In this step, we use an appropriate technique from Section 9.4 to optimize the decision variables that affect the costs but not the exergetic efficiency. At the end of this step the cost-optimal design is obtained.

7. Finally a parametric study may be conducted to investigate the effect on the optimization results of some parameters and/or assumptions made in the optimization procedure.

The applications of steps 1, 2, 4, and 5 to the cogeneration system case study is presented in Reference 7 and is not repeated here. Reference 7 is one of four papers [30] dealing with thermoeconomic optimization techniques applied to the cogeneration system considered throughout this book. Additional examples of thermoeconomic evaluations of complex thermal systems are discussed in References 28, 29, 31–34.

9.7 CLOSURE

To conduct a complete and successful optimization of a thermal system, engineers typically apply principles of engineering thermodynamics, fluid mechanics, heat and mass transfer, economics and mathematics, in addition to their experience and intuition. Engineers must also be thoroughly familiar with the system being optimized and should understand all technological options available and the interactions both among the system components and between thermodynamics and economics. Often a bit of luck can be helpful in optimization studies.

The impact of an effective mathematical model on the optimization outcome cannot be overemphasized: A good model can make optimization almost easy, whereas a poor one can make *correct* optimization difficult or impossible [3]. In any case, the quality of the optimization results cannot be better than the quality of the mathematical model used to obtain the results.

The analytical and numerical optimization techniques discussed in Section 9.4 should be used if the formulation of the optimization problem allows their successful application. If not, the thermoeconomic techniques presented in Chapter 8 and Sections 9.2, 9.3, and 9.6 may provide effective assistance. Since the application of the thermoeconomic evaluation techniques improves the engineer's understanding of the interactions among the system variables, and generally reveals opportunities for design improvements that might not be detected by other methods, these techniques are always recommended regardless of the optimization approach, however.

Much can be learned by reviewing case studies on optimal design applications. Complete case studies other than the cogeneration system have not been included in the present book, however, to keep its size within reasonable bounds. The interested reader can find additional optimization studies reported in various technical journals and in design optimization textbooks. The references listed below provide a starting point for individual study.

Finally, it should be emphasized that the optimization of a thermal system seldom leads to a unique solution corresponding to a global mathematical optimum. Rather, acceptable alternative solutions may be feasible. Different solutions developed by different design teams may be equally acceptable and nearly equally cost effective. As a rule, the more complex the system being optimized, the larger the number of acceptable solutions. The methods discussed in this chapter enhance the knowledge, experience, and intuition of design engineers, but do not substitute for engineering creativity.

REFERENCES

1. G. V. Reklaitis, A. Ravindran, and K. M. Ragsdell, *Engineering Optimization,* Wiley, New York, 1983.

2. T. F. Edgar and D. M. Himmelblau, *Optimization of Chemical Processes,* McGraw-Hill, New York, 1988.

3. P. Y. Papalambros and D. J. Wilde, *Principles of Optimal Design—Modeling and Computation,* Cambridge University Press, Cambridge, 1988.

4. M. Zeleny, *Multiple Criteria Decision Making,* McGraw-Hill, New York, 1982.

5. V. Chankong and Y. Y. Haimes, *Multiobjective Decision Making: Theory and Methodology,* North-Holland, New York, 1983.

6. C. Carlsson and Y. Kochetkov, eds, *Theory and Practice of Multiple Criteria Decision Making,* Elsevier, New York, 1983.

7. G. Tsatsaronis and J. Pisa, Exergoeconomic evaluation and optimization of energy systems—application to the CGAM problem, *Energy—Int. J.,* Vol. 19, No. 3, 1994, pp. 287–321.

8. G. Tsatsaronis, Thermoeconomic analysis and optimization of energy systems, *Prog. Energy Combustion Sci.,* Vol. 19, 1993, pp. 227–257.

9. J. Szargut, Application of exergy to the approximate economic optimization (in German), *Brennstoff-Wärme-Kraft,* Vol. 23, 1971, pp. 516–519.

10. T. Gundersen and L. Naess, The synthesis of cost optimal heat exchanger networks—an industrial review of the state of the art, *Comput. Chem. Eng.,* Vol. 12, 1988, pp. 503–530.

11. E. C. Hohmann, Optimum networks for heat exchange, Ph.D. thesis, University of Southern California, 1971.

12. N. Nishida, Y. A. Liu, and L. Lapidus, A simple and practical approach to the optimal synthesis of heat exchanger networks, *AIChE J.,* Vol. 23, No. 1, 1977, pp. 77–93.

13. T. F. Yee, I. E. Grossmann, and Z. Kravanja, Simultaneous optimization models for heat integration, *Comput. Chem. Eng.*, Vol. 14, Part I: No. 10, 1990, pp. 1151–1164; Part II: No. 10, 1990, pp. 1165–1184; Part III: No. 11, 1990, pp. 1185–1200.

14. B. Linnhoff, et al., *A User Guide on Process Integration for the Efficient Use of Energy*, Institution of Chemical Engineers, Rugby, Warks, England, 1982.

15. B. Linnhoff, Pinch technology for the synthesis of optimal heat and power systems, *J. Energy Res. Tech.*, Vol. 111, 1989, pp. 137–148.

16. A. R.Ciric and C. A. Floudas, Heat exchanger network synthesis without decomposition, *Comput. Chem. Eng.*, Vol. 15, No. 6, 1991, pp. 385–396.

17. J. M. Douglas, *Conceptual Design of Chemical Processes*, McGraw-Hill, New York, 1988.

18. K. Lucas and H. Roth, *Strategies and Systems for Rational Energy Use* (in German), Lehrgangshandbuch, Institut für Umwelttechnologien und Umweltanalytik, Duisburg, Germany, 1993.

19. B. Linnhoff and J. R. Flower, Synthesis of heat exchanger networks, *AIChE J.*, Vol. 24, 1978, pp. 633–654.

20. D. A. Sama, The use of the second law of thermodynamics in the design of heat exchangers, heat exchanger networks and processes, in J. Szargut, Z. Kolenda, G. Tsatsaronis, and A. Ziebik, eds., *Proceedings of the International Conference on Energy Systems and Ecology*, Cracow, Poland, July 5–9, American Society of Mechanical Engineers, New York, 1993, pp. 53–76.

21. D. W. Townsend and B. Linnhoff, Heat and power networks in process design; Parts I and II, *ACIhE J.*, Vol. 29, 1983, pp. 742–771.

22. B. Linnhoff, H. Dunford, and R. Smith, Heat integration of distillation columns into overall processes, *Chem. Eng. Sci.*, Vol. 38, 1983, pp. 1175–1188.

23. B. Linnhoff and F. J. Alanis, Integration of a new process into an existing site: a case study in the application of pinch technology, *J. Eng. Gas Turbines Power*, Vol. 113, 1991, pp. 159–169.

24. R. A. Gaggioli, D. A. Sama, S. Qian, and Y. M. El-Sayed, Integration of a new process into an existing site: a case study in the application of exergy analysis, *J. Eng. Gas Turbines Power*, Vol. 113, 1991, pp. 170–183.

25. G. S. G. Beveridge and R. S. Schlechter, *Optimization: Theory and Practice*, McGraw-Hill, New York, 1970.

26. R. W. Pike, *Optimization for Engineering Systems*, van Nostrand Reinhold, New York, 1986.

27. J. K. Kuester and J. H. Mize, *Optimization Techniques with Fortran*, McGraw-Hill, New York, 1973.

28. G. Tsatsaronis, L. Lin, J. Pisa, and T. Tawfik, Thermoeconomic Design Optimization of a KRW-Based IGCC Power Plant, Final Report, prepared for the U.S. Department of Energy, Morgantown Energy Technology Center, DE-FC21-89MC26019, November 1991.

29. G. Tsatsaronis, J. Pisa, L. Lin, and T. Tawfik, Optimization of an IGCC power plant—Parts I and II in R. F. Boehm et al., eds., *Thermodynamics and the Design, Analysis and Improvement of Energy Systems*, American Society of Mechanical Engineers, New York, AES-Vol. 27, 1992, pp. 37–67.

30. S. S. Penner and G. Tsatsaronis, eds., Invited Papers on Exergoeconomics, *Energy—Int. J.,* Special Issue, Vol. 19, No. 3, 1994.

31. G. Tsatsaronis, L. Lin, and T. Tawfik, Exergoeconomic evaluation of a KRW-based IGCC power plant, *J. Eng. Gas Turbines Power,* Vol. 116, 1994, pp. 300–306.

32. G. Tsatsaronis, A. Krause, T. Tawfik, and L. Lin, Thermoeconomic Evaluation of the Design of a Pressurized Fluidized-Bed Hydroretorting Plant, Final Report submitted to the Institute of Gas Technology and the Department of Energy, DE-AC21-87MC11089, May 1992.

33. L. Lin and G. Tsatsaronis, Analysis and improvement of an advanced concept for electric power generation, in J. Szargut, Z. Kolenda, G. Tsatsaronis, and A. Ziebik, eds., *Proceedings of the International Conference on Energy Systems and Ecology,* Cracow, Poland, July 5–9, 1993, pp. 557–566.

34. L. Lin and G. Tsatsaronis, Cost optimization of an advanced IGCC power plant concept design, in H. J. Richter, ed., *Thermodynamics, and the Design, Analysis and Improvement of Energy Systems—1993,* American Society of Mechanical Engineers, New York, AES-Vol. 30/HTD-Vol. 266, 1993, pp. 156–166.

PROBLEMS

9.1 Prepare a brief report describing a numerical optimization method involving nonlinear programming with constraints. Compare this method with similar methods.

9.2 At the end of Example 8.2, new values for the decision variables of the cogeneration system case study are suggested. Conduct a detailed thermoeconomic evaluation of the system using these values. Suggest changes in the decision variables to reduce the value of the objective function.

9.3 Develop a workable design for the cogeneration system case study different from the base design reported in Tables 1.2, 3.1 and 3.2. Conduct a detailed exergy analysis and thermoeconomic evaluation of this new design. Iteratively apply steps 2, 4, and 5 from Section 9.6 to improve the cost effectiveness of this design. The values of the parameter B_k and the exponents n_k and m_k that can be used in Equation 9.4 are given in Table P9.3:

Table P9.3

Component	B	n	m
Air compressor	58.135	3.0613	1.0
Air preheater	18816.1	1.2753	0.6
Combustion chamber	$294.03(1.0 + \exp(19.359 \cdot 10^{-3}T_4 - 28.854))$	1.0	1.0
Gas turbine	$169.949(1.0 + \exp(42.134 \cdot 10^{-3}T_4 - 64.362))$	1.8133	1.0
Heat-recovery steam generator	80119.8	0.24774	0.9

9.4 Develop a computer simulation for the cogeneration system case study based on the thermodynamic model presented in Section 2.5. Use the computer program to study the effect of the changes in the decision variables on the exergetic efficiencies of the overall system and the system components.

9.5 Use the computer simulation developed in Problem 9.4 to calculate iteratively the thermodynamic optimum of the cogeneration system. Discuss the procedure and the difficulties you encounter, if any.

9.6 Apply a numerical optimization technique to calculate the cost-optimal values of the decision variables for the cogeneration system case study using the levelized costs in current dollars for a 20-year period from Table 7.12. Discuss the differences between the values obtained here and the values reported in Table 9.1.

9.7 Prepare a report describing a design optimization case study from the literature. Discuss the objectives of the study, the assumptions made, the mathematical model used, the optimization technique applied, and the final results obtained in the case study.

9.8 Calculate the cost-optimal values of the decision variables η_{cs}, η_{ts}, p_2/p_1, and T_4 (temperature at the gas turbine inlet) for the cogeneration system of Figure 1.7 without the air preheater (see also Figure 1.6b). Your instructor will specify additional assumptions to be made. The thermodynamic model of Section 2.5 and the cost model of Appendix B are applicable.

9.9 Solve Problem 9.3 for the design configuration of the cogeneration system of Problem 9.8.

9.10 Using Equation 9.10, derive Equation 9.11.

Appendix **A**

VARIATIONAL CALCULUS

The basic problem in variational calculus consists of determining, from among functions possessing certain properties, that function for which a given integral (functional) assumes its maximum or minimum value. The integrand of the integral in question depends on the function and its derivatives. Consider the many values of the integral

$$I = \int_a^b F(x, y, y') \, dx \tag{A.1}$$

where $y(x)$ is unknown and $y' = dy/dx$. The special function y for which the integral I reaches an extremum satisfies the *Euler equation:*

$$\frac{\partial F}{\partial y} - \frac{d}{dx}\left(\frac{\partial F}{\partial y'}\right) = 0 \tag{A.2}$$

If, in addition to minimizing or maximizing I, the wanted function $y(x)$ satisfies an *integral constraint* of the type

$$C = \int_a^b G(x, y, y') \, dx \tag{A.3}$$

then $y(x, \lambda)$ satisfies the new Euler equation:

$$\frac{\partial H}{\partial y} - \frac{d}{dx}\left(\frac{\partial H}{\partial y'}\right) = 0 \tag{A.4}$$

where

$$H = F + \lambda G \tag{A.5}$$

Note that H is a linear combination of the integrands of I and C. The constant λ is a *Lagrange multiplier;* its value is determined by substituting the $y(x, \lambda)$ solution into the integral constraint given by Equation A.3.

ECONOMIC MODEL
OF THE
COGENERATION SYSTEM

Table B.1 shows the purchase costs of the cogeneration system components as a function of thermodynamic parameters.[1,2] The values of the constants C_{11} through C_{53} used in Table B.1 are given in Table B.2. The purchased-equipment costs obtained from Tables B.1 and B.2 are given in mid-1994 dollars.

Table B.1 Equations for calculating the purchased-equipment costs (PEC) for the components of the cogeneration system case study.[a]

Compressor	$$\mathrm{PEC_{ac}} = \left(\frac{C_{11}\dot{m}_a}{C_{12} - \eta_{sc}}\right)\left(\frac{p_2}{p_1}\right)\ln\left(\frac{p_2}{p_1}\right)$$
Combustion chamber	$$\mathrm{PEC_{cc}} = \left(\frac{C_{21}\dot{m}_a}{C_{22} - \dfrac{p_4}{p_3}}\right)[1 + \exp(C_{23}T_4 - C_{24})]$$
Turbine	$$\mathrm{PEC_{gt}} = \left(\frac{C_{31}\dot{m}_g}{C_{32} - \eta_{st}}\right)\ln\left(\frac{p_4}{p_5}\right)$$ $$\times\,[1 + \exp(C_{33}T_4 - C_{34})]$$

[1]In the design of thermal systems, cost functions such as these are usually not available. Thus, in the discussions of thermoeconomic evaluation (Section 8.3) and iterative optimization (Section 9.6) we recognize that after each design modification the new purchased-equipment costs would be calculated by a cost engineer. For simplicity of presentation, however, we assume that the cost values provided by the cost engineer are in full agreement with the corresponding values calculated from the cost functions and constants in Tables B.1 and B.2.

[2]Source: A. Valero, et al., "CGAM Problem: Definition and Conventional Solution," *Energy— Int. J.*, Vol. 19, 1994, pp. 268–279.

Table B.1 (*continued*)

Air preheater	$\mathrm{PEC}_{\mathrm{aph}} = C_{41} \left(\dfrac{\dot{m}_g (h_5 - h_6)}{U\,\Delta T_{\mathrm{lm,aph}}} \right)^{0.6}$
Heat-recovery steam generator	$\mathrm{PEC}_{\mathrm{hrsg}} = C_{51} \left[\left(\dfrac{\dot{Q}_{\mathrm{ec}}}{\Delta T_{\mathrm{lm,ec}}} \right)^{0.8} + \left(\dfrac{\dot{Q}_{\mathrm{ev}}}{\Delta T_{\mathrm{lm,ev}}} \right)^{0.8} \right]$ $+\, C_{52}\dot{m}_{\mathrm{st}} + C_{53}\dot{m}_g^{1.2}$

[a]$\dot{m}_a$, $\dot{m}_g$, $\dot{m}_{\mathrm{st}}$ are the mass flow rates of air, gas, and steam, respectively; η_{sc} and η_{st} are the isentropic compressor and turbine efficiency, respectively; h_5 and h_6 are the specific enthalpies of streams 5 and 6; ΔT_{lm} denotes a log mean temperature difference; $\dot{Q}_{\mathrm{ec}}$ and $\dot{Q}_{\mathrm{ev}}$ represent the rates of heat transfer in the economizer and evaporator, respectively.

Table B.2   Constants used in the equations of Table B.1 for the purchase cost of the components

Compressor	$C_{11} = 71.10 \text{ \$/(kg/s)}$, $C_{12} = 0.9$
Combustion chamber	$C_{21} = 46.08 \text{ \$/(kg/s)}$, $C_{22} = 0.995$ $C_{23} = 0.018 \text{ (K}^{-1})$, $C_{24} = 26.4$
Gas turbine	$C_{31} = 479.34 \text{ \$/(kg/s)}$, $C_{32} = 0.92$ $C_{33} = 0.036 \text{ (K}^{-1})$, $C_{34} = 54.4$
Air preheater	$C_{41} = 4122 \text{ \$/(m}^{1.2})$, $U = 18 \text{ W/(m}^2\text{K)}$
Heat-recovery steam generator	$C_{51} = 6570 \text{ \$/(kW/K)}^{0.8}$ $C_{52} = 21276 \text{ \$/(kg/s)}$ $C_{53} = 1184.4 \text{ \$/(kg/s)}^{1.2}$

Appendix C

TABLES OF PROPERTY DATA

Table C.1 Variation of specific heat, enthalpy, absolute entropy, and Gibbs function with temperature at 1 bar for various substances in units of kJ/kmol or kJ/kmol·K[a]

1. At T_{ref} = 298.15 K (25°C), p_{ref} = 1 bar

Substance	Formula	$\bar{c}_p^\circ$	$\bar{h}^\circ$	$\bar{s}^\circ$	$\bar{g}^\circ$
Carbon (graphite)	C(s)	8.53	0	5.740	−1711
Sulfur (rhombic)	S(s)	22.77	0	32.058	−9558
Nitrogen	N_2(g)	28.49	0	191.610	−57128
Oxygen	O_2(g)	28.92	0	205.146	−61164
Hydrogen	H_2(g)	29.13	0	130.679	−38961
Carbon monoxide	CO(g)	28.54	−110528	197.648	−169457
Carbon dioxide	CO_2(g)	35.91	−393521	213.794	−457264
Water	H_2O(g)	31.96	−241856	188.824	−298153
Water	H_2O(l)	75.79	−285829	69.948	−306685
Methane	CH_4(g)	35.05	−74872	186.251	−130403
Sulfur dioxide	SO_2(g)	39.59	−296833	284.094	−370803
Hydrogen sulfide	H_2S(g)	33.06	−20501	205.757	−81847
Ammonia	NH_3(g)	35.59	−46111	192.451	−103491

2. For $298.15 < T \leq T_{max}$, p_{ref} = 1 bar, with $y = 10^{-3}T$

$$\bar{c}_p^\circ = a + by + cy^{-2} + dy^2 \tag{1}$$

$$\bar{h}^\circ = 10^3 \left[H^+ + ay + \frac{b}{2}y^2 - cy^{-1} + \frac{d}{3}y^3 \right] \tag{2}$$

$$\bar{s}^\circ = S^+ + a \ln T + by - \frac{c}{2}y^{-2} + \frac{d}{2}y^2 \tag{3}$$

$$\bar{g}^\circ = \bar{h}^\circ - T\bar{s}^\circ \tag{4}$$

The constants H^+, S^+, a, b, c, and d required by Equations (1)–(4) are given for selected substances in the table below. The maximum temperature, T_{max}, is 1100 K for C(s), 368 K for S(s), 500 K for $H_2O(l)$, 2000 K for $CH_4(g)$, $SO_2(g)$ and $H_2S(g)$, 1500 K for $NH_3(g)$, and 3000 K for all remaining substances. To evaluate the absolute entropy at states where the pressure p differs from $p_{ref} = 1$ bar, Equations 2.71 and 2.72 should be applied, as appropriate. The same caution applies when using absolute entropy data from other references. Also, owing to different data sources and roundoff, the absolute entropy values from other references may differ slightly from those of the current table.

Substance	Formula	H^+	S^+	a	b	c	d
Carbon (graphite)	C(s)	−2.101	−6.540	0.109	38.940	−0.146	−17.385
Sulfur (rhombic)	S(s)	−5.242	−59.014	14.795	24.075	0.071	0
Nitrogen[b]	$N_2(g)$	−9.982	16.203	30.418	2.544	−0.238	0
Oxygen	$O_2(g)$	−9.589	36.116	29.154	6.477	−0.184	−1.017
Hydrogen	$H_2(g)$	−7.823	−22.966	26.882	3.586	0.105	0
Carbon monoxide	CO(g)	−120.809	18.937	30.962	2.439	−0.280	0
Carbon dioxide	$CO_2(g)$	−413.886	−87.078	51.128	4.368	−1.469	0
Water	$H_2O(g)$	−253.871	−11.750	34.376	7.841	−0.423	0
Water	$H_2O(l)$	−289.932	−67.147	20.355	109.198	2.033	0
Methane	$CH_4(g)$	−81.242	96.731	11.933	77.647	0.142	−18.414
Sulfur dioxide	$SO_2(g)$	−315.422	−43.725	49.936	4.766	−1.046	0
Hydrogen sulfide	$H_2S(g)$	−32.887	1.142	34.911	10.686	−0.448	0
Ammonia	$NH_3(g)$	−60.244	−29.402	37.321	18.661	−0.649	0

Special Note: Since the reference state used in the *steam tables* differs from that of the present table, care must be exercised to ensure that steam table data are used consistently with values from this table. Although the reference states differ, each of these sources must yield the same values for the *changes* in property values between any two states. Thus, for specific enthalpy and entropy we have

$$h(T, p) - h(25°C, 1 \text{ bar}) = h^* (T, p) - h^* (25°C, 1 \text{ bar}) \tag{5}$$

$$s(T, p) - s(25°C, 1 \text{ bar}) = s^* (T, p) - s^* (25°C, 1 \text{ bar}) \tag{6}$$

where the terms on the left side are obtained from the present table and the terms on the right side denoted by superscript * are obtained from the steam tables.

Liquid water is a special case of interest. Thus at 25°C we have from the present table, $h = -15866.2$ kJ/kg ($\bar{h} = -285,829$ kJ/kmol), $s = 3.88276$ kJ/kg·K ($\bar{s} = 69.948$ kJ/kmol·K). Applying Equations 2.42c and 2.42d at 25°C, 1 bar we have from the steam tables, $h^* = 104.85$ kJ/kg, $s^* = 0.3670$ kJ/kg·K. Inserting values in Equations (5), (6)

$$h(T, p) = h^* (T, p) - 15{,}971 \text{ kJ/kg} \tag{7a}$$

$$s(T, p) = s^* (T, p) + 3.51576 \text{ kJ/kg·K} \tag{8a}$$

On a molar basis

$$\bar{h}(T, p) = \bar{h}^* (T, p) - 287718 \text{ kJ/kmol} \tag{7b}$$

$$\bar{s}(T, p) = \bar{s}^* \, (T, p) + 63.3365 \text{ kJ/kmol·K} \qquad (8b)$$

Using these expressions, steam table values of the specific enthalpy and entropy of liquid water can be made consistent with enthalpy and entropy data obtained from the present table.

---

[a] $\bar{c}_p^{\,\circ}$, $\bar{s}^{\,\circ}$, and S^+ in kJ/kmol·K; $\bar{h}^{\,\circ}$, $\bar{g}^{\,\circ}$, and H^+ in kJ/kmol; T in Kelvin.

[b] Table values for nitrogen are shown as reported in the source given below. Corrected values for H^+, S^+, a, b, c, d are, respectively, -7.069, 51.539, 24.229, 10.521, 0.180, -2.315.

Source: O. Knacke, O. Kubaschewski, and K. Hesselmann, *Thermochemical Properties of Inorganic Substances*, 2nd ed., Springer-Verlag, Berlin, 1991.

Table C.2. Standard molar chemical exergy, $\bar{e}^{CH}$ (kJ/kmol), of various substances at 298.15 K and p_0

Substance	Formula	Model I[a]	Model II[b]
Nitrogen	$N_2(g)$	639	720
Oxygen	$O_2(g)$	3,951	3,970
Carbon dioxide	$CO_2(g)$	14,176	19,870
Water	$H_2O(g)$	8,636	9,500
Water	$H_2O(l)$	45	900
Carbon (graphite)	$C(s)$	404,589	410,260
Hydrogen	$H_2(g)$	235,249	236,100
Sulfur	$S(s)$	598,158	609,600
Carbon monoxide	$CO(g)$	269,412	275,100
Sulfur dioxide	$SO_2(g)$	301,939	313,400
Nitrogen monoxide	$NO(g)$	88,851	88,900
Nitrogen dioxide	$NO_2(g)$	55,565	55,600
Hydrogen peroxide	$H_2O_2(g)$	133,587	—
Hydrogen sulfide	H_2S	799,890	812,000
Ammonia	$NH_3(g)$	336,684	337,900
Oxygen	$O(g)$	231,968	233,700
Hydrogen	$H(g)$	320,822	331,300
Nitrogen	$N(g)$	453,821	—
Methane	$CH_4(g)$	824,348	831,650
Acetylene	$C_2H_2(g)$	—	1,265,800
Ethylene	$C_2H_4(g)$	—	1,361,100
Ethane	$C_2H_6(g)$	1,482,033	1,495,840
Propylene	$C_3H_6(g)$	—	2,003,900
Propane	$C_3H_8(g)$	—	2,154,000
n-Butane	$C_4H_{10}(g)$	—	2,805,800
n-Pentane	$C_5H_{12}(g)$	—	3,463,300
Benzene	$C_6H_6(g)$	—	3,303,600
Octane	$C_8H_{18}(l)$	—	5,413,100
Methanol	$CH_3OH(g)$	715,069	722,300
Methanol	$CH_3OH(l)$	710,747	718,000
Ethyl alcohol	$C_2H_5OH(g)$	1,348,328	1,363,900
Ethyl alcohol	$C_2H_5OH(l)$	1,342,086	1,375,700

[a] J. Ahrendts, "Die Exergie chemisch reaktionsfähiger Systeme," *VDI-Forschungsheft,* 579, VDI-Verlag, Düsseldorf, 1977, pp. 26–33. Also see, "Reference States," *Energy—Int. J.,* Vol. 5, 1980, pp. 667–677. In this model, $p_0 = 1.019$ atm.

[b] From J. Szargut, D. R. Morris, and F. R. Steward, *Exergy Analysis of Thermal, Chemical, and Metallurgical Processes,* Hemisphere, New York, 1988, pp. 297–309. In this model, $p_0 = 1.0$ atm.

Appendix D

SYMBOLS

a	dimensionless area (Equation 6.47)
A	area (m²); annuity ($) (Chapter 7)
B	condensation driving parameter (Equation 4.65); heat transfer duty parameter (Equation 6.30); transverse conductance parameter (Equation 5.49)
Bi	Biot number (Table 4.1)
B_k	constant (Equation 9.4)
Bo	Boussinesq number (Table 4.1)
c	cost per unit of exergy ($/GJ)
c^*	cost per unit of mass ($/kg)
c_p, c_v, c	specific heats (kJ/kg·K)
C	capacity rate (W/K); cost ($); constant (Tables B.1, B.2)
$\dot{C}$	cost rate associated with exergy transfer ($/h)
C_D	drag coefficient
COP	coefficient of performance
d	differential of a property
D	diameter (m)
D_h	hydraulic diameter (m)
DFX	design for X (Chapter 1)
E	emissive power (W/m²) (Chapter 4)
$\dot{E}_i, \dot{m}_i e_i$	time rate of exergy transfer at inlet i (MW)
$\dot{E}_e, \dot{m}_e e_e$	time rate of exergy transfer at outlet e (MW)
$e^{CH}, \bar{e}^{CH}$	chemical exergy (MJ/kg, MJ/kmol)
e^{PH}	physical exergy of a stream of matter (MJ/kg)
$\dot{E}_D$	time rate of exergy destruction (MW)
$\dot{E}_L$	time rate of exergy loss (MW)

$\dot{E}_w$	time rate of exergy transfer associated with work (MW)
$\dot{E}_q$	time rate of exergy transfer associated with heat (MW)
E, e, E^{PH}, e^{PH}	total exergy and physical exergy of a system (MJ, MJ/kg)
Ec	Eckert number (Table 4.1)
f	Fanning friction factor; factor (Figure 4.5), factor used for purchased-equipment cost calculations (Chapter 7); exergoeconomic factor (Chapters 8 and 9)
F	future value ($) (Chapter 7), dimensionless variable used for component optimization (Chapter 9)
F_D	drag force (N)
F_{12}	geometric view factor (Equation 4.80)
Fo	Fourier number (Table 4.1)
g	dimensionless mass velocity (Equation 6.45); gravitational acceleration (m^2/s)
g	denotes gas (or vapor)
G, $\bar{g}$	Gibbs function (MJ), molar Gibbs function (MJ/kmol)
$\bar{g}_f^\circ$	molar Gibbs function of formation (MJ/kmol)
G	mass velocity ($kg/m^2\cdot s$); total irradiation (W/m^2)
Gr	Grashof number (Table 4.1)
Gz	Graetz number (Table 4.1)
h	convective heat transfer coefficient ($W/m^2\cdot K$)
h_{fg}, h'_{fg}	enthalpy of vaporization (kJ/kg); augmented enthalpy of vaporization (kJ/kg)
h_l	head loss (kJ/kg)
H, h	enthalpy (MJ), specific enthalpy (MJ/kg)
$\bar{h}_f^\circ$	molar enthalpy of formation (MJ/kmol)
$\bar{h}_{RP}$	enthalpy of combustion (MJ/kmol)
H	height (m)
HEN	heat exchanger network
HHV	higher heating value (MJ/kmol)
i	cost of money (rate of return)
i^*	internal rate of return
j_H	Colburn j_H factor (Equation 6.21)
J	radiosity (Equation 4.91)
k	specific heat ratio c_p/c_v; thermal conductivity ($W/m\cdot K$); variable used in levelized-cost calculations (Chapter 7); average cost per unit of energy ($/MJ) (Chapter 8)
k_s	wall roughness (mm)
K	loss coefficient for a fitting (Equation 2.31)
K_c, K_e	loss coefficients for contraction or enlargement (Equations 6.8 and 6.9)
KE	kinetic energy (MJ)
l	denotes liquid
L	length (m); also denotes liquid phase
Le	Lewis number (Table 4.1)

$\overline{LHV}$	lower heating value (MJ/kmol)
m	exponent in cost equations
$m, \dot{m}$	mass, mass flow rate (kg, kg/s)
M	molecular weight (kg/kmol)
n	number of moles (kmol); number of plates (Equation 5.20); polytropic exponent; exponent in cost equations; number of time periods
N	number of mixture components; number of system components, streams, or utilities
N, NTU	number of heat transfer units (Equations 6.3 and 6.97)
N_S	entropy generation number (Equation 6.36)
Nu	Nusselt number (Table 4.1)
p, p_k	pressure; partial pressure (bars)
p	number of interest compoundings per year (Chapter 7); perimeter (m)
P	present value ($)
Pe	Peclet number (Table 4.1)
Pr	Prandtl number (Table 4.1)
PE	gravitational potential energy (MJ)
PEC	purchased-equipment cost ($)
q'	heat transfer rate per unit of length (W/m)
q''	heat flux (W/m^2)
$Q, \dot{Q}$	heat transfer (MJ); rate of heat transfer (MW)
Q	volumetric flow rate (m^3/s)
r	escalation rate (Chapter 7); relative cost difference (Chapters 8 and 9)
r	radial position (m)
r_i	general inflation rate
R	annual operating and maintenance costs that are independent of investment costs and actual production rate ($) (Chapter 9)
$\overline{R}, R$	universal gas constant (kJ/kmol·K); specific gas constant (kJ/kg·K)
R_t	thermal resistance (K/W)
Ra	Rayleigh number (Table 4.1)
Ra*	flux Rayleigh number (Table 4.1)
Ra$_m$	mass transfer Rayleigh number (Table 4.1)
Re$_D$	Reynolds number based on diameter
S, s	entropy (MJ/K); specific entropy (kJ/kg·K)
$\overline{s}^\circ$	molar absolute entropy at temperature T and reference pressure p_{ref} (kJ/kmol·K)
s	denotes solid
S	salvage value ($) (Chapter 7); shape factor (m) (Equation 4.5)
S_{gen}	entropy generation (MJ/K)

$\dot{S}_{gen}$	entropy generation rate (W/K)
$\dot{S}'_{gen}$	entropy generation rate per unit length (W/m·K)
$\dot{S}'''_{gen}$	entropy generation rate per unit volume (W/m³·K)
Sc	Schmidt number (Table 4.1)
St	Stanton number (Table 4.1)
St_m	mass transfer Stanton number (Table 4.1)
Ste	Stefan number (Table 4.1)
t	income tax rate; thickness (m); time (h)
T	temperature (K)
T_a	thermodynamic average temperature (K)
TCI	total capital investment ($)
U, u	internal energy (MJ); specific internal energy (MJ/kg)
U	overall heat transfer coefficient (W/m²·K)
U	velocity (m/s) (Chapters 4–6)
v	specific volume (m³/kg); also denotes vapor or gas; dimensionless volume (Equation 6.50)
V	velocity (m/s)
V	volume (m³); also denotes vapor phase
W	width (m)
$W, \dot{W}$	work (MJ); work rate (MW)
x, x_k	quality (kg/kg); mole fraction (kmol/kmol)
$X, \bar{X}_k$	extensive property; partial molal extensive property
X	variable expressing capacity of a component (Chapter 7)
X, X_T	flow and thermal entrance lengths (m) (Chapter 4)
y_D, y_D^*	exergy destruction ratio
y_L	exergy loss ratio
Y	cash flow ($)
z	elevation above some datum (m); attained age of property (years) (Chapter 7)
Z	compressibility factor (Chapter 2)
$\dot{Z}$	non-exergy-related cost rate associated with a plant component or a system ($) (Chapters 8 and 9)

Greek letters

α	area per unit volume (m⁻¹) (Equation 6.11); capacity exponent (Chapter 7); thermal diffusivity (m²/s); total absorptivity (Chapter 4)
β	capital-recovery factor; coefficient of volumetric thermal expansion (K⁻¹)
γ	coefficient expressing the part of fixed operating and maintenance costs that depend on the net investment expenditure for a plant component; overall surface efficiency factor (Equation 6.2)
Γ	condensation flow rate (kg/m·s) (Equation 4.66)
δ	differential of a nonproperty

Δ	difference
ε	effectiveness (Equation 6.4); exergetic efficiency; total hemispherical emissivity (Chapter 4)
η	cycle thermal efficiency; fin efficiency (Equation 4.10)
η_{st}, η_{sc}	isentropic turbine efficiency; isentropic compressor efficiency
λ	wavelength (m)
$\overline{\lambda}$	fuel–air ratio on a molar basis
μ	viscosity (kg/s·m); chemical potential (MJ/kmol)
ν	kinematic viscosity (m^2/s)
Π	pressure difference number (Equation 5.38)
ρ	density (kg/m^3); total reflectivity (Chapter 4)
σ	area ratio (Equation 6.8); Stefan–Boltzmann constant; surface tension (N/m)
τ	total transmissivity (Chapter 4); dimensionless temperature difference (Equation 6.41); shear stress (N/m^2); average annual time of plant operation at nominal capacity (Equation 9.5)
φ	relative humidity (kmol/kmol); irreversibility distribution ratio (Chapter 6)
Φ	integral (Equation 5.4); viscous dissipation function (s^{-2}) (Equation 6.23)
χ	factor (Figure 4.5)
ω	humidity ratio (kg/kg); constant used in calculating variable operating and maintenance costs ($/MJ) (Chapter 9); ratio of capacity rates (Equation 6.55)

Subscripts

a	air; dry air; average
ac	air compressor
ad	adiabatic
ah	ash handling
amb	ambient
aph	air preheater
AR	area ratio
at	after taxes
avg	average
b	base; blackbody; boundary; boiler
B	base cost
BL	book life
BM	base bare module
c	critical pressure or temperature (Chapter 2); combined system (Chapter 3); compressor (Figure 8.5), cold stream (Chapter 9); composite curve (Chapter 9); cross section

C	cold reservoir
cc	combustion chamber; cold composite curve
ce	common equity
cogen	cogeneration system
cs	value expressed in constant dollars (Chapter 7)
cu	value expressed in current dollars (Chapter 7); cold utility (Chapter 9)
cv	control volume
cycle	thermodynamic cycle
d	debt; design factor
ds	flue-gas desulfurization
DAF	dry and ash free
e	outlet stream
eff	effective
ec	economizer
ev	evaporator
f	fin
f	saturated liquid
fr	frontal
F	fuel
FC	fixed-capital cost
g	gas, gray medium
g	saturated vapor
gr	larger temperature difference
gt	gas turbine
h	hot stream
hc	hot composite curve
hen	heat exchanger network
hu	hot utility
hx	heat exchanger
H	high temperature
H	hot reservoir
hrsg	heat-recovery steam generator
i	inlet stream; inner; temperature interval number (Chapter 9)
int	internal
int rev, rev	internally reversible; reversible
isol	isolated
j	portion of boundary; stream of matter; year
k	mixture component
k	plant component; year
K	capitalized cost
lm	log-mean temperature difference
L	levelized value; low temperature
m	material factor; mean, melting

max, min	maximum; minimum
M	module cost
MACRS	modified accelerated cost recovery system
o	outer
OM	operating and maintenance
OTXI	other taxes and insurance
p	pressure
pinch	pinch point
P	product
PB	payback
PE	purchased-equipment cost
ph	preheater (economizer)
ps	preferred stock
r	real escalation
ref	reference state
R	rejected
s	isentropic
sat	saturated
sm	smaller temperature difference
st	steam
t	turbine
TL	tax life
tot	overall system
v	vapor
w	wall
W, Y	equipment item (Equation 7.3)
x	type of financing ($=$ce, d, or ps)
z	year
Δp	fluid flow irreversibility
ΔT	heat transfer irreversibility
λ	monochromatic
0	restricted dead state (Chapter 3); beginning of first time interval in time-value-of-money calculations (Chapter 7)
1, 2	initial and final state; control volume inlet and exit
∞	free stream

Superscripts

CH	chemical exergy
CI	capital investment
e	environment
KN	kinetic exergy
M	mechanical exergy
N	nonreactive exergy
OM	operating and maintenance costs

OPT	optimal value
PH	physical exergy
PT	potential exergy
R	reactive exergy
T	thermal exergy
TOT	total exergy
°	standard state value
0	indicates that only fuel costs are considered in thermoeconomic calculations

Appendix E

CONVERSION FACTORS

Mass and Density	1 kg = 2.2046 lb	1 lb = 0.4536 kg
	1 g/cm³ = 62.428 lb/ft³	1 lb/ft³ = 16.018 kg/m³

Length	1 cm = 0.3937 in.	1 in. = 2.54 cm
	1 m = 3.2808 ft	1 ft = 0.3048 m

Velocity	1 km/h = 0.62137 mile/h	1 mile/h = 1.6093 km/h

Volume	1 cm³ = 0.061024 in.³	1 in.³ = 16.387 cm³
	1 m³ = 35.315 ft³	1 ft³ = 0.028317 m³
	1 l = 0.0353 ft³	1 gal = 3.7854 × 10⁻³ m³

Force	1 N = 1 kg·m/s²	1 lbf = 32.174 lb·ft/s²
	1 N = 0.22481 lbf	1 lbf = 4.4482 N

Pressure	1 Pa = 1 N/m² = 1.4504 × 10⁻⁴ lbf/in.²	1 lbf/in.² = 6894.8 Pa
	1 atm = 1.01325 bars	1 atm = 14.696 lbf/in.²

Energy and Specific Energy	1 J = 1 N·m = 0.73756 ft·lbf	1 ft·lbf = 1.35582 J
	1 kJ = 737.56 ft·lbf	1 Btu = 778.17 ft·lbf
	1 kJ = 0.9478 Btu	1 Btu = 1.0551 kJ
	1 kJ/kg = 0.42992 Btu/lb	1 Btu/lb = 2.326 kJ/kg
		1 kcal = 4.1868 kJ

Energy Transfer Rate

1 W = 1 J/s = 3.413 Btu/h

1 kW = 1.341 hp

1 Btu/h = 0.293 W

1 hp = 2545 Btu/h

1 hp = 550 ft·lbf/s

1 hp = 0.7457 kW

Specific Heat

1 kJ/kg·K = 0.238846 Btu/lb·°R

1 kcal/kg·K = 1 Btu/lb·°R

1 Btu/lb·°R = 4.1868 kJ/kg·K

INDEX